The Inheritance of Personality and Ability
RESEARCH METHODS AND FINDINGS

This is a volume in
PERSONALITY AND PSYCHOPATHOLOGY
A Series of Monographs, Texts, and Treatises

Under the Editorship of David T. Lykken

A complete list of titles in this series appears at the end of this volume.

The Inheritance of Personality and Ability

RESEARCH METHODS AND FINDINGS

Raymond B. Cattell

Distinguished Research Professor Emeritus
Department of Psychology
University of Illinois
Urbana, Illinois

ACADEMIC PRESS   1982
A Subsidiary of Harcourt Brace Jovanovich, Publishers
New York   London
Paris   San Diego   San Francisco   São Paulo   Sydney   Tokyo   Toronto

ACADEMIC PRESS, INC.
111 Fifth Avenue, New York, New York 10003

United Kingdom Edition published by
ACADEMIC PRESS, INC. (LONDON) LTD.
24/28 Oval Road, London NW1 7DX

Library of Congress Cataloging in Publication Data

Cattell, Raymond Bernard, Date.
    The inheritance of personality and ability.

    (Personality and psychopathology)
    Bibliography: p.
    Includes indexes.
    1. Personality. 2. Ability. 3. Human genetics.
4. Psychological research--Methodology. 5. Genetic
research--Methodology. I. Title. II. Series.
BF698.C314   155.2'34   80-70667
ISBN 0-12-164260-7   AACR2

PRINTED IN THE UNITED STATES OF AMERICA

82 83 84 85   9 8 7 6 5 4 3 2 1

To
R. A. Fisher
J.B.S. Haldane
C. L. Burt
L. Darwin
Who pioneered, inspired,
and gave generous aid

Contents

1

Scientific and Social Issues
in the Advance of Behavior Genetics

2

Methods and Models Available
for Research in Behavior Genetics   19

3

The Twin Method of Investigation
with Illustrative Findings   55

4

The Genesis of the MAVA Model
and Its Solutions   89

5

Further Designs for Determining Genetic, Threptic, and Heritability Values

6

Models of Interaction of Learning and Genetic Processes

7

Evaluating Interactions:
Path Coefficients and
Diverse Heritabilities   **241**

8

The Inheritance of Abilities:
Some Psychometric Requirements   **275**

9

The Heritability of Nine Primary and Five Secondary Source Traits in Q-Data · · · 325

10

Heritability and Conceptual Advances for Source Traits in T-Data · · · 359

Preface

This book attempts to heal what is practically speaking a dangerous and intellectually deplorable scotoma in the eyes of many recently trained psychologists. This is the blindspot wherein the hereditary aspects of personality and motivation remain invisible and where consequently many phenomena remain inexplicable.

Several excellent books have recently appeared on the already quite advanced science of genetics, but few are written for psychologists. I have aimed here at a new and comprehensive treatment specifically kept in perspective for students and practitioners in the area of human personality, ability, and motivational psychology.

A problem facing the writer and reader in this field is that even if we set aside the more complex Mendelian, molecular, population genetic, and other very technically developed provinces of genetics, the necessary psychological treatment still bristles with fairly complex mathematical–statistical concepts, which psychology students seem untrained to handle. Psychology, as a *science,* is itself beginning to face complex models of this kind. But, to be frank, there are still many departments where students may "qualify" without mastering what is necessary for potency of practice or theory, and which still exist as a last roost for those who falter at the discipline they have met in such sciences as, say, chemistry or physics. I will add, nevertheless, that I am deeply sympathetic with young students centrally and primarily interested in psychology who are taught statistics so formally, and with such divorce from a rich field of multivariate theory in personality and motivation problems, that it becomes for them an arid and irrelevant exercise.

In spite of the fact that my aim here has been to bypass specialized purely genetic fields (those of the farm and the animal laboratory) and get to the substantive psychological findings on the interaction of heredity and environment that concern the practitioner, the diagnostician, and the researcher on personality theory, I have had to compromise. In order successfully to integrate with quantitative scientific psychology and psychometry, it is necessary that the first half of the book be given to methods, genetic concepts, and formulas demanded both by genetics and psychometrics. However, I have interspersed this material with illustrative psychological findings at points where these prove to be discussable.

The teacher or reader must decide where to skip. The essential psychological findings belong to the last three chapters, and some readers may turn to these and refer back to earlier chapters for whatever methodological definition they find they lack. To write within the compass of one book that which the psychologist needs to know inevitably means that several provinces belonging to *genetics as such* have to be taken for granted. Fortunately, today one can refer the reader to the several readable and technically impeccable texts in those areas. The text and the bibliography have been specially designed to guide the reader who lacks that background of purely genetic information in such matters as chromosome structure, the process of cell division, molecular genetics, and what might be called Mendelian algebra. These are the standard genetic substance of botany, biology, and animal husbandry, and in order to proceed insightfully in this book, those matters must be understood at least at an elementary level.

A second domain of genetics to which we give only fringe attention here as it briefly touches our problems is that of animal behavior genetics. Whereas each species is a world of genes to itself, *principles* are translatable, and where they are contributory I have used them.

Even without getting into genetic mechanisms, and the pursuit of effects into population genetics, students who wish to understand how the twin, the family adoption, the MAVA, and other methods yield the substantive results discussed must be prepared for as much statistical argument as they would meet in psychometry and some branches of learning theory. It has seemed best to clear this before coming to substantive findings, and so the first six or seven chapters are largely devoted to concepts and methods.

A good deal of planning has gone into making the presentation not only follow an order of necessary logical and factual dependence but also be effective by educational principles and didactive devices. For example, at the cost of a little mathematical elegance in the formulations of correlation, variance analysis, probability, factor analysis, and path coefficients, I have, where space and pace permitted, set out *in words,* alongside each expression, what the formulas tell us. Again, it is part of the didactic design to introduce concepts *developmentally*. That is to say, a concept may initially be presented in rough and simple form, and only get polished to a defensibly precise form in a second or third encounter. This means of course, that there will be repetition—a requirement in learning.

Finally, I have used here the device—which seems to have been much appreciated in my other books—of making a 10- or a 12-point summary at the end of each chapter. This, I am told, is of great help to the student who has to read without a teacher—or, alternatively, who has a teacher keen on quizzes!

Because this book is cut essentially to the needs of the psychologist and the personality student, it will necessarily lead up to certain frontiers which, as mentioned earlier, can be glanced over but not crossed. These adjoining areas include the Mendelian bases, molecular genetics, animal and plant breeding, population genetics, and more advanced mathematical analysis of some formulas. At all points "where the pavement ends" in the unavoidably short treatment here, I have given the student guidance to the central texts in the area. This practice, together with the comprehensiveness of the psychological coverage has resulted in an unusually large bibliography. So also has the fact that an attempt has been made to guide researchers to the latest findings in a subject—behavior genetics—that has experienced a "publication explosion" in the last decade.

A final feature of design of which the reader should be made aware is that although the book has the purpose of a general textbook, embedded in it are several quite new methodological and substantive research contributions, perhaps only half of which have as yet been published. That readers may be more aware of where they have to pass judgment on the new, it seems desirable to list these by concepts and method, and by substantive psychological findings.

New Propositions in Concepts and Methods

1. Some extensions and improvements (e.g., in response to comments of Loehlin, and Jinks and Fulker) in the multiple abstract variance analysis (MAVA) method, notably in the equation combining covariance and interaction (p. 67).
2. A formula for relating genotypic to phenotypic assortive mating, with some empirical support for it (pp. 137, 330).
3. A methodological suggestion to use race differences as an initial device for tracing individual genes, and doing so in factorial design with the new measures of cultural dimensions (p. 227).
4. Directing the choice of variables in human genetics in relation to

three theories concerning the mechanisms in the rise of unitary structures (pp. 234, 382).

5. Relating twin, sibs apart, adoptive family, parent–child correlation methods as partial, fragmentary methods within MAVA, in order to extract more unknowns, especially with reference to using known genetic correlations as an ancillary "lever" (p. 183).

6. Recognition of genetics as a necessary adjunct to learning theory, with provision of matrix (p. 195), *comparative MAVA* (p. 237), and regression (p. 225) methods for relating pure learning (threptic) gain to environmental experience.

7. Attacking behavior genetics by analysis of process curves into genetic and threptic curves, by four new methods (p. 213).

8. Entering path coefficient designs with a more comprehensive evaluation (from a basis of the obtainable correlations) of sources conceptually important to psychologists (p. 271).

9. Presentation of a model for the relation of primary to secondary trait structures that will lead to understanding of the genetic and environmental influences operating separately in each (p. 383).

10. The recognition and definition of *six* different heritability coefficients and of the contributions to clinical and social psychology especially from comparison of within- and between-family heritabilities (p. 269).

11. A schema for recognizing the relations of the subdivision methods —notably convarkin, MAVA, twin, and adoptive designs—within biometric genetics, to the advantage of productivity in design (p. 21).

12. The recognition of human behavior genetics as an integral part of personality and social psychology, and a supplement to learning theory, exemplified at the applied level by two new formulas for evaluating the genetic endowment of a particular individual in a particular trait (p. 226).

13. An empirical comparison of the agreement of three methods— OSES, least squares, and maximum likelihood—of analysis applied to the same raw variance MAVA data (pp. 330, 336, 340, 366, 370, 372, 376).

14. Trial of a model of production of crystallized intelligence by investment of fluid intelligence, with a formula giving a tolerable fit to data, and exemplifying splitting a life development curve into maturational and acquired components.

15. The proposal of a theory governing changing heritability values with age, to be checked by comparative MAVA. Three different principles are considered to be at work, the combined result of which is likely to be an increase of H with age (after early imprinting effects).

16. An addition, to existing formulas for sampling error, of further formulas for correction for measurement error, covering unreliability, invalidity, and function fluctuation (p. 287).

New Substantive Additions

1. Determination of a typical value by which within-family environmental variance for identical twins falls short of that for sibs (p. 356).
2. New data on significant physical linkage of trait measures (p. 39).
3. Determination of separate heritability values for fluid and crystallized intelligence (p. 322).
4. Determination and check in MAVA techniques of the heritabilities of nine primaries (C, E, F, G, H, Q_1, Q_2, Q_3, Q_4) and five secondaries (QI, QII, $QIII$, QIV, and $QVIII$) in personality as measured by questionnaire (p. 356).
5. Determination of the heritabilities of 11 personality factors measured by the new objective (T-data) test O–A Battery, namely, of U.I. 16, 17, 19, 21, 23, 24, 25, 26, 28, 32, and 33 (p. 399), with use of the new correction for reliability of measurement.
6. Fairly good agreement found between heritabilities of previously matched secondary Q-data and primary T-data factors, except for extraversion, suggesting an especially high heritability in the "stubs" of the primaries.
7. New support for (a) significant genothreptic correlations in about half the factors examined; (b) greater between- than within-family correlations; and (c) predominantly *negative* correlations of genetic and threptic deviations *between* families, fitting the theory of coercion to the biosocial norm.
8. In questionnaire primaries and secondaries we conclude that F (surgency), Q_3 (self-sentiment), I (premsia), and QIV (independence) are relatively highly inherited, whereas G (superego), E (dominance), O (inadequacy, guilt), Q_4 (ergic tension), and $QVIII$ (inhibitory control) have very low inheritance. If some estimate of bias is made, one could conclude that the former are patterns largely from genetic sources and the latter practically entirely learned.
9. In objective tests, high heritabilities are found for U.I. 16, 19, 24, and 25, and low for U.I. 17, 26, 32, and 21; and, again, it is conceivable on the new evidence that these extreme cases could be mainly hereditary and mainly learned patterns of behavior, respectively.

The general reader—student or professional psychologist—needs to be warned (if that is an acceptable term) that (a) as mentioned earlier, alongside the usual points covered in a textbook, the reader will encounter new ideas and facts, at present not much subjected to debate, but well defined and given clear-cut new labels appropriate to their definition; and (*b*) this is not a substitute for the regular genetics or behavior genetics text, but presents its own special comprehensiveness. That comprehensiveness consists of following the connections of genetics into the domains of personality, motivation, developmental, and social psychology, that is, into the circle of interests of the psychologist as such.

Author's Notes and Acknowledgments

For the reader who likes to find a more personal note concerning the book's production, I would add that I have been trying to produce this book for over 10 years, as it has seemed an organic necessity in my general approach to personality. A first version was actually typed in 1968, another in 1975, and the present version was completed in 1980. The first version owed a good deal to lengthy discussions with Professor Nesselroade, as well as to more brief ones with Professor Fisher and C. R. Rao. The explanation of the ridiculous delay is sadly familiar to many writers. As I approached completion of that first version, I was interrupted by an avalanche of earlier commitments to publishers, in regard to *Abilities* (1971), *Beyondism* (1972), and *Personality and Mood by Questionnaire* (1973). Meanwhile, behavior genetics as a research field had picked up momentum so fast that by 1975 the first draft needed to be rewritten and rearranged. In 1975, I was pressed to finish my two-volume work on *Personality and Learning Theory,* which went to the publisher in 1978 and 1979. By 1979, a whole new harvest of substantive behavior genetic findings had appeared, including those from the large study by Schuerger, Klein, Barton, and myself at the University of Illinois. These are written up in greater detail in several articles in press, but appear here for the first time in integrated form.

As for the roots of personal interest in this field, they may be traced to the 1930s, when I started the Child Guidance Clinic in Leicester. I became

intrigued, as all careful human observers must be, by the long arm of heredity in human behavior. I have borne with some impatience the 40 years of
ignoring of this field by psychologists, who have been scared away from a
rational handling of the problems by Hitler, on the one hand, and by doctri-
naire egalitarians, on the other. Meanwhile I have been encouraged and inspired by the remarkable contributions of a small group of men who truly
deserve the title of genius. I refer especially to Fisher, Wright, and Haldane,
but I am dedicating this book to four with whose rich personalities I had
benefits of personal contact.

To Fisher, I owe the ironing out, through the generous bestowal of his
time, of some kinks in my first development of the MAVA model (which
grew within the metric ideals of biometric genetics, but transcended the
latter's initial restriction to the nonpsychological world of the farmer). To
Burt, I am indebted for his lucid, creative, and socially courageous discussions on the importance of genetics to the science of psychology. That this
great man has come under much attack since his death, because of having
confused falsely remembered with real data, does not lessen my sense of
gratitude for his great positive contributions to our subject. That far lesser
men have presumed to tack an official censure on his tombstone cannot deny
his contributions at his prime when his mind scintillated with concepts as
deep as they were original. To Leonard Darwin, I am indebted not only for
the strange experience—itself an effect of heredity—of feeling myself un-
cannily in the presence of his father—in manner, voice, and eye—but also
for some very practical financial help. This he gave me, in the days before
"grants," in 1935–1937, to enable me to take time out to investigate the social question of the inheritance of intelligence in relation to birth rates. (The
provocative results of this study—in which culture-fair tests were used for
the first time—were written up in *The Fight for Our National Intelligence,*
1937.) Last but not least in providing the stimulation that has carried me to
this work was that intractable genius J.B.S. Haldane, who taught me that
mathematics is the highest form of imagination (though probably the weak-
est in myself). My first contact, when he was still in his Communist, pre-
Lysenko phase, was in a devastating criticism he wrote of my calculation of
the production rate of mental defectives. My last was when I said goodbye to
him, as he stood in a saffron robe by the Ganges, after a talk I gave on factor
analysis at a meeting at the Indian Statistical Institute. He was dying of cancer and, rebel against "conventional good taste" to the last, he had just
begun a verse, "Oh that I had the skills of Homer, to sing the epic of carci-
noma" (as I recall the opening stanza).

These men—so dramatically diverse in background, personality, and
contributions—put behavior genetics on the map in the first half of the twen-
tieth century. Nowadays I would not find it difficult to name a dozen others
of equal eminence to whom I also feel deeply indebted. For this fascinating
field seems to have attracted, by its inherent complexities, many brilliant

persons. These later leaders need no reference here for their references abound in the main text.

Finally, I wish to express more specific indebtness to John Loehlin and C. R. Rao for their critical aid with the MAVA method; to Donald Swan for bringing physical-linkage data to my notice; to John Nesselroade for his help in an early draft of certain chapters; to Frank Ahern, Velma Kameoka, Tom Klein, D. C. Rao, Newton Morton, J. Brennan, and James Schuerger for quite indispensable help in sorting, checking and rechecking, scoring, and analyzing the data. (Rechecking the twin and other family connections was no minor task. The degrees of freedom were threatened by such reports as that of John X who declared he had $3\frac{1}{2}$ brothers [2 brothers and 3 half brothers, as it transpired!].) I am indebted also to John Campbell and David Vaughan for checking formulas and for programs handling certain complex calculations.

In the scientific product of this kind, with so many formulas, tables, and figures to go wrong, perhaps the most important contributor is a secretary of enormous capability and patience. I count myself very fortunate to have had the help—through the various revisions—of such a secretary in Lucille Metcalf, who, in addition to the already mentioned virtues, happens to be one of the few living persons still able to read my handwriting.

Raymond B. Cattell

The Inheritance of
Personality and Ability
RESEARCH METHODS AND FINDINGS

1

Scientific and Social Issues in the Advance of Behavior Genetics

1. Historical Roots of Genetics Prior to 1800

Every human trait is due in part to heredity and in part to learning experience. That is to say the rank of a person's measurement on intelligence, ego strength, or extraversion—relative to other people—is fixed partly by what that person inherits from his or her parents and partly by what has happened to that individual since being conceived.

The fact that heredity and environment have interacted and merged in the final behavior has led some psychologists to declare the futility of trying to separate them. But science can assign to them relative quantities, as it does for percentages of alcohol and water totally mixed in wine. Furthermore, it can describe what properties will result from particular combinations, as it does for a given mixture of iron and nickel in steel.

Physicians and psychotherapists have naturally noticed the role of heredity in diseases more than in health, just as one pays more attention to the holes in a sock than to its remaining functionality. With thoughtless people this has given heredity a bad name. Some people deficient in biological education and perhaps in emotional maturity are horrified that heredity "dooms" them to some resemblance to their parents. Learning about laws of inheritance is apt to be as appealing to them as learning about the laws of

1

arithmetic would be to a spendthrift or learning about the law of gravity would be to a man who has carelessly fallen off a skyscraper. A scientific viewpoint and the truths of biology are fortunately becoming more widely accepted, but the words of the grand old man of biology, Thomas Huxley, are still much needed.[1]

Probably the vague antipathy to genetics that the psychology student senses is today to be found less in the young than in middle-aged teachers, journalists, and politicians. Some of the attitudes of the latter hinge on such crude thinking as confuses biological with social inheritance. That these have no necessary connection is shown by fact that aristocratic title can be abolished by a stroke of the pen, but biological heredity will go on as long as life. Nevertheless, in Russia, the inheritance of intelligence is flatly denied (at least it was when I was in Moscow). Possibly another source of antipathy among the middle-aged group is the memory of Hitler—suggestibility and contrasuggestibility being equally effective in shutting out objective discussion. On prejudice it may be necessary to comment later, as a more specific issue arises, if the innocent student is to get his or her bearings. But meanwhile let us be clear that heredity simply *is*, and that we make it good or bad according to our purposes. On its "bad" side, it is scarcely ever as inflexible as the Calvinistic predestination with which some of our forefathers learned to live. On its "good" side, it is gratifying to know that the genes of our newborn offspring will not allow that infant, however much we neglect his or her environment, to grow up into a gorilla.

Although various fashions, faiths, and "isms" may teach ill-balanced views—in either direction—regarding the interaction of heredity and environment, the improvement of mankind requires attention to both factors, and wise men have known that down through the ages. There is sensitive appreciation of heredity in, for example, Homer and the Old Testament. European plant and animal breeders in the seventeenth century, not to mention Jefferson in the eighteenth, and such geniuses as Lyell, Lamarck, Haeckel, Burbank, and the creators of hybrid corn who followed, knew enough at a practical level to work wonders with corn, cows, and collie dogs. Equally practical family doctors noted that disorders such as hemophilia, galactosemia, colorblindness, phenylketonuria, gout, diabetes, sickle cell anemia, and Huntington's chorea came out by heredity, appearing even when the child was reared away from the parent. They also noted such things as the differ-

---

[1] Huxley observed:

> The life, the fortune and the happiness of every one of us, depend upon our knowing something of the rules of a game infinitely more difficult and complicated than chess. The chessboard is the world, the pieces are the phenomena of the universe, the rules of the game are what we call the laws of Nature. The player on the other side is hidden from us. We know that his play is always fair, just and patient. But we also know, to our cost, that he never overlooks a mistake, or makes the smallest allowance for ignorance. To the man who plays well, the highest stakes are paid, with that sort of overflowing generosity with which, the strong shows delight in strength. And one who plays ill is check-mated—without haste, but without remorse. . . . What I mean by education is learning the rules of this mighty game. [Thomas Huxley, 1894, p. 117].

ence between disorders that marched without interruption from generation to generation and those that skipped one or two generations. And they noted that special inbred lines (e.g., of roses) "went back to the wild" when their special breeding was relaxed.

2. Modern Growth of Interest in Human Psychological Heredity

It was on the growing foundation provided by these observations of clever and enterprising farmers and doctors—and on his own shrewd observations during the voyage of the *Beagle* in 1832–1836—that Darwin developed the theory of evolution. As the Victorians painfully found out, this "put the fat in the fire," especially in regard to man's mentality as descended from that of the apes (though man doubtless stood, as Shakespeare's Hamlet said, only "a little lower than the angels"). With that barrier down, Spencer, Haeckel, and, in our own time, William James, McDougall, and Freud could (with due intellectual awareness of the superficiality of Watsonian reflexology) search for early forms of human instincts in the primates. Within that same span of a century, the partial dependence of intelligence on inherited brain (cranial) size became apparent from the unremitting increase in human skull sizes from 5 million years ago to the present. By the beginning of this century, psychologists comprehensively surveying the evidence had concluded that at least predispositions to the drives seen in primates existed in man; that several abilities, including intelligence, were likely to be partly inherited, and that forms of insanity had some degree of inheritance as specific predispositions.

The classical work which focused inquiry still more closely on psychological inheritance was Sir Francis Galton's *Hereditary Genius* (1869). The discussion around Galton's work brought forth a galaxy of researchers in this field in the next half century. Most—like Beloff, Burks, Burt, Cattell, De Fries, Eells, Erlenmyer-Kimmling, Hebb, Holzinger, Horn, Hurst, Jensen, H. E. Jones, McGurk, McClearn, Skeels, Skodak, Stice, Thurstone, Tryon and Vernon—dealt with inheritance of ability, but others—like Beloff, Cattell, Eaves, Eysenck, Gottesman, Kallmann, Klein, Loehlin, Nichols, Rosenthal, Royce, Scarr, Schuerger, Sheldon, and Vandenberg—began research on temperament and personality.

Inevitably, the social implications that some people drew from these findings, regarding race and class, generated more heat than light. This was the period when the dismal histories of the Jukes (Dugdale, 1877) and Kallikak families (Goddard, 1912) were cited to show the overweening influence of bad and good heredity. On the other hand, the student was reminded of the Darwin–Wedgewood–Galton–Huxley families to show the persistence of good heredity; and of Mozart; father and son, and the generations of Bachs, to show how even special talents were passed on. The socially outstanding instances in either direction were obviously not proofs of heredity alone, for the Jukes (Dugdale, 1877) were plagued by entanglement in a degenerate social milieu, and the Darwins and Bachs were given a good chance by their families maintaining a high level of achievement and educational stimulation. Lesser public visibility leaves out of our experience the probably well-talented individual brought up in a mediocre setting or those with moderate hereditary defects who are looked after well. It would be a mistake, however, to deny some hereditary action in historical instances of this kind. What they show is that good heredity needs generally good education to hit the really top performances, though there are many exceptions, in which great abilities transcended poor environmental beginnings.

3. Emotional Problems in Absorbing the Findings of Human Genetics

From the beginning, as shown by the insulting exchange over evolution that occurred between Bishop Wilberforce and Professor Huxley at the British Association for the Advancement of Science, genetics has met socioemotional prejudices even among well-educated individuals (at least well educated in a literary or classical sense). Conservatism has played its part, but so have the attacks and repressive restrictions on applications of genetics that have been initiated by social egalitarians, overtly in Russia, obliquely in the United States and Britain. (Although the term "liberal" is sometimes used by journalists for this position, it is important to distinguish this form of "wishful thinking" from the tradition of rational liberalism in Huxley, Darwin, Mill, James, Dewey, and others. [See Scarr, 1979].) It is deplorable that in this day and age one should have to digress to consider this intrusion into a scientific area, but the student, at least, deserves to be reminded of some points to help him get his bearings in this uproar—for as recently as 1975–1979 there was such a street parade over the case of Sir Cyril Burt. Both scientists—like Eysenck (1971, 1973), Herrnstein (1973), Hooton (1959), Jensen (1972, 1973), Kamin (1974), Lerner (1968), and Wilson (1975), to name a few—and social writers—like Hardin (1980), Marx (1890), Shaw (1947, 1965), Wells (1903), Williams (1953, 1956) and Young (1958)—have given frank and explicit discussion to this politicosocial side of genetics. And, in my book on morality and science (1972a), I have endeavored to show that the trouble in focusing the social aspects of human hereditary differences arises basically from a confusion of clear ethical ideals—the result of a farrago of various dogmatic religious viewpoints and political "isms."

As three errors seem dismally to repeat themselves in public discussion it would be well to pinpoint them, as follows:

1. That belief in biological inequality and individuality is incompatible with the democratic belief in equal dignity and right to opportunity.

2. That recognition of the effect of hereditary differences in, say, intelligence or school achievement is relatively pessimistic. (The converse might be better argued, since if intelligence were raised genetically in one generation, achievement would tend to stay high, whereas a merely environmental attack would require a great expenditure repeated every generation.)
3. That belief in completely environmental causation is somehow politically progressive, whereas belief in some influences from heredity is reactionary. (This last is a hangover from the *prebiological*, fanatical "rationalism" of the French Encyclopedists.)

A combination of these errors showed up in the left-wing attack of the 1950s on the British educational system. This had been designed to offer more prolonged and higher education to the more able, and it did so efficiently, on a basis of scholarship selection, with intelligence tests, and without regard to parental status or capacity to pay. It put into effect the socialist ideal of "career open to talent" rather than privilege and resulted in 60–70% of students in the "privileged" universities of Oxford and Cambridge being there on competitive scholarship grants. The nature of and arguments for promotion by capacity in a meritocracy have been admirably stated and analyzed in this generation by Young (1958) and by Herrnstein (1973).

Other correctives of the journalistic view[2] that attention to genetics is socially "conservative" (unless progressive conservatism is meant) could be cited. Plato's republic recognized innate differences, but was anything but conservative! Jefferson and Lincoln did, too. Throughout the socialist writings of Wells, and, especially, Bernard Shaw, the theme of greater attention to heredity and eugenics—in a national concern for the quality of the next generation—is brilliantly argued. And if Marxists fancy that Marx was blind to genetic influences they should read the analysis of his letters by Nathaniel Weyl (1979).[3]

---

[2] Journalism, presumably catering to what was believed to be public opinion, brought about the character assassination of Sir Cyril Burt, in the interests of the view that heredity has no role in intelligence. Burt's slip in data recording was trivial compared to his vast lifelong contributions to psychology and in fact did not affect the expert consensus on the heritability ratio one jot. What exposes the bias is that there have been far more gross claims for environmental influence on IQ, based on bogus experiment, the exposure of which was completely ignored by the mass media (see D. V. Glass's "Educational Piltdown Man," 1968). Incidentally, as Rowe and Plomin (1978) document, Burt's estimate was about the *average* of indpendent investigations to that date.

[3] Despite being ejected from Germany, Marx retained a conviction about the superiority of its aristocratic culture, and married a German aristocrat. By comparison he considered the Slavs (in a purely racist sense) incapable of matching such a culture. Believing Darwin's work on the evolution of man gave support to his views, he asked permission to dedicate *Das Kapital* (1890) to him; but Darwin, with the uncanny discretion that marked so much of his life, sensing an oblique intention, declined. It is at least of interest as showing historically that prejudices in favor of heredity are not intrinsically and necessarily "right wing."

An initial apology was made earlier to the purely scientific reader for this intrusion of discussion on social, political, ethical, and emotional issues, but politics has intruded uninvited into the domain, and the student needs to be warned that even some leading sociologists will tell him that mental heredity cannot exist. He must also get used to finding those psychology textbooks written over the last 30 years, in the heyday of Watsonian reflexology, either completely blank on heredity or suggesting that is is intellectually "passé." Psychology is now rubbing its eyes and awakening, through behavior-genetic findings, to a greater power of explanation and prediction where it was formerly impotent. But far too many citizens, limited to the mass media, still live under what has been in effect a censorship.

Some of those who enter universities in Britain, the United States, France, and so on in this decade may be inclined to find this discussion of the dangers of suppression and distortion exaggerated, but the experiences of older geneticists, if read in detail,[4] show that, if anything, it is understated.

[4] The writer can cite both personal experience and that of more eminent scientists on a grander scale. The history of the political demolition of genetics in Russia has been told in *Death of a Science* and in Medvedev's *The Rise and Fall of R. D. Lysenko*. J.B.S. Haldane and the Nobel Prize winner H. Muller both settled for a time in Russia with high hopes and sympathies, but found such political pressure against free genetic conclusions that they modified their views of the "democratic" process there and escaped, in one case with great difficulty.

It is scarcely believable that in Britain, the homeland of Darwin and Huxley, the Nobel Prize winner Shockley was forbidden to speak to a university audience, and Eysenck was knocked down by a mob at the London School of Economics when speaking on intelligence inheritance. As Herrnstein's uncontradicted account (1973) of the behavior of "educated" students (and some faculty) at Harvard shows, and the rowdyism at universities where Jensen has lectured demonstrates, there *is* a psychological–emotional problem still, of such magnitude that in some educational institutions rational discussion cannot even be begun.

My own experience is far milder, but equally clear on the direction of pressure. Heron's demonstration around 1900 of an inverse birth rate relation to social status and education justifiably raised the possibility in the more farsighted of socially concerned people that a dysgenic trend existed. However, I felt one could not infer intelligence from social status and so, for the first time in the area of population genetics, I administered in 1935 a culture-fair intelligence test to all 10-year-olds in a city of about 250,000 and in the rural areas of Devonshire. The result was clearly that families were larger that produced children of lower intelligence, and I predicted that unless the marriage rate was lower, and the death rate higher, at lower intelligence levels, about a 1 point drop of IQ could be expected in a generation (*The Fight for Our National Intelligence*, 1937). I returned half a generation (13 years) later to the new set of 10-year-olds and found that in fact no fall had occurred, thus indicating that, at least in that period, some compensating influences of lower marriage rates, more completely childless families, or higher death rates in the less intelligent had occurred.

In observing newspaper and other reactions to this prewar book and to my postwar articles appealing for further checks on my analysis of the factors that affected the mean intelligence level, I became aware of three problems in the interaction of science and society. First, the belief that progress is "built in" to society and calls for no examination of possible regression seems so essential to the public peace of mind that my questioning this birthright made me as unpopular as Cassandra. Second, any reference to individual differences of biological potential or concern with eugenics was automatically interpreted by many as a "right-wing" view. Now it is obvious to any social psychologist that Hitler with his weird superstitions about "blood"

Perhaps the best concluding comment on the present transition is that of a group of leading researchers (Eaves, Last, Young, & Marten, 1978): ''A psychology which ignores man's evolutionary past and the biological basis of his present differences is barren.''

Although one regrets that the student and the researcher in behavior genetics have been in this generation denied the privilege that Pasteur asked for scientists (''Vivez dans le calme des laboratoires''), those in this field are not worse off than those in, say, nuclear physics or recombinant DNA research. The prospect is that the changes in human life and organization due to the physical sciences in the last two centuries, are likely to be at least matched in the next two centuries by the applications of genetics. For example, our sheer level of technical control of the physical world is likely to be

---

and genetics was a worse catastrophe to a science of behavior genetics, first inside Germany, and since then, outside, than was the simple-minded egalitarian Russian opposition to genetics. But Hitler was a socialist. Nevertheless, an outdated view persists in many places that concern with eugenics is somehow ''right wing,'' whereas, as far as logic goes, the betterment of mankind must proceed equally environmentally and genetically, and all political parties should be interested in both. My personal experience of being accused of being right wing had its amusing sequel. A journalist, Ritchie Calder, immediately and vehemently attacked the conclusion of my book, not with fact or logic, but through guilt by association, inasmuch as I had three introductory contributions, one by Lord Horder (then the king's physician), one by a distinguished director of education, E. A. Armitage, and one by Leonard Darwin. The journalist lumped all together as ''old school tie'' rightists, and proceeded to ridicule intelligence tests and any concept of inherited individual dfferences. Some 30 or 40 years later I was interested to read that Mr. Calder's lifelong preaching of his ''uniformity'' misconception of democracy had, through a Labor peerage list, earned him the title of Lord Somewhere-or-other (I forget the locality)!

A third discovery, which I made with youthful surprise, was that among scientists interested in the field, division of evaluation also followed to some extent political affiliations and the less educated tendency to wishful thinking. Even comparatively recently Bodmer (see Cavalli-Sforza & Bodmer, 1971) insists there is no need to get concerned with eugenics, since work following my own (this subsequent data not being quoted) shows no differential birth rate by intelligence. The work referred to is based on smaller samples than my Leicester and Devonshire cross section of whole regions, did not use culture-fair tests, did not come back a dozen years later to check conclusions, and had as subjects an/atypical, unusually socially responsible midwestern group, by no means representative of the world generally, or even of ethnically diverse America.

These remarks are here confined to a footnote because this book is not concerned with social applications. But while on this matter I wish to express my concern about neglect of this social issue. In 1980 that study of 1937 still remains the only one using culture-fair tests, factored for a pure intelligence factor, on a large, complete, 10-year-old cross section, and repeated (with confirmation as to the size–intelligence differential) in the same population after half a generation. Over the years, results on large samples have nevertheless accumulated—such as those by Lentz, Thomson, Osborne, and others—and they agree with the Leicester 1936–1959 study in showing lower intelligence associated with larger families. Until adequate samples with death rates, childless family rates, assortive mating (for intergeneration regression values), marriage rates, and culture-fair tests of known heritability are taken in various subcultures, no exact conclusions are possible. In this matter I wonder whether the remedy I have suggested (1948, 1972a) of separate statements by social scientists of the research result and the ethical or other value, as premises prefixed to any applied technological advice, might help.

raised as much in the next century by genetic improvement in human beings themselves as it was in the last century by educational improvement. Even without any manipulation of genetic bases of mentality, our gain in insights on how social and political movements depend in part on genetic foundations and distributions may change our whole philosophy. But what new concepts of social organization and values will emerge no person experienced in history and the possibilities of science would venture to say, any more than a cautious scientist in 1900 would have predicted nuclear energy stations or men on the moon.

4. The Warring Ancestries of Genetic Methods and Concepts

Having digressed—hopefully not without some benefit to untrammeled further investigation—into socio-politico-ethical aspects of the science of behavior genetics, this book will henceforth leave that field to the reader's own ruminations, and concentrate on the science as such.

In any case, there are purely methodological internal debates within behavior genetics lively enough to keep us fully occupied. A major problem—and valuable heritage—is its growth from quite diverse scientific specialties. Genetics itself has grown up in botany, as in Mendel's classic experiments, but also from animal husbandry, farming, and the practical demands and experiences in the breeding of domesticated animals for diverse purposes. It has developed almost independently in the study of medicine, and obliquely from population and migration studies. As we shall see in the next chapter on methods, each field has developed its own adapted, but often limited, methodology. For example, in what we shall describe as the biometrical and the MAVA method, it is necessary to recognize that the former grew up in general biological genetics, where there are neither human cultural stratifications nor family domesticities to bother about. On the other hand, the MAVA method was propounded in psychology, where the cultural effect of the family is important, where psychometric measurement is different from physical measurement, and where the unknowns we seek are somewhat different. Both biometrical and MAVA methods however, deal with continuous, graded measures, in contrast to medicine and psychiatry, which have generally dealt with all-or-none syndromes.

All differences of origin tend to produce differences of language, of method of analysis, and even of concepts. For example, in MAVA and biometrical genetics we tend to deal with variances, whereas in medical genetics analysis has often been by "concordance" and "manifestation rate." In animal behavior genetics, much work has been done on "backcrosses," a term unknown to human genetics. In molecular genetics, a substantial development of chromosome Mendelian mechanics has developed (as in the recent work on "recombinant DNA") from which our behavior-genetic methods are still too remote to establish useful contact. And so on, through many mansions.

What one has to guard against is distortion of approach by terms and concepts that are safe in one area but subtly misleading in another. An instance is the classical geneticists' use of *genotype* and *phenotype* (see Glossary). In geneticists' physical data it is commonly safe to talk about *the* phenotype, but there is scarcely such a thing as *the* phenotype in human mental manifestations, so fundamentally does the genotypic expression vary with the pattern of environment. In Mendel's fortunate first choice of experiment, the genotype was a definite model, as it always is, but so was the phenotype of "tall" and "dwarf" peas, because soil and climate were the same for both. Yet, had some climatic and chemical conditions been applied to the tall plants, they might have been shorter than the dwarfs. The psychologist's approach through graded trait measures and diverse cultural environments makes the original "absolute" use of a fixed phenotype misleading, though to speak of the phenotype as what we literally observe in a given *individual* is another matter. Incidentally, analogous problems of historical origin occur elsewhere in science. The statistics that psychologists use in one part of their work (e.g., learning) derive largely from ANOVA, which derives from agriculture, whereas those used in the other part (e.g., personality and individual dfferences), derive largely from CORAN (correlational analysis methods), stemming from Galton and Pearson's concerns with personality structure.

Incidentally, a clear example of how different historical origins may produce difficulties in communication is seen in CORAN concepts themselves, in the realm of factor analysis. As shown elsewhere (Cattell, 1978a), the mathematical origin of "principal components" developed by Jacobi and others led many psychologists to unrotated and orthogonal factoring, quite unsuitable to psychology, whereas a separate historical origin in psychology, in the problems attacked by Spearman (1904) and Thurstone (1938), resulted in rotation, simple structure, higher-order factors, etc., and a different set of key terms and concepts. Something analogous to this, but less emphatic, occurred in the development of biometrical genetics concepts in plant and animal work, on the one hand, and MAVA concepts in psychological genetics, on the other. These differences in developmental origin show themselves in some new terms to which the reader needs to become accustomed. For example, we now recognize a whole class of approaches as the *convarkin* methods. Concern with learning theory in psychology also calls for our use of *threptic* as distinct from *environmental*. Human marriage introduces in MAVA a greater concern with *assortive mating,* which for no good etymological reason has often been written "assortative." Social and cultural psychology explain MAVA's greater concern in focusing on what we now call *genothreptic correlations,* as well as on effects of particular cultural life epochs, in *epogenic* curves, these terms are properly defined as we proceed, but convarkin, it may be pointed out at once, derives from *contrasting variances of kin*ship constellations, and embraces the twin method, MAVA, the study of adoptive families, and so on. Much of the convarkin developments are peculiar to the human branch of biometrical genetics and represent new concepts as well as methods.

    Threptic variance refers to *variance in a trait* that is produced by *environmental* variance. (The classical origin of *threptic* is described on p. 59.) To the psychologist, the separation of these two is important, for the whole of learning theory develops in relations between them, and behavior-genetic finding give a new depth of meaning to their separation.

5. Behavior Genetics as a Vital Part of Learning Theory and Social Psychology

    A point we shall develop in this book is that many psychologists do not sufficiently realize how much research on heredity at the same time throws light on environmental learning. It does so partly because there are such powerful interactions of heredity and experience that it is pointless to talk about learning theory without knowing the laws of joint action of personality structures and conditioning experiences. learning theorists have been prone to establish laws relating change to experience, ignoring the inexactness that arises through omitting the genetic *maturation* effect in, for example, personality change. Indeed certain advances in learning theory are impossible to reach without an understanding of the maturational findings of behavior genetics. Social psychology also is helped in several directions by behavior genetics toward laws it would otherwise have difficulty in pinning down. For example, we shall see that the higher heritability of crystallized general intelligence found *between* rather than *within* families gives us a measure of the extent of environmental social similarity (in cognitive stimulation) within families compared to that between families. And the significant negative correlation found by genetic research, between genetic deviations on dominance and environmentally produced effects on dominance, has led to the behavioral *law of coercion to the biosocial norm,* the pervasive action of which social psychology has yet neither measured nor analyzed as a dynamic influence.

    Anyone familiar with the general range and outcome of psychological research will soon recognize that the areas of personality and ability, of learning, and of the individual in the setting of social psychology, are relatively infertile without union with behavior-genetic findings and concepts. However, as mentioned in the Preface, the psychologist so enterprising as to seek to master behavior genetics will find some difficult mathematical models, which (unfortunately for the less psychometrically educated) one cannot really afford to neglect. But any science requires mathematics, and the ultimate rewards therefrom in building a science of psychology are great. Moreover, as indicated in the Preface, this book has sought to make most of those formulae clear to the less mathematically trained, by appropriate verbal discussion.

6. The Need for Precise Nomenclature

Many areas of psychology could reduce difficulties and avoid confusion by more explicit attention to a defined terminology and the avoidance of a clinging heritage of loose popular terms. With this in mind, we shall pause at suitable points for brief definitions, later held for reference in the Glossary. At the outset here let us define at least enough to carry us through the next chapter, beginning with such common terms as *innate, inherited, congenital,* and *constitutional.* Their relations are most readily perceived by the overlaps in Figure 1.1.

The difference between *innate* and *congenital* is that the latter covers what is given at *birth,* thus *including the effects of gestation* along with the innate properties inherent in the fertilized ovum. The difference *can* be considerable, as in those unfortunate children misshapen at birth through drugs or disease affecting the mother. Any substantial difference, on the other hand, between what is innate and what is *inherited* is very common. Though

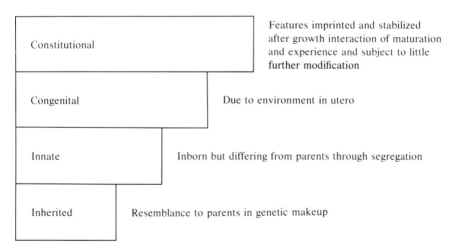

Figure 1.1. Relation of concepts of constitutional, congenital, innate, and inherited. These four terms are used approximately in popular speech but can be given precise meanings as in this illustration. The term *constitutional* lacks the operational precision of the others and requires an arbitrary drawing of limits.

infrequent, such a difference is quite real. It arises when the mutation of a gene occurs "spontaneously" or through exposure to X rays (as Muller, 1927, demonstrated) or chemicals. Such a mutation produces genetic effects in the child that are not in the gene heredity that would have been predictably handed on by the two parents. Rare though mutations are, they have been responsible, over millions of years, for what we see as evolution—though at a cost to individuals, since roughly 999 out of 1000 mutations cause defects rather than advantages, and consequently come to be eliminated by natural selection. Practically all our concern here will be with *heredity:* the degree and manner in which the child genetically resembles the parent—except for our additional study of interaction with environment.

The term *constitutional* seems to have settled down to a more popular meaning too vague to be really of any strictly scientific use. It apparently implies to most people a person's matured genetic make-up at a given time, in the sense of something largely inherited, or, if partially modified by earlier environment, now little likely to alter. We shall try to avoid it.

As we proceed we shall be distinguishing *sibs* (short for *siblings,* i.e., brothers and sisters) from *uniovular* and *biovular* twins. Synonyms for uniovular are *monozygotic* or *identical,* and synonyms for biovular are *dizygotic* or *fraternal.* In the first type of twin, a *single* ovum is fertilized but separates into two at the first cell multiplication step so that one organism grows from each of the identically structured but separate cells, thus giving genetic identity. In the dizygotic or fraternal case, *two* ova happen to descend into the womb at the same time and are separately fertilized. In that case we have two individuals, in two chorionic membranes, growing separately in the same gestation environment. Except for possible gestation period similarities, and the fact that they are born at the same parturition, they are no more and no less alike than ordinary brothers and sisters. Thus biovular twins can be of the same or opposite sex, but uniovular are always of the same sex. The occurrence of these two types of twins is, in the Caucasian race about 7.44 fraternal and 3.85 identical per 1000 births, in Japan 2.72 and 4.25, respectively, and somewhat different in other racial groups (Lerner, 1968). The existence of these two types of twins has provided a great opportunity for studying human heredity, as we shall see.

As the use of *zygotic* in the preceding paragraph indicates, a zygote is a single fertilized ovum. The student will encounter the terms *homozygotic* and *heterozygotic* in some later discussions. A fertilized cell (and the resulting individual) is said to be homozygotic *with respect to a certain gene* if the father and mother present the same gene form, and heterozygotic if the ovum and the sperm contribute differently (as to alleles) on that particular gene. A *gene* for the moment we must define simply as a unitary genetic influence affecting a certain character or set of characters in the final organism.

By two "forms" of a gene we mean what are technically called *alleles* (rhymes with eels). A gene commonly has two, but may have more, allelic forms. In some way the two alleles affecting the same organ or organs each set a train of physiological construction in motion that ends in a different somatic product. In humans, the distinction of light and dark eyes expresses such alternative alleles. In Mendel's peas, one allele determined tall and the other dwarf plants. It was Mendel's discovery, confirmed and extended since, that one allelic form is usually more or less dominant to the other, so that if one parent gives the dominant allele and the other the recessive, the offspring will show the character only of the parent with the dominant allele or will tend toward a more predominant expression of that character. For example, dark eyes are dominant to light, so that if one parent has entirely brown-eyed genes (both alleles brown) and the other blue-gray, the offspring will·be brown-eyed. (Some "shades" come from secondary genes, however.) Incidentally, one must not think that what appears as a Mendelian dominant necessarily implies a superior character. Some quite unpleasant diseases come with a dominant allele, though it is true that in a race long-evolved in a given situation there is a tendency for undesirable dominants to be more readily eliminated so that less desirable traits can and do "hide" longer as recessives. that is to say, a recessive, good or bad, is exposed to less efficient spreading or elimination by natural selection than is a dominant trait.

As indicated, in the Preface, the scope of this book does not extend to the physiological aspects of chromosome action and Mendelian mechanisms, though they are relevant. The student should read in any of the many excellent texts available about "Mendelian algebra" and *chromosomes*. The latter are the strings of genes, in the double helix, present in each cell, which contain virtually all of the cell that controls heredity. Parenthetically, not only the *gametes* (the cells—ova and sperm—concerned with reproduction) but every cell (except the nonnucleated red blood cells) in a person's body contains his or her unique chromosome pattern.

Probably the most important concept for the student to grasp as we glance over the physiology of the chromosomes is that the ordinary chromosomes as in the body cells divide down the middle to produce the gametes. Since one gamete comes in the ovum from the mother and one in the sperm from the father the child's chromosomes are composed of one-half of the gene alleles of the father and one-half of those of the mother. The number of chromosome elements is different in different species; in the human species it is 46, so that the gametes of each parent, after the "reduction division" to produce the germ cells, have 23, but their union restores the full 46 to the child. There are many consequences of this process of cell division to form gametes, followed by reunion, which must be read in the appropriate texts, but which we may refer to here briefly at appropriate points. An obvious consequence is that it is incorrect to say "Bill resembles the mother more, but Merle the father" except for some one quite specific quality. In toto, each child resembles both parents equally—except for some dominance effects and the X-chromosome.

It is now possible to photograph the elements forming the chromosome chain, and one fact that is evident in this resulting *karyotype* map, as it is called, is that the female has one substantial chromosome—the X-chromosome—which is larger in size and possesses more genes than the corresponding but diminutive Y-chromosome in the male. (Other insights given by photographically studying the karyotype are into the origin of Down's and Klinefelter's syndromes, as due to abnormal reproductions of chromosomes.) If a gene for a certain disorder appears only on the X- or Y-chromosome (as hemophilia does) it is said to be *sex-linked*. Among abilities there has recently been much discussion as to whether spatial ability (Thurstone's *S*) is sex-linked.

Discussion of whether a certain gene is on the X- or Y-chromosome, rather than those of the total *soma* (the chromosomes determining the organism other than in its sexual traits), leads to the general concept of the *locus* of a gene. That concerns not only whether it is on the sex chromosome or in the somatic range, but also in what part of the latter it stands. A gene's position along the chain has relation to what we shall later encounter as *linkage* effects. What the gene is chemically still baffles us in most instances (except that it acts through enzymes, and is stored in the "memory" of the chromosome by a pattern of amino acids and organic bases). But appreciable progress *has* been made in discovering the loci of a number of genes. They are then at least given what Shakespeare called in *A Midsummer Night's Dream* "a local habitation and a name" as a position on a chromosome. To explain how this is done by "linkage" and "cross-over" evidence is beyond our present domain, but it can be said to depend on accidents in which, in the reduction division, a gene tends to move with the genes immediately around it rather than with those more remote.

Two other terms, besides dominance and recessiveness of alleles, that a student needs to note while we are taking stock briefly of the biology of genetics are *epistasis* and *linkage*. The first refers to an interaction effect occuring between two or more quite different genes; the second, to genes in close proximity tending to move over together in the division process.

If the student reads further into the physiological basis of Mendelian laws, he will also encounter such terms as *haploid, polyploid, genome,* and *polymorphism* of alleles. The *genome* is virtually synonymous with the genotype, in the sense that it is the physical substrate of the genotype, which is an abstraction. It refers to a particular chromosomal make-up of a given individual. *Polyploidy* arises when the set of genes forming the usual genome form instead a new "multiple" genome by doubling, trebling, and so forth. This typically produces a stable or unstable new, but related, species. A *haploid* genome is one split by the meiotic division as in forming a gamete, thus having *half* the normal number in the parent genome. *Polymorphic al-*

lelism is when there are not just *two* alternative forms for the given gene to take, but several.

Most of these terms in the physiology of the chromosomes and their action we shall rarely need to deal with, because the inheritance of personality and ability is as yet mostly too imprecise quantitatively to permit inferences on gene action. But before leaving this area we should tighten up further the implications of the terms phenotype and genotype as they apply to this psychological area. The genotype refers, as stated earlier, to the effect of a certain pattern of genes (for an individual or a type) in terms of the organism produced. It is important to keep in mind that the genotype—what the organism would be with no environmental intrusion—is never actually seen, but is an inference from the only observations possible—those on the actual phenotype. The term *gene*—or *genome* for a collection of genes—is definite enough: It refers to what is known to exist in the karyotype. The genome is thus the *total* gene structure in the individual's chromosomes. In growth it is the number and nature of the individual's genes and the relations among them that determine how they will *begin* to "produce" the individual. In much simpler organisms, like the fruit fly, the efforts of Morgan and many since can give a useful map of what the genotype, as based on the genome, is like.

By contrast, phenotype, long treated by geneticists in botany and animal husbandry as a term as precise as genotype, is in fact always an uncertain, conditionally defined entity, though this is perhaps not clearly brought home to us until we get into behavior genetics. Mendel, as we pointed out, could safely call tall and dwarf peas two phenotypes, because there was nothing in the range of environment to which he exposed them that produced middling-tall specimens difficult to classify. By contrast, even the simplest theory of intelligence inheritance is such that some eight or nine genes, of different size effects, would be involved, and remoteness of the phenotype is made still greater by the fact that the performances which rest on a particular genome are always expressed through a considerable variety of culturally acquired skills. Presumably, the various combinations of the genes in the genotype would produce many small steplike discrete increments in a histogram of genotypic distributions, but the added environmental influences smooth these out into the normal distribution curve that we actually see in the phenotypes.

In short, whenever the range of environment is such that its interaction with heredity produces appreciable effects there is no such thing as *the* phenotype corresponding to *the* genotype. There is, of course, for an individual, at a given moment, *a* phenotype, namely, what he or she literally is at that moment; but there is no life-long phenotype for an individual, and *no fixed phenotype for a given typical genotype* such as the botanist or animal breeder can often safely talk about. We set out to define it by *the average phenotype, for a given genotype, in an absolutely defined and controlled en-*

vironment. A phenotype must therefore always be defined along with, and carry a tag for, a particular environment. And if by a genotype we mean not a particular genome—in actual physical pattern of genes—but some idea of what the genome would produce as an organism if "on its own," then we are again defining an abstraction. For the organism we see is never produced "on its own," so the genotype, like the standard phenotype, also needs a definition of the nature of the environment in which its properties are inferred.

7. Summary

1. Genetics is a science with diverse roots in animal and plant breeding, medicine, and general observation. As a science with clear theoretical underpinnings it is scarcely more than a century old.

2. Where human heredity is concerned there have been rash social, emotional, religious, and political assumptions which have interfered with support, growth and application of the science. They cannot be relegated to history because—especially where behavioral heredity is concerned—they are still active, though subsiding. Consequently the student needs to be aware of these sources of distortion, and three of the main misunderstandings are discussed. However, there is perhaps a fourth and deeper opposition in many people to the notion that their freedom is to some extent constrained by hereditary potentials. It arises from what Freud (1924) called the "infantile omnipotence" of the id struggling with the reality principle. At times we would all wish that the law of gravity, the laws of thermodynamics, and the law of compound interest, could be suspended. As A. E. Housman wrote:

> To think that two and two are four
> And neither five nor three
> The heart of man has long been sore
> And long 'tis like to be.

The more appropriate attitude is that the laws of heredity, like scientific laws generally, may be understood and managed toward human good: They can be denied and ignored only at our cost.

3. The diverse ancestries of genetic research have produced differences of concept and nomenclature (e.g., those between medical, genealogical, botanical, biometrical, and animal husbandry approaches), of which the student should be aware and which call for a clarifying synthesis, with some standardization terms, and adaptation to psychological genetics. Some preliminary discussion is given to clarifying genome, genotype, and phenotype.

It is shown that convarkin methods are a special human development of methodology within biometrical methods, and that twin, MAVA, adoptive family, etc. designs are divisions of the convarkin method.

4. Over the last half century there has existed what has amounted almost to a suppression of incorporation of genetics in psychological theories. This has been partly due to the social atmosphere and partly to the popular appeal of simplistic Watsonian reflexology. Both personality and learning theory are due for a substantal advance in potency and precision as the findings of behavior genetics flow into the neglected areas.

5. It appears not to be fully realized that behavior genetics can contribute substantially to learning theory. Learning theory has usually been calculated as if observed change is due to learning experience and thus has appreciable error, since only an unknown part of the change is actually learning. The division of variance into maturational and learning components enables more precise law to be developed regarding the latter. Furthermore, there are interaction effects from these two sources of change which remain to be analyzed out. The new *structured learning theory* (Cattell, 1979a, 1980a) develops the learning aspects beyond the range of the indications in this book.

6. The psychologist who studies behavior genetics must be prepared to handle some of the rigorous mathematical concepts in Mendelian and biometrical genetics, as well as to have at least a working familiarity with biological roots. The former deal largely with variance and correlation analyses; as these are already familiar in form to the psychometrist, they probably will not necessitate special studies. But introductory genetics reading may be needed to amplify the sketch given here of such concepts as the gamete, monozygotic and dizygotic twins, reduction division, chromosome locus of a gene, dominance and recessiveness, X- and Y-chromosomes, alleles, karyotype, heterozygotic, epistasis, linkage, phenotype, and genotype.

2

Methods and Models Available for Research in Behavior Genetics

1. Overview of Methods

At this stage of our science a compromise has to be made between the ambitious technical goals we would like to set ourselves and the practical availability of methods. For example, one would doubtless like to know how many genes account for the growth of the genetic part of intelligence, and on which chromosomes they have their loci. But, in fact, the most we can hope to find out is what fractions of the variance of intelligence, in a given ethnic (racial and cultural) group, are associated, respectively, with genetic and environmental causes.

Or, in the domain of pathological behavior, we might like to know the placement of genes connected with proneness to manic-depressive disorder, and what other characteristics they pleiotropically[1] affect, and whether epis-

---

[1] *Pleiotrophy* is the tendency of a gene to affect more than one characteristic in the organism's development. Naturally, with an observer interested in one effect it took some time before other effects of the same gene were noticed. Consequently, whereas a pleiotrophic gene was for some time thought of as unusual, it is now realized that most genes are probably to some extent pleiotropic, the connections between the diverse manifestations sometimes appearing initially strange.

TABLE 2.1
List of Principal Methods

1. *The clinical syndrome, genealogical, pedigree method.*
   This observes the qualitative, all-or-nothing presence of a trait or pathological syndrome, noting relatives, usually ancestors, in whom it appeared.
2. *The animal, inheritance-manipulating method.*
   This measures performances or syndromes in animals, who can be bred in ways to bring out Mendelian connections.
3. *The physical-linkage method.*
   This seeks to establish a statistically significant relation between a behavioral measure and a physical feature already of established heredity.
4. *The biometrical, convarkin methods.*
   These use correlational and variance analysis methods on measurements (continuous) on subgroups of plants, animals, or people of various degrees of genetic relationship and shared environment.
5. *The longitudinal growth and learning analysis method.*
   This attempts to analyze change curves, with maturational and experiential effects variously combined, to determine the quantitative role of each in development.

tasy is involved, and how the genes act in terms of affecting enzymes and hormones as intermediate steps in neurological action (see Sutton, 1961). What we are actually likely to settle for is values for the increased probability of manic-depressive disorder in a person born of one and of two parents who are manic-depressives.

At this stage it is good strategy not to "stand on ceremony" in method, but to garner evidence by whatever method we can, even though the technical standing of methods may differ a good deal. For example, seeking behavioral links with anatomical features deemed to be genetic is one of our poorer methods, and genealogical, "pedigree" results over only one or two generations are not always dependable. But they give us leads. Let it be understood, therefore, that we apply different standards of precision to different methods, and attach different levels of confidence and generalizability to the results.

In Table 2.1 we list five main methods that have been used at various times and in various areas.

Only a brief descriptive sentence is given to each because each of the following sections will enlarge on one method. In the first three—clinical genealogical, animal manipulative, and physical linkage—we also propose to give here whatever research results need to be given. In the convarkin methods (twin, MAVA, etc.), on the other hand, since the findings are alto-

TABLE 2.2
Broader View of Possibilities in Genetic Investigation Methods

---

1. The "classical," CSG, clinical syndrome genealogical approach (pedigree method). It might also be called the "synkin" method, tracing *syn*dromes in *kin*.
2. Manipulative animal breeding, with controlled, quantitative experiment.
3. Somatic tie search. Attempting correlation of physchological traits with physical traits of known heritability.
4. Convarkin methods, comparing variances in kin by biometric genetic methods. These include
   a. Animal biometric genetics
   b. Human family and culture biometric genetics, constituted by the MAVA model and method
   c. The twin method, which can be regarded logically as a restricted MAVA method used before the appearance of the latter
5. Analysis of volution and learning components out of life development curves.
6. Populating genetics. Examination of trait distribution in large groups, showing various degrees of homogamy (inbreeding, assortive mating, etc.).
7. Molecular genetics. Examination of chromosome structures by microscopic means and chemical and other experiments.
8. Studies of physiological paths of gene action, in the embryo and later, by enzymes, connecting ultimate somatic characters with genes.
9. Embryological development analysis.
10. Ethological observation of animal behavior in situ.

---

gether more extensive and more important, all setting out of findings is deferred to the later chapters of the book.

An overview of a domain more extended than that of Table 2.1—and than that which we plan to cover in this book—is given in Table 2.2.

2. The Clinical Syndrome, Genealogical, or Pedigree Method (CSG Method)

This method is not confined to the pathological, though most instances come from psychiatry and medicine.

Essentially this is the method embedded also in the long *literary* history of "personality study," which, incidentally is not to be despised, since it contains many shrewd observations—unfortunately not separated from unproved generalizations of folklore. Homer and the Old Testament contain recognitions of trait heredity as does Plutarch's *Lives of the Noble Grecians and Romans,* written about A.D. 100. The Renaissance writings abound in comments on heredity in Shakespeare (where "good wombs have born bad sons [Macbeth]" suggests recognition of Mendelian recessives), in Sir Francis Bacon, Comenius, and in the medical writers of the day.

In modern times, the characteristic feature of this method has been that it rightly does not attempt to handle that which depends on measuring some continuously distributed parameter of the organism (like stature, intelligence, or reaction time), but fastens on some syndrome (usually a "bunch" of variables) that has an all-or-nothing character. Such are eye color, schizophrenia, Huntington's chorea, a five-fingered hand, epilepsy, and Down's syndrome.

The observation of such all-or-nothing appearances has characteristically been combined with genealogical observations, going back to grandparents and great-grandparents, since such features are usually obvious enough to have been recorded. Family physicians have been the greatest contributors by this method. The syndromes examined have commonly been of a type of which no environmental explanation is easily possible and are so little prone to varying degrees of manifestation that no dispute arises in recording them. However, there are borderline instances where the concept of a "manifestation rate" has been brought in, admitting that in certain environments the expected hereditary effect does not appear. This amounts to drawing up a Mendelian model of dominance or recessiveness on part of the evidence and seeing how it fits the rest. If the fit is good, but with a few "holes," the latter are considered cases where unusual environment prevented the emergence of the syndrome, and so a manifestation rate is calculated. That amounts to the probability, in the given cultural environment, that the genetic endowment will express itself. "Concordance rates" which we use in what follows (Table 2.4) are a particular form of manifestation rate.

Many quite firm Mendelian forms of heredity have been established by this method. McKusick (1964) has provided genetics with over 2000 instances. The great majority are not psychological. They begin with such easy physical observables as eye color (light recessive to dark), albinism, brachydactyly and proceed to diseases which family doctors observe, such as Anderson's disease (failure in glycogen storage), Bloom's syndrome, Cooley's anaemia, Fabry's disease, galactosemia, cystic fibrosis, and so on. Some psychological inheritances so established are those of myoclonic epilepsy (an autosomal recessive), phenylketonuric mental deficiency, and Huntington's chorea (a simple dominant with 100% manifestation).

It has likewise been possible by this method to recognize sex-linked traits (i.e., where the gene is on the X- or Y-chromosome), such as hemophilia, which affected Queen Victoria's family, vitamin D-resistant rickets, and (as some psychologists think) a gene in visual or spatial ability. Finally, the method has been successful in recognizing new mutations, which appear spontaneously and then breed true. It is believed that a variation to five fingers is such an instance.

However, it is now evident that two kinds of "mutations" must be considered. In true mutations, a gene "flips" to an allelic (alternative) form, producing quite new features, usually in the same bodily area. For example, achondroplastic dwarfism occurs in about 10 in 94,000 births (McKusick, 1964), which is typical of the normally low rate of reversive chromosome abnormality. At the present, with more radiation exposure and chemical provocation, such rates may be higher. In the second kind of mutation, we find something appearing that is not in the parental line, as a possible allele, at all (i.e., it produces something inborn but not inherited). Or, it may involve some radical change in the form of doubling or new linkage in genes that are themselves unchanged. This is recognizable by microscopic cell photographing of the karyotype—the layout of the distinguishable elements in the human chromosome—revealing change in the systematic number that normally exists. A well-known and distressing instance is the birth of a child with Down's syndrome (which used to be called mongolism because some physical features, for example, the eye fold, resemble the racial Mongoloid type). There, with mental defect at the imbecile IQ level, susceptibility to pulmonary infections, lack of sexual maturation, etc., we have a fairly broad syndrome which proves to be due in most cases to an "accidental" triple representation of chromosome 21 in the genome.

The great majority of mutations are defects, so the path of selective evolution is a steep and thorny one, in which the few rare progressive mutations need to be preserved while a throng of lethal or disadvantageous ones have to be eliminated by better survival rates for the better adapted.

The clinical-genealogical method may be said to have had its heyday in the first half of this century, in such classical writings and summaries as those of Blacker (1934), Blakeslee and Fox (1932), Bleuler (1933), Huntington (1927), and, later, Kallmann (1938, 1950), Ruggles-Gates (1946), Slater (1936), and Slater and Cowie (1971). It has advanced greatly in method and knowledge through such workers as McKusick (1964) and Li (1961). It has been particularly productive in Scandinavian, Swiss, and scientifically oriented cultures because of more popular interest and the relative completeness of genealogical records there obtainable. It is still contributing steadily, but less spectacularly, to our body of knowledge, with some special developments in connection with population genetics, as in Cavalli-Sforza and Bodmer (1971), Crow and Kimura (1970), Darlington (1969), Lerner (1968), Malécot (1948), Osborne, Noble, and Weyl (1978), and Waddington (1953).

Among the present generation of behavior geneticists one senses an apparent recession in the prominence of this genealogical method—partly because of more alert criticisms concerning syndrome definition and the methods of locating "cases," and partly because of the attraction of researchers to the newer, "more mathematical" biometrical methods with continuous variables that we have still to discuss. Among legitimate criticisms, we would recognize the charge that certain syndromes—like "inheritance of hot temper" and others in, for example, Ruggles-Gates's survey

(1946)—were too vague or subjective in definition. In many earlier studies—
and this would include some parts of Kallmann's work on schizophrenia—
no allowance had been made for persisting environmental family atmospheres. Less legitimate criticisms—as Slater, Eysenck, and others have
brought out—spring from a more emotional, popular, ostrichlike desire to
deny that heredity could have such potential potency as facts show it to
have. In this connection, a particular revulsion greeted Lange's book, *Crime
as Destiny* (1931), which was unpopular with sociologists because, like Hooton's (1959) researches it continued the Lombroso tradition, focusing on heredity to the exclusion of environment. Lombroso's work was crude and is
probably best dropped from modern evaluations, but Hooton is full of leads
to the behavior-genetics researcher, and Lange's findings, puzzling though
they may be, are well supported by later studies.

Lange found out which men in prison had a twin. He then ascertained
whether the twin was identical or fraternal (i.e., no more related than an ordinary sib). He discovered (Table 2.3) that if the twin was identical he was
much more likely also to be in prison than if he was a fraternal twin. The results were so disturbing that others critically repeated the study relating to a
much better overall sample than Lange's meager 30 cases, as shown in Table
2.3. Stumpfle's results also showed a greater concordance of identical twins
in a class defined as serious crimes.

Such findings will be examined more critically in Chapter 3 on the twin
method. Here we may note, methodologically, that when similarities are
counted in this way, the CSG (clinical syndrome genealogical) method is
moving part way toward the convarkin methods, described in what follows,
which contrast measured *continuous* variables for large samples from different kinship groups. Indeed, convarkin may be thought of as a historical
growth out of the CSG approach toward continuous quantitative characters.

The net result of the clinical-genealogical approach is a considerable

TABLE 2.3
Incidence of Recorded Crime for Identical and Fraternal Twins of
Propositi

	Identical twins		Fraternal twins	
Investigator	Number of cases	Percentage also criminals	Number of cases	Percentage also criminals
Lange	13	76.9	17	11.8
Legras	4	100.0	5	.0
Rosanoff	37	67.6	28	17.9
Stumpfle	18	72.2	19	36.8
Kranz	31	64.4	43	53.5
Total	101	69.9	102	33.0

harvest of dependable findings, and the methodology has been improved by the technical Mendelian action concepts that one can read about in Cavalli-Sforza and Bodmer (1971), Falconer (1960, 1965), Li (1961), McClearn and DeFries (1973), McKusick (1964, 1968), Sturtevant and Beadle (1964), Vandenberg (1965b), and Waddington (1953, 1967). A good summary of contributions has been made by McKusick (1964), who lists over 2000 syndromes with presumed known Mendelian mechanisms. The great majority, as stated above, are physical and physiological, but the various forms of mental defect especially are relevant to our present study of personality inheritance.

It remains true to say, however, that though the CSG method has contributed enormously to knowledge of specific Mendelian chromosomal action, it has been incapable of any comparable contribution to the heredity of the broad *normal ranges* of continuous ability and personality source traits in which we, as personality theorists, are most interested. However, let us consider the syndrome evidence that has been most valuable to psychiatrists and clinical psychologists. Although some minor revisions have been made in the programmatic research findings of Kallmann (1938, 1950) and his associates, his samples remain among the largest available and his diagrammatic representations, presented in Figure 2.1, give the essential picture on the heredity of schizophrenia.

This condensation nevertheles presents some degree of overstatement of the role of heredity because schizophrenia was then considered so much a definite disease entity that the environmental effect of ordinary family atmospheres did not enter into the analysis. Also, although the psychiatrists concerned agreed on the schizophrenic diagnosis, agreement is known to be poor in the designation of "schizoid personality."

Let us therefore consider results that bring some of the precision of the twin method into the CSG. Table 2.4, from Shields (1962), summarizes a good range of independent studies, some, however, on small samples.

Other studies (see McClearn & DeFries, 1973) center on a figure of 46% for identical co-twins of schizophrenics being schizophrenic, which a more weighted average in Table 2.4 might also roughly indicate, whereas, for fraternal twins the average, in Table 2.4, is 9%. From this approach the heritability is obviously substantial, but let us look also at a "syndrome" use of the adoptive-family method which, in itself, is studied further in what follows (p. 170). The sample so far available—as given in Table 2.5—is small, but it agrees in indicating (at a .024 probability) that biological parentage is more important than home atmosphere in determining schizophrenia. It also offers an interesting suggestion that the biological inadequacy is such that it contributes also to antisocial personality, felonies, and neurosis.

Although the subject of schizophrenia is so immense in literature that we cannot begin to point here to explanations or implications of the obvious high heredity, let it be said that when we get to personality factors as such, in Chapters 9 and 10, we find five primary factors in Q data—$C(-)$, $F(-)$,

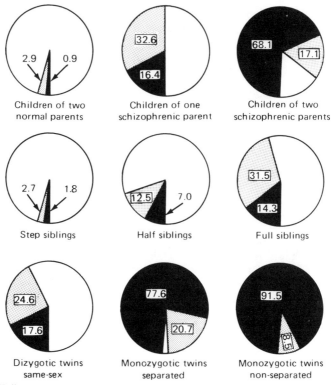

Figure 2.1. Kallmann's findings on frequency of schizophrenia and schizoid personalities among relatives of schizophrenics; black areas indicate schizophrenia; lightly shaded areas indicate schizoid personality.

$H(-)$, $I(+)$, and $O(+)$ and five in T data (U.I. 19($-$), 21($-$), 23($-$), 25($-$), and 32($-$)—that repeatedly and significantly distinguish schizophrenics from normals, and in four of each they also distinguish neurotic and antisocial personality. The batteries for these show appreciable similarity, in tests involving higher susceptibility to threat—$H(-)$, $F(-)$, U.I. 19($-$), U.I. 32($-$), to the GSR and other overreaction measures noted in experiments by Mednick and Schulsinger (1965) to distinguish schizophrenics. The reader will note later the substantial heritability (Chapters 9 and 10) of C, F, H, and I, and of U.I. 19, 21, 23, and 25. Our tentative hypothesis would be, therefore, that, genetically, schizophrenia is not a single entity but an unfortunate coincidence of high inheritance on each of some four or more personality factors. However, this is very different from Heston's (1966, 1970) hypothesis that it is a single entity determined by a single-locus, dominant, autosomal gene allele.

Before leaving the genetics of schizophrenia, let us summarize "risk" at three important degrees of relation by adding some more recent data

TABLE 2.4
Concordances for Schizophrenia in Identical and Fraternal Twins of Diagnosed Individuals

(a) *Percentages by direct count*

Research	Country	Identicals		Fraternals	
		Sample fraction affected	Percentage	Sample fraction affected	Percentage
Kollmann (1946)	United States	120/174	69	53/517	10
Slater (1968)	United Kingdom	24/37	65	10/112	9
Inouye (1961)	Japan	33/55	60	2/17	12
Tienari (1963)	Finland	0/16	0	2/21	10
Kringlen (1966)	Norway	19/50	38	13/94	14
Gottesman and Shields (1966)	United Kingdom	10/24	42	3/33	9
Fischer (1966)	Denmark	3/10	30	0/8	0
Unweighted mean			43		9

(b) *After Gottesman (1978), corrected for proband choice effect (more recent studies, 1963–1973, only)*

	Kringlen	Fischer	Tienari	Gottesman and Shields	Pollin
MZ pairs					
Pairwise range (investigator)	25–38	24–48	0–36	40–50	14–27
No. of pairs (used for "consensus")	55	21	17	22	95
Probandwise concordance (our "consensus")	45%	56%	35%	58%	43%
DZ pairs					
Pairwise range	4–10	10–16	5–14	9–10	4–5
No. of pairs	90	41	20	33	125
Probandwise concordance	15%	26%	13%	12%	9%

Source: Courtesy of Gottesman and Shields (1976c) and Academic Press.

(Elässer, 1952) to our introductory data from Kallmann (1946), in Table 2.6 (from Heston, 1970).

The other great class of psychoses—the manic-depressive disorders—show equally definite heredity. Today, with finer discrimination of varieties of depression, one would like to see separate data for these, but Table 2.7, on a large London sample, gives the essence of relations for manic-depressive disorder diagnosed as such.

Again, the weakness is in the vagueness of "mood disorders of a lesser degree," because families with parents who have not been definitely certified as having such "disorders" *might* also be considered to show such when

TABLE 2.5
Children of Schizophrenic and of Normal (Control) Mothers Who
Showed Abnormality When Raised in Adopting Homes

Abnormality	Control ($N = 19$)	Schizophrenic mother ($N = 22$)	Significance (Fisher's test)
Schizophrenic	0	5	.024
Antisocial personalities	2	9	.017
Neurotic diagnosis	7	13	.052
Felons	2	7	.054

Source: Heston, "The genetics of schizophrenic and schizoid disease,"
Science, 1970, *167,* 249–256. Copyright by the American Association for the
Advancement of Science.

more specifically brought under scrutiny by an investigator. The marked difference of the "offspring unaffected" rate between those having one and two parents affected is striking, and, along with similar results, has led to some promising Mendelian models for manic-depressive disorders, especially in the depressive form.

A useful check on the preceding is provided by Kringlen's (1966) study by the twin-concordance method, as in Table 2.8.

An interesting variant is Schulz's study (1939) of families with one schizophrenic and one manic-depressive parent. In early life the children showed 8% schizophrenic and 19% manic-depressive, but in later life (the manic-depressive syndrome typically shows later) the percentage of the latter rose. This may indicate somewhat higher heritability for manic-depressive behavior. Penrose (1942; see also 1959) found more inheritance in the same-sex child, and argued that auxiliary genes for the major mental disorders are probably to be found on the sex chromosome.

As far as psychology is concerned there can be no doubt whatsoever that a substantial degree of heredity exists for the psychoses, largely specific to the two major forms, and with manic-depressive disorder showing the higher heritability. Regarding this comparison, we note that with both parents affected, 34% of offspring are affected in schizophrenia (Table 2.6) and 67% in manic-depression (Table 2.7), and in identical twins 43% in the former (Table 2.4) and 66% in the latter (Table 2.8). Kallmann's results (Figure 2.1) would place schizophrenia inheritance much higher, and on a par with manic-depression, but, for the reasons already given, these older results are commonly regarded as too high. The general impression of clinical psychology, from Kraepelin, Bleuler, and Kretschmer onward, that the psychoses are to a substantial extent heritable, and physiological, but the neuroses only slightly so, is thus born out by the CSG method. Furthermore, it is recog-

TABLE 2.6
Percentages of First Degree Relatives Found to Be Schizophrenic or Schizoid

(1)	(2)	(3)	(4)	
Relationship	Sample (N)	With schizophrenia (%)	Schizoid (%)	Total (3) + (4) (%)
Children[a]	1000	16.4	32.6	49.0
Siblings[a]	1191	14.3	31.5	45.8
Parents	2741	9.2	34.8	44.0
Children of two schizophrenics[b]	171	33.9	32.2	66.1

More recent and complete summarization by Gottesman

		Schizophrenic			
		Number[c]		Percentage[c]	
	Total relatives	(a)	(b)	(a)	(b)
Parents	7675	336	423	4.4	5.5
Sibs (all	8505	724	865	8.5	10.2
Sibs (neither parent schizophrenic)	7335	621	731	8.2	9.7
Sibs (one parent schizophrenic)	675	93	116	13.8	17.2
Children	1227	151	170	12.3	13.9
Children of mating schizophrenic × schizophrenic	134	49	62	36.6	46.3
Half-sibs	311	10	11	3.2	3.5
Uncles and aunts	3376	68	123	2.0	3.6
Nephews and nieces	2315	52	61	2.2	2.6
Grandchildren	713	20	25	2.8	3.5
First cousins	2438	71	85	2.9	3.5

Source: Heston, "The genetics of schizophrenic and schizoid disease," *Science,* 1970, *167,* 249–256. Copyright by the American Association for the Advancement of Science.

[a] From Kallmann (1938, 1946).

[b] From Kallmann (1938), Kahn (1923), Schulz (1940), and Elässer (1952).

[c] Figures in Column (a) include diagnostically certain cases only; Column (b) also includes probable schizophrenics. Both columns of risk figures are age-corrected. From F. I. Gottesman (1978).

nized that several other mental abnormalities—Huntington's chorea, myoclonic epilepsy, and several forms of mental defect—have specific gene heritabilities.

There are some confusions of nomenclature and problems of method in the pedigree approach, which we shall touch on only briefly, because personality and ability study in the normal range here will depend much more on other methods. The terms *penetrance, expressivity,* and *manifestation rate* have been given somewhat overlapping use in this domain. Partly

TABLE 2.7
Inheritance of Manic-Depressive Disorder

	Manifestation in parents	
	One parent	Both parents
Percentage of children affected	33	67
Percentage showing lesser mood disorders	17	33
Percentage unaffected	50	0

Source: A. Lewis (1933).

through differences of life environment, and partly through the background of other genes in the genome, persons "affected" by a particular allele known to produce, say, a certain disease syndrome, will in fact show it to varying degrees, as in a normal curve or some approach thereto. *Expressivity* refers to this property of a gene of having a range of phenotypic expression.

If one has ulterior, Mendelian evidence as to the invididuals who must possess a certain gene than one can calculate a manifestation rate for that gene in the given environment. When the gene manifests itself it is said to have penetrance, and, for most syndromes, this is discernible and definable in all-or-nothing terms. When we take the frequency in the population with which penetrance occurs we have the distribution curve defining expressivity, at the lower limit of which there are individuals showing low or nonpene-

TABLE 2.8
Incidence of Manic-Depressive Psychosis in Identical and Fraternal Twins of Diagnosed Cases

	Population	Identical twins		Fraternal twins	
		Number of pairs	Percentage same	Number of pairs	Percentage same
Luxenburger (1928)	Germany	4	75	13	0
Rosonoff *et al.* (1935)	United States	23	70	67	16
Kallmann (1950)	United States	23	96	52	26
Slater (1953)	United Kingdom	8	50	30	23
da Fonseca (1959)	United Kingdom	21	75	39	39
Harvald and Hauge (1963)	Denmark	15	60	40	5
Kringlen (1967)	Norway	6	33	9	0
Unweighted mean			66		16

Source: Kringlen (1967), pp. 73 and 93.

trance. The relation of those pedigree syndrome values in expressivity and manifestation rate to the alternative expression (primarily for continuous variables) in *variances,* respectively, from heredity and environment, as done later here, need not be pursued into its mathematical forms here, but the reader will find further discussion thereon in McKusick (1964, p. 58), Falconer (1960, 1965, 1967), and others.

A problem in the genealogical, pedigree method is that one unavoidably makes some selection, relative to a chance sample from the population, when one *begins* by taking affected cases. The case taken is called the *proband* (or *propositus*) and at present most investigators are alert to certain precautions needed in the *proband method.* What expectations one has when examining the rest of the family and remoter relatives will depend on the genetic hypothesis. If the assumption is that the effect comes from a single gene, say a Mendelian recessive and that parents in the population are heterozygous for this gene, then normally we should expect 25% of the collected offspring to possess the genetic property, whether they manifest it (from environmental permissiveness) or not. The rules of the proband method, however, tell us that the expectation would actually be 57%. In short the approach by clinical selection of cases (the proband method) biases the statistical outcome toward a decidedly higher expectation than the general population sampling expression of Mendelian laws would indicate.

In retrospect, over uses of the clinical syndrome genealogical (CSG) method we must accord to it a substantial contribution to behavior genetics and a still more substantial one to medical genetics, especially when aided by twin-method data.

3. Experimental, Inheritance-Manipulating Animal Research

Whereas the proportion of researchers working on the clinical-genealogical approach has somewhat declined (though many prizes in new findings are still to be gained), experimental work on animals has rapidly increased. This mode of research covers, on the one hand, specific syndromes (e.g., the behavior of "waltzing mice") and, on the other, measured traits of performance on continuous scales (e.g., rate of running a maze, and number of errors in a perceptual-discrimination task).

The considerable advantage of the animal method is that matings can be arranged—various inbreedings, such as backcrosses, can be carried out, and in sufficient numbers to give reliable percentages. A second advantage over research with humans is that the environmental experiences can be fully controlled. A third advantage is that such studies can be combined with brain operations to determine the hypothesized locus of the indicated neurological basis, as Lashley (1963) did with rats and many others have done since with rats and apes.

Basically, animal behavior genetics is concerned with the same targets as human behavior genetics—namely, such findings as the CSG method brings, the breakdown of phenotypic into genetic and threptic variances, the discovery of physiological intermediates, and gene identification. However, despite its advantages, animal research has for our present book one severely disabling disadvantage: that each species has its own gene system and that, consequently, one cannot reason precisely from animal to human findings. For that reason, important though the subject is in itself, we shall allow it only one section here. The justification for giving it even one section is that there are Mendelian laws that transcend any particular gene pool, and there are findings which, at least by analogy, give leads for human genetics. For example, if associations of innate behavior components with physical traits are found in animals—as they are—it suggests this route will not be barren of promise in humans.

Perhaps the social psychologist—and certainly the sociologist—have tended to deprive us of such leads by insisting too much on the irrelevance of animal research. "What," they ask us, "is ambition in a rat or guilt proneness in a monkey?" But perhaps there could be similar genetic mechanisms in the analogues, as when we compare patterns of inheritance in a shy person and a shy horse, although the actual behaviors measured are necessarily different. The sociologists who are inclined to criticize comparison of traits even between different human cultures, on the grounds that, say, filial piety in India is very different from what would be called filial piety in Japan, are at any rate consistent in objecting to reasoning to human genetics from that of mice.

To these objections of, especially, the sociologist, one naturally replies that medical animal research—for example, on cancer producing and curing compounds—habitually reasons "by analogy." It does so because mammalian physiology has been found to keep to a fairly common denominator. But clearly the social customs and cultures of rat, dog, and human are something else. Furthermore, one may reasonably hypothesize that if a seemingly genetic trait is physiological, and the same in meaning across species (e.g., autonomic reactivity, or a sexual or fear response), it is by that very breadth of phylogenetic occurrence something so largely innate and inheritable as scarcely to require any confirmatory experiment.

Dogmatic methodological statements are inappropriate here; the true scientists have to "play by ear." When Williams, McClearn, and others found differences in the capacity of breeds of rats to become addicted to alcohol, they started a valuable new theory in human alcohol addiction. When Feuer and Broadhurst (1962) found that measured "emotionally more active" rats tended to be more hyperthyroid, they called attention to a possibly useful link in human behavior genetics. When Tryon (1940) showed that in ten generations of rat breeding a good learning substrain could be so genetically separated from a poor learning substrain that their performances scarcely overlapped at all, he made the arguments for the inheritance of human intelligence more palpable to many who had mistrusted the bare statistical arguments of test similarities in relatives. Indeed, we would argue that in not too dissimilar species there are traits—such as pugnacity, anxiety, intelligence (relation-perceiving capacity), and inhibitory capacity—and certain behavior disorders—such as ataxias, epilepsies, muscular-neural dystrophies, audiogenic and other induced seizures—that are sufficiently alike (e.g., in chimpanzee and human) to offer conclusions from animal behavior genetics that are profitable for human research. This matter is discussed more closely (with some differences of conclusions!) by Broadhurst (1971) and Royce (1966).

What is scientifically beyond debate is that certain genetic principles of wide generality will carry across species—for example, any further findings in Mendelian principles as such. Thus Fuller (1965) found in certain traits, with the precision which animal research permits, evidence of decidedly more dominant and epistatic action than has been *assumed* in the largely hypothetical additive-action models in human inheritance of roughly similar traits. He also brings supportive evidence for our argument (p. 219) that the effects of heredity may become more, not less, prominent with age; for in dogs he found the genetic resemblances greater in an older set of animal relatives.

Probably one of the most vital contributions of animal research, in these laws of wide generalizability, is that for the fullest development of an ability or emotional attachment, the environmental stimulus must come at the right time, as fixed by the inner maturational "clock" of the species concerned. These phenomena, which we may cover by the term *imprinting,* have been widely studied in regard to visual perception in chimpanzees, sexual behavior, attachment of offspring to a parent object (Lorenz & Leyhausen, 1970), succorant behavior to the young, etc. The involvement of the imprinting effect with genetic processes was noted by Fuller (1965) in regard to genetic differences of two dog breeds, subjected to "isolation procedures" at critical maturation periods. He records "the measurements on which the beagles and terriers are most unlike are the ones which are most modified by isolation procedures [p. 252]."

The reader who wishes to pursue the behavior genetics of particular species will find a rich literature in Broadhurst (1971), DeFries, Weir, and Hegmann (1967), Fuller (1965), Fulker (1966), Hirsch (1967), Lush (1940, 1968), McClearn, Wilson, and Meredith (1970), Royce (1955, 1966), Royce and Covington (1960), Scott and Fuller (1965), Tolman (1924), Tryon (1940), and others.

4. The Physical-Linkage Method

It happens that knowledge of inheritance of obvious *physical,* and researchable *physiological,* heritabilities has moved ahead more rapidly than behavior genetics. Indeed, especially in animal breeding, the mode of inheritance of coat color, short and long ears, hornedness, color of eggs, etc. is widely known. And, in humans, the heritability of eye color, crinkliness of hair, red hair, blood groups, sickle cell erythrocytes, form of eye, mid-digital hair, ear lobe form—and many other things subject to close everyday human observation—has become textbook property.

Research therefore becomes possible, at least as a preliminary reconnaissance, in designs aimed at discovering some behavior traits that can be *significantly tagged to some physical trait already known to be highly heritable.* Investigation of physical–behavioral correlations thus throws light on the degree and manner of heritability of the associated behavioral tendency. Much of the evidence in this field is at present confused, inchoate, and mixed with superstition. Folklore abounds with relevant but doubtful observations. For example, do we put any weight on the English folk comment "Ginger for pluck" which supposes reckless courage in redheads? Or the belief that men of Falstaffian build are especially genial and sociable; or that Scottish sheep dogs can be more readily trained than any others to round up sheep? Shakespeare's contrast of the genial Falstaff with the double-crossing "lean and hungry" Cassius receives some support from Kretschmer's (1929) researches, but one doubts that Becky Sharp's ambition had anything to do with her green eyes. And one shudders at the 24 hours when the fate of biology and the theory of evolution hung in the balance while Captain Fitzroy debated whether to take Darwin on the *Beagle,* because his snub nose suggested to Fitzroy that he might not have enough character to withstand the hardships! Yet in this garbage heap of folklore there may well be a few jewels, as medicine found when it tried the old wives' brew of foxglove for heart disease.

Attaching mental traits to physical features is logically in the same class as raciology and physical anthropology. And, unfortunately, research along these lines with humans is thoughtlessly attacked by certain groups who mistake (sometimes deliberately) raciology for racism. Although some single physical elements have been found, with real but rather low statistical significances, to correlate with psychological measurements, associations with the whole physical pattern and gene pool that defines a race have been matters of hot debate. In short, we have to recognize that research on racial contrasts even when carefully and scientifically pursued (Baker, 1974; Eysenck, 1971; Hooton, 1959; Huntington, 1927; Jensen, 1971; Kuttner, 1967; Loehlin, Lindzey, & Spuhler, 1975; Lynn Dziobin, 1980; McGurk, 1967; Porteus, 1967) has not often produced results that are beyond explanation—if zealously sought—on other grounds.

TABLE 2.9
"Blind" Separation of Dog Breeds by Behavior Dimensions Alone,
Using the Taxonome Type-Seeking Program[a]

Group segregated by Taxonome	Actual breed of dog			
	Basenji	Beagle	Cocker spaniel	Sheep dog
Group 1	10	0	0	1
Group 2	0	14	3	0
Group 3	0	0	5	1
Group 4	1	0	0	10

[a] The chi square for this agreement is 116.09, which is significant at $p < .001$. An alternative "cut" in the Taxonome program, which included (but showed difficulty in separating) the fox terriers, classified 90% of the animals correctly, for a ψ^2 of 234.41, also significant at $p < .001$ (Cattell, Bolz, & Korth, 1973).

That within a single genus or species such as man or dog there can be significant differences of psychological predispositions associated systematically with physical features is nevertheless beyond question. Scot and Fuller (1965) found different breeds of dogs, not in any way trained differently, to differ significantly on 49 out of 50 measures of behavioral response.

Cattell, Bolz, and Korth (1973) took Brace's 42 measures on five different breeds—basenji, beagle, cocker spaniel, sheep dog, and fox terrier—factor-analyzed to 16 "personality" factors. Using profiles on 15 traits covering behavior alone (i.e., dropping one with physical variables) and applying the Taxonome computer "typing" program, they obtained positive separations as shown in Table 2.9.

As regards human races, there have been allegations that the difficulties in objectively making clear separations, even on physical bases, invalidates the concepts. Several leading physical anthropologists (Baker, 1974; Coon, 1958, 1963; Hooton, 1959) have perhaps sufficiently answered this, but at present in any case, there are objective sorting and typing methods and computer programs like Taxonome (Campbell, 1979; Cattell, Coulter, & Tsujioka 1966; Sokal & Sneath, 1963) to answer the problem.[2]

We do not propose here to get involved in the morass of arguments and counterarguments that needs to be passed if any succinct conclusion on racial–psychological trait differences is to be properly supported. The reader is referred to the excellent modern critical evaluations of Baker (1974), Ey-

[2] These difficulties are routinely exaggerated by critics. The methods of Wright (1934, 1968), Fisher (1930), Haldane (1932), Simpson (1961), and Dobzhansky (1962), and the applications by Baker (1974), Coon (1963), and Hooton (1959) can lead to operationally definite concepts of race. More recently, as stated, computer programs of Cattell, Coulter, and Tsujioka (1966), Sokal and Sneath (1963), and Campbell (1979) have provided comparatively *objective* sorting into types.

senck (1973), Freedman and Freedman (1969), Jensen (1970b, 1973), Loehlin *et al.* (1975), and by Kuttner (1967), and Osborne *et al.* (1978). As Kuttner (1968) points out, Sir Ronald Fisher has commented: "Available scientific knowledge provides a firm basis for believing that the groups of mankind differ in their innate capacity for intellectual and emotional development, seeing that such groups do differ undoubtedly in a very large number of genes [p. xxiv]." To which comment on "the myth of race" (Montagu), Nobel Prize laureate Muller[3] has added a calm scientific "Amen." The question remains, of course, whether this "large number of genes" in which they *differ* is large *relative* to the number in which they are *similar*. Lewontin (1970) has contended that it is relatively small. On the phenotypic level it is true that every behavioral variable yet measured across races shows a difference of mean which is small relative to the range within each race. (This might be said even of the 15 points of IQ mean difference found between blacks and whites in the United States.)

That significant mean differences are found *phenotypically,* on personality as well as ability measures, among blacks, whites, Japanese, Chinese, and Hindus, and even between British, United States, and New Zealand (Vaughan & Cattell, 1976) populations cannot be gainsaid. The evidence on *Q*-data measures will be found in Cattell, Eber, and Tatsuoka (1970), Lynn (1977), and Meredith (1965), as well as in the references cited earlier. More recently it has been shown that objective *T*-data (O-A Battery) measures support this (Cartwright, Tomson, & Schwartz, 1975). Interesting though these results[4] are, we cannot seriously analyze them because we lack the factorial design for separating races and cultures that Table 5.21 (p. 180) shows to be necessary.

---

[3] H. J. Muller writes,

> To the great majority of geneticists it seems absurd to suppose that psychological characteristics are subject to entirely different laws of heredity or development than other biological characteristics. . . . Since now there *are* these very abundant *individual* differences affecting psychological traits it would be extremely strange if there were not also differences, in the frequencies of such genes, between one major race and another, in view of the fact that there are such pronounced differences in the frequencies of genes affecting physically and chemically expressed traits. That would surely be the attitude of the great majority of geneticists [Muller, quoted by Kuttner, 1967. p. xxvi].

[4] Cartwright *et al.* (1975) found, in Chicago young males (age 11–24), 242 black and 78 white, all of the same socioeconomic status, differences of means on seven of the personality factors in objective tests (as studied in Chapter 10 here) to be significant at the $p < .01$ level and one at $p < .05$. The whites were higher ($p < .01$) on: U.I. 16, ego assertion; U.I. 22, cortertia; U.I. 25, reality contact; U.I. 26, self-sentiment; and lower on: U.I. 18, hypomania; and U.I. 33, sanguineness. There were no significant differences on anxiety U.I. 24; asthenia, U.I. 28; and dissofrustance, U.I. 30. Except for U.I. 26 and U.I. 18 (the heritability of which is unknown), the factors with substantial difference are also those with appreciable ("within-white") heritability (see Chapter 10), though one trait with fair heritability, U.I. 24, anxiety, shows no race difference here. Inasmuch as quite a number of anatomical, blood-group, and other differences are known between these racial groups, it is reasonable to see if differences on 16, 22, 25, 26, and 33 could be attached to these by pleiotropy.

The racial-difference approach, when culture is not grossly dissimilar, is, however, one from which the behavior geneticist may yet hope to glean leads to *specific* gene associations. For it is now clear that certain genes are so much more prevalent in one race than another that the establishment of a potential racial–behavioral difference is a good clue for eventually tracking down the gene responsible. The much higher incidence of sickle cell anemia in blacks is an example of such a lead, and there are others.

Another difficulty, which we discuss elsewhere, is that a certain phenotypic trait may prove to be accountable for by different genes in different races. This need not cause confusion because though both gene A and gene B account for the same phenotypic trait X, yet, because of pleiotropic effects, A is likely also to account for Y, whereas B may account for Z but not Y; so that such false instances of "understudy" can eventually be detected.

Leaving, for reasons given, the search for racial associates to much more extensive scholarly works (Baker, 1974; Loehlin *et al.*, 1975), we shall turn instead to any evidence of ties with quite specific physical and physiological features. Such evidence is still sparse, but gradually increasing and improving in its methodological underpinnings. It includes observations on blood group types (Cattell, Young, & Hundleby, 1964; Mai & Beal, 1967; Osborne & Suddick, 1971; Swan, Hawkins and Douglas, 1980), skeletal build (Kretschemer, 1929), fingerprints (Swan, 1981, in press) and eye color (Bernhard, 1965; Cattell & Malteno, 1940; Gary and Glover (1976).

Some of the findings obtained in the statistically crude work of Sheldon and Stevens (1942), associating personality factors with body type, deserve further, more systematic study.[5]

Even associations with cranial capacity deserve renewed attention with modern methods. After years of skepticism (since Pearson's initial [1906] low but positive values of head size and estimated intelligence), it has become clear that the skepticism is unwarranted, though the correlations remain quite low. Table 2.10 gives correlations obtained by Valen (1968), Suzanne (1979), Swan, Hawkins, and Douglas (in press), and other sources.

For a thorough discussion of the many factors attenuating or otherwise influencing such correlations, and the influence from head size to cortical weight, see Gellis (1977) and Suzanne (1979). Conceivably cranial size is not

---

[5] Kretschmer's (1929) observation of more leptosomatic body build in schizophrenics, contrasted with broad (pyknic) body build in manic-depressives, was in several ways methodologically sounder than later work. Sheldon and Stevens (1942) essentially repeated Kretschmer with more refined body build indexing (endomorphs, mesomorphs, ectomorphs) and less pathological temperament descriptions. I have criticized that work on statistical grounds, such as inadequate size of samples and use of correlation clusters instead of factors for both physical and behavioral parameters. But in the crude ore of these studies there are bright gleams of significant relations which behavior genetics would do well to separate from the dross by better statistical treatment and improved personality measurement.

TABLE 2.10
Cranial Size Correlated with Intelligence

Researcher	Year	Sample and size	Measure	Correlations
Pearson	1906	Random, 4486	Rating	.11 ± .015
Murdoch and Sullivan	1923	Random, 595	IQ test	.22 ± .041
Sommerville	1924	University, 105	IQ test	.10 ± .099
Valen	1968	Random, mean of 80 and 71	IQ test	.10 ± .11
Suzanne[a]	1979	Random, 2071	IQ test	.19 ± .001
Swan, Haskins and Douglas	In press	School children—age constant, 547	IQ test	.11 ± .04

[a] Head perimeter in inches. Falls to .14 instead of rising (as might be expected) with stature partialed out.

entirely an inborn characteristic, but it is sufficiently so for the relation to illustrate the present methodological approach.

A physical characteristic which we know, through the work of Landsteiner and his successors, to be entirely hereditary, is the blood group to which an individual belongs, and this basis has the advantage that its modes of Mendelian inheritance are well known. It has been questioned whether blood type is an ''adaptive trait'', if it is not and is spread only by ''genetic drift,'' it would be unlikely to relate to much of psychological importance. Mourant (1954) who wrote the classical work on blood-group distribution argues that there are adaptive features, for example, the O blood group in Europe is more prevalent in fringe and mountain retreat areas where competition might be less. His survey shows the A group most prominent in Europe, the B among Mongoloids, the O in American Indian and Negroid races. Within Western Europe, A decreases and O increases as one goes north and west, the latter being highest in Ireland. But despite some such signs of order, the world population is a crazy quilt of these and the dozen other blood groups that have since been discovered. The possibility that one might nevertheless find some temperament or ability connections was suggested by the finding of a significantly higher stomach cancer incidence in group A than B or O, and of higher rates of duodenal ulcers and high blood pressure in O, descending through A, B, and AB. In the syndrome approach, the A group has shown a higher incidence of schizophrenia and obsessional-compulsive disorders, and the O group more depressives (Masters, 1967; Nance et al., 1965; Parker, Thielie, & Spielberger, 1961). For further documentation, the reader is referred to an excellent survey by Swan, Hawkins, and Douglas (1980) and to the article by Cattell, Brackenridge, Case, Propert, and Sheehy (1980).

Attempts to relate blood groups to ability measures have been made by Osborne and Suddick (1971), Swan (1980), and others with both some success and some confusion. The former found six genes—Le, K, Fy, Jk, A, and Ab—with sufficient relation to give a significant multiple R for general intelligence, but with markedly different weights in white and nonwhite samples. Swan, with six primary abilities, found a negative relation Rh (Rhesus) at $p < .05$ with spatial ability and at close to $p < .01$ with reasoning (Thurstone primaries). Cattell, Brackenridge, Case, Propert, and Sheehy (1980) found a $p < .05$ relation of intelligence to blood groups 6-PGD and BGP.

When personality relations are examined not through syndromes, but through measured source traits, significant correlations are found, but they are low and, furthermore, tend to change with different populations (as with Osborne's finding). Cattell, Young, and Hundleby (1964) found ($p < .05$) that type A is higher on I factor (premsia) than are O, B, and AB (in that order), and (just short of $p < .05$) found personality factor J higher for AB. A more extensive study (Cattell, Brackenridge, Case, Propert, & Sheehy, 1980) failed to support this significantly, but gave decidedly more significances than chance overall expectations, particularly (a) tying factor A, at the $p < .01$ level to Fh(C) and MN, and Q_3 (self-sentiment) to ABO; and (b) showing a highly significant relation ($p < .001$) of the second-order anxiety factor to the P blood group. The latter finding is particularly convincing because significant P blood-group correlations run very regularly through the primaries that normally enter into the second-order anxiety factor. Eysenck (1977) mentions data association ABO with introversion–extraversion, but our data showed association only with the P group and emanating particularly ($p < .01$) from the H (parmia) primary component in exvia. Finally, the work of Swan *et al.* (1980) shows a significant and clear relation of O blood group with Q_4 (ergic tension) and a suggestion ($p < .07$) of a relation of B, A, O, and AB, in that order with declining H (parmia). Their more recent, unpublished work approaches significance in the associations of Rh with A (affectia) and ABO with I (premsia). The research of Mai and his colleagues (Mai & Beal, 1967; Mai & Pike, 1970) gives indications in this area of other faint associations. The personality factors with degrees of blood-group association worth following up are thus A (affectia), H (parmia), I (premsia), J (zeppia), Q_4 (ergic tension), and the second-order anxiety factor, QII. (May and Stirrup [1967] claim some findings of O blood group association with neurosis which may represent this association with anxiety.) Some convergence to a definite hypothesis is surely indicated in the association of the O blood group with (a) duodenal ulcers; (b) ergic tension, Q_4, in the 16 P.F.; and (c) the relative association of O (Mourant) with retreat to less competitive "backwater" regions. It has been suggested that some of the instability of blood-group findings may be due to the personality effects not being "pleiotropic" from those genes, but due to linkage (genes in one chromosome *tending* to shift together) of the blood-group genes with genes more directly concerned with personality. Another source of possible initial contradictions is the possible different gene derivation of the same phenotypic trait in long-separated populations.

In general, though the fraction of variance accounted for by the blood-group differences has been small in spite of significance, it is noteworthy that the particular personality factors affected are those that later, by independent methods (Chapter 9), have been found most substantially inherited. Thus, besides the abilities—general intelligence and spatial ability—we find primary personality factors A, H, I, J, Q_3, and Q_4 among the more highly inherited. The defects in the fit are that F is missing from the blood-group associations and that the heritability of Q_4 is not outstanding. At the second order of personality factors, anxiety proneness, QII, was found (Chapter 9) to have appreciable heritability, and its strong association with the blood gene P is in accord.

Although results in blood-group relations have been small and sometimes erratic—as indeed one would expect them to be considering their remoteness in human evolution and their questionable adaptiveness—they are a part of the behavior-genetic jigsaw puzzle, and justify the labor spent. This position is also that of the leading physical anthropologist Rife (1961) in his plea for a systematic study of blood-group relations.

Eye color, because it is so easy to observe rather than because of any suspected significance, has been the subject of a large number of psychological studies. It is methodologically worth discussing as a representative "low significance" but suggestive instance of linkage to a known Mendelian physical determiner. In virtually all cases the following studies were (a) confined to Caucasians (Alpine, Nordid, Mediterranid mixtures); and (b) to subjects within the same culture.

There is no support for any correlation with intelligence, but there is support for some specific abilities and for temperament traits. Swan's recent results (739 cases) give a significant eye-color relation only to premsia (I factor) ($p < .03$) among primary personality factors. This shows dark eye color associating with premsia, and blue-gray with the opposite, harria. Premsia approaches significance also with dark hair coloring and (inverted) with Nordid-versus-Mediterranid racial diagnosis (using Baker's [1974] definitions). No other trait on the HSPQ, CPQ, etc., came anywhere near significance with eye color within Swan's group of 739 white Americans.

In the ability field, Stafford (1972) has indicated some relation of eye color to handling quantitative reasoning as "mental arithmetic," as shown in his diagram in Figure 2.2. His work must be read for due support for this possibly improbable-seeming result. However, it is supported by another study in Jordan's (1980) survey, and since then Swan, Hawkins, and Douglas's (in press) results with the primary ability measures on 834 cases show a correlation significant at $p < .01$ of the PMA numerical factor N with light eye color. Bernhard (1965), on a large German population sample, found small but statistically significant associations with eye and hair color consistent with these indications, which in his data he sees as inexplicable by any cultural–environmental differences. The small but significant correlations found by Swan, Hawkins, and Douglas (in press) are also hard to ac-

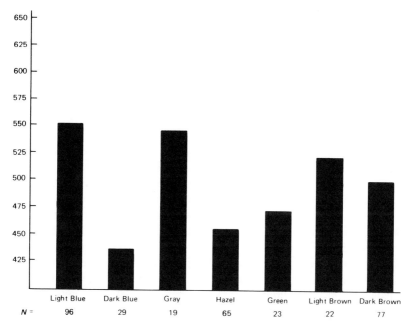

Figure 2.2. Relation of quantitative reasoning ("mental arithmetic") to eye color. Graph indicates mean scores on the Mental Arithmetic Problems of quantitative reasoning by eye color; $F = 2.221$, $p < .05$. From Stafford (1972). By permission of R. E. Stafford and the *Review of Educational Research*.

count for by cultural differences, since they used a socially and culturally very uniform private school white American population sample.

The organization of promising hypotheses in eye color and other physical-linkage areas is rendered difficult less by uncertainties about the physical inheritance than by (a) the unavoidable use in data of 10 to 20 years ago of poor psychological measures, unrelated to ability and temperament factors as now determined; and (b) the pleiotrophy of genes which undoubtedly ties some different modality expressions—abilities and temperament—to the same gene sources, as suggested in the following. If we sought to generalize across races we should also encounter the well-known phenomenon of isoallelism, in the broader sense, where different genes, especially in different gene pools, can produce the same (or practically indistinguishable) results.

Nevertheless we shall attempt some integration. Thus the finding in darker-eyed people of quicker reaction time (Gary & Glover, 1976; Landers, Obermeier, & Wolf, 1977; Tedford, Hill, & Hensley, 1978; Markle, 1975), lower flicker fusion speed (Smith & Misiak, 1973; see Jordan, 1980), and what Markle calls dexterity in reaction to external stimuli rather than skill in self-paced tasks (light eyes) suggests the pattern in the personality factor of *cortertia* in objective temperament tests, indexed U.I. 22.

Similarly, the supported evidence for significantly higher field independence, and some "recognition perception" in light-eyed subjects (Lane & Mendelsohn, 1977; Nisbet & Timoshok, 1977) suggests it would be worth testing the hypothesis that light-eyed subjects are higher on U.I. 19, independence, which Chapter 10 shows to have appreciable inheritance.

One notes also that the syndrome of higher metabolic rate, quicker pulse, quicker respiration, and higher oral temperature found in autonomic factoring (Cattell, 1950) and reported by several investigators (See Cattell, 1946, pp. 444–448) as a correlation cluster associated with sanguine, sociable, and energetic leadership behavior (Herrington, 1942), was found to have a significant association with blue eye color (Lester, 1974). In psychiatric patients and a smaller sample, however, Happy and Collins (1972), found blue eyed more "introverted" (desurgent?) and prone to autism.

A broader, vaguer, yet perhaps deeply temperamentally important syndrome associated with eye color and nigrescence is that which begins with the neglected clue—neglected perhaps because of its age—in Havelock Ellis's (1926) survey of the British National Portrait Gallery, in Trafalgar Square, seeking and finding associations between area of eminence and eye and hair coloring. His finding of more mathematical and science performance in the fair, and more language and religious eminence in the dark, checked well with Bramwell's (1923) finding on choice of subjects by Cambridge students and studies by Onslow (1920) and McDougall (1926). These studies in a culturally uniform British population with intelligence controlled (by confining to high intelligence) point to some as yet not adequately definable psychologically but repeatedly indicated association in the European gene pool of light eye and fairness of complexion with objective, quantitative thinking and of the converse with personal emotional involvment and interest in subjects rich in emotional content. Gary and Glover (1976) interpret, by ratings, the dark eyed as more impulsive and imaginative, which might be called more emotional. Markle (1975), using Rorschach responses, found a difference in light eyed making more *form* responses and dark more *color,* which, by Rorschach empiricism, he interprets as less and more emotional. He also found more marked physiological disturbance from stimuli in the dark eyed, and Sutton (1959) found dark eyed giving stronger reactions to pain. Happy and Collins (1972), in attitude studies with retests, came across a relation of dark eyed showing fewer spontaneous attitude changes and a greater tendency to like people with similar attitudes to their own. This kind of emotionality with rigidity (and religious interest) begins to look like the emotionality of the *protension* factor, *L,* in personality primaries.

Since higher motor-perceptual rigidity has been found correlated with more emotionality on the one hand, factor U.I. 23 (Cattell & Schuerger, 1978; Hundleby, Pawlik, & Cattell, 1965) and dark eyedness on the other (Cattell & Malteno, 1940) the theory has been put forward (Cattell, 1946, p. 480) that we have a genetic plexus here of higher rigidity, dark eyedness, deeper interest in emotional matters, less interest in "dry," quantitative, scientific (and handicraft) matters.

It may be, therefore, that the findings by Stafford (1972) and several other investigators, of quicker and more accurate quantitative performance,[6] is the outcropping in the ability area of a more fundamental temperament pattern (possibly L) having among its genetic determiners a gene either itself producing, or linked with other genes producing, the light–dark eye-color physical difference. The reader will recognize that the hypotheses in this section are quite speculative, and that the concrete relations, though statistically significant, are far lower in accounting for variance than those we shall deal with by later methods. But it is the task of an alert science to attempt to put clues together.

Finally we come to physical features of bodily structure, concerning which reference has been made to the less precise and adequate approaches of Lombroso (1895) and Sheldon and Stevens (1942), and the more adequate of Kretschmer (1929) on mental disorders and Hooton (1959) on the diverse physiques of diverse types of criminal. In the domain of defined source trait measurement, and in abilities, Swan *et al.* (1981) found a significant positive r of lighter hair color with verbal ability and with perceptual ability ($p < .05$ and $< .01$). The latter had significant positive correlation also with frontal-zygo-nasal index (lower) and a number of other skull features. The finding of an association of a fingerprint characteristic with intelligence by these investigators (Swan & Hawkins, 1978) is possibly not statistically noteworthy in view of the number of features examined.

Among personality primaries the number of physique associations found by Swan is very striking, for F (surgency) and J (zeppia). Surgency is positively related—often at $p < .01$—to head breadth, facial length, nasal length, head circumference, stature, and chest girth. The breadth of face and chest girth were also found positively related to surgency in unpublished work of our own, but less significantly. Factor J had $p < .01$ correlations in the main with characteristics (head breadth, jaw breadth, facial length) similar to those found to correlate with with F. (It should be noted that F and J are prominent in the second-order exvia factor.) Factor H, parmia, was significantly positive with the Nordid physical type assignment in Swan's subjects. So far it is noteworthy that the factors with most numerous physical associates are those with highest heritability, as found by later variance anal-

---

[6] Among the main primary abilities, most attention has been given to a possible genetic influence in spatial ability, but, if one is guided by associations with physique, it is numerical and, even more, perceptual ability that have the most evidence of a specific inheritance beyond that of general intelligence, as later paragraphs will show.

Moreover, the findings of Sanders, Mefford, and Brown (1960), showed that students with relatively high *numerical* performance on the WAIS had significant departures in physiology, namely, a higher rate of excretion of urine, nor-epinephrin, arginine, and glutamine.

ysis methods. An exception is Q_3, which variance analysis shows to have high heritability but which here shows a $p < .04$ relation only with nasal index and morphological facial index. An exception in the opposite direction is the existence of negative r's of superego strength, G, with several indices of bodily size, though we find it of low heritability by MAVA. Possibly this is an environmental effect through the larger children tending to be more "above the law!"

The limitations of the physical-linkage method will be obvious from the preceding discussion. First, the correlations are almost always small (most of those discussed here would be about .2, even with attenuation correction) and become checkable only with large samples. Second, the physical feature itself needs checking for degree of innateness. And, third, the possibility of production of the association by an environmental interaction cannot be ruled out, as in the instance of smaller children showing more superego development.

Nevertheless, the physical-linkage method has value as a completely independent approach relative to the other four main methods. And, whatever qualifying conditions the modern scientist may want to put on it, shrewd observers down through the ages have noted the joint inheritance of physical and temperamental features.[7]

5. The Convarkin Methods: Initial Survey

The methods of variance analysis on continuous variables, and analysis of continuous developmental process, will occupy the rest of this book. For that reason we have included in this chapter the findings on the CSG and physical-linkage methods (and as much as we plan to mention from animal experimentation). But the two remaining methods we shall only sketch in as principles here, leaving findings and finer issues for expansions in later chapters.

With the exception of certain mental-disorder syndromes, the phenotypes of which are stable enough to be individually recognized, the bulk of human behavior-genetic research has to deal with behavioral traits that are *continuously graded in their distribution*. Of course, the Mendelian genet-

---

[7] Thomas Hardy, in *The Dynasts* gives tribute to the power of heredity.

> *The years-heired feature that can*
> *In curve and voice and eye,*
> *Despise the human span*
> *Of durance—that is I.*

One sometimes meets a widow startled into commenting on something just observed in the son who never saw his father. Most often it is a combination of a physical feature with a behavior, for example, a specially long-fingered hand and a particular way of gesturing with it.

icist hopes someday to break that continuity down to particulate gene effects that have become smoothed over by environment, but for that we must all be patient. This emphasis is not unique to human genetics, with it restrictions on manipulation; the biometrical study of plants and animals has increasingly had to direct itself to continuous variables—for example, analyzing yields of corn and records of milk production by pedigree cows. This character of the data in fact provoked introduction of the methods of *biometrical genetics*. Because of the the further special character of human psychological-genetic data, biometrics was also developed in that area, but somewhat differently, resulting in the MAVA and the twin methods. Although these fall into the same statistical framework as the biometrical methods applied to animals, historically they have grown up independently, and in several respects biometrical methods applied to humans show developments quite specific to psychological needs. This justifies human twin, MAVA, and family-adoption designs being considered a distinct methodology. Thus, the reader may find that the MAVA method—to be discussed in what follows—has sometimes been classified simply as "biometrical genetics," whereas, though it is generically in the latter, it has many concepts and terms—notably about psychological interaction within families and correlations of genetic gifts with social impacts—that were not developed in biometrical methods with rats, mice, and cows.

As stated earlier, the essence of the methods of *contrasting variance of kin* (convarkin) is that where relatives can be found who combine genetic and environmental influences in different proportions, there is a hope of separating these influences by algebraic methods. Nature has been kind to geneticists by providing them with an extreme case of this difference in combinations in identical twins as contrasted with sibs or fraternal twins. These constellations lead immediately to a separation of the genetic from the environmental variance *within* families. Today we recognize it as the basis of the convenient and widely used twin method within the large realm of convarkin methods, of which MAVA is another example.

These methods operate by measuring the *variation* of trait scores. Statistically this means, technically, finding the *variance*, σ^2, which is the square of the standard deviation[8] about the mean. Thus $\sigma^2 = (\Sigma d^2)/(N - 1)$. One finds the variance for different types of parent and offspring groupings. For example, we might take the variance of IQ among children of ordinary marriages and compare it with that of children of first cousins; or take the variance of twins and compare it with that of ordinary sibs; or the variance of say, surgency, in pairs of adopted children and contrast it with that of children born in their own homes. By contrasting such variance values, using

[8] Since we have no need to get involved in sampling problems until much later we shall use σ^2, that is, the *population* value, by the unbiased estimate, using $N - 1$. As a simple descriptive statistic of the variance *of a given sample,* we would use $s^2 = \Sigma d^2/N$. Note some writers switch the symbols σ^2 and s^2.

algebraic methods that are described in what follows, it is possible to reach conclusions about the relative magnitude of genetic and environmental influences on a given trait.

Figure 2.3 indicates the main subdivisions of the biometrical and convarkin methods. We shall first briefly decribe the nature of MAVA (for *m*ultiple *a*bstract *v*ariance *a*nalysis).

The history of science is replete with methods that began in some relatively narrow usage and were later perceived to permit development in broader ways—both in logical extension and in encompassing new phenomena. (An instance in psychology is the simple test–person–occasion "covariation chart" which later became comprehensive as the basic data relation matrix.) Each special development usually holds to the basic principle but adds its own unique properties and powers. In its potentially *broadest, logically inclusive* sense, biometrical genetics is *genetic analysis based on measurement and statistics* (but *no longer necessarily confined* to continuous variables, to additive heritability, to the concept of an "ideal phenotype," etc. [see pp. 15, 149] with which it naturally began in its botanical–zoological origin). In its title MAVA indicates the fact that it belongs to the convarkin and biometrical methodological stem through the words *variance analysis* (VA). But the "multiple" in *multiple abstract* (MA) betokens its difference from, for example, the twin method, in that it can extend to as many distinct familial constellations (genetic–environmental groups) as an

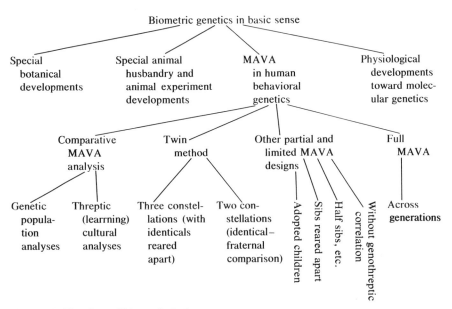

Figure 2.3. The class of biometric designs.

experimentor may be able to find. The *abstract* betokens its introducing an extensive psychological model of conceptual abstractions from concrete variances not fully present in biological biometrics. As a human, social, behavior-genetic research design, and as a model for theoretical analysis, MAVA covers three or more submodels, as shown in Figure 2.3 (e.g., models with and without recognition of possible interaction).

The MAVA method logically subsumes the twin method, since it becomes the twin method when reduced to relations between those two constellations. It also contains, as we shall see later, the adoptive-family, the half-sib, and biological sibs reared apart designs. MAVA has a distinct branch beyond these, however, in *comparative MAVA*, which extracts higher-order psychological relationships concerning the relation of environmental parameters to threptic (acquired trait) parameters and of genetic variance to genetic structure, by use of MAVA analysis repeated on different populations.

Although Figure 2.3 represents the logical structure of the relation of methods, historically the twin method began sooner and froze a number of concepts and statistical approaches. An interesting parallel in scientific history is the way the triads of Newlands, in the study of the chemical elements, arrested attention and were later explained and extended in the full recognition of a periodic table by Mendeleyev. The twin method has in practice essentially rested on two constellations—identical and fraternal twins—with rare instances of a sufficient sample of identical twins reared apart. It is inevitable that the broader MAVA development has bought some difference in notation from the animal biometrical tradition and the twin method in human psychology. As will be seen in Chapter 4, these new MAVA concepts and formulae are shaped more to the conveniences and standard notations of the psychologist.

If we pause to look at the history rather than the logical form of methodology, we recognize the original inspiration of the twin method in Sir Francis Galton's enquiries into inheritance. In his thinking, it came ideally to embrace the variance of identical twins reared together, fraternal twins reared together, and identical twins reared apart. The third category is rarely encompassed in practice because of the rarity of statistically sufficient samples of twins reared apart *from birth*. Mostly, what the twin method has done has been to compare the variance of pairs of identical twins living together, which is assumed to be due purely to environment, with that found for fraternals living together, which must be due to some composite of environmental and genetic variances. By subtraction (see Chapter 3), the two influences are separable.

The MAVA method, on the other hand, takes *all* kinds of child and parent "constellations"—ordinary sibs, unrelated children reared together, sibs reared apart, parents and children together and apart, and so on. The twin approach we shall present in the next chapter, but the more complex MAVA design will require two or three succeeding chapters. In an overview we should recognize that the different *methods* of analysis in all cases imply *models* of what is happening and also indicate what *kinds of empirical values* we need to gather in order to check them. So methods and models must be studied together in the subsequent chapters.

6. Maturation–Learning Process Analysis

We examine here a proposed methodology that has not yet been clearly worked out, still less ever effectively used. First, let us recognize that what is inherited is more than that which is phenomenally present at birth, as the growth of a beard may remind us. Consequently, the behavioral and other acquisitions of a growing organism cannot all be put down to current learning, but rather represent learning interwoven with maturation. We know that physiologically the action of the genes proceeds through protein production, and, in the absence of an uncooperative environment, in processes that are as regular as clockwork. But the educator's emphasis on youth inclines him to see *maturation* (evolving) of properties as the essence of gene action and to forget that *involution* in later life (the refusal of cells to grow further or replace themselves) is also part of the genetic clock. To avoid this we propose to call the process of gene action *volution,* a term covering both unfolding and folding processes, which in fact are closely related in mechanism. Thus genetic *volution* will be taken etymologically to cover both *maturation* (or evolving) and *involution.*

The central and never-to-be-forgotten principle, therefore, is that the *phenotypic outcome* at any moment is the cumulative result of processes of *learning* (including unlearning) and of *volution.* Through use of appropriate models and research methods as discussed in the preceding section, the main initial aim of behavior genetics is to find out how much of any trait of the phenotype, at a given stage and point in time, is due to past, accumulated genetic and learning processes in interaction. The difference of the present proposed methodology from the biometrical methods is that instead of working only cross sectionally, by variance determinations at different ages, we attempt to evaluate genetic and environmental influence through examining the *ongoing* process itself.

In its simplest form, an idea justifiably common among psychologists is that something not strongly inherited can be made to change rapidly and considerably with teaching and environmental impact, whereas something highly heritable will not do so. Psychology has debated this particularly in regard to innate "instincts" (or *ergs* as we should more technically call those reactivities evidenced by factorial experiment [Cattell & Child, 1975]). The notion of innately preferred responses occurred in Freud (e.g., in his stages of the sex erg) and in more cognitive form in Jung's archetypal ideas. McDougall's use of "propensities" to "attend, perceive and react" involved this same cautious notion of only inclinations or preferences being inherited, and not specific behaviors. To concepts of this kind, which are now emerging with good support from the general biological approaches of such ethologists as Tinbergen and Lorenz, the psychometrist has added concepts and factorial findings of unitary ability, temperament, and dynamic traits, with expectations that research will show them to have genetic components. Probably every observant psychologist has ideas of how various abilities and skills will differ significantly in heritability. For example, he observes that training in "thinking" affects powers of intelligent discrimination relatively little, whereas training, say, in shooting or skiing or in acquiring a new language produces great increase in accuracy and speed scores.

But, beyond the simple concept, in the actual processes of measuring the relative "trainability" or "cultivatability" of different abilities and skills, there lie many complications. Even in the simpler cross-sectional research methods one soon finds that the intention of getting some meaningful value for the relative importance of genetics and environment (or nature and nurture), whether by manifestation rate of a syndrome, or by biometric and MAVA analyses of heritability coefficients of continuous traits, is not as conceptually simple as it first appears. For example, it will be shown later that the heritability index reached will depend, on the one hand, on the level and range of the cultural influences and, on the other, on the racial–genetic endowments (the gene pool) of the given racial population.

Among other qualifying values there is, for example, a "ceiling" effect in that if, on the cultural side, through some cultural urgency to acquire a certain skill, *all* people are "trained" to their highest limits (e.g., in needed word symbols for communication, or in driving fast vehicles in heavy traffic as most commuters must, or in remembering the names of a large number of people—the "gift of kings," which some kings have unfortunately not had!), individual differences in performance will become largely hereditarily determined, that is, H will be large. Quite a different heritability value might be obtained in some skill more casually picked up among the different members of the population, in which individual differences of interest determine training level. This does not mean that H (heritability) is not a value worth seeking. Along with other data, the H's of different performances tell us a lot about the race *and* the culture involved, and also about how much change we can, on an average, hope to produce by classroom and clinical intervention.

An experimenter accustomed to traditional experimental methods might, by his usual use of simultaneously controlled and manipulated designs, think to get around those uncertainties by *controlling* both the amount of training and the genetic range of genomes, as can be done in animal experimentation. But this is out of the question with humans except on quite trivial things that we can—ethically, practically, and socially—manipulate. When it comes to effects on major traits of personality, one cannot manipulate the number of life disasters or the size of the circle of congenial friends that conceivably affect such traits as ego strength or introversion. Multivariate experimental analysis of what life itself does is the only answer where major personality traits are concerned. We have recognized this in the cross-sectional approach, as in MAVA, and it will have to be recognized similarly in designs of process analysis, when applicable to human beings.

The third complication in obtaining and understanding heritability values through process study is that the helical course of intertwining maturational and learning processes is likely to yield different values according to the age at which it is cut for examination of the numerical values. Environmental influences are more potent in early life, which might lead us to expect that *H* would be low from, say, 1 to 5 years and rise, say, at 20, after differences in environmental influences have receded to lesser potency. But when *cumulative* effects are considered, as they must be, the analysis of what happens at different stages of a process calls for more complex mathematical consideration than we have yet needed.

The scanty data (largely of high-school and undergraduate student subjects) on which human behavior geneticists and personality theorists must form their hypotheses today urgently needs to be augmented by (*a*) comparative MAVA results cross sectionally at each of several different ages; and (*b*) analyses of the learning–maturational curves themselves. At this point we have made no tangible suggestions as to how the latter can be accomplished, but we shall do so in Chapter 6. Meanwhile, psychologists could at least prepare themselves by accumulating much of the needed material for the latter, namely, accurate age development curves for the principal source traits, beyond intelligence, of which we have data for only about 20 (Cattell, 1978b; Cattell, Eber, & Tatsuoka, 1970) at present.

One dependable feature on which to base analysis can fairly confidently be anticipated: that the separated age curve of volution (maturation and involution) will be relatively fixed for a given racial population. The threptic curve—the impacts of environment on a given trait—will differ enough to require that local *H* values be "corrected" or "standardized" according to the intensity of "teaching" or cultural pressure. This will require objective standards for comparing the magnitude of external environmental impacts. Those questions of concept and method will be discussed later, as this chapter is intended to be only a brief survey of the chief available methods. Those methods we have cut, as regards any further intensive study, to the six essential designs at the foot of Figure 2.3.

7. Summary

1. Although one can see broadly ten methods by which knowledge of genetic influences can be pursued, only five of them are of a precision or relevance suitable for human behavior genetics, and it is these that are surveyed in this chapter. The first of them, which is of greatest antiquity—going back to Galen and Hippocrates, and powerfully carried forward since, mainly by doctors and animal breeders in the eighteenth and nineteenth centuries—has been called the *clinical syndrome genealogical* (CSG) method. It depends on being able to recognize in the phenotype a sufficiently defined syndrome pattern, representing a genotype not obscured by environmental influences, and noting its recurrence in relatives and over generations. Though unrefined in its evaluation of the magnitude of influence of common environmental effects (except by such concepts as *manifestation rate* or *concordance* index), it has successfully established the existence of Mendelian mechanisms for close to 2000 heritable human syndromes, mainly pathological and largely physical, but including several psychological disorders that are neurologically based.

2. The CSG method, sometimes combined obliquely with the twin method, has been successful in demonstrating a high heritability of the two main forms of psychosis—schizophrenia and manic-depression—and suggesting Mendelian gene mechanisms involved in them. It has revealed a far lower degree of inheritance in neurosis, but quite specific dominant and recessive genetic origins for several forms of mental defect, Huntington's chorea, certain ataxias, etc. It has contributed nothing of certainty regarding normal, continuous human traits, but some suggestions (e.g., of inheritance of violent temper, excessive sexuality) which better methods are needed to clarify.

3. A second important approach is *experimental manipulation of animal matings* usually with biometrical measurement of performances. This method is being pursued vigorously in many laboratories, but, insofar as any attempt to identify homologous human trait behavior is concerned, it cannot be depended upon. What *is* valuable for human genetics is that through this approach it is possible to establish—more easily than can be done with humans—general *principles* and laws applicable to human genetics, concerning frequency of dominance and epistasis, intermediate physiological processes in behavior inheritance, effects of imprinting and age on inheritance, and so on. Its unquestionable contribution is a knowledge of heritabilities tied to *the gene pool of each particular* species investigated.

4. The *physical-linkage* method also has antiquity, but mainly in doubtful folklore stereotypes. It takes physical features of proven heritability and seeks to discover statistically significant associations of them with mental

traits. Some physical traits (e.g., facial beauty in a woman, small size in a boy) can evoke social environment responses which create a psychological–physical trait association that is really environmentally acquired. Caution on causality is therefore needed, but with blood groups, fingerprints, and other such "hidden" physical genetic features, the method can be trusted, and some positive findings have resulted. The science of *raciology*, seeking trait difference levels in physically defined races within the same culture can give leads to associations of individual physical features with particular genes. For although within-race genetic variations are decidedly greater than between, any race tends to have, on certain genes, a much greater preponderance than another, the study of which could lead eventually to gene isolation.

5. So far, the physical-linkage method, though ably pursued by several investigators, has revealed only low correlations—but statistically significant ones—between (*a*) body build and relative liability to manic-depressive versus schizophrenic disorders; (*b*) blood groups A, B, O, AB, Rh, and P and certain personality source traits—affectia, anxiety, parmia, premsia, and ergic tension; (*c*) eye color and certain temperamental traits; (*d*) surgency (F), zeppia (I), and intelligence (g); and a series of skull and skeletal ratios. The associations fall where they might be expected on other evidence, but are small and vary, presumably according to gene frequencies, with the population samples used.

6. The *convarkin* methods, which arise within biometrical genetics, include mainly the twin method and MAVA. The acronym designates that they operate by *con*trasting the *var*iances of family *kin* constellations, comparing those constellations that combine hereditary and threptic (environmental) components in different proportions. From ensuing calculations the varying weights of heredity and environment in various traits can be mathematically worked out. These methods began with continuous, graded traits, but can also be adapted to all-or-nothing incidences of syndromes. The convarkin methods are giving the psychologist the most extensive information available from any methods developed at the present time.

7. The MAVA method and model, developed within convarkin and general biometrical methods, is a special refinement in psychology, for human behavior-genetic investigation. Its special concepts, terms, and notation are adaptations beyond the basic biometrical designs to the special family and cultural influences involved in (*a*) the psychology of human families; (*b*) the cultural interests of social psychology; and (*c*) the special psychometric problems of human behavioral measurement. MAVA subsumes the twin method, adoption studies, the design for sibs reared apart, etc., as logically special subdivisions, though in terms of historical emergence the twin method is the oldest.

8. *Maturation–learning process analysis* is as yet somewhat inchoate and unexploited. It essentially evaluates heredity by determining how much

behavior can be modified by standard amounts of learning experience and finding how age development curves modify in different cultural settings. Since any course of learning extends over a time interval, in which innate volution (maturation or involution) can also occur, the problem is one of analyzing a curve of, say, measured gain on a trait, into separate but interwoven and interacting process curves for maturation and learning. The method inherently requires, if comparisons are to be made, the quantification and standardization of environmental "presses" important in the culture.

9. Use of any of these experiment and analysis methods implies that we have to start with one or more exact theoretical *models* of how environment and heredity interact. The first three methods have proceeded satisfactorily with little more than acceptance of Mendelian mechanisms and its associated algebra of alleles and gene frequencies, as established by physical-trait research. The last two, however, require construction of new models for polygenic genetic action and forms of interaction with environment, and these models will be fully studied in ensuing chapters.

3

The Twin Method of Investigation with Illustrative Findings

1. Comparisons of Variances as a First Step in Behavior Genetics Research in the Convarkin Methods

A gap must exist for some years to come between the genetic study of personality development and that allotment of the basis of specific phenotypic traits to specific genes which has long been the aim of general genetics. If we needed a word for that knowledge of the gene locations themselves we might call it *genomics,* and today genomics has many successes (McKusick, 1964). It has defined the inheritances of certain physical features, certain enzymes in human physiology, and even of a few definite pathological behavior syndromes, as we saw in the preceding chapter. However, there is every reason to believe that the major personality source traits—intelligence, surgency, anxiety proneness, etc.—have their genetic part determined by *many* genes. Such polygenic determination at present baffles genomic analysis. We must content ourselves with analyzing the genetic and environmental contribution in terms of functions of the observed variance. However, our treatment will consider later some possible methods by which behavior genetics in the future can take steps toward finding genetic locations.

In the general convarkin methods, which aim at division of genetic and environmental sources of variance by contrasting the observed variances of well-defined kin groups in different environments, the two main developments have been the twin method and the MAVA method, both biometrical. The former, which has historically been the nursery for the latter, is best studied first, because it provides a simpler introduction to principles that are more complex in the latter.

2. The Nature of the Complete and Partial Twin Methods

At an anecodotal level the twin method is quite ancient. There are instances of identical twins reared apart who showed the same special talents, married at close to the same age, committed the same crimes, suffered a certain disease at the same age, and even had the same accidents. Current studies by Bouchard at Minnesota University repeat such striking observations. Naturally, it is the more spectacular instances that are remembered, but quite a number have been verified. It was such observations that caused Lange to write *Crime as Destiny*. However, the first attempt at a scientific and quantitative treatment was made in 1883 by Sir Francis Galton, in pursuit of his interest in the degree to which high intelligence is inherited.

The method hinges on the fact that there are definitely two types of twins: fraternal twins and identical twins.

1. *Fraternal, dizygotic, or biovular twins*. These twins arise from two ova happening to be available at the same time for fertilization in the mother, and, of course, the ova are fertilized by different sperm cells. Fraternal twins commonly grow in the womb in two different chorionic sacs. They can be of different sexes and, all in all, are no more alike than ordinary sibs. We say "all in all" because finer analysis might show that being subject to the same health state, etc. in the mother's womb throughout gestation could produce some extra similarities at birth (what we called *congenital* effects p. 11), absent in sibs born at different ages and states of the mother.

2. *Identical, monozygotic, or monovular twins*. These twins arise from the fertilization of a single ovum by a single sperm, but at the first cell division the two cells produced happen to fall apart and go on to produce two individuals, who are necessarily identical in their chromosome structure and are, of course, always of the same sex.

The *practical* separation of identicals from fraternals is not always so easy, and so certain, as their real differences might lead one to expect. Every now and then one meets a pair of fraternal twins who are are so alike in appearance that they get mistaken for identicals. In one direction the choice is certain—if they are of different blood type, eye color, or sex they are fraternals; however, if they appear the same on such genetic features they are not necessarily identicals. Today it is realized that a majority but not all biovular twins are born from separate chorionic sacs. And among twins the first-born twin tends to be larger than the second-born, which sometimes affects decisions based on obstetric evidence. Even fingerprints are normally slightly different between identicals. In the end, the distinction in most research has to rest on an accumulation of probabilities, with the possibility that some fraternals may "contaminate" the group of identicals.

Inferences from twin study as to the general population inheritance variances are liable also to a second source of "error": Because twinning is itself an inherited tendency, twins may not be representative, in their means and variances, of the general population. The incidence of fraternal twins is about 1 in 135 of all births and, of identicals, about 1 in 260, but different races have different incidences. In the Japanese, for instance, it becomes 1 in 368 and 1 in 236, respectively (Lefner, 1968). So far, it is true, no marked divergence from the general population has been found, except, perhaps for *slightly* lower *average* intelligence in twins. In the next chapter we shall argue that the MAVA method is superior to the twin method, but not so much because of these error sources, which are trivial, as because of the greater "power" of the MAVA method (e.g., in giving correlations of hereditary and environment deviations within and between families).

Granted adequate separation of twin types, the inferences from the twin method (so long as we aim at analysis of within-family variance) rest on the following:

1. Variance in the trait in identical twins reared together. This we symbolize as σ^2_{ITT} (Between *I*dentical *T*wins *T*ogether).
2. Variance in the trait in identical twins reared apart: σ^2_{ITA} in our mnemonic notation.
3. Variance in the trait in fraternal twins reared together: σ^2_{FTT}.
4. Variance in the trait of fraternal twins reared apart: σ^2_{FTA}.

The variance we speak of is primarily what is called *within-family variance*, obtained, (as will be described regarding degrees of freedom, etc. later) in calculations based on the difference of score *within* each pair, that is between the two members of the pair.

Although the four sources of concrete, empirical variances listed here are theoretically possible for the *complete twin method*, the great majority of twin-method research have been by the *partial twin method* which omits numbers 2 and 4: the twins reared apart. Indeed, the psychologist seeking genetic perspective needs to be warned that probably 95% of twin researches have been made by the partial method. For cases of twins reared apart from birth have proved impossible to obtain in samples large enough to meet statistical requirements.

In addition to the within-family variances we can calculate the between-family (twin) variances, i.e., that of the means of pairs from the grand mean.

However, the inferences therefrom in calculating general heritabilities are (a) possibly biased by twin families being biologically or socially biased selections relative to the general population; and (b) demanding of some special, essentially incorrect assumptions which will become evident when we get to contrasting the twin with the MAVA model.

3. The Breakdown of Concrete into Abstract Variances

Although most psychologists will be familiar with variance calculations, it seems desirable to start here from the very ground in calculating the concrete variance. For the within variance we subtract the mean of each pair, from each of the twins' scores, obtaining deviations. We then square and sum all the deviation measures and divide by the number of families, N, which defines the degrees of freedom in this case. If n—the number within each family—is 2, as here,

$$\sigma_w^2 = \frac{\Sigma^{i=2n} d_i^2}{N},\tag{3.1}$$

there is only one degree of freedom $(n - 1)$ in each family. If D_f is the difference of the twins in family f $(D_f = 2d_i)$ we can calculate σ_w^2 perhaps more conveniently as $\sigma_w^2 = (\Sigma ND^2)/2N$ which is equivalent to (3.1).

The between-family variance, again using degrees freedom to estimate the best value for the population, is:

$$\sigma_b^2 = \frac{\Sigma^N \bar{d}^2}{N - 1},\tag{3.2}$$

where $\bar{d}$ is the deviation of the family mean [used to get (3.1)] from the grand mean of all families.

Now variances have the property of being additive, which enables us to set up a theoretical model that *the observed, concrete variance is the sum of two abstract contributory variances: that due to the genetic variance between fraternal twins and that due to the environmental variance between them.* It will be recognized that other models can (and will) be considered, for example, those involving interaction effects too, but this is a beginning and is indeed what the twin model has habitually been restricted to. Incidentally, one should not be surprised that although they have the same parents, the *fraternals* nevertheless do not have the same genomes (genetic make-up). Each is a different juggling of parental genes, from different splits of the parental chromosomes.

We are dealing now with two variance terms which may better be called, *abstract*, as in the preceding, than hypothetical. There *can* be no other hypotheses about the primary sources themselves, for environment is operationally defined as that which is not the genetic part, and there is thus no room for any third term. Nevertheless, these variances are abstract, because we cannot directly calculate them like the concrete variances. If we omit possible interaction effects we are saying:

$$\sigma^2_{\text{FTT}} = \sigma^2_{\text{wg}} + \sigma^2_{\text{wt}}, \qquad (3.3)$$

where wg stands for "within-family genetic" and wt for "within-family threptic (environmental)."

In using the term *threptic* we employ a word which only a minority of researchers have yet used, but which assists clarity of thought. Threptic means *that part of the variance of a trait which is the result of variance in the environment.* To call this "environmental variance," which has often been done, is to confuse the variance *in the environment itself* (e.g., in the social status of the subjects, their years of training, and the number of accidents they have suffered) with the variance *in the person,* that is, in the individual's *traits, due* to the environment. Later, as we get to the interaction of learning with maturation, we shall need to be very clear about the difference of threptic and environmental, because we shall be looking for laws relating the two, that is, concerning how much threptic[1] variance results from a given amount of cultural, environmental variance, objectively measured.

The typical, most basic, and most frequently used procedure in the twin method is to compare the concrete variances of identical and fraternal twins, in both cases as raised together. With identicals there can be no genetic difference so the concrete within-family variance, calculated as before, is

$$\sigma^2_{\text{ITT}} = \sigma^2_{\text{wt}}. \qquad (3.4)$$

The solutions for the abstract variances here can be readily seen, though the more complicated MAVA equations we shall see later really rest on just the same principle. Subtracting the given concrete values of (3.4) from (3.3) we get

$$\sigma^2_{\text{wg}} = \sigma^2_{\text{FTT}} - \sigma^2_{\text{ITT}}. \qquad (3.5)$$

Knowing the genetic variance, σ^2_{wg}, and the corresponding threptic variance, σ^2_{wt}, we have the basis for the heritability calculations shortly to be described.

---

[1] Some 20 years ago I chose the term *threptic* from a careful study of many Greek roots that might convey the exact meaning here needed. Τροφή has the meaning merely of nourishment, as in the biologists' term *trophogenic*, and Τυγχάνεν, "to happen," suggests too much fortuitous luck in environment. What we have in mind psychologically is the growth-promoting effect of environment upon a genetic tendency. For this meaning, Θρεπτική, as used by Aristotle in *De Anima* (and later by Galen), meaning "nourishing and promoting growth," seems most apt. I have accordingly, here and elsewhere, used *threptic* to mean the "nourishing" effect of environment. I can only hope that behavior geneticists will find it useful to clarify discussion and measurement by using *threptic* variance for *the trait in the organism* and *environmental* variance for variance in those *environmental features,* such as years of schooling, which the learning theorist will want to relate to the threptic component. For, in terms of this example, the most educated person is not necessarily the one who has the longest education.

4. Some Qualifying Considerations in Method and Model

The thoughtful student may have some doubts regarding the fit of procedures to concepts in the model. For a psychologist, a fairly obvious misgiving concerns the equating of the within-family environmental variance for identical and fraternal twins; that is, calling both simply σ^2_{wt}. Almost certainly, the fact that everyone recognizes the identical twins as being more alike results in their being *treated* as more alike. On the other hand, psychoanalysts have speculated that this situation causes a problem of identity for identicals and that they struggle, almost desperately, to pull apart from each other insofar as they can mold themselves environmentally. Several writers have addressed themselves to this problem (e.g., Scarr, 1968; Shields, 1954; Smith, 1965). The matter is doubtlessly complex, but in the MAVA equations we can do what we cannot do in the twin method—namely, solve for *two, different* abstract variances: $\sigma^2_{wt.t}$ for identicals, and $\sigma^2_{wt.s}$ for fraternals and ordinary sibs. When this is done we generally find $\sigma^2_{wt.t}$ is smaller (Chapters 8, 9, and 10). Making an approximation by the assumption that they are the same is a major shortcoming of the twin method.[2]

[2] That the environmental differences of twins and sibs are different has long been questioned and debated (see Scarr, 1968). A good survey of medium amplitude is given in the book by Claridge *et al.* (1937), which, incidentally, is an excellent overview of the whole twin method and its results. The student needing to delve more fully into the twin method would be well advised to read it, along with the other scholarly surveys, such as Lytton (1977), Metheny, (1976), Mittler (1971), Eaves (1976), and the discussion in Vandenberg (1965a, 1965b, 1969).

On the side of hypotheses from observation Shields (1954) argues that emotional attachments and identifications are much greater between identical twins than fraternals or sibs. This would not necessarily argue for greater threptic similarity on *all* traits. On such a trait as dominance (*E* factor) the assumed greater willingness of twins to come into harmony might lead one to be dominant and one submissive, whereas, by contrast, two sibs, with age difference, might obstinately preserve high levels of individual dominance. Whereas psychological observers on the whole have favored hypotheses of greater threptic-environmental similarity of twins than sibs, and of identical than fraternal twins, psychoanalysts have put forward the opposing and subtle theory that the need for establishing different identity is so great among identical twins that they make abnormally great efforts to be different. This might be one more instance of psychoanalysis taking the abnormal as a legitimate account of normal and average behavior.

Everything at a direct observational and quantitative level points to identical-twin environment being of smaller variance than for fraternals or sibs. They are commonly dressed more alike, go about together more, and are treated more alike by parents and by peers. Smith (1965) irrefutably shows that on a wide span of interests, habits, values, and viewpoints identical twins are more similar than sibs, and the nature of some of the traits, which were chosen to represent those most likely to be largely acquired, suggests that the greater resemblance (relative to sibs) is in the environmental sector. Scarr (968) contributes to the same view.

Nevertheless, this position has been questioned by Loehlin (1980) who writes:

> A number of studies evaluating different aspects of this assumption [of equality of environmental difference] have been reported in recent years (e.g., Scarr, 1968; Scarr & Carter-Saltzman, 1979; Lytton, 1977; Matheny *et al.*, 1967; Loehlin & Nichols, 1976; Plomin, Willerman, & Loehlin, 1976). On the whole, these studies have tended to sustain the viability of the equal-environments assumption for personality [p. 18].

There will be plenty of opportunity in the course of this book for the reader to weigh evidence and consider special aspects. But I shall say here that I consider the new evidence—presented here in Chapters 8, 9, and 10—to point to appreciable reduction of environmental difference in identical twins relative to sibs. So long as one stays within the twin method of research alone, attempting, as the writers cited by Loehlin have done, to get slants from bits of evidence in that field, the conclusion of the twin researchers may be tenable. But the inference amounts, methodologically, to trying to pull oneself up by one's bootstraps. As soon as a broader basis of MAVA evidence became available the independent solutions for $\sigma^2_{wt.i}$ and $\sigma^2_{wt.s}$ (identicals and sibs) the comparison pointed to a decidedly greater variance for the latter. These studies, it is true, have mainly compared a solution for twins as a pooled type ($\sigma^2_{wt.i} + \sigma^2_{wt.f} = 2\sigma^2_{wt.t}$) with sibs as a type ($\sigma^2_{wt.s}$) and have found a ratio of the first to the second around 1.25–1.30. But there is other evidence suggesting that further allowance be made on the ground that contamination of the identicals' variance with that of fraternals has lowered the above ratio, and that in *some* traits it could rise to 2.0. That is to say, as an educated guess, we might find a series beginning with 1.0 representing σ^2_{wt} for identicals, going to 1.25 for fraternals and 1.50 or more for sibs (same sex). Our data is based on sib scores corrected to the same age, for twin comparability, and their greater variance therefore arises from being born at different epochs of parental life, having different peer groups, etc.

Whatever the causes of difference may be, the assumption of equality, we argue here, is quite wrong, and constitutes one of several crudities that render twin method results approximate in value and limited in the extent of inferences that can be drawn from them.

A lesser source of possible error in interpretation—and again one that can effect other methods too—is the finding that essentially genetic effects of a minor kind can come from some sources in the body of the zygotic cells outside the nucleus and the chromosome structure. All that is known about these effects, possibly not handed on beyond a generation, is that they are quite small and unlikely to affect the main Mendelian interpretations.

Finally, there is a real limitation to the twin method in that twin families, in their biological means and sigmas, may be an atypical sample of the general population. As stated, twinning is itself heritable, and comparisons of races, for example, show significant differences in liability to multiple births (see p. 57). Enough traits have been examined to show that there are in fact *some* significant differences in trait *means,* at least, and probably also in interfamily sigmas on traits, between twins on the one hand and sibs on the other. Our discussion has so far only been on the correctness of assuming that $\sigma^2_{wt.i} = \sigma^2_{wt.f} = \sigma^{22}_{wt.s}$; but it seems there may also be some problem with assuming that σ^2_{bg} for, say, fraternal twin families equals σ^2_{bg} for ordinary sib families.

Another question that arises in using the twin method is whether, since identicals are bound to be of like sex, the fraternals and sibs compared with them in experiments should also be of like sex, and of the same sex as the identicals used. From a cautious belief in the desirability of maximum control, I have practiced this step-by-step advance, beginning with all male pairs and trusting others will research the H values with female pairs. Actually, provided researchers are explicit in going through these stages comparison of results would permit certain further conclusions to be drawn, on sex-linked genes and the effects of environmentally demanded sex roles in the σ^2_{wt} and σ^2_{bt} variances. For example, Eaves, Last, Young, and Martin (1978) point out that working with unlike sex fraternals will throw light on the question of whether the same gene loci are involved in determining the same measured trait in men and women. Stafford (1961) and others (but see disagreement by DeFries *et al.,* 1979) argued for evidence that the visualization ability factor is differently determined in men and women and there are, of course, clear evidence of the *same* genotypically determined phenotypic features, such as chest girth, being determined by different chromosome loci, namely the X and Y chromosomes. Eaves, Last, Young, and Marten (1978) give the calculation to proceed from mixed-sex comparisons in order to check this.

The twin method is normally confined to reporting relative genetic and threptic variances *within* families, but, as Loehlin (1977) points out, and as shown in Eq. (3.2) above, it can be extended to getting between-family vari-- ances, thus:

$$\sigma_{wt}^2 = (1 - k_1 k_2 r_{ITT}) + (1 - k_2) r_{FTT}, \qquad (3.6)$$

$$\sigma_{wg}^2 = (1 - k_2 r_{FTT}) - \sigma_{wt}^2, \qquad (3.7)$$

$$\sigma_{bg}^2 = k_3 \sigma_{wg}^2, \qquad (3.8)$$

$$\sigma_{bt}^2 = k_2 r_{FTT} - \sigma_{bg}^2. \qquad (3.9)$$

These more complicated-looking formulas look so primarily because they represent a shift from statements of variances [as in (3.1)–(3.5)] to statements of (intraclass) correlation. This transformation we shall tackle as a general problem in the next section. For the moment the first two formulas are to be viewed as statements of (3.4) and (3.5), in which, however, k_1 is introduced as a correction for the otherwise assumed exact equality of $\sigma_{wt.t}^2$ and $\sigma_{wt.s}^2$. It is a ratio of the identical- to fraternal-twin threptic variance. This value has to be gained from somewhere outside the twin-method data (actually from MAVA). The correction k_2 is a ratio of between- to within-family genetic variance, which also could come, if we want it to be empirically based, from MAVA; but, if certain assumptions are made, it may come theoretically from within the twin method itself. Correction k_3 is for assortive mating—like marrying like—etc., with which we shall deal more fully later. The major weaknesses of this design therefore are (*a*) that k_1 cannot be obtained in a twin study itself; (*b*) that k_2 has to be assumed on other grounds; and (*c*) that Eq. (3.9) supposes that the genetic variance among twin families is the same as for ordinary families in the population. As to the last, it is well known (Lerner, 1968; McClearn & DeFries, 1973) that twinning families are *not* completely representative of the general population. Ordinary families are different, and since the general population is composed of about, say, 90% of sibs reared together (the rest being twins and foster children), the sum of the within- and between-family variances obtained for sibs will alone (following ANOVA rules) come close to restoring (estimating) the variance of the general population. Since the corresponding "within plus between" calculations for twins will give us the population values *as if the population were wholly twins*, the heritability calculations employing these variances cannot give us a dependable estimate for the general population.

Even at this introductory stage it is advisable to point out that σ_{wt}^2 (regardless of whether it is $\sigma_{wt.t}^2$ or $\sigma_{wt.s}^2$), though *called* "within-family variance due to environment," is by no means entirely due to differences of environment occurring *within* the family; it is also due to certain *outside* influences. The former would arise from differences in handling of the two children by the parents, and especially to one twin or sib being the environment of the other. ("A pessimist is someone who has lived with an optimist.") Statements of mutual influence sometimes go so far as to say that the genetic make-up of one is the environment of the other. This is true, but only at one or two removes in the causal chain (i.e., through the phenotype), and that part is perhaps so small an arc of the other's environment as to be trivial. However, our later results show that sources of variance within the family environment are substantial and commonly greater than those between families. Particulars can be studied later, but we can also note examples here of influences outside the family that contribute to within-family variance among offspring, such as the children going to different schools and having different peer experiences. Or, again, one may be brought up with two parents for the first 4 years of life and the other orphaned from birth, and so on. The σ_{wt}^2 value thus has to be further analyzed when we proceed to causes, and meanwhile we must not think of it as entirely explicable psychologically by *what happens within the family*. The intrusion of some outer world influences into the within-family threptic variance probably evens up to some extent $\sigma_{wt.t}^2$ and $\sigma_{wt.s}^2$ and accounts for their not being such radically different values as they might otherwise be.

The procedure of addition and subtraction of concrete, observed variances to get abstract variances, which runs through convarkin methods, raises other basic questions that are best answered in this introductory phase, within the twin method. A first question that will occur to the reader at this point is one that mixes statistical and psychological considerations. The reader will ask, "Is it fair to estimate the within-family concrete (total) variance from only two children when, in the case of sibs, there may be several more?" If families of different sizes were in each case a random selection from a potentially large number, the statistical answer is straightforward: By adjusting the within-family degrees of freedom $(n-1)$ to the number, n, of children measured we are, of course, aiming at the very same population estimate as when we take families of different sizes (though the same number of large families will give us a better estimate.)[3] But it is really

---

[3] For those who are unfamiliar with derivation of variances from scores themselves let us go down to bedrock here and show the steps as given in other terms in (3.3) etc. First:

$$a_{oi} = a_{gi} + a_{ti},$$

where a_{oi} is the *observed* score of person i, and where a_{gi} and a_{ti} are, respectively, his genetic and his threptic score components, all in *standard scores*. Then

$$\sum_{i=1}^{i=N} a_{oi}^2 = \sum_{i=1}^{i=N} (a_{gi} + a_{ti})^2 = \sum_{i=1}^{i=N} a_{gi}^2 + \sum_{i=1}^{i=N} a_{ti}^2 + 2 \sum_{i=1}^{i=N} a_{gi} a_{ti}.$$

Bringing these to variances, by dividing by N—or $(N - 1)$ if estimating)—we have (since $r_{gt} = \Sigma^{i=N} a_{gi} a_{ti} / \sigma_g \sigma_t$)

$$\sum_{i=1}^{i=N} a_{oi}^2 / N = \sigma_o^2 = \sigma_g^2 + \sigma_t^2 + 2 r_{gi} \sigma_g \sigma_t.$$

more than a statistical question. It could well be that, psychologically, the measurable environmental variance (though not necessarily the threptic variance) *is* different with different family sizes. Therefore, when research in this area is sufficiently endowed, psychologists should seek to establish σ^2_{wt} values (σ^2_{wg} values do not alter) for families of varying size, by taking samples in which all families are of one size in each given investigation, and then comparing results.

To a psychometrist it will go without saying that when we deal with sibs, whatever trait is being scored should be age corrected—that is, members of each pair of sibs, and all sib pairs, should be brought to the same developmental age, if their variances are to be compared with those of twins, and the *pairs* among whom should also be corrected to a common population age. This has typically been done in studies on intelligence, where the IQ is automatically used; but it seems to have been overlooked in most published personality studies. The existence of this adjustment therefore needs to be remembered in comparing most past personality inheritance results with the present age-corrected results, in Chapters 8, 9, and 10. Correction is carried out by using normal standard age trend curves, which are available for both the primary and secondary personality traits in the factored 16 P.F., HSPQ, and O-A batteries. Even *with* this correction, the psychological realities would still expect $\sigma^2_{wt.s}$ to be greater than $\sigma^2_{wt.f}$, for fraternal twins, and $\sigma^2_{wt.f}$, in turn, to be greater than $\sigma^2_{wt.i}$, for identicals. Age correction is also necessary *among* pairs of twins, as well as of sibs, as stated, primarily because we deal with between-family variance too. Being of different ages also means being born into different epochs, different stages of development of the parents, etc., and these differences beyond normal age trends should ideally be taken into account as discussed in Chapter 6.

A more general consideration that is likely to occur to anyone entering on the convarkin methods is that of why we are, so far, operating with the children and leaving parents out. Is not the relation of parent to child the very essence of what we mean by inheritance? There are some subtle problems here that call for postponing (Chapter 5) that mode of attack on general kin relations. It is true that basic knowledge of the chromosome-splitting process enables us to say exactly how much children will resemble parents in their purely genetic make-up, but the same is also true of resemblance of sibs. The difficulty that would arise in considering sib and parent equations in the same system is that the personality-forming environment of the parent and of the child cannot be considered the same, even though they both are presently in one family. Clearly, most of the formative environmental influences of the parents resided in the grandparents' home, and the parents' parents and peers were different for the two parents. However, let us not forget also that today children do to some extent bring up parents (if not considered too old to be educable!). These complications in the meaning of family environmental variance we will consider at a suitable later stage, in Chapter 5.

5. Interaction and Correlation of Genetic and Threptic Effects

Effects from interaction and correlation are not peculiar to the twin method, but their consideration naturally intrudes at this point and is best handled here, for future reference. As a first, elemental attack on definition, we have stated that genetic and environmental effects are sufficiently defined in the total observed variance by saying that what is not one is the other. This is true in the sense that what does not come with the genetic endowment, from within the skin, must come, environmentally, from outside. But, as statistical analysis may show, the final effect on behavior may involve some causal interaction of the two, thus throwing into the observed score range measurable variance that is more than the result of simple addition. For example, if we have, first, two vocabulary-improving books and, then, four books, read by two children, A and B, whose personal genetic gifts in intelligence measure, say, 3 and 5 (components in mental age), and consider vocabulary, in a first model, derived as a simple sum of books read and intelligence, then we have, in terms of level reached by each on a vocabulary score:

	Two books	Four books
Additive model	Vocabulary $A_2 = 3 + 2 = 5$	Vocabulary $A_4 = 3 + 4 = 7$
	Vocabulary $B_2 = 5 + 2 = 7$	Vocabulary $B_4 = 5 + 4 = 9$

(the subscript being the number of books—the "environment"). But we know from educational data, that, as in the parable of the sower, the bright child will get more out of the books, and this effect we might represent by introducing a product interaction term thus:

	Two books	Four books
Interactive model	Vocabulary $A_2 = 3 + (3 \times 2) + 2 = 11$	Vocabulary $A_4 = 3 + (3 \times 4) + 4 = 19$
	Vocabulary $B_2 = 5 + (5 \times 2) + 2 = 17$	Vocabulary $B_4 = 5 + (5 \times 4) + 4 = 29$

Whereas in the additive model B's level after an "environment" of two books reaches 1.4 times that of A, in the interactive model it reaches 1.55 times that of A; and the difference is even greater with the four-book environment.

In pondering revisions on the basic formula underlying (3.3), one may also wonder whether another algebraic change should be introduced by giving different weights, v_g and v_t, to g and t scores, thus

$$a_i = v_g g_i + v_t t_i \qquad (3.10)$$

(using a_i for act, response, performance, or trait of individual i, and a, g, and t being in standard [deviation] scores). If we had some extraneous basis for assigning different units of particular size to the g and t scores, such a "scaling" procedure by weights might be required to bring them to their true contributions. But, since we do not possess any already inherent scale for the genetic and threptic effects, it is simplest to bypass (3.10) and let the sizes of g and t be expressed in the units of the particular trait or performance being measured.

Interaction, on the other hand, is on scientific grounds a real possibility, and, of course, it could theoretically take more complex forms than a simple product as illustrated earlier, such as g^2t, $(g + t)^2$, kg^2t^3, and so on indefinitely. To state an interactive model in simplest form, to which reference may be made for illustration, we will indicate interaction merely as a weighted (v) product thus:

$$a_i = g_i + v_{gt}(g_i t_i) + t_i. \tag{3.11}$$

Let it be stated here and now, however, that our experimental sources of data and our analytical methods of solution are scarcely ripe for handling this. It would in effect be asking us to run before we can walk, and, since there is proof that the simpler additive model *in most cases* comes close to fitting the expectancies, we shall for the present stay with the additive form. Nevertheless, later in this book, we shall indicate paths to follow for interactive solutions.

The psychologist familiar with ANOVA will remember, however, that interaction and correlation (covariance) are two different things. In ANOVA, even though two "effects" are correlated in the population from which the cases are taken, in standard experiments we so choose the cases in the cells (equal numbers) that the variables are not correlated. This is a necessary basis for testing for true interaction. For, if the effects are actually correlated in the population, we would not know how much of the departure from zero interaction is due respectively to true interaction [as in (3.11)] and how much to correlation.

In the investigation of genetic and environmental influences, it is widely recognized by investigators that in many cases these two influences (and therefore the genetic and threptic deviations from the mean among individuals) *are* correlated. In educational systems with effective scholarship selection (for more advanced classes or entry to universities), the more gifted child is likely to get into an environment with more scope and stimulation. In this case there could be a positive correlation of children's deviations from the average in the genetic and the threptic components as shown by the size of the positive correlation of fluid and crystallized intelligence (Horn, 1972b). (The adolescent-age r of about $+ .5$ of crystallized and fluid intelligence found by Horn and Cattell [1966a] is consistent with this, though it could have other causes.) On the other hand, in school systems that expend much more on the backward than the bright, the *genothreptic correlation* (as we shall call it) could be negative.

If the genetic and threptic deviations should be correlated, as r_{wgwt}, within the family (i.e., in terms of deviations from the family mean), then the observed concrete variance of, say, fraternal twins would have a covariance term added[4] into Eq. (3.3), thus

$$\sigma^2_{FTT} = \sigma^2_{wg} + \sigma^2_{wt.f} + 2r_{wgwt.f}\sigma_{wg}\sigma_{wt.f}. \tag{3.12}$$

(Remember that subscript f means fraternal, or dizygotic.) If one admits that there may be interaction in the production of the final score, as set out in Eq. (3.11)—that is, with interaction as a simple product $\sigma_{wg}\sigma_{wt}$ (weighted v), *and* intercorrelation—the observed within-family variance becomes

$$\sigma^2_{FTT} = \sigma^2_{wg} + \sigma^2_{wt} + v^2\sigma^2_{wg}\sigma^2_{wt} + 2r_{wgwt}\sigma_{wg}\sigma_{wt}$$
$$+ 2vr_{wgwt}\sigma^2_{wg}\sigma_{wt} + 2vr_{wgwt}\sigma_{wg}\sigma^2_{wt}. \tag{3.13}$$

Since the product of two noncorrelating variances centers on zero, this would simplify considerably, but with correlation, so that $r_{wgwt} \neq 0$, an r term must enter all products as shown in (3.13). Then the solution of a whole series of equations—for σ^2_{BITT}, σ^2_{BST}, etc. would become formidable (though not necessarily impossible, since the only new unknown is v).

Ultimately a form of the MAVA solution may reach out to handle this, but at present we leave such a model aside. For, as a trial of the available simultaneous equations will show, the twin method is not even able to separate the covariance from the genetic variance, when Eq. (3.3) is subtracted from (3.12).

In the rare case where we have enough identical twins reared in different families a solution for the between-family threptic variance becomes possible because *their* variance would be

$$\sigma^2_{ITA} = \sigma^2_{wt.s} + \sigma^2_{bt}, \tag{3.14}$$

since they would differ also by being in different cultural families (the σ^2_{bt} term), while the within-family σ^2_{wt} would be the same as for sibs. Since we know the general population concrete variance, the value for σ^2_{bg} could be obtained as shown in the next chapter, namely, by subtraction of the three known variances from total. But even this "full resource" twin solution, that is, with data on twins apart, still limps because it makes the incorrect assumption that $\sigma^2_{wt.i} = \sigma^2_{wt.s}$, and it still lacks solutions for covariance.

---

[4] The covariance could be expressed *without* going through the correlation coefficient, of course, simply as $2\Sigma z_g z_t / N$, where the zs are the deviations in standard scores of the g and the t components. These are, of course, not directly measurable, being "abstract components," and we actually proceed by solving for r_{wgwt} and σ^2_{wg} and $\sigma^2_{wt.f}$. Note that although a positive sign precedes the r in (3.12), the r itself can be negative and the whole covariance, therefore, negative. In fact, in quite a lot of psychological data (see Chapters 8, 9, and 10) it *is* negative.

6. The Calculation of Heritability, H, and the Nature–Nurture Ratio, N

Despite the weakness occasioned by its approximations and gaps, the twin method, in the usual short form (i.e., lacking twins reared apart), has been the main workhorse of medical and psychological genetics from the beginning of the century, and many areas up to the present conclusions have had to depend entirely on the results obtained by it.

The reader will find that the actual calculation of heritabilities in twin research has often proceeded by a different route from the direct calculation of variances as described here. The analyses are then presented as correlations between members of pairs of, for example, identical twins (ITT), fraternal twins (FTT), and so on. There is a simple equivalence of these two approaches with which the reader should become familiar, involving an understanding of the *intraclass correlation*. The ordinary correlation, with which all students are familiar, is one between two different kinds of test scores on one series of persons. By contrast the intraclass correlation, as used here, is between two different, matched people on one kind of test score. Since there is no general logical basis for deciding which twin should go in the first and which in the second column (as there is with two different tests), the solution is to achieve symmetry by entering each twin pair— Joyce and Jane—twice, once with Jane in the first column, once with Jane in the second, and conversely for Joyce.

The intraclass correlation comes out essentially the same as the ordinary correlation, but (*a*) it avoids a bias, by the double entry; and (*b*) it derives, conceptually, in a different way. Here we happen to be dealing with *pairs,* but the intraclass correlation, r_i, was first conceived as applying to *groups* ("classes") of any size, and works on a comparison of variance within groups with that between groups (see Haggard, 1958). It expresses the total population variance less the within-twin variance, divided by the total variance. This works out as

$$r_i = \frac{\Sigma^N t_1 t_2 - NM}{N\sigma_p^2}, \tag{3.15}$$

where M is the population mean, σ_p^2 is the population variance, and t_1 and t_2 are the scores on any pair of twins. It can also be thought of as the analogue of the ordinary correlation when the latter is used as a reliability coefficient, that is, when two scores on the same test from two occasions are correlated.

Now it is a basic habit of thought in psychometry that if two tests correlate r, the proportion of variance of one accounted for by the other is r^2. But if the two are symmetrical, in having the same amount of a common factor (as in a test–retest, reliability coefficient, that is, the *dependability* coefficient) in them, then the proper inference to be made is that that amount (proportion) of *common variance literally equals* r (not r^2). [The factor analyst will recognize this as an aspect of the basic factorial proposition that $r_{ab} = r_{ag}r_{bg}$, where g is the common factor. The amount of g variance in a is r_{ag}^2. If a and b are the *same* in common factor variance, $r_{ab} = r_{ag}r_{bg} = r_{ag}^2$, that is, r_{ab} ($\times 100$) is the amount of common variance in a and b.]

Now, in two children from the same family, we suppose that their resemblance is based on a *symmetry* of what is common. Therefore, if the total variance is called σ_p^2 for the total population variance in a trait,

$$r_{\text{ITT}} = (\sigma_p^2 - \sigma_{\text{wt.i}}^2)/\sigma_p^2, \qquad (3.16)$$

$$r_{\text{FTT}} = (\sigma_p^2 - \sigma_{\text{wt.f}}^2 - \sigma_{\text{wg}}^2)/\sigma_p^2, \qquad (3.17)$$

$$r_{\text{ITT}} - r_{\text{FTT}} = \sigma_{\text{wg}}^2/\sigma_p^2. \qquad (3.18)$$

The last holds if we permit ourselves to continue with the assumption that $\sigma_{\text{wt.i}}^2 = \sigma_{\text{wt.f}}^2 = \sigma_{\text{wt.t}}^2$, that is, that we may speak of a within-twin-family threptic variance, $\sigma_{\text{wt.t}}^2$, that is the same for both. [Equation (3.18) is *not* a standard H.)

Now heritability—which we will write H (following the early use of Holzinger)—is the fraction of the *total* concrete variance between fraternal twins that is ascribable to the genetic component, so that the within-family heritability, H_{w} is obtained as follows:

$$H_{\text{w}} = \sigma_{\text{wg}}^2/(\sigma_{\text{wg}}^2 + \sigma_{\text{wt}}^2). \qquad (3.19)$$

We can calculate this either from the concrete variances, coming through Eq. (3.4) and (3.5), thus:

$$H_{\text{w}} = (\sigma_{\text{FTT}}^2 - \sigma_{\text{ITT}}^2)/\sigma_{\text{FTT}}^2 \qquad (3.20)$$

or from the correlations, as indicated in (3.16), (3.17), and (3.18), thus:

$$H_{\text{w}} = (r_{\text{ITT}} - r_{\text{FTT}})/(1 - r_{\text{FTT}}). \qquad (3.21)$$

Incidentally, given the awkwardness of the subscripts FTT, etc. in writing out algebraic manipulations, it might be asked why we do not use subscripts D and M, or B and M for dizygotic, monovular, biovular, etc. One reason is that we need to keep perspective on the fact that these variances are *within-family* variances, and later, in the MAVA method, we shall need to use BITTF, BFTTF, etc., to designate *between-family* variances, which might cause confusion with B for biovular. There exists in our notation also a mnemonic advantage in that subscripts immediately recall the proper terms.

A way to calculate between-family heritability exists without the assumption of Eq. (3.8) if one can get a sufficient sample of identicals reared apart to get a reliable r_{ITA} (see the end of Table 3.1). Writing the ITA correlation, as illustrated in Eq. (3.16) for twins together, but now for twins apart, as shared, common variance divided by the total twin population variance, $\sigma_{p'}^2$, we have

$$r_{ITA} = \sigma_{bg}^2/\sigma_{p'}^2 = [\sigma_{p'}^2 - (\sigma_{bt}^2 + \sigma_{wt.i}^2)]/\sigma_{p'}^2 \qquad (3.22)$$

[since, in Eq. (3.14), $\sigma_{bt}^2 + \sigma_{wt.i}^2$ is the variance *between* them, that is, the part that is *not* common]. However, the "population variance" by taking one member of each twin pair is written p′ because it is not the same as that of sibs. It is $(\sigma_{wt.i}^2 + \sigma_{bg}^2 + \sigma_{bt}^2)$ not $(\sigma_{wg}^2 + \sigma_{wt.s}^2 + \sigma_{bg}^2 + \sigma_{bt}^2)$. Consequently we must calculate $\sigma_{p'}^2$ by taking one twin from each pair, as something different from σ_p^2 in order to get σ_b^2 from (3.22). Knowing now σ_{wg}^2 from (3.18), σ_{bg}^2 from (3.22), and $\sigma_{wt.i}^2$ from (3.16), we can get σ_{bt}^2 by subtracting these from σ_p^2. This gives us the basis for calculating H_b, the between-family heritability, as well as the total population heritability, H_p. Expressed as derivations from correlations, these calculations would be complex and are best left as the calculations from the variances in H_w and σ_{bg}^2 in Eqs. (3.19) and (3.22). It should be noted that though avoiding the assumption of values for k_2 and k_3 of Eqs. (3.6)–(3.9), this extension of the usual limited twin method, by adding identicals apart to the experiment, still involves (a) assumption of k_1; and (b) the assumption of no covariance of genetic and threptic deviations. But even this much of a breakthrough to H_b in the twin method is usually precluded by the rarity of adequate size of the sample of identicals reared apart.

Although H is not the only way in which the degree of heritability has been expressed (in Chapter 7 we shall discuss others, including those of Falconer, Jensen, and Vandenberg), it is the most generally useful. However, later we shall deal with some variants of these three subscripts to its present forms—w for within the family, b for between families, and p for the total population. What we have just written [Eq. (3.21)] is, of course, H_w, and it gives the ratio specifically for the range of variation *within* the family.

As indicated, the between-family heritability, H_b, and the total population heritability, H_p, are not usually calculated in twin studies (see results offered in Table 3.1 and subsequently) because of the various extra assumptions involved. In general (Chapters 8, 9, and 10) H_w and H_b can have values appreciably different for the very same trait. The total population value, H_p, can be different again from these, though it must tend to be *between* H_w and H_b for purely arithmetic reasons. The differences of H_w and H_b values, as we shall see later, are almost entirely due to situational differences in the threptic, not the genetic, component. Some environmental effects are greater within families; others are greater with the social status and atmosphere differences between families.

In popular discussions, a more favored index than heritability, H, has been the *nature–nurture* ratio, N, for this fitted the era when public debate thought of them as opposites, and always set nature (genetics) against nurture (environment). The expression is the simplest possible contrast of variances.

$$N = \sigma_g^2/\sigma_t^2. \tag{3.23}$$

Today we are more deeply interested in how heredity and environment interact. However, as the arithmetical relation of H and N is very simple, if one wants to obtain N's from the H's given throughout this book he has only to calculate

$$N = H/(1 - H). \tag{3.24}$$

Note that H can never exceed 1.0; N can have any value.

7. A Brief Survey of Ability Heritabilities by the Twin Method on within-Family Variance

If we include the medical and physiological research accumulating by the twin method, the existing literature is truly enormous. There are impressive volumes even on specialized aspects, such as the diagnosis of monozygosity and dizygosity in itself (see Claridge, Canter, & Hume, 1973, p. 103);[5] the effects of sharing the same gestation situation, as well as the resulting interaction of hormones when one twin is a boy and one a girl; the genetic roots of twinning, as such, in families, and so on. The student surveying the twin method results, however, will find most that is likely to interest him in the valuable summaries, largely from the psychological standpoint, by Mittler (1971), Claridge *et al.*, Eysenck (1967), Shields (1954, 1962), Nichols (1969a, 1979), Vandenberg (1965b, 1968) and others. Some of the best work on ability, incidentally, was done quite early (Newman, Freeman, & Holzinger, 1937). Indeed, Burks's (1928), Shield's (1962), and Roe's (1954) perseverance in augmenting the usual twin data with a sample of iden-

---

[5] The problem in the twin method of reliable separation of identicals and fraternals enters also into MAVA if twins are included but its effect is then much less. Separation rests on general physique, hair and eye color and form, fingerprints, blood types, and account of the birth (single or double chorionic sacs). The physiological measures are probably the most important and blood types can at least tell us when twins are definitely *not* identicals. Chorionic sac record is not alone final since a significant percentage of dizygotics manage to grow in a single sac. A fuller account of this problem and further sources will be found in Mittler (1971), Claridge *et al.* (1973), Osborne (1980), and other twin texts. As Muller (1925) pointed out long ago, the decision is ultimately based on a weighted total of many points, in effect a multiple regression equation, and this total, with modern observations, can lead to virtually no overlap between the two distributions. Nevertheless, most existing studies would have to admit to a real likelihood of minor confusion of the two types, which in the main form of calculation must have led to some slight underestimation of the heritabilities.

tical twins reared apart has apparently not been equaled since. (Burt's "apart" data have been questioned.)

Our glance at findings here will explicitly accept the H values with the understanding that they are *those of the twin method*, with the limitations examined earlier. In personality, especially, they do not always agree well with heritabilities of the latter MAVA method (Chapter 8), nor is there really good agreement among the different twin studies themselves. The student should therefore not "make up his mind" and store in his memory the inheritance values from twin data until these have been put in the perspective of agreement or disagreement with the MAVA method. The student should also preserve due alertness to comparative sizes of samples, and to the fact that the reference is usually only to the *within*-family partitioning—though, as we have seen in Eqs.(3.18) and (3.22), it can be extended, with further assumptions, to between and population H's. However, he should not fall into undue skepticism about the method itself merely because the smallness of samples in past research necessarily gives sometimes fairly substantial differences of H values.

In the interests of continuity with personality and ability structure research generally, we shall set aside from serious consideration results obtained on merely arbitrary "personality-trait" scales, of unknown factor composition. Behavior genetics has sufficient problems without obscurities from subjective factor-confused scales. In personality we shall concentrate first on the *primary* personality factors themselves—affectia (A), ego strength (C), surgency (F), superego strength (G), premsia (protected emotional sensitivity, or sensitivity in Eysenck) (I), strength of self-sentiment (Q_3), etc. After that we shall turn to secondaries—a QI, exvia–invia (popularly, extraversion–introversion), QII, anxiety (so defined on the 16 P.F., HSPQ, etc., but as "neuroticism" on Eysenck scales), and so on. In abilities there are Thurstone's (1938) primaries, now extended by Hakstian and Cattell (1974, 1975) to 20 functional unities, and the secondaries g_f (fluid intelligence), g_c (crystallized intelligence), retrievability, g_r; speed, g_s; memory, g_m, etc. (Horn, 1978). Unfortunately, the theory of factorial distinction and the associated test batteries for fluid and crystallized intelligence were not available when most of the twin work now surveyed was done. This is partly responsible for the debates over inheritance of intelligence extending over an unnecessarily wide range of disagreement. The Binet, Wechsler, etc. tests used have commonly contained g_f and g_c in undefined mixtures, though generally (in the Stanford, Binet, WISC, and WAIS) more toward the crystallized intelligence factor.

Table 3.1 summarizes the evidence on inheritance of intelligence. It omits the Burt data since the dust has not yet settled on the dispute of how much of it is dependable. It should be reiterated that we are separating out here exclusively the *twin-study* data, and we have, for example, taken out of the spectrum of constellations in the well-known Erlenmyer-Kimmling survey the twins only (the rest is left until the analysis of ability by MAVA in Chapter 8).

In Table 3.1 we have kept to the standard procedure of reporting the identical- and fraternal-twin correlations, deriving the H by Eq. (3.15), and giving an F value when offered by the investigator. The F value is the usual check on the significance of a variance ratio, in the light of the sample sizes, and in this case is simply $\sigma^2_{\text{FTT}}/\sigma^2_{\text{ITT}}$, i.e., it simply examines whether there *is* a significant difference between identicals and fraternals.

It must be kept in mind in later discussion that the mean value for ordinary intelligence tests, $H = .66$, is the value for *within-family* variance

TABLE 3.1
Values Obtained for the Heritability of Intelligence by Twin Method[a]

	r_{ITT}	r_{FTT}	H_{w}	F	Size of samples
I. Older studies					
(a) Crystallized intelligence (g_c)					
Newman, Freeman, and Holzinger (1937)	.88	.63	.68		50 + 50
Binet	.91	.83	.47		50 + 50
Otis	.92	.62	.79		50 + 50
Achievement Stanford[b]	.96	.88	.67		50 + 50
Shields (1962)	.76	.51	.51		44 + 44
Canter (1969) Short test on 16 P.F.	.23	.13	.12	1.84*	39 + 44
Osborne (1980) Short test HSPQ[c]	.54	.44	.17		82 + 61
Eysenck (1959) (See 1971)	.82	.38	.71		40 + 45
Blewett (1954)	.75	.39	.59		26 + 26
Husén (1959) S.M.I.T.[b]	.90	.70	.67		416
Nichols (1965) N.M.S.Q.T.[b]	.87	.63	.65		482
(b) Fluid intelligence (g_f) (But matrices (Raven) only)					
Canter (1973)	.68	.46	.41	3.04	170

Mean value of above 12 studies, weighed by size of sample $H_{\text{w}} = .57$
Mean values in Erlenmeyer-Kimling survey[d] (1963) of 14 studies
        .87   .63 $H_{\text{w}} = .72$
[The values of r in all these studies ranged over (.76–.95) for identicals and (.45–.88) for fraternals.]
Mean H_{w} of the above and the Erlenmeyer-Kimmling and Jarvik (1963) H_{w} survey (which deals with crystallized intelligence) gives $H_{\text{w}} = .645$
The above values are all on American, British and Swedish data. It is of interest to compare Halperin, Rao, and Morton (1975) on Russian data (146 MZ; 155 DZ) yielding; $H_{\text{w}} = .49$

(continued)

TABLE 3.1 (*continued*)
Values Obtained for the Heritability of Intelligence by Twin Method[a]

	r_{ITT}	r_{FTT}	H_w	F	Size of samples
II. Recent studies[e]					

(a) Crystallized intelligence (g_c)
1. Cattell (1980) (Analyzing MAVA data of Chapter 8 purely by twin method on 94 MZ and 124 DZ) H_w = **.53** (See Chapter 8, p. 314.)
2. Osborne (1980) (MZ 141 white, 70 black; DZ 115 white, 46 black)

	r_{ITT}	r_{FTT}	H_w	F
White	.85	.60	.62	2.36
Black	.80	.34	.70	3.22
Male	.85	.55	.66	3.23*
Female	.83	.54	.62	2.22*
		Mean H_w = **.65**		

(b) Fluid intelligence[f] Osborne (1980)

			H_w	F
White	.79	.41	.65	1.28
Black	.76	.38	.38	1.02
		Mean **.52**		

* p < .01.

[a] The twin method can be extended by utilizing data for twins reared apart, and in that case it permits a solution also for H_b and H_p (between-family and general population heritabilities). However, the sample yet available is too small. Jensen (1972, p. 313) lists obtained values .78 (Shields, N = 38), .67 (Newman *et al.*, N = 19) .68 (Juel-Nielsen, N = 12).

It should be kept in mind in comparing with MAVA results that all the above heritabilities are within family heritabilities

$$H_w = \frac{\text{Within family genetic variance}}{\text{Within family genetic and within family threptic}}$$

[b] Husén's Swedish Military Induction Test and the Stanford Achievement are accepted as close to g_c for the social groups involved. The National Merit Scholarship (Nichols, 1965) can also be accepted.

[c] Osborne's study is on a mixed white and black group.

[d] For a more detailed lay-out of these results see Cattell (1977, p. 265).

[e] Separating fluid and crystallized intelligence, 1980.

partition. Further, since it assumes $\sigma^2_{wt.t}$ and σ^2_{wt} are equal, when the latter may well typically be twice as big, the numerator in H is exaggerated, and H is thus an overestimate even in regard to the within-family ratio.[6] Another conclusion from Table 3.1 that needs watching is the indication that g_c, crystallized general intelligence, has higher heritability than g_f, fluid intelligence,

[6] The MAVA method gives for *sib* families H_w = .59 and .62 for g_f and g_c, respectively (see Table 8.8), but changed to fit the twin method assumptions, .72 and .74. If we assume the traditional tests used in twin work ere 50/50 g_f and g_c we get .73, very close to the *largest* twin survey by Erlenmeyer-Kimmling and Jarvik(1963)in Table 3.1.(But higher than the unweighted average of all, at .66.) It would seem that .70 is the best rounded final estimate today of $H_{w.t}$, that is, within-family crystallized intelligence heritability, by the *twin* assumptions.

TABLE 3.2
Heritability Values:
Developments over Early Ages

Age	r_{BITT}	r_{BFTT}	H	Size of samples
4	.81	.64	.47	51 + 54
5	.81	.62	.50	62 + 58
6	.85	.59	.63	70 + 64

Source: Wilson (1971).

which goes against all evidence of the greater physiological determination of the latter. This oddity might be explicable by (*a*) the usual variation over samples (the .41 comes from a single sample, the .52 from only two; but the .66 from a dozen); or (*b*) the Raven matrices not being the best measure of g_f (relative to a culture-fair intelligence test, which, unlike the Raven is an individual *or* group test spreading over four or more *diverse* subtests): That is, a large specific factor is probably contaminating the Raven test.

An interesting further development here is the determination by Wilson (1972), p. 583) of *H* at yearly intervals. We drop years 1, 2, and 3 from his table, as leaning on undemonstrated *g* validity of tests at that early age, but at 4, 5, and 6 the values are shown in Table 3.2. With the moderate size samples one cannot infer a definite trend, but there are perhaps indications that the genetic component plays a lesser part in younger children (compare Table 3.1). This is an illustration supporting our criticism elsewhere (p. 219) of the popular view that what is inborn must necessarily show itself most in the early years.

In regard to definite primary abilities much less evidence has yet accumulated, though there is a fair amount on cognitive performances of mere "face validity," arbitrarily named categories, for example, "word association" and "card sorting." Some of these latter are couched in Guilford's special abilities, which, as Horn (1978) has pointed out, cannot stand conceptually on the firm functional basis of maximally-simple-structure-rotated factors. However, it is of some practical interest to know also the heritability of such quite specific performances, notably of particular *scholastic* divisions—English, mathematics, physics, history, geography, etc. Canter (1973) and Nichols (1965) give good summaries of what is available there. The vocabulary result in Canter deals with a test that is virtually Thurstone's *V*, and is included in Table 3.3 on primaries, for which we are indebted, however, largely to Vandenberg (1967c).

There is evidently both an appreciable degree of inheritance of several special primary abilities and quite a lot of difference among them. From other sources we would expect considerable heritability for spatial ability,

TABLE 3.3
Heritabilities of Primary Mental Abilities[a]

Ability (Thurstone primaries)	Blewett (1954)	Thurstone et al. (1955)	Vandenberg (1962)	Vandenberg (1967c, 1968)	Osborne[b] (1980)	Canter[c] (1973)	Mean
Verbal (V)	.68	.64	.62	.43	.29	.53	.53
Reasoning (R)	.64	.26	.28	.09	.48	—	.35
Word fluency (W)	.64	.59	.61	.55	—	—	.60
Spatial (S)	.51	.76	.59	.72	.35	—	.59
Number (N)	.07	.34	.61	.56	.29	—	.37
Memory (M) (short distance)	—	.39	.20	—	—	—	.30

[a] As indicated in various special articles, the spatial, fluency and verbal primaries seem to have highest heritabilities. The verbal is pulled down a little here by Osborne's mixed black and white samples, between which groups verbal environment is probably very different.

[b] Mixed black and white sample (80B, 63W).

[c] See Claridge et al., 1973 (Value from a vocabulary score, yielding $r_{ITT} = .85$, $r_{FTT} = .68$, $F = 2.02$.)

and for fluency[7]—the latter as part of the temperament factor U.I. 21, exuberance. The comparatively high value for verbal ability should be seen in the light of (a) its strong loading in the g_c factor, that is, it is a consequence of the comparatively high inheritance of intelligence; and (b) the fact that environmental-verbal contact (within a subculture) tends to be relatively uniformly available. Nevertheless, there *is* a specific primary in verbal, and in this connection it has been known for several years that V is higher in females at all ages and S in males, the differences being highly significant. Naturally one must look partly to sex roles (although girl babies begin to speak before boys, and it is interesting to note that Einstein, who could think readily in four dimensions, was very late in learning to talk). Stafford (1961) has developed an interesting theory about S falling on the sex-differentiating chromosome, and there is growing support for it. Possibly V will be found also to depend in part on a gene locating on the X chromosome.

[7] The theory of the fluency factor (Cattell, 1979b, p. 263) is that it is a function of (a) the size of the stored reservoir of vocabulary, etc.; and (b) of low inhibition (or high retrieval rate) in expression. Probably, when fluency is verbally measured it should have as high a heritability as V. But U.I. 21, low inhibition (Chapter 10), has a lower heritability, which might account for the fact that in Table 3.3 W is .60 and V .65.

8. A Brief Survey of Personality and Psychopathology Heritabilities by the Twin Method on within-Family Variance

Considerable evidence has appeared just in the last few years on personality factors by the MAVA method. But at this point we are confining ourselves to the somewhat less internally consistent values emerging from the twin method. On primary personality factors we have mainly the work of Canter (1973), Gottesman (1963), Osborne (1980), and Vandenberg (1967b).

The Canter and Osborne studies are summarized in Table 3.4. The Gottesman and Vandenberg results are best discussed in more detail and in relation to congruence with those in Table 3.4. The latter disagreed by finding lower heritability (than Klein and Cattell, 1978) in dominance (E factor), affectia (A), superego (G), ergic tension (Q_4), and premsia (I). They agreed with Osborne in finding high heredity for surgency (F) and with all in Table 3.4 in high values (relatively to others) for self-sentiment (Q_3), and for parmia (H). Klein's analysis of Cattell's data by the twin method (ITT and FTT only) led to partly congruent results, namely, agreeing roughly with these investigators' findings of a significant value for affectia A (.42), a moderate low for premsia I (.16) and self-sufficiency Q_2 (.30), and high values for sur-

TABLE 3.4
Heritability of Primary Personality Factors by Twin Method

Trait		Canter[c] (1973)	Osborne (1980)	Klein and Cattell (1978)	Mean
A	Affectia versus sizia	.35	.43	.44	.41
C	Ego strength	.26	.20	.40	.29
D	Excitability	(—)	(0)	.29	.15
E	Dominance	(0)	(0)	.20	.07
F	Surgency	(0)	.44	.29	.24
G	Super ego	(0)	.29	.53	.27
H	Parmia versus threctia	.40	.23	.37	.33
I	Premsia versus harria	.57	(0)	.30	.29
J	Zeppia	(—)	(0)	.63	.32
L	Protension	.31	(—)	(—)	.31
O	Guilt proneness	.34	(0)	.35	.23
Q_2	Self-sufficiency	.38	.14	.10	.21
Q_3	Self-sentiment	(0)	.23	.59	.27
Q_4	Ergic tension	(0)	.24	.25	.16

Note: The intelligence results on the HSPQ-16 PF, namely, factor B, are transposed to Table 3.7. A (0) is entered where result falls below significance or .14 in Osborne's data; a (—) when the factor is not included.

Canter's results are from Claridge *et al.* (1973, p. 31), the others from Osborne (1980) and Klein and Cattell (1978). The ITT and FTT samples are, respectively, 39 and 44, 82 and 61 (combined black and white), and 43 and 55. Details of confidence limits are to be found in the sources given.

gency F (.62), self-sentiment Q_3 (1.0), and parmia H, (.75). The Klein twin analysis of Cattell's data disagrees with the other results here and with Cattell's MAVA analysis in giving high values for G, and higher than middling values for ego strength (C) and ergic tension (Q_4). The last are central to anxiety, so it is not surprising that at the second order he obtained for anxiety (QII) the unusual value of .91, whereas for exvia (QI) obtained a more usual value of .43, and for cortertia ($QIII$) .60, and independence (QIV) .63. These values have the usual augmentation of twin studies above the MAVA sib values (Chapter 9). But it is possible that the unusual high value for H for anxiety is due to a true psychological effect in which identical twins tend to an unusual degree to share the anxiety level within the family.

Loehlin (1977) is impressed in personality scale results generally more by the lack of differentiation among them in H values than by any lack of average real heritability. He therefore cites (1977) the median result among 12 researches as a identical twin r of .48 and a fraternal twin r of .24, which would give a twin-style heritability of $H_{w.t} = .32$. Osborne (1980) reacts somewhat similarly, arguing that the discrepancies are "not so much a weakness of the twin method of research as of the tests used to measure personality. When personality factors are identified and measured with the same dependability as IQ, then [characteristic heritabilities may appear] [p. 137]." With this judgment I feel bound to agree: a confused spread of essentially different heritabilities into indistinguishable values, such as Loehlin observes, is precisely what I predicted would occur so long as test users failed to distinguish between ad hoc personality scales on subjective concepts and those scales targeted on personality structures emerging from several decades of carefully replicated unique factorial resolutions.

The numerous available subjectively-named-and-conceived personality scales present us with variables mixing several more factor-pure personality source traits. If the primary colors on an artist's palette are washed together it is not surprising that the areas of the palette no longer show different properties. A case can be made (consider Loehlin, 1977, p. 336, Table 1) for the conclusion that thoroughly factored tests, such as Thurstone's (1938) and Cattell's primaries (1973a) and Eysenck's secondaries (1959), do show more diverse and distinctive heritabilities. They constitute, moreover, five of the six cases in Loehlin's table where the data refused to fit the false assumptions of twin and sib σ^2_{wt} being identical, or of σ^2_{wg} simply equaling σ^2_{bg}.

As psychologists, let us note in passing, however, that even Loehlin's conservative conclusion that scale differences are not clear, but that a mean heritability of .32 is established for them as a whole, has psychological importance. There are still enough teachers in personality with Watsonian–Skinnerian explanations of personality traits (or nontraits) who need to recognize that this much heritability (which is likely, through error, to be an underestimate) completely precludes a purely reflexological theory of human personality differences.

Although agreeing with that aspect of Osborne's comment which traces much vagueness of results to factorial confusion, I would (as a psychometrist) have to ascribe an appreciable part of the poor consistency to nothing more than the brevity—and therefore the unreliability—of the scales used. Osborne's comparison of personality with intelligence heritabilities must be considered with recognition that Form A of the 16 P.F., HSPQ, CAQ, etc. has 8–12 items and about 3–4 min of testing for each personality factor, compared to 30–60 min for intelligence tests. The constructors of the 16 P.F., etc., have gone to the trouble of producing (and the publishers to the trouble of standardizing) no fewer than five equivalent forms (A, B, C, D, and E) of the 16 P.F., so that the measurement of any factor can now rest on some 40–50 items and thus on about 20–25 min of testing. But the exhortation constantly given to utilize *all* forms in serious research is not much heeded. Although the relation of the H values to test reliability and length is somewhat complex (see Chapter 7) lower reliability in general tends to give lower H. This is easily seen empirically in that in the tables from which Table 3.4 is abstracted (it omits intelligence, B) the short (10-item) intelligence test in the 16 P.F. and HSPQ gives lower H than the longer tests. Similarly in Osborne's tables (from which the intelligence value in Table 3.4 is taken) higher heritabilities are found (mean of three = .67) when both Forms A and B of the culture-fair test are used than when either is used alone (mean of six = .51).

The weaknesses of personality-trait heritability estimates discussed in this chapter in general arise from two causes: (a) tests too short for reliability; and (b) the faulty nature of the assumptions and procedures in the twin method itself. In some studies not covered here there is a third weakness, just discussed; the use of scales never thoroughly factored and replicated as primary source traits, across ages and cultures. The need for allowances in results of the twin method will be brought to light more as we proceed, and despite improvements examined by several writers (e.g., Hasemann & Elston, 1971), its basic weakness of considering $\sigma^2_{wt.i} = \sigma^2_{wt.s}$ stays with it. The addition of data for twins reared apart, as in four researches in Table 3.1, does not obviate this, though it would help in other ways if these samples were adequate, which so far they have never been.

Despite these shortcomings, if we take the overlap of the five existing researches (including Gottesman and Vandenberg) on primary personality factors cited earlier, there is agreement in all five that H has moderately high heritability, in four out of five that F and Q_3 are high, in three that values are substantial for A and Q_2, in two that C (ego strength) and I (premsia) have appreciable heritability, with only one proposing that G and Q_4 may have some significant heritability. Moreover, none of the five argues that $D, E, L, M, N, O, Q_1,$ or Q_4 have heritability exceeding the moderate level of .35. It will be found when we get to the MAVA results that the agreement with this twin consensus is high, except for premsia, I, receiving a somewhat higher rank (close to Canter's .57). Our psychological hypotheses on the nature of primary personality factors must definitely henceforth take into account the notion that surgency (F) and self-sentiment (Q_3) are to a considerable extent genetically determined; that affectia-sizia (A) and premsia (I) follow fairly close, and that any explanation of ego strength (C) must allow for a moderate genetic component. On twin evidence (as on MAVA), dominance (E), super ego (G), radicalism (Q_1), and ergic tension (Q_4) are quite low in heritability, and several other source traits are predominantly environmentally formed.

Eight second-order factors are well replicated in Q-data (see Nichols in Cattell, 1973), but only QI, exvia (extraversion–introversion) and QII, anxiety, have been examined for heritability by the twin method. Confusion arises because Eysenck names his EPI scale for QII "neuroticism," whereas it has been shown that as measured by actual neurotic–normal differences (discriminant function) (Cattell & Scheier, 1961) a *neuroticism* scale for maximal separation has to contain, *besides* the major variance in QII, anxiety, certain characterological contributions, notably of high I, premsia, and low E and F. For uniformity, and also because Canter's results do properly distinguish QII, anxiety, from the IPAT neuroticism scale, giving results for both, it has seemed desirable to put the EPI so-called neuroticism scale results[8] along with other measures of QII under the pure *anxiety* factor, as in Table 3.5.

The disagreements among experimenters on H values are here quite marked. Some of this could be due to different scales with the same label or testing conditions. The report of a large *negative* correlation of fraternals in Eysenck's 1956 study is out of line with any other study. However, as a consideration of the standard error of an r—on 40 cases tells us that r has to reach approximately .4 to be significant ($p < .01$)—variations of this magnitude could appear in published studies that have stopped after collecting no more than 40 pairs of each kind of twin. Work in this area definitely calls for greater endowment. However, the central values of Table 3.5 are not so far from the MAVA values in Chapter 9, Table 9.7, namely, .44 and .38. Nevertheless, perhaps the conservative conclusion from the twin method in view of the variability in Table 3.5 is that there is significant and appreciable inheritance in both exvia and anxiety. No significantly greater mean value for one than the other can be concluded.

Canter has pursued the interesting question, with twins who are reared

---

[8] As has been pointed out for some years (Cattell, 1964b), a masterpiece of confusion confronts the trusting test user here. In objective test, T-data, personality factors, Eysenck called the U.I. 23 pattern *neuroticism*. It is quite distinct from U.I. 24, which by every known association is anxiety, and which falls exactly on the same axis as the second-order anxiety factor, QII, from the 16 P.F. Eysenck's Q-data neuroticism scale, however, in PEN, etc. does not correlate with his neuroticism factor in T data, U.I. 23, but with U.I. 24, the Q "neuroticism" scale being, as stated earlier, an anxiety scale. The structure has been clear for years; but the verbal confusion meanwhile has been costly to theory and experimentation (see Cattell, 1964b).

TABLE 3.5
Heritability of the First Two Broad Secondaries in Personality

	r_{BITT}	r_{BFTT}	H	F	Size of sample
QI extraversion–intraversion (exvia)					
Canter (1969) (EPI scale)	.34	.29	.07	1.41	40 + 45
Canter (1969) 16 P.F., second order factor	.56	.33	.36	1.36	39 + 44
Eysenck (1956) Eysenck scale	.50	−.33	.62		
Canter (1973) 16 P.F. second order	.43	.08		1.36	
Shields (1962) Eysenck scale	.42	−.17	.50		
Mean H			**.39**		
QII anxiety					
Canter (1969) 16 P.F. second order	.43	.08	.38	1.34	39 + 44
Canter (1969) EPI scale N	.37	.23	.18		40 + 45
Eysenck (1956) Eysenck scale					
Canter (1973)	.56	.33	.37	1.34	
Shields (1962) Eysenck N	.38	.11	.30		
Eysenck and Prell (1951) Eysenck N	.85	.22	.81		
Mean H			**.41**		
Neuroticism[a]					
Canter (1969)	.36	.06	**.32**	1.30	

[a] This is anxiety plus other 16 P.F. factors differentiating neurotics.

apart for various numbers of years, of what the length of separation does to their resemblance. For those factors in which indications of some trends appeared the results are set out in Table 3.6.

In most traits—notably, ego strength (C), shrewdness (N), guilt proneness (O), self-sentiment strength (Q_3), ergic tension (Q_4), and anxiety (QII)—more years of separation reduce the correlations. One must conclude that environment—specifically the environmental difference of (adopting) families, that is, σ^2_{bt}—has an appreciably greater role in these factors than in those not listed. By contrast, we find the correlations are *increased* with increased length of separation in the case of dominance (E), surgency (F), parmia (H), and exvia–invia (QI) (also slightly in premsia, I). This shows either that in these traits earlier environment is not so much more important, or that when sibs are genetically high on the traits a common family situation is mutually "abrasive." This I have elsewhere called the "Brazil nut effect" within a family environment. Just as a bump on one Brazil nut coincides with a dent in its companion within the outer tight container, so a dominant or a surgent twin might force his mate to be relatively submissive or desurgent. (Surgency loads "talkativeness" highly. In an hour, more talk by one could mean more silent habits in the other.) Released into different families, the natural inherited level of both is more likely to come to expression. Canter undertook this age-change study presumably as a demonstration, since the

TABLE 3.6
Effects of Greater and Lesser Separation on Correlations within
Identical and Fraternal Twin Pairs

	Separated less than 5 years		Separated more than 5 years	
	BIT ($N = 23$)	BFT ($N = 28$)	BIT ($N = 15$)	BFT ($N = 16$)
Primary factors				
C Ego strength	.50	.22	.22	−.03
E Dominance	.13	.18	.37	.37
F Surgency	.42	.24	.79	.60
H Parmia (shy)	.57	.28	.63	.35
I Premsia	.67	.20	.70	.36
N Shrewdness	.34	.41	.03	.07
O Guilt proneness	.50	.18	.27	−.19
Q_3 Self-sentiment	.33	.22	.11	.07
Q_4 Ergic tension	.49	.23	−.11	−.06
Secondaries				
QI Exvia	.29	−.65	.85	.50
QII Anxiety	.56	.44	.27	−.53

Source: Canter (1973, p. 41).

samples are quite small, but the indications are interesting enough to justify follow-up with larger samples.

In the domain of personality pathology, since psychiatric syndromes have not been brought into measurement or the form of a continuum—at least in these studies—behavior genetics falls back on somewhat cruder forms of calculation. The first involves the concept of "manifestation rates," and the second of "concordance ratios," as already mentioned. The manifestation rate is the percentage of those certainly possessing the genetic potential, according to a Mendelian model, for a certain disorder or phenotypic character who actually show it in a given environment. For example, if two dark-eyed parents are themselves each born of a light- and a dark-eyed parent combination we know they must be heterozygous for eye color, and since dark is dominant a large number of such parents should approach a total of 75% dark-eyed children. If only 50% showed, we should say that the dark-eyed gene has a 66.6% manifestation rate. In real data, in this eye-color example, the manifestation rate is close to 100%; but, conceivably, embryological environments could exist in which it would be less.

The notion of concordance ratios can best be seen by the following arrangements in which it is supposed that we have 100 identical (reared together: ITT) and 100 fraternal (reared together: FTT) twins, and we enter the percentage that are alike on the give character—say, schizophrenia—for the ITT's and FTT's. (The reader should be reminded that most tables in genetics books use MZ for monozygotic, and DZ, for dizygotic, where we have ITT and FTT. As has already been mentioned, the latter symbolism is preferred throughout here because it conveys, *also,* whether they were reared together or not. The symbols MZ and DZ are perfectly descriptive of zygosity but of nothing more. In data where the environment is "understood" MZ and DZ will cause no confusion, but here, and still more when we come to MAVA, it is necessary to distinguish the resemblances of environment too.)

	ITT	FTT	
Concordant	70	40	Both members show it
Discordant	30	60	Only one member shows it

The concordance–discordance ratio here would show substantial inheritance. It will be noted that it is possible to transform this into correlations and also into heritabilities (see Shields, Gottesman, & Slater, 1967).

The most extensive surveys of schizophrenia, by Kallmann and his associates 1938, 1950) gave results (also for nontwin relations) most readily shown by Figure 2.1, p. 26. Shields *et al.* (1967) surveyed and averaged 12 studies (including Kallmann's) from several countries (Mittler, 1971, p. 124), giving concordances and an H value as follows:

ITT	FTT	H
53%	11%	.47

On manic–depressive disorder, Lewis's (1933) London survey gave estimates little altered by later work, namely:

	If one parent affected	If both parents affected
Percentage children affected	33	67
Percentage showing lesser mood disorders	17	33
Percentage unaffected	50	0

It will be seen that this cannot be directly integrated with the environment-separating twin method because the parental atmosphere would also be involved in these results. Direct twin data (Kallmann, 1950, 1953; Rosanoff, Handy, & Plesset, 1935) give the following concordances:

	ITT	FTT
Kallmann	96	26
Rosanoff *et al.*	70	16

If these data stand they indicate a heritability for manic–depressive even higher than the rather high value for schizophrenia.

The twin method has also been applied to less-defined behavior disorders and to neurosis. Lange's initially surprising data on "crime as destiny"

dy been set out in Table 2.2. Eysenck (1970) adds to this, the following:

	ITT	FTT
Adult crime	71%	34%
Juvenile delinquency	85%	75%[9]

In neurosis, if we average percentages over nine studies (Mosher, Pollin, & Stabenau, 1971—but neglecting the study with only 16 cases) we get

		ITT/FTT concordance
ITT	FTT	ratio
51	29	1.76

(See also Claridge, *et al.*, 1973.) This indicates lower inheritance than for the psychoses (and also than for crime), which fits most of what is known about cause and cure of these personality abnormalities.

Mental defect has also been treated in a psychiatric context (Penrose, 1963; Prehm, Hamerlynck, & Crosson, 1968) as a syndrome, and there is appreciable data (separating out also the fraction that is congenital but not innate) but it is too well known for us to extend into it here. However, it would be a useful contribution to methodology to compare the H values for this all-or-nothing categorical approach with those from intelligence test continuous measures, as in the tables of the previous section.

9. Summary

1. Although attachment of genotypic features to genes is the ultimate goal of behavior genetic research, the first step toward it, in multigene-determined, continuous traits, has to be the determination of relative variance contributions of genetic and environmental influences to the observed phenotypic variances. A major class of methods presently available for this purpose are the convarkin methods (*con*trast of *var*iance across *kin*ships), of which the twin method was historically the first, and is so studied here.

2. Ideally, the twin method compares three kinds of concrete, observed variances; those of identical twins reared together; of fraternal twins reared together (or sibs), and of identical twins reared apart. But *adequate* samples of the last, *known to be separated from birth,* are rare, and, as the main workhorse in nearly a century of human genetic research, the method has mainly rested on comparison of identical and fraternal twins in ordinary family situations.

[9] This covers only 25 cases.

3. This comparison, operating on a model that "reconstitutes" the observed, concrete variances from abstract, "hypothetical" genetic and threptic variances and covariances, is used here as a good introduction to the problems of models and modes of solution for genetic data. The calculations are described, from variances and from intraclass correlations.

4. All considerations point to the fact that it is not enough to consider the observed concrete variance as only a *sum* of genetic and threptic variances. A fully satisfactory model must have terms for both interaction and covariance as well. The difference, conceptually, between covariance and interaction is pointed out, but, unfortunately, although it is simple to introduce terms for the former, presently used behavior-genetic equations have not yet incorporated the latter, or led to experiments on that model.

5. The twin method has certain limitations which it is important to keep in mind in interpreting its contributions and heritability values. These limitations are the following:

(i) Incompleteness, in most studies, of the separation of identical and fraternal twins. The problem in the twin method of reliable separation of identicals and fraternals enters also into MAVA if twins are included but its effect is much less. Separation rests on general physique, hair and eye color and form, fingerprints, blood types, and account of the birth (single or double chorionic sacs). The physiological measures are probably the most important, and blood types can at least tell us when twins are definitely *not* identicals. Chorionic sac record is not in itself final since a significant percentage of dizygotics manage to grow in a single sac. A fuller account of this problem and further sources will be found in Mittler (1971), Claridge *et al.* (1973), Goldberger (1977), and other twin texts. As Muller (1925) pointed out long ago, the decision is ultimately based on a weighted total of many points, in effect a multiple regression equation, and this total, with modern observations, can lead to virtually no overlap between the two distributions. Nevertheless, most existing studies would have to admit to a real likelihood of minor confusion of the two types, which in the main form of calculation must have led to some slight underestimation of the heritabilities.

(ii) Restriction to same-sex pairs. Through contrast of different-sex pairs, it is true, some useful further results can be obtained. Eaves *et al.* (1978) point out that working with unlike sex fraternals will throw light on the question of whether the same gene loci are involved in determining the same measured trait in men and women. Stafford (1961) and others (but see DeFries *et al.*, 1979) cite evidence that the spatial visualization ability factor is differently determined in men and women and there are, of course, clear evidences of phenotypic features determined by X or Y chromosomes differently in men and women, for example, chest girth. Eaves *et al.* (1978) give the calculation to proceed from mixed-sex comparisons in order to check this.

(iii) Its use of the assumption that the within-family environmental variance of identical twins is the same as that of fraternals and sibs—which we know to be incorrect (see Note 2 in this chapter, and Plomin, Willerman, & Loehlin, 1976).

(iv) Its assumption that twin families, though rare, are representative in their mean and variance of ordinary families that yield the community population variance. This also is incorrect. Twinning is itself heritable, and comparisons of races, for example, show significant differences in liability to multiple births. Enough traits have been examined to show that there are in fact some significant differences in trait *means,* at least, and probably interfamily difference of environmental variance between twins and sibs. We may have to take some account also of genetic variance differences, of a slight magnitude, between fraternal twins and sibs.

(v) In analysis, its inability to separate genothreptic covariance terms.

(vi) In analysis, its inability to yield other than within family heritability values (H_w) without further assumption. (Notably that $\sigma^2_{bg} = k\sigma^2_{wg}$)

(vii) As shown later the $H_{w.t}$ values themselves are spuriously high —($H_{w.t}$ (using $\sigma^2_{wt.t} = \sigma^2_{wt.s}$) equals about 1.2 times $H_{w.s}$ (sib within family).

(viii) A lesser source of possible error in interpretation—and again one that can affect other methods too—is the finding that essentially genetic effects of a minor kind can come from some sources in the body of the zygotic cells outside the nucleus and the chromosome structure. All that is known about these effects, possibly not handed on beyond a generation, is that they are quite small and unlikely to affect the main Mendelian interpretations.

6. It may seem ungracious to those who have labored through the twin method in behavior genetics to list eight shortcomings, but when scientific issues arise on the relative weight to be given to twin results vis-à-vis the MAVA and other biometric, convarkin methods they must be faced. To the extent that the MAVA method, in the next chapter, *may* include a twin constellation, it also is vulnerable to defects (i), (ii), (iv), and (viii). As Chapters 8, 9, and 10 show, its sampling errors produce variations of result that are at first more openly revealed than in any single, uncompared twin study.

But the susceptibility of the latter in this respect is equally obvious when, as in tables here, we set different twin studies side by side.

As we shall see in the next chapter, the substantial superiority of the MAVA method consists in avoiding the error and bias of the twin method with respect to (iii) by accepting different identical, fraternal, and sib within-family threptic variances, (v) by handling covariance, (vi) by getting a more assumption-free value for between family heritability, and (vii) by avoiding spuriously high H_w values, i.e, the $H_{w.t}$ values.

7. The same heritabilities are implicitly aimed at by the convarkin methods (twin and MAVA), namely, H_w, within family, H_b between family, and H_p for the total population range. These values are the ratio of the genetic variance to the total variance, the latter being the simple sum of genetic and threptic variance. [The nature–nurture ratio, N, is genetic divided by threptic only, and is equal to $H/(1 - H)$.] However, most twin studies have stopped at H_w results, and it is to these that we give serious attention, remembering, in comparing with MAVA results, that they will be inflated by a multiple of approximately 1.2.

8. In the ability field, the heritabilities of both general intelligence and primary abilities have been studied, though studies of the former have involved, with one exception, traditional rather than culture-fair tests, thus covering g_c or a mixture of g_c and g_f, rather than g_f. The H's of primary abilities vary widely, with some—like spatial, verbal, and word fluency— being as high as general intelligence, but others much lower. The H_w for intelligence, averaging .65, is close (if corrected for the twin basis bias), to that obtained later by MAVA.

9. Primary personality factors studied by questionnaire scales show an average heritability (about .4, if Klein's results are included) that makes a Watsonian—Skinnerian view of personality development quite untenable. There is tolerably good agreement among available results that F (surgency), H (parmia), J (zeppia), and Q_3, (self-sentiment) have high heritability; and that D (excitement), E (dominance), L (protension), M (autia), N (shrewdness), Q_1 (radicalism), Q_2 (self-sufficiency) and Q_4 (ergic tension) have low heritability. There is some disagreement on A (affectia), C (ego strength), I (premsia), and G (superego), but Canter's results tend to agree in those cases with later MAVA data placing all but G with appreciable heritability. The secondary personality factors also show middling heritabilities (.39 and .41) for exvia and anxiety, giving approximate agreement with later MAVA findings.

10. Personality inheritance in the pathological range has been investigated in terms of all-or-nothing psychiatric syndrome diagnoses, using analytical methods yielding evidence as manifestation rates and concordance values (which with some assumption can be converted to H's). Many years of extensive data gathering leave no doubt of a rather high inheritance of manic-depressive disorder and of schizophrenia (probably higher than for most normal traits), but quite low heritability in regard to neurosis.

4

The Genesis of the MAVA Model
and Its Solutions

1. The Intrinsic Potency of Using Many Family
Constellations in MAVA

When traits that yield continuous measures are involved—as in most normal personality studies—the indispensable preliminary for any further theorizing, in learning or in gene structure, is to obtain the variance contribution and covariance contributions from genetic and threptic sources. As we have seen, the twin method does this for a limited number of such influences. The MAVA method, which we have so far barely sketched in its role within convarkin methods, uses the same algebraic principles for breaking down concrete, observed variances into abstract, "hidden," hypothetically underlying, genetic and threptic sources, but does so in a broader context.

By taking all feasible family *constellations*, as we may call them, together (e.g., twins, sibs, unrelated children raised together), we are able, as we shall see, to solve for more diverse abstract contributing variances. Before describing it let us head off criticisms sometimes voiced that obtaining contributory variance values is not really to reach the Mendelian structures that geneticists want. It is quite true that these variance approaches do not *immediately* do so; but they are a necessary step on the way. Furthermore, the psychologist should notice that the analysis leads not to one highway but to two: that to *genetic* understanding and that to *learning theory*—the latter in terms of the acquisition of threptic increments from cultural environments.

It is because of this integration of genetic and learning interests that the new term *genothreptics* might well have been given as the subtitle of this book. The term implies an equal concern with genetic and threptic influences and an especial concentration on their mode of interaction.

The rationale of the multiple abstract variance analysis (MAVA) method is that a model of sources of variance can be constructed, proceeding logically from certain commonsense assumptions, for any family constellation whatever—sibs reared together, sibs reared apart, unrelated children reared together, twins together, and so on. After finding out by actual testing what these observable variances are, we can then solve for the contributory

TABLE 4.1
Constellations Providing Concrete Variances as a Basis for Analyzing into Abstract Variances

---

1. *Measurable within-family concrete variances*

   σ^2_{ITT} = Within-family variance of identical twins raised together

   σ^2_{ITA} = Within-pair variance of identical twins raised apart

   σ^2_{FTT} = Within-family variance of fraternal twins raised together

   σ^2_{FTA} = Within-pair variance of fraternal twins raised apart

   σ^2_{ST} = Within-family variance of sibs raised together

   σ^2_{SA} = Within-pair variance of sibs raised apart

   σ^2_{UT} = Within-pair variance of unrelated children raised together

   $\sigma^2_{UA\ or\ GP}$ = Differences of pairs of unrelated persons reared apart. This is initially called UA here to show its relation, but henceforth is GP because it is the variance of the general population.

   Less practicable within-family constellations

   σ^2_{HST} = Within-family variance of half sibs raised together

   σ^2_{HSA} = Within-pair variance of half sibs raised apart

   σ^2_{CT} = Within-family variance of cousins raised together

   σ^2_{CA} = Within-pair variance of cousins raised apart

   There could follow a number of raised together and apart constellations in which the genetic variance is unusual, as in cousin marriages, uncle–niece marriages, father–daughter incest offspring, sib incest offspring, and so on. Whatever we may feel about some of these socially and ethically, they do provide extra sources of solution in genetic analysis!

2. *Measurable between-family concrete variances* (among means of children in each family)

   σ^2_{BITTF} = Between identical twin raised together families

   σ^2_{BITAF} = Between identical twin raised apart families

   These simply continue, case for case, the list given in (1), adding in the mnemonic form of notation here used, simple B for between and F for family, in BFTTF, BFTAF, BSTF (the symbol BNF, between normal—or natural—families has been used in several studies instead of BSTF).

   Several of the constellations listed can be further subdivided. For example, same-sex sibs and fraternal twins conceivably have a different within-family variance from opposite sex members. Half sibs raised together could be raised all their lives with the mother or the father. The genetic components could also differ slightly for same-sex and different-sex pairs, because of the role of the sex chromosomes, X and Y.

---

unknown genetic and threptic sources by the same mathematical approach —the solution of simultaneous equations—as in the twin method. Thus MAVA is distinguished from the twin method by—among other things—(a) use of more numerous and varied family *constellations,* as we shall call them; (b) a more *comprehensive* model of what is actually happening, including covariation of genetic and threptic influences; and (c) ultimately more sophisticated methods of solution. Incidentally, MAVA is in principle a model applicable to other units than the family, for example to orphanages, but these are beyond practicality at present.

A list of some of the main constellations that could conceivably be used to provide the varied combinations of genetic and environmental terms needed to solve for such separate sources is given in Table 4.1.

Actually each constellation is *repeated* in Table 4.1 in the form of one equation for the *within*-family and one for the *between*-family variance (the latter being that among the means of child families, about the grand population mean.) Both within and between can be simultaneously used in equation solution calculations because, with one exception, they are statistically independent. If the within and between added up to the full population variance they could *not* be used, however, along with the variance for the total population (σ^2_{GP}). But only in the case of sibs does this dependency arise through the sum of squares for the within and between adding to that for the actual population. This last point is justified later, when we note that a population of identical-twin families, for instance, is not representative of the actual world population; but it will also be explained that by special, iterative methods "within," between," and "population" *can* be used together.

2. The Principles for Relating Abstract to Concrete Variances

It should be noted that genetic and threptic sources of variance differ in one important characteristic. Granted certain reasonably firm genetic and other assumptions, we can depend on fixed relations among the various genetic contributions. For example, with random mating assumed, (and no dominance) the within-family genetic variance of sibs and fraternal twins, σ^2_{wg}, is exactly equal to the between-family genetic variance, σ^2_{bg}. Or again, the genetic variances in the "less practicable samples" in Table 4.1—half sibs, cousins, etc.,—bear a fixed relation to the ordinary between sib, σ^2_{wg}. One is fortunate when such knowledge can appropriately be applied, for in research one generally has fewer equations (concrete variances and constellations) than are needed to solve for all the unknown abstract variances one would like to know.

Let us now exemplify the logic and assumptions in building up MAVA equations. So far, in the twin method, we have assumed, in the basic twin-method equations—(3.3) and (3.4)—only that the within-family threptic variance, σ^2_{wt}, equals σ^2_{ITT}, since identical twins cannot differ in a genetic component, whereas fraternals will have a variance equal to the sum of σ^2_{wt} and σ^2_{wg}. Now we shall go beyond that, in the twin case, by (a) considering that σ^2_{wt} splits into an environment for fraternals differing from that of identicals, so that $\sigma^2_{wt.t} \neq \sigma^2_{wt.f}$; and (b) accepting that there will be correlations of genetic and threptic deviations so that

$$\sigma^2_{FTT} = \sigma^2_{wg} + \sigma^2_{wt.f} + 2r_{wgwt}\sigma_{wg}\sigma_{wt.f}. \tag{4.1}$$

Similarly, for ordinary sibs raised together [refraining, as in (4.1), from the further refinement of sex-differing pairs] we have

$$\sigma^2_{ST} = \sigma^2_{wg} + \sigma^2_{wt.s} + 2r_{wgwt}\sigma_{wg}\sigma_{wt.s}. \tag{4.2}$$

(This derives from the common statistical proposition for the summing of variance with correlation of the deviations.) Again, refinements occur to one, for example, that r_{wgwt} might differ for same-age fraternals and different-age sibs, so that one might ultimately have to assume $r_{wgwt.f} \neq r_{wgwt.s}$. However, let us not be spendthrifts for luxuries of special abstract variances till we see how many concrete variances can be made available in actual experiment to pay for their discovery.[1]

Turning to the newer field of between-family variances, which was essentially set aside from twin analysis, and taking ordinary families, we shall first have to introduce two new terms, σ^2_{bg}, the genetic variance between families, and σ^2_{bt}, the threptic variance between families. And because the status of a family in the culture, and its internal atmosphere, may have some relation to its genetic average make-up (e.g., as in the Bach family an apparently natural talent got tied to being brought up in a culturally musical

---

[1] There are actually three kinds of within-family environmental (and therefore threptic) variances that can and should be psychologically distinguished among related individuals raised together:

  1. $\sigma^2_{wt.i}$, for identical twins, same age, same genomes
  2. $\sigma^2_{wt.f}$, for fraternal twins, same age, different genomes
  3. $\sigma^2_{wt.s}$, for sibs, different ages, different genomes

These can be multiplied further if we consider differences along parameters of (a) absolute age; (b) male or female pairs; (c) same or different sex, for (2) and (3); and (d) two only or embedded in larger family of offspring.

The possibility also exists that r_{wgwt} is different for fraternals, $r_{wg.wt.f}$, and for sibs, $r_{wg.wt.s}$, because the age difference in sibs could magnify the association of environmental differences with genetic differences. As will be clear later in this chapter, we must be content, until research endowments permit the full MAVA to be used, to compromise by letting one term, r_{wgwt}, stand for both correlations and by letting one term, $\sigma^2_{wt.t}$ (t for twin), stand for both kinds of twins. From this page on the symbolism will be so simplified.

family), there will be covariance, positive or negative, represented by $r_{bgbt}\sigma_{bg}\sigma_{bt}$, analogous to $r_{wgwt}\sigma_{wg}\sigma_{wt}$ within families. These are components of the observed concrete variance σ^2_{BNF} (between natural families: sibs raised together) calculated, as explained earlier from the deviations of the means of pairs of sibs about the grand mean for all families.

The basic manner of simultaneous equation solution in this field has been introduced by Eqs. (3.3) through (3.5), though it should be noted that in the twin method we contented ourselves with a simpler model in which there was no covariance term such as will be used here. Incidentally, as Table 4.1 reminds us, the mnemonic subscripts for within family variances will remain as in the twin illustration (e.g., σ^2_{ITT} the within variance for *identical twins raised together*), but now a *between* family series is added (e.g., σ^2_{BITTF}, for *between identical twin (raised together) families*).

The student's usual background in analysis of variance will initially be of assistance in following the present development. For example the reader will note that the variance among groups (families in this case) plus the variance within groups (the within variance being simply that calculated as in Eq. (3.1) earlier or Eq. (4.5) here, equals the population variance). Though by our mode of calculation in Table 4.2 their sum is actually just *double* the population variance. (In equation numbers in Table 4.2, $(7) = \frac{1}{2}[(3) + (6)]$.) Thus what may not at first seem logically necessary is the intrusion of within-family variance terms into Eq. (4.3), given in what follows. One way of perceiving this requirement is to recognize that if the between-family variance were calculated from σ^2_{bg}, plus σ^2_{bt} suppose that each mean is fixed by a family of infinite size (a parent's nightmare!), whereas in fact it is but a few, indeed two, so that the means vary also by the within-family variance.

If the reader will patiently proceed with deriving Eq. (4.3) from the first principles, as in Loehlin (1965b) using the deviations as starting point, he will see that

$$\sigma^2_{BNF} = \sigma^2_{wg} + \sigma^2_{wt.s} + 2r_{wgwt}\sigma_{wg}\sigma_{wt.s} + 2\sigma^2_{bg} + 2\sigma^2_{bt}$$
$$+ 4r_{bgbt}\sigma_{bg}\sigma_{bt} \qquad (4.3)$$

though essentially correct contains a special assumption, as become evident from Eq. (4.4)

However, in the MAVA equations several much more subtle issues arise than in straightforward ANOVA, having to do with the differences of twin and ordinary sib population and especially the covariances of various kinds. The general behavior genetics student may not need to face them all, and we shall turn to footnotes for the more recondite issues, but the researcher will need to follow the debates we shall describe around these issues. The original sets of MAVA equations (Cattell, 1953, 1960, 1963b) constituted a model based on first principles in genetics and psychology, but a valuable critique by Loehlin (1965b) has checked it against algebraic derivation as follows.

Let us write the deviations of sib_1 and sib_2 from the *population* mean, as x_1 and x_2. Then

$$x_1 = x_{wg_1} + x_{wt_1} + x_{bg_1} + x_{bt_1}$$
$$x_2 = x_{wg_2} + x_{wt_2} + x_{bg_2} + x_{bt_2}. \tag{4.4}$$

This simply says there are four additive parts to the total deviation, two (genetic and threptic) from the family means and two of the family means from the grand (population) mean. Now if we consider calculating the concrete variance σ^2_{ST} and put the formula in the more intelligible form from first principles we have

$$\sigma^2_{ST} = \frac{\sum_{f=1}^{N}\sum_{i=1}^{2}(x_{if} - \bar{x}_f)^2}{N}, \tag{4.5a}$$

where $\bar{x}_f$ is the mean of the two sibs for a given family and i is either of the two sibs in turn. For simpler calculation from the difference of the two sibs we can write this as

$$\sigma^2_{ST} = \frac{\sum_{f=1}^{N}(x_{1f} - x_{2f})^2}{2N}, \tag{4.5b}$$

where 1 and 2 are the two sibs.

Correspondingly, in calculating the between-family variance we have

$$\sigma^2_{BNF} = \frac{\sum_{f=1}^{N}\{[(x_{1f} - x_{2f})/2] - \bar{x}\}^2}{N - 1}, \tag{4.6}$$

where $\bar{x}$ is the grand mean of the pairs of all families. There are $N - 1$ degrees of freedom for estimating the variance around a mean of N families. Note we are dealing so far with summed squares.

Looking first at Eq. (4.5b) and substituting for x_1 and x_2 as in (4.4) we have

$$\sigma^2_{ST} = (2N)^{-1} \sum_{1}^{N} (x_{wg_1} + x_{wt_1} + x_{bg_1} + x_{bt_1} - x_{wg_2} - x_{wt_2}$$
$$- x_{bg_2} - x_{bt_2})^2 \tag{4.7}$$

which, expanded, becomes

$$\sigma^2_{ST} = (2N)^{-1} \sum_{1}^{N} (x^2_{wg_1} + x^2_{wt_1} + \cdots + x^2_{bt_2} + 2x_{wg_1}x_{wt_1}$$
$$+ 2x_{wg_1}x_{bg_1} + \cdots + 2x_{bg_2}x_{bt_2}) \tag{4.8}$$

and, expressed in variances and covariances, becomes

$$\sigma^2_{ST} = \tfrac{1}{2}\sigma^2_{wg_1} + \tfrac{1}{2}\sigma^2_{wt_1} + \cdots + \tfrac{1}{2}\sigma^2_{bt_2} + r_{wg_1wt_1}\sigma_{wg_1}\sigma_{wt_1}$$
$$+ r_{wg_1bg_1}\sigma_{wg_1}\sigma_{bg_1} + \cdots + r_{bg_2bt_2}\sigma_{bg_2}\sigma_{bt_2}. \tag{4.9}$$

Among four variances (or deviations) there are, algebraically, six possible combinations (covariances). But correlations of within and between deviations are impossible in the ordinary ANOVA conditions, so we can drop all terms with r_{wgbg}, r_{wtbt}, r_{wgbt}, and r_{wtbg} (see, however, page 244). To show where these are dropped from a systematic devolution of the last equation they are put in square brackets in Eq. (4.10).

$$\begin{aligned}
\sigma_{ST}^2 = {} & \sigma_{wg}^2 \, (1 - r_{wg_1wg_2}) + \sigma_{wt}^2 \, (1 - r_{wt_1wt_2}) \\
& + \sigma_{bg}^2 \, (1 - r_{bg_1bg_2}) + \sigma_{bt}^2 \, (1 - r_{bt_1bt_2}) \\
& + 2\sigma_{wg}\sigma_{wt} \, (r_{wg_1wt_1} - r_{wg_1wt_2}) + [2\sigma_{wg}\sigma_{bg} \, (r_{wg_1bg_1} - r_{wg_1bg_2})] \\
& + [2\sigma_{wg}\sigma_{bt} \, (r_{wg_1bt_1} - r_{wg_1bt_2})] + [2\sigma_{wt}\sigma_{bg} \, (r_{wt_1bg_1} - r_{wt_1bg_1})] \\
& + [2\sigma_{wt}\sigma_{bt} \, (r_{wt_1bt_1} - r_{wt_1bt_2})] + 2\sigma_{bg}\sigma_{bt} \, (r_{bg_1bt_1} - r_{bg_1bt_2}).
\end{aligned} \tag{4.10}$$

As Loehlin points out, by assuming particular values for the r's, the various within-family variance equations in the MAVA test (Table 4.2) can be derived. For example, Eq. (4.2)—appropriately making $\sigma_{wt} = \sigma_{wt.s}$, because we are dealing with sibs—appears when we assume the following:

1. That $r_{wg_1wg_2}$ and $r_{wt_1wt_2}$ are zero, that is, the sibs' deviations from the true family mean are unrelated. (The way we calculate from the mean of the two might seem to negate this. But if one thinks of a large family it is readily seen that the deviation of one from the mean approaches independence of another from the mean.) This converts the first row of Eq. (4.10) to $\sigma_{wg}^2 + \sigma_{wt}^2$.
2. That $r_{bt_1bt_2}$ and $r_{bg_1bg_2}$ are unity. This is true because both the sibs come from the same family and have the same family mean. This condition eliminates the second row of (4.10).
3. That all within–between correlations, like r_{wgbg}, are zero, by ANOVA rules, as stated earlier. This eliminates all terms in square brackets in Eq. (4.10).
4. That the between-family correlation $r_{bg_1bt_1}$ as calculated for one of the sibs is the same as that for the second sib, $r_{bg_1bt_2}$ because bt_1 and bt_2 are just the same for both inasmuch as they belong to the same family. This eliminates the last term.

We are left with the peculiar observation, to which Loehlin was the first to draw attention, that the covariance we have set down on "common-sense" grounds in various Table 4.2 equations as $2r_{wgwt}\sigma_{wg}\sigma_{wt}$ is actually algebraically a *difference*, namely, $2(r_{wg_1wt_1} - r_{wg_1wt_2}) \, \sigma_{wg}\sigma_{wt}$. This becomes the expression we have used only if $r_{wg_1wt_2}$ is zero (for there is no problem over $r_{wg_1wt_1} = r_{wg_2wt_2}$). Actually an argument can be reasonably made that though $r_{wg_1wt_2}$ is likely to be small it need *not* be zero. For the particular genetic endowment of sib 1 results in behavior which becomes part of the environment of sib 2. Indeed, later (Chapter 6), in considering causal relations within the whole family, some special discussion will be given to one member's inheritance being a functioning part of another's environment.

At this point two alternative courses will be described, and one followed here until the other is *developed later*.

1. The within-family environmental deviation of an individual, though *measured* in the family, actually arises substantially from forces outside the family (different schools, peer groups, accidents). Only a fraction of the variance comes from influences rooted in and varying within the family, and only a still smaller fraction of the forces arise from *genetic* influences in the family members. Consequently, we can approximate that $r_{wt_1wg_2}$ is zero, and simply proceed to use r_{wgwt} as before.

2. Alternatively, we can admit that $r_{wt_1wg_2}$ has a small real value and, by entering it as a new unknown, seek for further equations in MAVA from which to solve for it. This is a refinement to be left till later, but we put the equation that demands it on record, in (4.11), thus:

$$\sigma^2_{ST} = \sigma^2_{wg} + \sigma^2_{wt.s} + 2r_{wgwt}\sigma_{wg}\sigma_{wt.s} - 2r_{wg_1wt_2}\sigma_{wg}\sigma_{wt.s}. \quad (4.11)$$

Just as we derived the class of *within*-family variance equations from deviation scores, by Eqs. (4.7)–(4.10), so we can derive analogously the *between*-family equations recognizing now that we deal with values based on (4.6). This produces a final expansion that is completely isomorphous with (4.10), but in which *every negative value becomes positive*. Again, since $r_{wg_1wg_2}$ and $r_{wt_1wt_2}$ are zero, the first row reduces to $\sigma^2_{wg} + \sigma^2_{wt}$. But the second row, since $r_{bg_1bg_2}$ and $r_{bt_1bt_2}$ alike become 1, instead of vanishing, becomes $2\sigma^2_{bg} + 2\sigma^2_{bt}$, and so on through other rows transformed from (4.10). If we make the same assumptions as in handling (4.10), but again pause before venturing to assume $r_{wg_1wt_2}$ equal to zero, we have for the between-sib-family variance:

$$\sigma^2_{BNF} = \sigma^2_{wg} + \sigma^2_{wt.s} + 2\sigma^2_{bg} + 2\sigma^2_{bt} + 2r_{wgwt}\sigma_{wg}\sigma_{wt.s}$$
$$+ 2r_{wg_1wt_2}\sigma_{wg}\sigma_{wt.s} + 4r_{bgbt}\sigma_{bg}\sigma_{bt}. \quad (4.12)$$

Let us, for ease of discussion, call the sixth term in (4.12) and the fourth in (4.11) the *cross-genothreptic* or, *CG* term. Then we note that it enters as a negative in (4.11) and a positive in (4.12). If, as we shall do in subsequent equations (and have done in Table 4.2), we write the within-family covariance simply as $2r_{wgwt}\sigma_{wg}\sigma_{wt.s}$, we are assuming that $r_{wg_1wt_2}$, the CG term, is 0, and if it really is not we have an underestimate in one case and an overestimate in another, introducing some inconsistency of the model to the best view of the reality.

In the empirical solutions in Chapters 8, 9, and 10, the sets of equations used (Tables 4.2 and 4.6) have omitted the $r_{wg_1wt_2}$ term. Recently Loehlin (personal communication, 1981) has retracted emphasis on this covariance saying that the assumption that $r_{wg_1wt_2} = r_{wt_1wt_2} = 0$ (which holds for an infinite family) implies that $r_{wg_1wt_2} = 0$, but slight departures from independence would be compatible with appreciable values for the last term. I would base the argument that omitting $r_{wg_1wt_2}$ is an approximation, affecting the ultimate H very little, upon a psychological addition to this statistical argument, namely, that as the child family gets greater than two we should expect to find, in any two taken at random, that the genetic make-up of one has neglible relation to the threptic build-up of the other. A path coefficient analysis would show the causal roots of a child's threptic component to lie, first, in the genetic and threptic components largely of the parents, secondly of all other sibs, thirdly of peer groups, etc. The genetic component of the second sib (diluted in the phenotype) must be an extremely small fraction of the total environment of the first.[2] A second point of dispute in the MAVA model as we have used it (Table 4.6) concerns the terms $2\sqrt{2}r_{\text{wgwt}}\,\sigma_{\text{wg}}\sigma_{\text{wt.t}}$ and $2\sqrt{2}r_{\text{bgbt}}\sigma_{\text{bg}}\sigma_{\text{bt}}$ in equations (5) and (8)(Table 4.6). Loehlin, Fulker and others have argued for a coefficient of 4 rather than $2\sqrt{2}$ in these terms. We are dealing here with a covariance between $2\sigma_{\text{wg}}^2$ and $\sigma_{\text{wt.t}}^2$ (or a similar situation of one variance being twice the usual size, in the $2\sigma_{\text{bt}}^2$ and σ_{bg}^2). If the variance is twice as great the σ is $\sqrt{2}$ times as great, and if r stays the same the covariance becomes $r_{\text{wgwt.t}}(\sqrt{2}\sigma_{\text{wg}})(\sigma_{\text{wt.t}})$—to emphasize its make-up by parens. The usual term for twice the covariance then becomes $2\sqrt{2}, r_{\text{wgwt}}\sigma_{\text{wg}}\sigma_{\text{wt.t}}$. There are, however, subtleties in this issue, such as whether the *correlation* or the *regression* value should remain constant. These complications are best relegated to a footnote.[3]

[2] If there were just two children it can be shown by the *correlation of sums* formula that $r_{\text{wg}_1\text{wt}_2} = -r_{\text{wg}_1\text{wt}_1}$. Thus with this change to the literal pair and no inference to a population a solution could be obtained for it, though MAVA equations would alter. The derivation of σ_{ST}^2 for example by 4.10, for example, by the formulas above would then become $2\sigma_{\text{wg}}^2 + 2\sigma_{\text{wt}}^2 + 4r_{\text{wgwt}}\sigma_{\text{wg}}\sigma_{\text{wt}}$. Since we do not wish to shift the whole model from infinite families to two's (that is to say to abandon use of degrees of freedom) we do not adopt this solution. Incidentally we shall discuss later the general question of family size.

[3] A literal covariance is a product of deviations summed and divided by N, that is, $\Sigma d_g d_t / N$ in this case. The correlation is $\Sigma d_g d_t / \Sigma d_g^2 \Sigma_t^2$. If we double the variance of d_g it will cancel in the latter but not the former and the covariance $r_{\text{gt}}\sigma_g\sigma_t$ in the first case will become $r_{\text{gt}}\sqrt{2}\sigma_g\sigma_t$ in the doubled case.

It should be noted that this is not a case for the familiar "restriction of range" effect and resulting formula concerning a correlation coefficient. There is no *selection* within a population: We have simply a different population. If the correlation is held constant in the above doubling the regression coefficient will *not* remain constant. In the doubled g variance case $b_{\text{gt}} = r\sqrt{2}\sigma_g/\sigma_t$. Psychologically we might call this the relativistic or family "gestalt" hypothesis. It says the environmental treatment begetting the threptic deviation depends on the child's *relative* not *absolute* genetic deviation. The genetically most assertive child, Bill, in a spread-out adoptive family in this model is treated like the most assertive child, John, in an ordinary family, though the deviation of the former is *absolutely* greater.

As stated in the text, the alternative "absolute" theory is that *for every given increment in the genetic component there is a constant increment in the threptic*. In that case our model can either (a) still use σ_{wt} as the environmental variance of the adoptive family; or (b) conclude the the environmental variance itself gets stretched to keep up with the increased genetic extensic The calculation of this stretching would be difficult, for the environmental (strictly the thre' variance in a family, σ_{wt}^2 is not *just* a function of the children's genetic variance, but of other preexisting influences. $\sigma_{\text{wt.s}}^2$ would require a new and different term from $\sigma_{\text{wt.s}}^2$ ' elsewhere and it would require that we assume r_{wgwt} changes in value when the genetic is doubled. Thus if $b_{\text{gt}_1} = r_{\text{gt}_1}\sigma_g/\sigma t$ must equal $r_{\text{gt}_2}\sqrt{2}g/\sigma_t = b_{\text{gt}_2}$ then $r_{\text{gt}_2} = r_{\text{gt}_1}/\sqrt{}$

So far we have used the first, *relativistic* model, assuming r to remain the s' change of range. The second, *absolute* model would have some disturbing effects ' of the equation. It would require not only that a different r be introduced in (4) ar different $\sigma_{\text{wt.s}}^2$—two new unknowns.

The third and last point of dispute in the model concerns the term $2r_{\text{wgwt.t}}\sigma_{\text{bg}}\sigma_{\text{wt.s}}$ in equations (4a) and (8a) (Table 4.6). This well illustrates the clash of psychological and traditional ANOVA thinking habits in new model construction; for by the latter it is unthinkable to get a covariance by linking a correlation r_{wgwt} to variance terms one of which, σ_{bg}^2, is *not* involved in the correlation. Equations (4) and (8) are those for within- and between- *adopting* families (i.e., families in which all the children come from different outside biological families). The genetic variance of these children will be the full genetic variance of the population (assuming no adopting agency selection). That is to say it will be $(\sigma_{\text{wg}}^2 + \sigma_{\text{bg}}^2)$, including within-family and between-family sources of variance, whereas in the ordinary sib or fraternal twin family it is only σ_{wg}^2.

Now, as will be discussed more later (p. 248) the causes for the correlation r_{wgwt} are several, but mainly they can only be through the genetic make-up of the individual affecting (*a*) the situations he gets into; and (*b*) the way people react to him, not by threptic affecting genetic. The most important, if personality is most formed in early years, is the way parents react to him. ("John is more impulsive than Bill; he must be taught restraint.") But his sibs, his peers, and his teachers also play a part. The correlation $r_{\text{wgwt.s}}$ represents a final "exchange value," a rate of transformation of genetic into threptic increments from the totality of these influences. This could be expressed in the linear equation of regression of genetic on threptic measures, derived from $r_{\text{wgwt-s}}$, as $b_{\text{wgwt.s}} = r_{\text{wgwt.s}}\sigma_{\text{wg}}/\sigma_{\text{wt.s}}$.

Now in the adoptive family the genetic range just about doubles ($\sigma_{\text{bg}}^2 = 2\sigma_{\text{wg}}^2$ in the simplest random mating). My argument in setting up equations (4) and (8) in Table 4.6 is that the covariance will therefore also double. The assumption here is that the correlation remains the same (i.e., that we do not have to introduce a new correlation $r_{\text{bg.wt.s}}$ fitted to σ_{bg}^2 and $\sigma_{\text{wt.s}}^2$), but that the same *psychological* laws are in effect in transforming from genetic to threptic deviations, so that $r_{\text{wgwt.s}}$ suffices. The view of Loehlin, Fulker, and Rao, which would require a new unknown r here, depends on a different assumption, as discussed in Footnote 3 in this chapter, and in our present ignorance of the psychological mechanism by which the threptic deviation derives from the genetic deviation it deserves to be given a trial, though my own choice is as in Table 4.6.

The MAVA model as a whole of course does not stand or fall by these alternatives or a few others that could be mentioned. It requires progressive rectification of its details by successive experiments on goodness of fit of alternatives. When we come to discuss path coefficients (Chapter 7) it will be evident that the model may undergo appreciable complication through recursive effects (i.e., additional interactions among particular terms).

As we pass from the first presentation in Table 4.4 to the possibilities of getting more constellations and equations the reader should note that refinements of definition are made beyond those present here. For example $\sigma_{\text{wt.t}}^2$

for twins generally splits into $\sigma^2_{wt.i}$—the threptic variance within identical twin pairs—and $\sigma^2_{wt.f}$—that for fraternal twins, which is almost certainly larger.

One misunderstanding which has plagued some statisticians' views of the MAVA model is that in each constellation the within and between variances should add up to one and the same population variance [i.e., that in Table 4.6 equations (1) + (5), (2) + (6), (3) + (7), and (4) + (8) should all add exactly to (9)]. Loehlin and Fulker would argue this also for the population threptic part only [i.e., that $(\sigma^2_{wt.t} + \sigma^2_{bt.t})$ should equal $(\sigma^2_{wt.s} + \sigma^2_{bt.s})$]. But there is no inherent reason why the *total* environment range to which twins are exposed should equal the *total* environmental range to which sibs are exposed. They, to a small degree, live in different worlds within the same community and culture.

If we set covariance terms aside it will be true, as the above paired summations in Table 4.6 show, that the sums of within and between add to the total population Equation (9) variance (twice its value to be exact). This is exactly true of sibs raised together because we have assumed that there is so great a majority in our child population of sibs raised together that for all practical purposes the within (sib-together) and between (natural families) will together give the actual community population variance.

But even in the non-covariance terms there are exceptions to certain constellations having within and between variances adding to the community population variance. The total population threptic variance for identical twins will have $2\sigma^2_{wt.t}$ (ideally $2\sigma^2_{wt.i}$) in the total population estimate instead of $2\sigma^2_{wt.s}$, and such a population, because of the closeness of environment in a twin pair should (and does) literally have a lesser variance than for the general population. However, it is particularly the covariances, notably the peculiar covariance in adoptive families that break the simple expectation of all withins and betweens adding to the same values. However, these deviations are small and with some of our traits and samples as examined by least squares and maximum likelihood fits in Chapters 9 and 10 the covariances are declared nonsignificant, so that the question of their role does not arise.

It may perhaps be superfluous to point out that the relative importance of terms in the MAVA model in Tables 4.2, 4.6, and 4.8 will vary with ages of subjects, races, and cultures, but the reader should perhaps be reminded that what behavior geneticists sometimes call the "infinite family" is paradoxically really of a particular size. The MAVA method may work with two sibs from each family, as we have here for twin comparability, or three, or four, etc., but it makes an inference, by using degrees of freedom, $(n - 1)$, in each case, to an estimate of the variance for a population—an "infinite family" (only statistically possible!). But if we took our pairs always out of families of five children this infinite family would have the psychological characteristics of a family of five. That is to say the within-family threptic (environmental) variance, $\sigma^2_{wt.s}$, would be estimated as that of an infinite family having the psychological properties of a family of five. Recognizing this is particularly important in connection with our dropping the vexed term $r_{wg_1wt_2}$ since with five it is easy to consider it as zero, but perhaps not so easy with two.

In what follows we shall expand the MAVA model through three degrees of richness of posited unknowns, beginning with the limited and proceeding to the less limited and the full MAVA models.

3. The Unknowns That Genothreptics Seeks:
A Priority List

Even with some simplification of assumptions and restriction to fewer than the ideal number of constellations, MAVA remains a research design formidable in range and potency.

Let us first set a priority for the unknowns we would like to know. The core set is

1. σ^2_{wg} = within-family genetic variance
2. $\sigma^2_{wt.s}$ = within-family threptic variance for ordinary sibs
3. σ^2_{bg} = between-family genetic variance
4. σ^2_{bt} = between-family threptic variance

Since there are now substantial indications that $\sigma^2_{wt.s}$ is nearly twice as big as $\sigma^2_{wt.t}$ we would like next to avoid using a compromise single σ^2_{wt} for both, and add a further term instead:

5. $\sigma^2_{wt.t}$ = within-family threptic variance for twins

In the differentiation reached to this point we would use the same term for identical and fraternal twins, assuming same-age pairs are more alike than sibs in their environments. More urgent than splitting $\sigma^2_{wt.t}$ again into $^2_{wt.i}$ and $\sigma^2_{wt.f}$ for identicals and fraternals, respectively, is recognition of the reality of the *covariance* terms. For accumulated direct and indirect evidence suggests that effects of genetic deviation on threptic deviation and of outside influences simultaneously on both can be considerable (see Cattell, 1963b p. 199). Thus we next add to the priority list:

6. r_{wgwt} = correlation of within-family deviations
7. r_{bgbt} = correlation of between-family deviations

These seven make up what might be called the basic desired quorum (let us call it Quorum 1) for most genothreptic discussion. However, one hankers after certain further refinements. For example, of relevance to the category of unrelated children reared together (σ^2_{UT} in Table 4.1) are indications that (a) adoption agencies tend to place children in families that are culturally, and therefore to some extent genetically, like their biological parents; (b) children received for adoption may be from a genetically and socially somewhat selected subset of the general population; and (c) there may be some tendency, conscious or unconscious, for parents to treat the deviations of adopted children with less intuitive understanding, and perhaps to handle their problems with more conscientious sympathy (or guilt) than in the case of their own.[4] Consequently, in an adoptive family, where the genetic variance would normally be $(\sigma_{bg}^2 + \sigma_{wg}^2)$—because the adopted have both within- and between-genetic variances contributing—we may need to substitute for σ_{bg}^2 a new and probably larger value, thus:

8. $\sigma_{bg.a}^2$ = between-family genetic variances of children of parents whose children have to be adopted

Previously (1960), I also suggested a special correlation $r_{wgwt.a}$ within adoptive families. But an argument for simplicity can be made, as earlier, that in UT families the *rate* of response of environmental treatment to heredity remains the same, namely, that the coefficient, b_{wtwg} remains constant, but that it operates over a wider range. The alternative (see Footnote 3) is to change $r_{wg.wt}$ in equations (4) and (8) to a new unknown $r_{wgwt.a}$ the eleventh unknown. However, further down in the priorities we shall also invoke, as an unknown, special form of this correlation, r'_{wgwt}.

Our judgment is that the next needed priority is recognition of a distinction of the within-family threptic variances of identical and fraternal twins, by leaving the former $\sigma_{wt.t}^2$ and calling the latter

9. $\sigma_{wt.f}^2$ = within-family threptic variance of fraternals

Next we return to the adoption effect we called (a), above, in which placement action gives a special correlation of genetic and threptic due to placing children in culture families resembling those of their genetic origin, thus

10. $r_{bgbt.a}$ = between-family correlation due to placement and then to the possibility that the within-family correlation in adoptive families is intrinsically different, as in (c)

11. $r_{wgwt.a}$ = within-family correlation in adoptive, UT, families

It will be noted that the UT constellation—and the SA which involves some of the same issues—brings in more potential complications than one would

---

[4] One is reminded of Bernard Shaw's quixotic suggestion in "Parents and Children" (Preface to *Getting Married*) that children should not be brought up by their own biological parents because the inheritance of specific temperamental weaknesses means that parents react excessively, allergically, to those particular weaknesses!

like. The recent work on personality (see Chapters 9 and 10), however, unfortunately indicates that the simplified model does give a poorer fit of expectancies on σ_{UT}^2 than on any others, and that adoption relations are intrinsically complex.

Although, with 11 unknowns, we have already overspent the usually available number of equations (number of concrete constellation variances) for an ordinary simultaneous equation solution, we are still not at the end of all we would like to know. There is still the value which Loehlin's analysis suggests would be worth having, and which we may list as the twelfth abstract unknown:

12. $r_{wg_1wt_2}$ = correlation of one child's threptic with the other's genetic environment

and also a possibly different correlation of genetic and threptic deviations for fraternals from those for ordinary sibs. As explained later, there being presently no solution possible for $r_{wg_1wt_2}$, we accept the approximation already made of considering it equal to zero. The arguments made earlier for this amounting to a very small approximation are (*a*) that even if the family environment were only another sib this relation would be small, but actually it is usually several sibs and all the peer, parental, and school experiences that operate differentially on sibs; and (*b*) that in many instances the verdict of maximum likelihood has been that *no* within-family covariance of genetic and threptic deviation is significant.

13. $r_{wgwt.f}$ = within-family genothreptic correlation that might be specific to fraternal twins.

These are the chief unknowns if we keep to same-sex pairs, as we would normally do because we have to make comparisons with identicals. Conceivably, however, different values would be found for all boy, all girl, and mixed families (except for identicals). Other features of the family constellation, for example, being brought up by two parents or one, also ultimately deserve separate evaluation by studying special families. However, these do not require new unknowns in MAVA, but only a special choice of subjects, *and a tagging of our final results to the family or sex choice used.* This should be kept in mind when making inferences from comparative results by *all* convarkin methods, twin or MAVA. While there is no doubt that important inferences for personality and ability theory, and for social psychology, that can be reached in no other way, are attainable by comparisons of genothreptic values from MAVA applied to different sexes, ages, and family conditions, the task of investigating *one single variety* is so great that it will be many years before those further inferences can be approached. The opportunities for continued research are indeed great.

4. The OSES Solutions for the Most Limited MAVA Experiment

With certain family constellations being impracticable of access for certain investigators, and with some psychologists wanting a different order of priority of unknowns from that given earlier, it is obvious that the number of combinations, that is of experimental designs and analyses, could be great. We propose, therefore, to set out in this chapter three experimental designs of what are probably the most practicable and appropriate kind, in increasing order of ambitiousness, from the most limited through the limited, to the full MAVA. The first two are, in any case, those by which we shall produce existing findings on the main personality factors and intelligence; the third we leave to the better endowed researches of the future.

The first set or quorum of equations uses identical twins reared together, ITT, fraternal twins reared together, FTT, and ordinary sibs reared together, ST—plus, of course, the general population GP (or UA unrelated reared apart). These four constellations have always proved the easiest to get for a MAVA study. The model of unknown variances and the concrete variances used are as given in Table 4.2.

It will be seen that there is an algebraic dependency produced by the fact that equations (3) + (6) = (7), in terms on the right. Thus, except in special conditions mentioned later, it is pointless to use all three in a simultaneous equation solution. One would choose from these three either (3) and (7) or (6) and (7), because cases for (7) are easy to get and that sample will have

TABLE 4.2
The Quorum of Equations for the Most Limited Design[a]

(1) $\sigma^2_{\text{ITT}} = \sigma^2_{\text{wt.t}}$

(2) $\sigma^2_{\text{FTT}} = \sigma^2_{\text{wg}} + \sigma^2_{\text{wt.t}} + 2r_{\text{wgwt}}\sigma_{\text{wg}}\sigma_{\text{wt.t}}$

(3) $\sigma^2_{\text{ST}} = \sigma^2_{\text{wg}} + \sigma^2_{\text{wt.s}} + 2r_{\text{wgwt}}\sigma_{\text{wg}}\sigma_{\text{wt.s}}$

(4) $\sigma^2_{\text{BITTF}} = 2\sigma^2_{\text{wg}} + \sigma^2_{\text{wt.t}} + 2\sigma^2_{\text{bg}} + 2\sigma^2_{\text{bt}} + 2\sqrt{2}r_{\text{wgwt}}\sigma_{\text{wg}}\sigma_{\text{wt.t}}$
$\qquad + 4r_{\text{bgbt}}\sigma_{\text{bg}}\sigma_{\text{bt}}$

(5) $\sigma^2_{\text{BFTTF}} = \sigma^2_{\text{wg}} + \sigma^2_{\text{wt.t}} + 2\sigma^2_{\text{bg}} + 2\sigma^2_{\text{bt}} + 2r_{\text{wgwt}}\sigma_{\text{wg}}\sigma_{\text{wt.t}}$
$\qquad + 4r_{\text{bgbt}}\sigma_{\text{bg}}\sigma_{\text{bt}}$

(6)[b] $\sigma^2_{\text{BNF}} = \sigma^2_{\text{wg}} + \sigma^2_{\text{wt.s}} + 2\sigma^2_{\text{bg}} + 2\sigma^2_{\text{bt}} + 2r_{\text{wgwt}}\sigma_{\text{wg}}\sigma_{\text{wts}}$
$\qquad + 4r_{\text{bgbt}}\sigma_{\text{bg}}\sigma_{\text{bt}}$

(7) $\sigma^2_{\text{GP}} = \sigma^2_{\text{wg}} + \sigma^2_{\text{wt.s}} + \sigma^2_{\text{bg}} + \sigma^2_{\text{bt}} + 2r_{\text{wgwt}}\sigma_{\text{wg}}\sigma_{\text{wt.s}}$
$\qquad + 2r_{\text{bgbt}}\sigma_{\text{bg}}\sigma_{\text{bt}}$

[a] The symbol σ rather than s is used throughout because we are stating general principles, leaving sampling aside.

[b] Not usable because of dependency produced with (3) and (7).

a smaller standard error due to its larger size. However, it will be increasingly evident, especially as we meet larger sets of equations, that one cannot always solve for as many unknowns as there are equations. The reasons are

1. There are sometimes hidden dependencies, discovered as one proceeds to solutions. For example, sometimes it is hard to pull apart the terms in $\sigma^2_{wg} + 2r_{wgwt}\sigma_{wg}\sigma_{wt}$ (as Loehlin has pointed out), and sometimes $\sigma^2_{bg} + \sigma^2_{bt} + 2r_{bgbt}\sigma_{bgbt}$ moves as an unfissionable block, that is, one can only get a value for the whole.

2. These simultaneous equations are not linear but quadratic, which means that we may get two solutions for an abstract variance that are mathematically equally correct.

The fact that the equations are nonlinear, and complex, also in general precludes the ordinary computer matrix solution for linear simultaneous equations, as I have checked. And it must be said that even with the most limited MAVA seven equations in Table 4.2, and the less limited nine in Table 4.6, the solution by trial and error is a considerable undertaking. (For this reason I published [Cattell, 1960] the solutions reached, though these are now superseded by some changes in the model.) From the essentially six equations of Table 4.2 the set of six proves soluble as do four of the six possible sets of five. The identifying numbers of these are shown in Table 4.3, but a expert algebraist might find others.

The fact that only one six-equation solution is found suggests that we may have to forego one of our desired unknowns, for example, separate solutions for $\sigma^2_{wt.t}$ and $\sigma^2_{wt.s}$. Or, alternatively, we might abandon the search for an r_{bgbt}. However, by invoking two subsidiary assumptions [e.g., (6) in Table 4.4], we can get solutions for these, as shown in that table. These assumptions deserve thorough consideration because they will be encountered in several later limited designs used by behavior geneticists and are in any case essential to understanding the present solutions in Table 4.4, illustrating derivations from Set 3 and Set 4 of Table 4.3.

TABLE 4.3
Soluble Equations[a]

	Equation numbers in Table 4.2					
Set 1	1	2	3	4	5		7
Set 2	1	2		4	5		7
Set 3	1	2	3	4			7
Set 4	1		3	4	5		7
Extra set (six-set)	1	2	3	4	5		7
(used with C factor)							

[a] Four five-equations sets and one six-equation set for which OSES solutions for abstract variances were found and used.

The first is that the between-family genetic variance has a fixed relation to that existing within families. This means that if we can solve for the former we know the latter. Genetics has a firm law that with random mating and no genetic dominance in the population the value for σ_{bg}^2 will be exactly equal to that for σ_{wg}^2. However, human mating is not random, but characterized for most traits by a positive correlation of husband and wife (Cattell & Nesselroade, 1967; Vandenberg, 1972). If we know what the correlation is *for the genotypes* (not just the phenotypes) then we can reliably calculate what effect the nonrandom mating has on the relation. The calculation is

$$\sigma_{bg}^2 = \sigma_{wg}^2 \left(\frac{1 + r_{fm.g}}{1 - r_{fm.g}}\right), \tag{4.13}$$

where $r_{fm.g}$ is the correlation of father, f, and mother, m, in genetic terms, g.

In this initial perusal of the MAVA equations we shall not digress into determining $r_{fm.g}$ (leaving that for p. 144) but shall simply state that for Table 4.4 we accepted trial values of a higher one at $+.25$ and a lower value at $+.07$. These give by Eq. (4.13)

$$\sigma_{bg}^2 = 1.67\sigma_{wg}^2, \tag{4.14a}$$
$$\sigma_{bg}^2 = 1.15\sigma_{bg}^2. \tag{4.14b}$$

With this aid (see Table 4.4) the six-equation quorum will give a solution for r_{bgbt} too, as shown at the bottom of that table. But in the sets of only five equations we have to bring in auxiliary assumption number 2. Therein one gets a value for the block $\sigma_{bg}^2 + \sigma_{bt}^2 + 2r_{bgbt}\sigma_{bg}\sigma_{bt}$, which we may write B. It is obtainable from the concrete variances $(\sigma_{GP}^2 - \sigma_{ST}^2)$ and $(\sigma_{BFTTF}^2 - \sigma_{FTT}^2)/2$, duly averaged, to give B. In solving the quadratic $\sigma_{bg}^2 + \sigma_{bt}^2 + 2r_{bgbt}\sigma_{bg}\sigma_{bt} = B$, one gives r such a value as will not result in the square root of a negative quantity becoming involved for σ_{bt}^2, that is, the r is the lower limit of values giving nonimaginary solutions. This at least gives operationally a definite value to the r_{bgbt} correlation, though we must remember it is a limit only.

As indicated earlier, the quadratic solution still leaves us plagued with a choice of signs for r_{bgbt} and here there is nothing for it but to turn to extraneous evidence. That extraneous evidence is as follows:

1. Most of the influences that decide the correlation *within* family reside also in the *between*-family environment. It is therefore highly probable that r_{wgwt} and r_{bgbt} will have the same sign. The sign of r_{wgwt} also rests on a square root, but is in terms of a variance, and a negative variance is hard to conceive, so that the denominator of Eq. (3) has been taken as positive.
2. It is legitimate to seek guidance on the sign question from other researches—those with the less limited or the full MAVA design, increasingly available—which give us independent evidence on the

TABLE 4.4
Solutions for Most Limited Design

(a) *Solution for most limited design Set 4 in OSES (Table 4.3)*

(1) $\sigma^2_{wt.i} = \sigma^2_{ITT}$

(2) $\sigma^2_{wg} = .706\sigma^2_{ITT} - 1.412\sigma^2_{ST} + 1.706\sigma^2_{BITTF} - 2.412\sigma^2_{BFTTF} + 1.412\sigma^2_{GP}$

(3) $r_{wgwt} = \dfrac{-1.706\sigma^2_{ITT} + 3.412\sigma^2_{ST} - 1.706\sigma^2_{BITTF} + 3.412\sigma^2_{BFTTF} - 3.412\sigma^2_{GP}}{2[\sigma^2_{ITT}(.706\sigma^2_{ITT} - 1.412\sigma^2_{ST} + 1.706\sigma^2_{BITTF} - 2.412\sigma^2_{BFTTF} + 1.412\sigma^2_{GP})]^{1/2}}$

Entering existing 3 solutions for brevity in remaining equations:

(4) $\sigma^2_{wt.s} = [\pm(\sigma^2_{wg}r^2_{wgwt} + \sigma^2_{ST} - \sigma^2_{wg})^{1/2} - \sigma_{wg}\,r_{wgwt}]^2$

(5) $\sigma^2_{bt} = [\pm(1.15\sigma^2_{wg}r^2_{wgwt} + \sigma^2_{GP} - \sigma^2_{ST} - 1.15^a\sigma^2_{wg})^{1/2} - 1.07^b\sigma_{wg}r_{wgwt}]^2$

Assumptions in (5) are that

(6) $\sigma^2_{bg} = 1.15\sigma^2_{wg}$

(7) $r_{bgbt} = \pm\dfrac{(\sigma^2_{ST} - \sigma^2_{GP} + 1.15\sigma^2_{wg})^{1/2}}{1.15\sigma^2_{wg}}$ Substitute this for r_{wgwt} in (5) and solve for σ^2_{bt}.

(b) *Solution in most limited design (using Set 3 of OSES in Table 4.3)*

(1) $\sigma^2_{wt.t} = \sigma^2_{ITT}$

(2) $\sigma^2_{wg} = 3.416\sigma^2_{ST} + 1.707\sigma^2_{BITTF} - .707\sigma^2_{ITT} - 2.415\sigma^2_{FTT} - 3.416\sigma^2_{GP}$

(3) $r_{wgwt} = \dfrac{3.416\sigma^2_{FTT} - 1.707\sigma^2_{ITT} - 3.416\sigma^2_{ST} - 1.707\sigma^2_{BITTF} + 3.416\sigma^2_{GP}}{2[\sigma^2_{ITT}(3.416\sigma^2_{ST} + 1.707\sigma^2_{BITTF} + .707\sigma^2_{ITT} - 2.416\sigma^2_{FTT} - 3.416\sigma^2_{GP})]^{1/2}}$

(4) $\sigma^2_{wt.s} = [\pm(\sigma^2_{wg} \cdot r^2_{wgwt} + \sigma^2_{ST} - \sigma^2_{wg})^{1/2} - \sigma_{wg}r_{wgwt}]^2$

(Utilizing some of the above solutions as entries, for brevity.)

(5) $\sigma^2_{bt} = [\pm(1.15\sigma^2_{wg}r^2_{wgwt} + .354\sigma^2_{BITTF} - .5\sigma^2_{FTT} - .293\sigma^2_{ST} + .293\sigma^2_{GP}$
$\quad\quad + .146\sigma^2_{ITT} - 1.357\sigma^2_{wg})^{1/2} - 1.07\sigma_{wg}r_{wgwt}]^2$

Equations (2) through (5) employ the auxiliary assumption that $\sigma^2_{bg} = 1.15\sigma^2_{wg}$. However, parallel solutions are available if one chooses the second alternative in the likely assumptions in (6) following:

(6) $\sigma^2_{bg} = 1.15\sigma^2_{wg}$ or $1.67\sigma^2_{wg}$

(7) $r_{bgbt} = r_{wgwt}$, or lowest value for a real solution.

These solutions and those in other tables have been checked and rechecked by three co-researchers, but their complexity is such that an investigator planning to use them is recommended to make his own check.

---

[a] 1.67 in best estimate.
[b] 1.29 in best estimate.

sign. Thus the r_{bgbt} of the less limited design constitutes independent evidence on the magnitude but not on sign of this covariance.

After obtaining solutions for five (or any other number of) sets of equations, as listed in Table 4.3, the question arises as to how to put them together. Parenthetically, even the number of *algebraically* soluble equations is not always achieved also with real numerical data, the principal reason being that the vagaries of small sample variation may finish by demanding roots of negative quantities in some Table 4.4 equation. In the solutions here presented for psychological discussion later the values used are straight averages across the equation solutions that finally emerge from all sets. It might be reasonable, however, alternatively, to weight the solutions according to the total size of the samples accumulated in the particular concrete variances involved in a given set. For example, the values from Sets 2, 3, 4, and 5 in Table 4.3, involving the large *general* population sample might accordingly be weighted more. But if weighting were considered one would also want to entertain less objectively quantifiable weightings such as come from the usual doubts about whether the identical twin sample is as pure as it should be, or, in related children reared together, whether the assumption in the model that they do not suffer from selective placement is tenable, and so on. Consequently, straight averages were taken in our actual solutions in Chapters 8, 9, and 10.

It should be noted that in the approach described in this section—which may be designated the *overlapping simultaneous equation sets* (henceforth *OSES*) method—it is not possible to insert the averaged resultant abstract values in the Table 4.2 model and see, as a check, how closely one comes to the original concrete variances. This particular check is lacking because the algebraic solutions for abstract variances from the different sets of equations differ according to the number of equations involved and the different constellations operative in each set. The abstract variance estimate settled upon, however, has the advantage of being an average of several sets. In the next two methods of solution of MAVA simultaneous equations that we shall consider, namely, the least-squares and the maximum-likelihood methods a calculation back from the obtained abstract variances to the concrete variances appropriate to them can be made. From the closeness of fit of these to the actual concrete variances an evaluation of the confidence limits of the total solution can be reached. In the OSES method the confidence limits must be estimated by less adequate methods.

Until the recent least-squares and maximum-likelihood solutions by MAVA (Chapters 8, 9, and 10 here) virtually all twin-method investigations (and the two previous MAVA investigations) approached estimation of the standard error and confidence limits by more conservative statistics. Loehlin (1977) has handled the basic twin experiment need by the following formula for the standard error of the heritability coefficient:

$$\sigma_H^2 = \left[\frac{(1 - r_{\text{ITT}}^2)}{N_{\text{ITT}}} + \frac{(1 - r_{\text{FTT}}^2)}{N_{\text{FTT}}} \right]^{1/2}, \qquad (4.15)$$

where N_{ITT} is the number of identical twin pairs (raised together; hence ITT), N_{FTT} is the number of fraternals, and the r's are the intraclass correlations. This assumes, as the twin method usually does, that the purely genetic correlation of sibs is .5 and that there is no assortive mating, that is, that $r_{mf.g} = 0$ (m = mother, f = father, g = genetic).

To show what magnitudes of standard error might typically result, Loehlin, enters (4.15) with typical correlations of $r_{ITT} = .5$ and $r_{FTT} = .3$, and takes $N_{ITT} = N_{FTT}$ sample sizes with results shown in Table 4.5.

As Loehlin (1977) aptly says, "The many inconsistencies in the twin study literature are all too intelligible. Until recent years the typical such study has used on the order of 40 or 50 pairs of each kind." He notes that some recent studies (Husén, 1959; Nichols, 1965; Schoenfeldt, 1968) have obtained much larger samples. But one must, unfortunately, add that in most cases their psychological scales are ad hoc contrivances which have ignored the advances of the last 20 years in precisely factorially locating primary personality source traits. The value of the larger samples happens to have been lost by neglecting factored scales established across age development (the ESPQ, CPQ, HSPQ, EPI, CAQ, and 16 P.F.) and across cultures (United States, Great Britain, Germany, Italy, India, Japan, etc. [Cattell, 1973a]) as stable human basic traits.

The principle in approaching standard error estimation in MAVA is essentially that the sampling variance of a given abstract variance is the sum (or other derivative) of the sampling errors of the concrete variances that enter into the formula for it. Thus in Table 4.6 (given in what follows) the σ^2 of σ^2_{wg} is a function of the σ^2's of σ^2's for ITT, FTT, ST, BITTF, and GP (UA). Statisticians have moved slowly in tackling this problem but (assuming independence of the concrete variances, as we may) Daniels (1939) and Welch (1956) have proposed, for any abstract variance σ^2_a, the following:

$$\sigma^2_{(\sigma^2_a)} = \frac{2k_1^2\sigma_1^4}{n_1 - 1} + \cdots + \frac{2k_n^2\sigma_n^4}{n_n - 1}, \tag{4.16}$$

where the k's are coefficients for the concrete variance, σ^2_a is an abstract variance, n's are sizes of the particular concrete variance samples, and σ_n^4 is

TABLE 4.5
The Standard Error of the Heritability Coefficient
for Various Sample Sizes (after Loehlin, 1977)

$N_t{}^a$	Approximate standard error
25	.47
50	.33
100	.24
200	.17
400	.12
800	.08
1600	.06

^{a} N_t = number of pairs in each of the two (ITT and FTT) twin groups.

the square of the given concrete variance. With small samples (under 30), the degrees of freedom should be taken as $n + 1$.

Thus the confidence limits for a given abstract variance are

$$P_r \left\{ \left[\sigma_a^2 - Z_j \frac{(2\Sigma k_i \sigma_i^2)^{1/2}}{n_i - 1} \right] < \hat{\sigma}_a^2 < \left[\sigma_a^2 + Z_j \frac{(2\Sigma k_i \sigma_i^2)^{1/2}}{n_i - 1} \right] \right\} = P_j, \quad (4.17)$$

where Z_j is the normal standard deviate for a given probability, P_j, and σ_a^2 is the "true" value for the population variance, assuming a normal distribution of variances from the larger population.

The values in the preceding paragraph are discussed more fully in Chapter 8.

In current OSES solutions, I have not ventured to use these formulas because of continuing debates on the question of significance. The reader interested in pursuing this awkward issue is referred to the sources already cited, as well as to Cattell (1960) and to later statistical papers (e.g., Eaves [1972], Eaves et al. [1978]). In any case, after the standard errors of the variances themselves have been determined, it remains to calculate that of the H (heritability) ratio derived from them. To give a trial to the above formulas (in Cattell, 1960) I selected a trait at random (source trait U.I. 23, from the early study by Cattell, Stice, and Kristy, 1957) and found that the nature–nurture ratio, N, which had a value of 6.7, yielded $P < .05$ confidence limits of 1.9 and 32.0. A second trait at random—parmia versus threctia (Cattell, Blewett, & Beloff, 1955)—which yielded $N = .52$ showed $P < .05$ limits at .18 and 1.39. Our reluctant contingent conclusion has to be either that the formula has features that make it too exacting, or that the need for larger samples (Cattell, Stice, and Kristy [1957] had a total of 542 pairs and Cattell, Blewett, and Beloff [1955] a total of 481 pairs) is even more imperative than Loehlin has stated. Consequently, in twin studies available today, and even in the larger MAVA studies to be reported in Chapters 8, 9, and 10, we are compelled in the OSES, or where simultaneous equations are simply solved, largely to depend on the central tendency (and degrees of dispersion) of several independent studies ranged side by side. The least-squares and maximum-likelihood methods should theoretically give us a better basis for MAVA analysis than the OSES we are presently discussing; but later we shall see that their statistical cleanness may be counter balanced by some alternative advantage in the OSES.

5. The OSES Solutions for the Less Limited MAVA Model and Experimental Design

MAVA has been initially described in its "limited" form, with a core of familiar constellations, including those used in the twin design. But it can be considerably extended as to both concrete variances and the further unknowns one would like to derive. Probably the most practicable initial further extension is to bring in the concrete variances for unrelated children (adopted) reared together and for true sibs reared apart. To report practical realities we have to mention that most researchers have found the latter noticeably more difficult to get than the former. The probable reasons are that sibs apart are numerically less common and that both the original parents and the adopting agencies are (in principle or because of bureaucratic regulations) comparatively unwilling to disclose the location of one sib to the other or of either to investigators. (This fetish on the part of current "social work" may pass, but its effects are evident in our study, in that, restricting ourselves to males, we were able to obtain 132 unrelated children raised together but only 10 known sibs raised apart.) The σ_{SA}^2 will therefore be considered in the unlimited model (next section) but for practicality (and to show results) we shall keep the moderate, middle extension—which we will call the less limited model—as our main concern. It uses five concrete variances yielding, with the permissible between-family equations, eight or nine equations as shown in Table 4.6.

Table 4.6 introduces one or two considerations not encountered in the most limited design, in Table 4.2. The most important has to do with *unrelated children raised together*, σ_{UT}^2, in adopting homes. This will be given special consideration in Chapter 5, where "fragmentary" uses of MAVA are considered (designs using, separately, genesis groups as in the twin method, adopting homes, and parent-child correlations). The adoption design, as we shall see in Chapter 5, has considerable popular appeal (Scarr, 1979), because

TABLE 4.6

Equations Analyzing Concrete Variances into Abstract Variances:
The Less Limited MAVA Design[a]

(1)	σ_{ITT}^2	$= \sigma_{wt.t}^2$
(2)	σ_{FTT}^2	$= \sigma_{wg}^2 + \sigma_{wt.t}^2 + 2r_{wgwt}\sigma_{wg}\sigma_{wt.t}$
(3)	σ_{ST}^2	$= \sigma_{wg}^2 + \sigma_{wt.s}^2 + 2r_{wgwt}\sigma_{wg}\sigma_{wt.s}$
(4a)	σ_{UT}^2	$= \sigma_{wg}^2 + \sigma_{wt.s}^2 + \sigma_{bg}^2 + 2r_{wgwt}\partial\sigma_{wg}\sigma_{wt.s} + 2r_{wgwt}\sigma_{bg}\sigma_{wt.s}$
(4b)	σ_{UT}^2	$= \sigma_{wg}^2 + \sigma_{wt.s}^2 + \sigma_{bg}^2 + 2r_{wgwt}\sigma_{wg}\sigma_{wt.s} + 2r_{wgwt}\dfrac{(\sigma_{wg}^2 + \sigma_{bg}^2)}{\sigma_{wg}}\sigma_{wt}$
(5)	σ_{BITTF}^2	$= 2\sigma_{wg}^2 + \sigma_{wt.t}^2 + 2\sigma_{bg}^2 + 2\sigma_{bt}^2 + 2\sqrt{2}r_{wgwt}\sigma_{wg}\sigma_{wt.t} + 4r_{bgbt}\sigma_{bg}\sigma_{bt}$
(6)	σ_{BFTTF}^2	$= \sigma_{wg}^2 + \sigma_{wt.t}^2 + 2\sigma_{bg}^2 + 2\sigma_{bt}^2 + 2r_{wgwt}\sigma_{wg}\sigma_{wt.t} + 4r_{bgbt}\sigma_{bg}\sigma_{bt}$
(7)	σ_{BNF}^2	$= \sigma_{wg}^2 + \sigma_{wt.s}^2 + 2\sigma_{bg}^2 + 2\sigma_{bt}^2 + 2r_{wgwt}\sigma_{wg}\sigma_{wt.s} + 4r_{bgbt}\sigma_{bg}\sigma_{bt}$
(8a)	σ_{BSF}^2	$= \sigma_{wg}^2 + \sigma_{wt.s}^2 + \sigma_{bg}^2 + 2\sigma_{bt}^2 + 2r_{wgwt}\sigma_{wg}\sigma_{wt.s} + 2r_{wgwt}\sigma_{bg}\sigma_{wt.s} + 2\sqrt{2}r_{bgbt}\sigma_{bg}\sigma_{bt}$
(8b)	σ_{BSF}^2	$= \sigma_{wg}^2 + \sigma_{wt.s}^2 + \sigma_{bg}^2 + 2r\sigma_{bt}^2 + _{wgwt}\sigma_{wg}\sigma_{wt.s}$
		$+ 2r_{wgwt}\dfrac{(\sigma_{wg}^2 + \sigma_{bg}^2)}{\sigma_{wg}}\sigma_{wt} + 2\sqrt{2}r_{bgbt}\sigma_{bg}\sigma_{bt}$
(9)	σ_{UA}^2	$= \sigma_{wg}^2 + \sigma_{wt.s}^2 + \sigma_{bg}^2 + \sigma_{bt}^2 + 2r_{wgwt}\sigma_{wg}\sigma_{wt.t} + 2r_{bgbt}\sigma_{bg}\sigma_{bt}$

[a] Note we have to be satisfied in a nine-equation set with $\sigma_{wt.i}^2 = \sigma_{wt.f}^2 = \sigma_{wt.t}^2$, the last symbol being used for identicals *and* fraternals.

it seems to tell one directly the effect of a common home environment. But it is actually beset by some complications which reduce the accuracy of its contributions.

In this σ^2_{UT} constellation we encounter an unusual term for the within-family covariance, which has just been discussed (p. 97) in connection with some disputes between experts. We saw there that genetic variance within such an adopting family is that usual in families, σ^2_{wg}, *plus that between children from different families*, σ^2_{bg}. The question we have already discussed in introductory fashion (p. 97) about equations (4a) and (8a) in Table 4.6, is whether the magnitude of the correlation of genetic with threptic deviations remains the same for the σ_{wg} part, the σ_{bg} part of the deviation, and the sum of those parts, or not?

Our preliminary attack on the question in Footnote 3 (p. 97) has recognized that the alternatives hinge primarily on a question of psychological assumptions, between a "gestalt" (relativistic) view that a child is treated more according to his genetic *position* and an "absolute" view that his threptic increment is a fixed linear function of his *absolute* genetic increment. The latter we have seen requires that the *regression coefficient, b,* not the correlation, *r,* remain constant. If *b* is constant then the correlation over the new range will alter and become bigger, as follows:

$$r_{(wg+bg)wt} = r_{wgwt} \frac{\sqrt{\sigma^2_{wg} + \sigma^2_{bg}}}{\sigma_{wg}}. \tag{4.18a}$$

This follows from $b_{wtwg} = r_{wgwt}(\sigma_{wt}/\sigma_{wg})$ which must equal

$$b_{wt(wg+bg)} = r_{(wg+bg)wt} \frac{\sigma_{wt}}{\sqrt{\sigma^2_{wt} + \sigma^2_{wg}}}.$$

In that case the covariance within the adoptive family (UT) is

$$\text{Cov} = 2r_{(wg+bg)wt}\sigma_{(wg+bg)}\sigma_{wt}. \tag{4.18b}$$

If we wish to keep in terms of r_{wgwt}, because this is used in the other equations, we can substitute (4.18a) and get

$$\text{Cov} = 2r_{wgwt} \frac{\sigma^2_{wg} + \sigma^2_{bg}}{\sigma_{wg}} \cdot \sigma_{wt}, \tag{4.18c}$$

which is the term used in the alternative equations (4b) and (8b).

In the solutions we have used we have stood, instead, by the "gestalt" hypothesis, employing (4a) and (8a) rather than (4b) and (8b). However, as briefly mentioned in Footnote 3, since we do not know the relative importance of mechanisms involving parents, sibs, peers, etc. in producing the child's threptic deviation from his genetic deviation, a third possibility is that the increased genetic deviation in an adoptive family actually changes σ^2_{wt} to a larger threptic variance. This assumption would require a new σ_{wt} or a new r_{wgwt} in the 4.18 equations and we could no longer solve from the 4.6 set of equations and would have to seek more constellations and equations as in the full MAVA design.

Thus with the limited design set out in Table 4.6 we are committed, in addition to assumption (1) the "gestalt effect" just made, to (a) letting $\sigma_{wt.t}^2$ stand for both kinds of twins, (3) considering $r_{wg_1wt_2}$ to be zero, and (4) deriving the between-family genetic variance from the known within-family and two estimates of assortive mating. These are, of course, over and above those basic assumptions made from the beginning that though we do not admit *interaction* we do admit *covariance*.

It has just been noted that with eight equations we are nevertheless able to solve for only seven unknowns. As pointed out earlier, losses like this are not unusual and are due to hidden algebraic dependencies, and additionally, to mathematical imaginary solutions in certain cases where sampling errors lead to roots of negative quantities. With a fixed number of constellations it will be noted that as we reduce the number of equations in the sets used we increase the number of alternative set solutions that have to be averaged. For example, starting with eight equations more sets of six can be made from it than of sevens.

From the eight equations of Table 4.6 we found three soluble five-equation subsets, two six-equation subsets, and one seven-equation set. The solutions, like the particular example given from a seven-equation set in Table 4.7, are naturally diverse, and cannot all be tabulated and set out here. Table 4.7, involving use of what we called above a subsidiary solution, uses eight concrete variances from Table 4.6, dropping from the nine available only BNF because of its unquestionable dependence if used with ST and GP. By

TABLE 4.7
Algebraic Solutions of Equations for Seven Unknowns

(1) $\sigma_{wt.t}^2 = \sigma_{ITT}^2$

(2) $\sigma_{wg}^2 = .707\sigma_{ITT}^2 - 2.414\sigma_{FTT}^2 + 3.414\sigma_{ST}^2 + 1.707\sigma_{BITTF}^2 - 3.414\sigma_{UA}^2$

(3) $r_{wgwt} = \dfrac{1.707\sigma_{FTT}^2 - 1.707\sigma_{ST}^2 - .854\sigma_{ITT}^2 - .854\sigma_{BITTF}^2 + 1.707\sigma_{UA}^2}{[\sigma_{ITT}^2(.707\sigma_{ITT}^2 - 2.414\sigma_{FTT}^2 + 3.414\sigma_{ST}^2 + 1.707\sigma_{BITTF}^2 - 3.414\sigma_{UA}^2]^{1/2}}$

(4) $\sigma_{wt.s}^2 = \left\{ \pm \dfrac{\left[(1.707\sigma_{FTT}^2 - 1.707\sigma_{ITT}^2 - 1.707\sigma_{BITTF}^2 + 1.707\sigma_{BFTTF}^2)^2 + (\sigma_{ST}^2 - .707\sigma_{ITT}^2 + .707\sigma_{FTT}^2 - 1.707\sigma_{BITTF}^2 + 1.707\sigma_{BFTTF}^2) \right]^{1/2}}{4\sigma_{ITT}^2} \right.$

$\left. - \left(\dfrac{.853\sigma_{FTT}^2 - .853\sigma_{ITT}^2 - .853\sigma_{BITTF}^2 + .853\sigma_{BFTTF}^2}{\sigma_{ITT}} \right) \right\}^2$

(5) $\sigma_{bg}^2 = [\pm (\sigma_{wt.s}^2 \cdot r_{wgwt}^2 + \sigma_{UT}^2 - \sigma_{ST}^2)^{1/2} - \sigma_{wt.s}^2 s_{wgwt}]^2$

(6) $\sigma_{bt}^2 = 1.707\sigma_{BSF}^2 - 2.414\sigma_{UA}^2 - 1.707\sigma_{ST}^2 + 2.414\sigma_{bg}^2 + .707\sigma_{wt.s}^2 + 1.212r_{wgwt}\sigma_{wg}\sigma_{wt.s}$

(7) $r_{bgbt} = \dfrac{\sigma_{UA}^2 - \sigma_{ST}^2 - \sigma_{bg}^2 - \sigma_{bt}^2}{2\sigma_{bg}\sigma_{bt}}$

the laws of combinations there could be from Table 4.6 8 subsets of seven equations, 28 of six, 56 of five, and so on. But, as all these are not simply soluble on the computer by matrix methods, the number actually soluble can be found only by long exploration. Although 3–4 months were given to this it is possible that we have not found all. The practice of what we have here called the OSES analysis method proceeds by averaging the solutions from as many of the equations, embracing this required number of unknowns, as can be solved.

Incidentally, the solutions with subsets of only five and six equations yield values for the required unknowns only by bringing in the same auxiliary assumptions as we earlier used for the most restricted set: notably (a) the genetic law about the relation of σ_{bg}^2 to σ_{wg}^2; and (b) on the limiting value of r_{bgbt}, that avoids an imaginary value being reached for the σ_{bt}^2 term, as stated previously, and others listed above.

The seemingly large amount of basic mathematical and tactical discussion given here to the OSES method of analysis of MAVA design experiments seems justified (a) by the fact that the full MAVA, with adequate numbers of constellations for the number of unknowns, which may be practicable for investigators before long, need not have these complications; and (b) because the solution of the limited or less limited MAVA by least squares and maximum likelihood—as in Chapters 8, 9, and 10—has features that will not appeal to all psychologists, so that the present mode of attack is likely to be generally as much in steady demand as coal in a nuclear power age.

Since it seems inherent in the OSES approach that we shall always have to seek more equations than the desired number of unknowns, there will always be more algebraic labor in exploring for possible solutions than a mathematician welcomes; there will be alternative quadratic solutions needing quidance from ulterior evidence; and there will be need for auxiliary aids for solution from genetic and other assumptions. These circumstances require resources and flexibility of a fairly high order in the investigator. We have illustrated in a series of seven articles (including Cattell, 1960; Cattell, Klein, & Schuerger, in press; Cattell, Schuerger, & Klein, in press, 1981 a, b, and c; Cattell, Schuerger, Klein, & Kameoka, in press, 1981) at least some attempts to obtain maximally satisfactory solutions with different local sample conditions, etc.

Nevertheless OSES has certain advantages relative to the sweeping simplicity of least squares and maximum likelihood. It keeps closer to the ground of data, and though this imposes the laboriousness of pedestrian, compared to airplane, transit, it permits (a) getting an impression of which particular concrete variances are producing erratic results; (b) watching the effects of various assumptions (e.g., of degree of assortive mating), and adjusting them with respect to particular regions of consequences; and (c) evaluating the stability of results by comparing outcomes of different solution sets.

6. The Full MAVA as the Most Complete Convarkin
Design for Analyses through Offspring Data

Convarkin designs are, of course, not limited to measures on offspring only, but in principle apply to contrasts of all kin variances. The intergenerational relations are pursued in the next chapter. As it stands the MAVA design (and the twin method, which is logically contained in it, in Tables 4.2, 4.6, and 4.8) has been presented as a restricted form which operates only on variances measured in the offspring generation. We run into highly complex and different questions of what is a "family environmental variance," when we take parent–child comparisons, which properly require separate treatment in the next chapter.

Meanwhile let us consider the fullest possible development of the MAVA design in which all but quite rare and bizarre constellations are enlisted. This "full MAVA" rests on 10 basic constellations: identical twins together and apart, fraternal twins together and apart, sibs together and apart, unrelated children together and apart (general population), and half sibs together and apart. Each constellation, except GP, generates a within- and a between-family variance, so that there are 19 equations in all.

These equations were generated by a method independent of the original analysis from first principles (Cattell, 1960). They derive algebraically by Loehlin's (1965) method as set out earlier in this chapter (p. 94). Except for two terms now to be discussed, and a typographical error in the sixth term of Eq. (17) in the 1960 set, it is gratifying to discover that the two analyses fully check.

The first term that is new is the acceptance of $r_{wg_1wt_2}\sigma_{wg}\sigma_{wt}$ as nonzero, in all the "raised together" constellations. That is to say, with the extra constellations we can abandon the approximating assumption we made in Tables 4.2 and 4.6. A second improvement is the acceptance of nonzero placement correlations, in accordance with accumulating evidence discussed earlier and that of Scarr, Webber, Weinberg, and Wittig (1980) that a positive correlation exists between biological and adopting parents. Whether the practice of adopting agencies is conscious or unconscious, the relationship is statistically significant at least in our current society, and it has always existed where children are adopted by relatives of deceased parents. (Cases of the latter kind, however, are not usually included in behavior-genetic research.) Obviously whatever social-environmental similarity of adopting to original families exists will not enter as *threptic* placement selection so long as the shifts are at birth. But *genetic*-placement selection as demonstrated to exist will, through genetic similarity, contribute to the associations of genetic and threptic trait components that normally exist to some slight extent in most cultures. Both the within- and the between-family genetic deviations, which, together, define the genetic individuality of the child as seen, can contribute to genothreptic covariation. In some constellations —mainly unrelated children reared together—both genetic sources contrib-

TABLE 4.8
Equations for the Fully Extended MAVA Design[a]

1. *Within-family variance analyses*

(1) $\sigma^2_{ITT} = \sigma^2_{wt.i}$

(2) $\sigma^2_{ITA} = \sigma^2_{wt.s} + \sigma^2_{bt}$

(3) $\sigma^2_{FTT} = \sigma^2_{wg} + \sigma^2_{wt.f} + 2r_{wg_1wt.f}\sigma_{wg}\sigma_{wt.f} - 2r_{wg_1wt.f2}\sigma_{wg}\sigma_{wt.f}$

(4) $\sigma^2_{FTA} = \sigma^2_{wg} + \sigma^2_{wt.f} + \sigma^2_{bt} + 2r_{wgwt.f}\sigma_{wg}\sigma_{wt.f} + (2r'_{wgbt}\sigma_{wg}\sigma_{bt})$

(5) $\sigma^2_{ST} = \sigma^2_{wg} + \sigma^2_{wt.s} + 2r_{wgwt.s}\sigma_{wg}\sigma_{wt.s} - 2r_{wg_1wt.s2}\sigma_{wg}\sigma_{wt.s}$

(6) $\sigma^2_{SA} = \sigma^2_{wg} + \sigma^2_{wt.s} + \sigma^2_{bt} + 2r_{wgwt.s}\sigma_{wg}\sigma_{wt.s} + (2r'_{wgbt}\sigma_{wg}\sigma_{bt})$

(7) $\sigma^2_{UT} = \sigma^2_{wg} + \sigma^2_{wt.s} + \sigma^2_{bg} + 2r_{wgwt.s}\sigma_{wg}\sigma_{wt.s} - 2r_{wg_1wt.s2}\sigma_{wg}\sigma_{wt.s}$
$\qquad + 2r^b_{wgwt.s.a}\sigma_{wt.s}\sigma_{bg} - 2r_{wg_1wt.s2}\sigma_{wt.s}\sigma_{bg}$

(8) $\sigma^2_{HST} = \sigma^2_{wg} + \sigma^2_{wt.s} + \tfrac{1}{2}\sigma^2_{bg} + 2r_{wgwt.s}\sigma_{wg}\sigma_{wt.s} - 2r_{wgwt.s}\sigma_{wg}\sigma_{wt}$
$\qquad + \sqrt{2}r_{wgwt.s}\sigma_{wt.s}\sigma_{bg}$

(9) $\sigma^2_{HSA} = \sigma^2_{wg} + \sigma^2_{wt.s} + \tfrac{1}{2}\sigma^2_{bg} + \sigma^2_{bt} + 2r_{wgwt.s}\sigma_{wg}\sigma_{wt.s} + \sqrt{2}r_{bgbt}\sigma_{bg}\sigma_{bt} + (2r'_{wgbt}\sigma_{wg}\sigma_{bt})$

2. *Between-family variance analyses*

(10) $\sigma^2_{BITTF} = 2\sigma^2_{wg} + \sigma^2_{wt.i} + 2\sigma^2_{bg} + 2\sigma^2_{bt} + 2\sqrt{2}r_{wgwt.i}\sigma_{wg}\sigma_{wt.i}$
$\qquad + 2\sqrt{2}r_{wt_1wt.i2}\sigma_{wg}\sigma_{wt.i} + 4r_{bgbt}\sigma_{bg}\sigma_{bt}$

(11) $\sigma^2_{BFTTF} = \sigma^2_{wg} + \sigma^2_{wt.f} + 2\sigma^2_{bt} + 2\sigma^2_{bg} + 2r_{wgwt.f}\sigma_{wg}\sigma_{wt.f} + 2r_{wg_1wt.f2}\sigma_{wg}\sigma_{wt.f}$
$\qquad + 4r_{bgbt}\sigma_{bg}\sigma_{bt}$

(12) $\sigma^2_{BFTAF} = \sigma^2_{wg} + \sigma^2_{wt.f} + 2\sigma^2_{bg} + \sigma^2_{bt} + 2r_{wgwt.f}\sigma_{wg}\sigma_{wt.f} + 2\sqrt{2}r_{wt.fbg}\sigma_{wt.f}\sigma_{bg}$
$\qquad + 2\sqrt{2}r_{bgbt}\sigma_{bg}\sigma_{bt} + (2r'_{wgbt}\sigma_{wg}\sigma_{bt}) + (2r'_{bgbt}\sigma_{bg}\sigma_{bt})$

(13) $\sigma^2_{BNF} = \sigma^2_{wg} + \sigma^2_{wt.s} + 2\sigma^2_{bg} + 2\sigma^2_{bt} + 2r_{wgwt.s}\sigma_{wg}\sigma_{wt.s}$
$\qquad + 2r_{wg_1wt.s2}\sigma_{wg}\sigma_{wt.s} + 4r_{bgbt}\sigma_{bg}\sigma_{bt}$

(14) $\sigma^2_{BBF} = \sigma^2_{wg} + \sigma^2_{wt.s} + 2\sigma^2_{bg} + \sigma^2_{bt} + 2r_{wgwt.s}\sigma_{wg}\sigma_{wt.s} + 2\sqrt{2}r_{wt.sbg}\sigma_{wt.s}\sigma_{bg}$
$\qquad + 2\sqrt{2}r_{bgbt}\sigma_{bg}\sigma_{bt} + (2r_{wgwt}\sigma_{wg}\sigma_{bt}) + (2\sqrt{2}r_{bgbt}\sigma_{bg}\sigma_{bt})$

(15) $\sigma^2_{BSF} = \sigma^2_{wg} + \sigma^2_{wt.s} + \sigma^2_{bg} + 2\sigma^2_{bt} + 2r_{wg_1wt.s2}\sigma_{wg}\sigma_{wt.s} + 2r_{wg_1wt.s2}\sigma_{wg}\sigma_{wt.s}$
$\qquad + 2r_{wt.sbg}\sigma_{wt.s}\sigma_{bg} + 2r_{wt.s1bg}\sigma_{wt.s}\sigma_{bg} + 2\sqrt{2}r_{bgbt}\sigma_{bg}\sigma_{bt}$
$\qquad + (2\sqrt{2}r'_{wgbt}\sigma_{wg}\sigma_{bt}) + (2\sqrt{2}r'_{bgbt}\sigma_{bg}\sigma_{bt})$

(16) $\sigma^2_{BHSTF} = \sigma^2_{wg} + \sigma^2_{wt.s} + \tfrac{3}{2}\sigma^2_{bg} + 2\sigma^2_{bt} + 2r_{wgwt.s}\sigma_{wg}\sigma_{wt.s} + 2r_{wg_1wt.s2}\sigma_{wg}wt.s$
$\qquad + 2\sqrt{3}r_{bgbt}\sigma_{bg}\sigma_{bt}$

(17) $\sigma^2_{BHSAF} = \sigma^2_{wg} + \sigma^2_{wt.s} + \tfrac{3}{2}\sigma^2_{bg} + \sigma^2_{bt} + 2r_{wgwt}\sigma_{wg}\sigma_{wt} + \sqrt{6}r_{wt.sbg}\sigma_{wt.s}\sigma_{bg}$
$\qquad + \sqrt{6}r_{bgbt}\sigma_{bg}\sigma_{bt} + (2r'_{wgbt}\sigma_{wg}\sigma_{bt}) + (\sqrt{6}r'_{bgbt}\sigma_{bg}\sigma_{bt})$

(18) $\sigma^2_{BITAF} = 2\sigma^2_{wg} + \sigma^2_{wt.i} + 2\sigma^2_{bg} + \sigma^2_{bt} + 2\sqrt{2}r_{wgwt.i}\sigma_{wg}\sigma_{wt.i} + 2\sqrt{2}r_{wt.ibg}\sigma_{wt}\sigma_{bg}$
$\qquad + 2\sqrt{2}r_{bgbt}\sigma_{bg}\sigma_{bt} + (2\sqrt{2}r'_{bgbt}\sigma_{bg}\sigma_{bt})$

(19) $\sigma^2_{GP} = \sigma^2_{wg} + \sigma^2_{wt.s} + \sigma^2_{bg} + \sigma^2_{bt} + 2r_{wgwt.s}\sigma_{wg}\sigma_{wt.s} + 2r_{bgbt}\sigma_{bg}\sigma_{bt}$

[a] The concrete, experimentally obtainable variances in this table are given in the same mnemonic notation as before. For example, SA is *sibs (raised) apart*, HST is *half sibs (raised) together*, BITTF is *between identical twin (raised) together families*, BNF is *between normal families*, BBF is *between biological families* (i.e., sibs raised apart), BSF *between social families* (i.e., unrelateds brought up in families that adopt them), and GP is *general population* (i.e., unrelated children raised apart).

In equations the variances precede the covariances, and the placement terms come last, in parens. Within precedes between, and genetic precedes threptic. The equations for the more common and readily obtained constellations precede the more rare. The within-family analysis precedes the between in the whole table, and each "raised together" is immediately followed by the equivalent "reared apart."

[b] This is the "a"—"adoptive"—correlation special to the range of between family genetic variance as discussed on p. 97.

ute to genothreptic relations *within* the family. In previous, limited MAVA designs, since the effect is deemed small, these correlations were the first to be sacrificed, but until a complete MAVA is done, solving for these, we cannot really be sure how trivial they are; and so they are given place in Table 4.8.

The unknowns involved in these 19 equations are listed in Table 4.9, and since they are only 13 in number there is every hope of finding solutions for all of them, by the OSES, or the least-squares and maximum-likelihood methods to be described in the next section.

At last we are able with full MAVA to include three different within-family threptic variances, and along with them the correspondingly different genothreptic correlations, that is, it is not supposed that the relation of genetic and threptic deviations is the same in same-age fraternals as in different-age sibs.

Despite preliminary *negative* findings by investigators who have addressed themselves to the important question of whether means and sigmas in intelligence of children in adopted families differ from those of the general population (Scarr, 1968, 1981), the fact that the *majority* of adoptees are illegitimate makes this surprising. The only data, however, is on intelligence of mothers, and it is perhaps not so unexpected that nonconforming individuals are intelligent (though a bimodal or extended sigma might be finally more probable).

TABLE 4.9
List of Unknowns Involved in Table 4.8, with Codings

(1) σ_{wg}^2	=	Within-family genetic variance.
(2) $\sigma_{wt.i}^2$	=	Within-family threptic variance for identical twins.
(3) $\sigma_{wt.f}^2$	=	Within-family threptic variance for fraternal twins.
(4) $\sigma_{wt.s}^2$	=	Within-family threptic variance for ordinary sibs.
(5) σ_{bg}^2	=	Between-family genetic variance.
(6) σ_{bt}^2	=	Between-family threptic variance.
(7) $r_{wgwt.f}$	=	Within-family correlation of genetic and threptic deviations in fraternal twins.
(8) $r_{wgwt.s}$	=	Within-family correlation of genetic and threptic deviations in ordinary sibs.
(9) $r_{wg_1wt_2.f}$	=	Within-family correlation of one fraternal's threptic deviation with the genetic deviation of the other.
(10) $r_{wg_1wt_2.s}$	=	Within-family correlation of one sib's threptic deviation with the genetic deviation of the other.
(11) r_{bgbt}	=	Between-family (means) correlation of genetic and threptic deviations.
(12) r'_{wgbt}	=	Correlation of genetic deviation of adopted children from their own family genetic mean with the deviations of the cultural (threptic) means of families in which they are placed. This is one part of a "placement effect."
(13) r'_{bgbt}	=	Correlation of deviation of the family genetic mean of the family from which an adopted child comes, with the cultural (threptic) deviation of the family in which he or she is placed. This is a second part of a "placement effect."

Questionable additions

(14) $\sigma_{bg}'^2$	=	Between-family genetic variance of adopted children.
(15) $r_{wgwt.s.a}$	=	Within- family correlation in adoptive families adjusting (see p. 97) to the magnitude of between-family genetic variance.

In personality traits, however, such as superego (G), self-sentiment (Q_3), protension (L), and ego strength (C), it would be surprising to a clinical or social psychologist if there were not some differences. As far as I know no primary personality factor profile data exist comparing mothers or fathers of illegitimate children with the average parent, but Cattell, Eber, and Tatsuoka (1970, pp. 279–281) have sufficient 16 P.F. samples for sex crimes, sociopath, and addict groups, which *may* have some overlap with parents of illegitimate children. These deviant groups share low ego strength (C), low super ego (G), high protension (L), and high ergic tension (Q_4). An extension of inquiry to these traits is indicated. Of course, a direct examination merely of the *actual* mean and variance, of, say, adopted children, will not give an answer, because the measure mixes genetic and threptic contributions. But if the illegitimate children's genetic variance proves significantly different from average one would be tempted to protect the integrity of the MAVA solutions by adding to Tables 4.8 and 4.9 one more unknown, $\sigma_{bg}'^2$. Meanwhile, the question has been raised (p. 97) relative to adoptive families whether the *threptic* variance in the more diverse children reared together in an adoptive family is greater than in an ordinary family. The within-family variance as expressed in equations (7) and (15) in Table 4.8 should ideally be permitted to assume a different value from $\sigma_{wt.s}^2$, but under the demand for economy we have allowed this to be taken care of by the addition of the unusual convariance term $r_{wtbg}\sigma_{wt}\sigma_{bg}$. If this is shared between the genetic and threptic variance, it would give a larger than usual value to the latter, handling to some extent the effect upon the psychological diversity of atmosphere in an adoptive family of the greater diversity of heredity, due to children's varied origins (σ_{bg}^2).

In Table 4.8 there remain perhaps a few odd-looking values that call for justification. One might wonder, for example, about the $\sqrt{2}$ values, which, however, as mentioned earlier come from a root of a variance product, for example $(\sigma_{bg}^2 \times 2\sigma_{bt}^2)^{\frac{1}{2}}$, in getting a covariance. Again, in BITAF, one might wonder whether the term for covariance of within threptic and between genetic $r_{wt.ibg}$ is called for. However, one must recognize that each twin in a new home will differ from the other members due to his or her family genetic mean (as expressed in the σ_{bg}^2 variance). Such a twin will also have σ_{wg}^2 variance because he or she differs (as does his or her twin) genetically from some other twin pair that might have been born into that family. Some other unusual terms concern mainly the half-sib constellations. Thus, $\frac{1}{2}\sigma_{bg}^2$ in equations (8) and (9) and $\frac{3}{2}\sigma_{bg}^2$ in (16) and (17) derive from $\sigma_{bg}^2 (1 - r_{bg_1bg_2})$ and $\sigma_{bg}^2 (1 + r_{bg_1bg_2})$ in Loehlin's framework, where the genetic correlation of half sibs is .25 (from having half the genes from one parent in common).

Incidentally, in the half-sib constellations one can conceive of some possible subdivisions and elaborations regarding σ_{bt}^2. With a common father and two mothers the half sibs are likely to be reared in two different maternal homes. In that case certain environmental similarities from assortive mating are likely to make the two environments more alike than two homes at random. There are also possibilities of pursuing X- and Y-chromosome effects in contrasting one father and two mothers with one mother and two fathers. These open up research possibilities of locating X and Y chromosome effects but extend our model beyond present practicality.

Reasonable justification for holding to the same correlation—r_{wgwt}—when the genetic range is greater, thus writing $r_{wgwt}\sigma_{bg}\sigma_{bt}$, has already been given. But, if one has qualms about this and desires to introduce a different, new, unknown r, it seems it would now be possible to carry it in the full MAVA (as r_{wtbg}) and still solve for it. In fact, if we responded both to this and to the hypothesis of a special genetic variance for adopted children, we should still only be demanding 15 unknowns from 19 equations.

Various compromises between the full and the less limited MAVA are of course possible. In each reduction from the full model, dropping the concrete variance of a particular constellation will generally call for a compensating crossing out of an unknown abstract term. One of the most likely trimmings to meet availability of groups is dropping the rather scarce or unreliable category of half sibs. This loses 4 equations, but unfortunately, it gives no corresponding reduction in unknowns, leaving 14 equations but still 13 unknowns. In view of the gains in the least-squares or maximum-likelihood analysis methods from extra equations it would be good to extend experiment to these constellations. Because of cost, one should consider dropping identical twins reared apart, which in losing 2 equations still leaves demand for the same number of unknowns, but—since the full MAVA starts with an excess of equations—all unknowns should still be soluble.

Psychologists and behavior geneticists may think of other things beyond those obviously given in Table 4.8 that they would like to know. Especially one thinks of (a) the changes in variances as one moves from pairs that are both boy or both girl, to those that are one girl and one boy; (b) the differences in genothreptic correlations with these changes; (c) the differences of being raised with one parent only or both; (d) the effect on $\sigma_{wt.s}^2$ and σ_{bt}^2 of being raised in families of different sizes; and (e) the effect of different numbers of years of rearing apart upon the size of $\sigma_{wt.s}^2$ in SA equations. All these are readily handled by applying MAVA to the appropriate selections. There are, indeed, several others besides these four—for example, the variances at different ages of the child subjects—and some of them will be considered more carefully later.

7. MAVA Analyses by Least-Squares and Maximum-Likelihood Methods

It is important to distinguish among the concepts of (a) a theoretical model, with its built-in relational system; (b) an experimental design for investigating it; and (c) a statistical analysis method for examining the experimental results (Cattell, 1966, p. 52). MAVA is a theoretical model of how genetic and environmental influences are supposed to act, when we define them in particular ways. In conjunction with a proposed statistical analysis method it leads to an experimental design in which one gathers data in certain constellations of certain sample sizes, etc. The method of analyzing those data which we have so far considered has been by solving simultaneous equations in the manner described as the OSES (overlapping simultaneous equation sets).

Some advantages and disadvantages of this method have been discussed, and one must now add that the algebraic labor of solving for the last development—the full MAVA—will probably (we have not yet tried) demand heroic qualities in the investigator. Although we are convinced that it would be scientifically valuable to complete the application of OSES to full MAVA, we shall here consider two other methods of analysis applicable to all three levels of the MAVA model. These are the method of least squares and the method of maximum likelihood.

These, and two other methods (moments and minimum χ^2)—all of course, of wider application than to behavior genetics alone—are very usefully described for geneticists by Kempthorne (1957). As most psychological statisticians will recognize, least squares was the first general method of estimation of fit proposed, and it takes the form

$$x_i = E(x_i) + e_i. \tag{4.19}$$

Here an expected value $E(x_i)$ is defined and e_i is the deviation of the observed value from this expectation. The aim is to minimize the sum of squares of the deviations,

$$\Sigma[x_i - E(x_i)]^2, \tag{4.20}$$

in this case over the concrete variances of all equations, as a result of a given set of abstract variances.

In Chapters 8, 9, and 10 of this volume, I have used a least-squares computer program by D. C. Rao and J. M. Brennan. This iterates the starting approximate abstract variances to an ever better fit of the concrete variances (consistent with the iterated values to the observed concrete variances. The reader will find this and many other statistical needs in genetics comprehensively handled in Kempthorne's classical work (1957, 1966). The goodness of fit finally reached is evaluated in this program by a value Q, which represents the residual sum of squares divided by the square of the expected population variance. Although Q has relation to χ^2 its distribution presents problems, and our evaluation of significance of fit, in findings in Chapters 8, 9, and 10, is based on comparisons among the many solutions that we obtained, using different models. For example, in the intelligence investigation three models were tried: (a) the more limited MAVA; (b) the same with genetic terms omitted; and (c) the same with only genothreptic terms omitted.

Since the smallest Q value was found for the first (relative to degrees of freedom), it was accepted.

The general principles of maximum likelihood are as familiar to statisticians as least squares and have become well known to psychologists through Lawley and Maxwell(1963) and Sörbom and Jöreskog(1976), making it practicable by computer (see Cattell, 1978b, p. 395). Its logic is that the set of concrete variances reproduced from the obtained abstract variances can be considered as the population values of which the given concrete variances are the most likely sampling values. Like least squares it is iterative, and its function climbs to find a maximum on the multiparameter space. There can be more than one local maximum and so it is necessary to check by starting iteration from more than one set of guessed rough beginning values.

In the maximum likelihood method let us call any one of the concrete variances $E(s^2)$—short for, say, $E(\sigma^2_{\text{ITT}})$ when s^2_1 is used, and for $E(\sigma^2_{\text{FTT}})$ when s^2_2 is used, which will be continued here to a consideration of the nine equations in Table 4.6 used for the objective personality source traits. A log function $\ln s^2_i$ is now used, which in large samples is expected to follow a normal distribution with mean equal to $\ln E(s^2_i)$, and variance equal to $2/n_i$, where n_i equals the sample size. Therefore, the overall log-likelihood is given by

$$\ln L = -\chi^2/2 + \text{constant},$$

where

$$\chi^2 = \sum_{i=1}^{9} \left(\frac{n_i}{2}\right) [\ln s^2_i - \ln E(s^2_i)]^2. \tag{4.21}$$

This expression is derived under the assumption that all nine variance components, s^2_i, are independent. Even though these components, or at least some of them, are estimated from the same data, thus questioning our assumption, analyses with and without this assumption gave essentially identical results for similar but not identical problems (Gulbrandsen et al., 1979; Rice et al., 1979; Rao, Morton, Cloninger, Elston, & Yee, 1977). Since the alternative approach without the assumption is very complicated, and given the experiences cited earlier, we have chosen to retain the assumption.

We estimate the abstract variance components, σ^2_i, by maximizing $\ln L$, or equivalently, by minimizing χ^2. The residual χ^2 follows a χ^2 distribution with $(9 - K)$ degrees of freedom, where K equals the number of parameters estimated. Under a null hypothesis, we may fix some parameters and estimate only the remaining ones (less than seven). The residual χ^2 still follows a χ^2 distribution with appropriate degrees of freedom. In fact, if $\chi^2_{(9-K)}$ and $\chi^2_{(9-K-W)}$ denote residual χ^2 values after estimating K and $K + W$ parameters, respectively, where the K parameters are included in the $K + W$ parameters,

$$\chi^2_W = \chi^2_{(9-K)} - \chi^2_{(9-K-W)}, \tag{4.22}$$

provides the likelihood-ratio test for the null hypothesis on the W other parameters.

It should be noted that the ninth component, s^2_{GP}, denotes the general population variance whose expectation is not independent of the other eight expectations. This component does not improve the estimability of the seven abstract variances. On the other hand, the very large sample size on which s^2_{GP} is based contributes a big share to the χ^2 values. Therefore, the residual χ^2 values when all the seven parameters are estimated should not be overly emphasized as testing goodness of fit of the model. However, differences in χ^2 values provide adequate tests of null hypotheses. The above comment reminds us that in the least-squares and maximum-likelihood approaches, as distinct from the solution of simultaneous equations in the OSES method, all three of the interdependent equations, for σ^2_{ST}, σ^2_{BNF}, and σ^2_{GP}, can be used, since they are *experimentally* independent.

Parenthetically, it will be noted that the sum of within and between equations for the *same* constellation in any of the cases in Table 4.7 in general comes quite *close* to twice the population variance, just as the *terms* for σ^2_{ST} and σ^2_{BNF} sum exactly to twice the terms in σ^2_{GP}. But this closeness does not affect the independent use of the former algebraically. However, it is interesting to see that a population entirely of twin families, or of sibs raised apart, would not be expected to produce the population as we know it. For example, the sum of identical twins together families produces a population variance of

$$\sigma^2_{GP.IT} = 2\sigma^2_{wg} + 2\sigma^2_{wt.i} + 2\sigma^2_{bg} + \sigma^2_{bt} + 4r_{wgwt.i}\sigma_{wg}\sigma_{wt.i}$$
$$+ 4r_{wg_1wt.i_2}\sigma_{wg}\sigma_{wt} + 4r_{bgbt}\sigma_{bg}\sigma_{bt}, \qquad (4.23)$$

which differs by having wt.i rather than wt.s terms from the general population. However, if our aim were to estimate from abstract variances as exactly as possible the variance of any *literal* human population we should need to allow for the small percentages of identical and fraternal twins, adopted children, etc., relative to ordinary sibs and weight accordingly relative to $\sigma^2_{wt.s}$ etc. It is assumed, therefore, that in experimentally obtaining σ^2_{GP} we take, ideally, only from ordinary sib families, thus coming relatively very close to the literal population, having only a very small percentage of twins, etc.

8. Summary

1. The MAVA method achieves comprehensive solutions for the wider variety of genetic and threptic variances and covariances that the psychologist needs to know, and which he cannot reach by the twin method. It does so by embracing in its data-gathering a wide variety of family constellations

—sibs reared apart, unrelated children reared together, half sibs, etc. For each, a "concrete," experimentally measurable variance is determined as the basis for analysis.

2. By a model with elements that have proved effective concepts in psychology and genetics generally, the MAVA Method relates the conceptually abstract contributory variances one wishes to know to the obtained concrete variances. The general model, expressed in equations, assumes simple (noninteracting) addition of genetic and threptic variances within and between families, but accepts as possible "genothreptic" correlations of genetic and threptic deviations within and between families.

3. Each family constellation will yield two equations (within pairs and between families), and there must be enough constellations to yield at least as many simultaneous equations as desired unknowns. Because, in practice, data on enough constellations on sufficient sized samples is expensive or hard to get, the investigator is soon forced to give rank to priorities among the unknowns he desires, and drop the less important.

4. Such ranking usually places four basic abstract variances first—σ_{wg}^2, σ_{wt}^2, σ_{bg}^2, and σ_{bt}^2—followed first by covariances, then by recognition of three forms of σ_{wt}^2 and associated correlations, and, last, correlations due to placement selection effects in adoptive families. Faced with the differing difficulties of obtaining various constellations, and the different priorities of desired unknowns, one reaches quite a number of acceptable permutations of sets of equations to be combined and unknowns to be found. Three such sets of constellations and equations, chosen as most apt to theory and practice, are considered here: the *most limited* (7 equations), the *less limited* (9 equations), and the *full MAVA* (19 equations).

5. The notion from the elementary solution of linear equations that solution for n unknowns follows from n simultaneous equations does not hold in the present model, for reasons that are in part algebraic (quadratic terms, hidden dependencies denying splitting up a block), and in part arithmetical–statistical (sampling and measurement error together creating negative variances, etc.). Consequently, one uses such subsets of equations as prove soluble, and averages the estimations of each unknown from them. This is called the overlapping simultaneous equation sets or OSES method. It cannot be computerized, as with linear simultaneous equations, but involves appreciable skill and labor, for which reason it seems desirable to publish the algebraic solutions reached.

6. Two solutions for a four-constellation experiment (with five sets from seven equations) and one from a five-constellation experiment (eight or nine equations) are given here. To obtain the $\sigma_{wt.t}^2$, $\sigma_{wt.s}^2$, $r_{wgwt.}$, σ_{bg}^2, σ_{bt}^2, and r_{bgbt} abstract variance values from the shorter set, it is necessary to supplement the equations with subsidiary assumptions, notably (a) concerning the genetically known relation of σ_{bg}^2 to σ_{wg}^2; (b) an assortive mating r to determine (a); (c) the rejection of negative variances; (d) allowing the sign of r_{bgbt}

to agree with that found for r_{wgwt}; and (e) fitting an r_{bgbt} that avoids an imaginary number resulting for σ_{bt}^2.

7. Nineteen equations, based on 10 constellations, each (except general population) in a within and between analysis, are presented for the full MAVA. The constellations are the following: ITT, ITA, FTT, FTA, ST, SA, UT, UA (GP), HSTT, and HSTA. Except for some trivial, questionable, or impracticable alternatives, this is the fullest extent of the possible use of convarkin methods on one-generation data—that on offspring.

8. A list of suggested priorities in desired unknowns is presented, in regard to the utility of behavior genetic findings in clinical, social, and learning theory areas of psychology. One may anticipate that as far as OSES is concerned 13–15 unknowns might be solved for (though with heroic algebraic labor!). If the three most difficult-to-collect constellations are dropped— ITA, HST, and HSA—the equations fall to 14, but a solution of most of the desired abstract variances is still theoretically possible from these.

9. MAVA has the role of a precise theoretical model with the purpose of (a) giving the psychologist estimates of heritability of important primary source traits according to the presently most likely model; and (b) testing and tentatively modifying the model to get maximum goodness of fit to empirical data of diverse kinds.

Different statistical analytical methods may be applied to this model, the three used on subsequent data here being OSES, least squares, and maximum likelihood.

10. Each method has advantages and disadvantages. Although formulae are put forward for standard errors and significances by the OSES, they may not be adequate, and though it is possible to work back to a single set of concrete variances from averaged abstract variances no simple test of goodness of fit is possible. What is evident from standard errors, at least, is that existing published researches have been on samples decidedly too small.

Statisticians value the least-squares and maximum-likelihood methods as more elegant than OSES. These are clear in principle, but have to operate by brute computer iteration procedures, which can fail and yield absurd results if caught at some local maximum in the multidimensional distribution. The OSES, though pedestrian and clumsy in some ways (e.g., in blindly averaging estimates from rather arbitrary subsets of equations), has some advantage in permitting the investigator to stay near the data and see what assumptions immediately cause what effects. Given the present "state of the art," we have in this book applied all three methods to the same data, to gain experience and to make best possible contingent H estimates in personality and ability.

5

Further Designs for Determining Genetic, Threptic, and Heritability Values

1. The Incidence of Assortive Mating in Typical Populations Today

The breakdown of observed variances into genetic and threptic components which we call *genothreptics* has within the convarkin methods tools other than MAVA and the twin method. Indeed, so far we have confined ourselves to metric relations *among offspring*, whereas to the person without specialized knowledge the most obvious way to study heredity is to look at the relations of parents and children. It is true we *have* considered the line of descent in the genealogical, pedigree method, but that was concerned with all-or-nothing syndromes and not the biometric devices of the convarkin method. Accordingly, in this chapter we shall ask what independent illumination can be obtained from grandparent–parent–child measurements, and shall consider some further approaches, such as adoptive family studies, which are actually fragmented MAVA in design. We shall also glance briefly at mass racial-cultural comparison methods.

As we focus on parent–child relations it will become evident that two ambient circumstances plague the discussion: (*a*) the degree of assortive mating (i.e., of similarity, inbreeding, or homogamy in the parents); and (*b*) the extent to which Mendelian dominance and epistacy, rather than simple additive genetic action, operate in the genes affecting the trait under study.

In discussing the first issue, studied in this section, I use the word *assortive* rather than the more common *assortative* to describe similarity selection in mates, because, etymologically, there is no justification for using the more tongue-twisting form. In fact, at a meeting of leading behavior geneticists fairly recently, when I raised discussion on the matter, there seemed to be unanimous agreement that *assortive* should be preferred, but judging by current publications, few have yet had the innovative courage to adhere to this!

Naturally, we are concerned here with the degree of assortive mating in the societies from which we draw our experimental samples. A comparative cultural study of homogamy and social customs everywhere is another matter. One may note in passing, however, that brother–sister marriages in the royal family shocked the first missionaries to Hawaii, and that such marriages were the ideal of the Eighteenth Dynasty of the Pharoahs which, over 200 years, produced such cultural leaders as Tutankhamen and Akhenaton. What lesser degrees of assortiveness exist today and how they affect genetic relations we shall study in this and the next section.

Assortive mating, if positive, means that a person high on trait A tends to mate with another high on A. If negative, as in some popular beliefs that "opposites" attract, it means A − with A +. Among orderly generalizations in this area we must also include a broader "assortiveness" called cross-trait assortiveness in which a person plus on trait A is attracted to a mate plus on trait B. Since at least the time of Karl Pearson (1906–1907), sound statistical evidence has supported the existence of *positive* assortive mating. The sharp comment of the wife whose professor husband ran off with a chorus girl—"The only thing they have in common is a difference of sex"—does point, however, to some uninvestigated mysteries connected with some biological differences being mistaken for sex differences, and, by enhancement, producing negative assortiveness. However, as Pearson and others have shown, physical build, characteristics such as height, weight (and even depths of chest) correlate substantially (with a substantial difference of means, of course), and even eye color and hair features have some lesser positive correlation, as shown in Table 5.1. Naturally one wonders how much these would be reduced if the world were such that everyone met everybody. But, even in a metropolis, mating tends to be physically assortive, and a deeper principle, such as keeps sexual activity in nature largely within species, may be at work.

Assortiveness of mating is obviously something likely to alter with history and local culture. The advent of the steam engine almost certainly reduced "marrying the girl next door" and the inbreeding previously current in isolated villages. The circumstances that play their part are those of social stratification, age, education, race, religion, and ethnic background. As to the last, the rather extreme case of close propinquity of different ethnic groups in Hawaii shows that common racial background and ethnic culture still lead to considerable inbreeding (nonmiscegenation, from another view-

TABLE 5.1
Typical Evidence of Assortive Mating in Physical Traits[a]

Trait	Correlation	Source	Size of sample (in marital pairs)
Height	.28	Pearson & Lee (1903)	1000
Weight	.31	Burgess & Wallin (1944)	989
Hair color	.34	Harris (1912)	774
Eye color	.26	Pearson (1906)	774

[a] From Vandenberg (1972).

point). As shown in Table 5.2, by the large values down the diagonal, roughly five-sixths of children come from race-assortive marriages.

Because psychologists have had a greater interest in marriage psychology than in population genetics, they have given more attention to what may be considered "inherent" personal influences rather than those arising from population circumstances. Empirical evidence on what is actually happening in regard to measured ability and personality traits is beginning to stabilize, after 50 years, on ability measures, and after 20 years on factored personality traits, as shown in Tables 5.3, 5.4, and 5.5.

The mean value of intelligence r's of .52 in Table 5.3 and .51 in the first part of Table 5.4 are higher than those found in some subsequent studies not yet fully published, and in the second part of Table 5.4. The change might support the notion of some behavior geneticists (e.g., Johnson, Ahern, & Cole, 1980), that some reduction of assortiveness has occurred in the last generation. (Sociological changes, such as some breakdown of class distinction in England, more cross-ethnic marriage in America, or a general tendency to enter marriage in a lighter, more experimental spirit, could be invoked as causes for such a reduction, if it is confirmed.)

Since Table 5.3 was put together a review of classical thoroughness on assortive mating for intelligence has been contributed by Jensen (1978; see his Table 4.5, p. 71). Over 44 surveyed researches, in Britain and the United

TABLE 5.2
Mating Choice and Racioethnic Origin[a]

Father	Mother				
	Caucasian	Japanese	Filipino	Chinese	Hawaiian
Caucasian	41,939	2,877	1,002	688	558
Japanese	692	45,976	182	417	101
Filipino	697	962	12,033	209	658
Chinese	344	1,002	71	6,579	71
Hawaiian	219	179	53	78	1,176

[a] Taken from a total of 179,327 children born in Hawaii 1948–1959 (Morton et al., 1967).

TABLE 5.3
Typical Assortiveness of Marriages with Regard to Intelligence and Some Other Abilities

Ability	Source	Sample size	Phenotypic correlation $r_a{}^a$	Attenuation-corrected r_a
Intelligence				
Average of five verbal tests	Willoughby (1927, 1928)	90	.44 ± .08	.51
Average of six nonverbal tests	Willoughby (1927, 1928)	90	.44 ± .08	.51
Army Alpha	Jones (1928)	105	.60 ± .04	.70
Stanford Binet	Burks (1928)	174	.47 ± .04	.55
*Cattell Scales II and III	Cattell & Willson (1938)	101	.77 ± .04	.89
Progressive Matrices	Halperin (1946)	324	.76 ± .02	.88
Scale B, Verbal in 16 P.F.	Cattell & Nesselroade (1967)	102	.31 ± .09	.53
Scale B, Verbal in 16 P.F.	Barton & Cattell (1972)	171	.37 ± .07	.64
		Mean	**.52**	**.65**
Achievement				
Vocabulary	Carter (1932)	108	.21 ± .09	
Arithmetic	Carter (1932)	108	.03 ± .09	
Various achievements in total	Smith (1941)	433	.19 ± .04	
		Mean	**.14**	

^{a} This British population deliberately spanned upper and lower classes, giving a sigma half as large again as usual. It is therefore recognized to be higher than for a stratified sample (Note: r_a—assortiveness—is synonymous with r_{fm}—father–mother r—used elsewhere.)

States (largely the latter) he finds an average phenotypic husband–wife r of .43. It seems reasonable to conclude that in contemporary Western culture an assortive r between .4 and .5 prevails. (With corrections for attenuation from test unreliability the estimated "real" r would be .45 to .55.) Incidentally there is an interesting possible indication in Table 5.3, that (a) the correlation for intelligence is higher than for school achievement and (b) the correlation for fluid intelligence, g_f (in Matrices and Culture Fair tests) is higher than for crystallized intelligence, g_c (in traditional Binet and Wechsler tests). These would imply that the basis of spouse congeniality is one of "native wit" (g_f being more innate) rather than schooling level. This last psychological observation bears on the dispute largely between (as it happens) psychologists on the one hand, believing the r_{fm} (father–mother) is mainly genetic, and some geneticists, on the other, believing it to be mainly between the acquired traits of the spouses.

In support of the argument that assortiveness operates on genetic qualities are the findings that show assortiveness tends to be lower on traits that are indicated to be lower in heritability. Thus in Table 5.4, which extends the preliminary evidence in Table 5.3, we see (along with the lower part of Table

5.3) that more specific, narrow types of test performance, with lower heritability, tend to have lower assortiveness, whereas g_f (Progressive Matrices), g_c, crystallized intelligence in Table 5.3, and spatial, visual, and speed abilities, known to be unitary factors of appreciable heritability, have appreciable assortiveness.

This relation is suggested also by the results in Chapters 8, 9, and 10. The use of the maximum-likelihood and least-squares method in those chapters, however, gives best fit with $r_{fm.g}$ (genetic assortive mating) values ranging only from .0 to .25. The true value for genetic assortive mating is thus at present not as accurately estimatable as one would wish, and is almost certainly different for different traits and different cultures. Since the value enters into estimates of heritability, and is important for population genetics and social psychology, it is to be hoped that researchers will soon get more data, especially on clearly measured primary personality factors.

One could easily get the impression, on turning to assortiveness in the personality domain, that the correlations—be they genetic or threptic in

TABLE 5.4

Assortiveness on More Specific and More General Abilities

Assortiveness, unadjusted correlations
Spouse correlations and t values

Abilities	$r_a{}^a$	t
Vocabulary	.36	4.3*
Subtraction and multiplication	.11	1.2
Card rotations	.19	1.0**
Hidden patterns	.42	5.0*
Paper form board	.30	3.4**
Number comparisons	.18	1.9**
Social perception	.37	4.3*
Spatial	.32	3.7**
Speed	.05	0.5
Verbal	.42	5.1*
Memory	.35	4.1*
Intelligence (g_f?)	.51	6.3*
Intelligence (g_f?)	.49	6.1*

$*\ p < .001.$
$**\ p < .01.$
$***\ p < .05.$

Source: Modified from Zonderman, Vandenberg, Spuhler, & Fain (1977).

Note: Correcting $r_{g.a}$ for attenuation by a reliability of .85 we obtain an estimate of $r_{g.a} = .60$.

^{a} Standard errors of correlation are between .08 and .09 except for age where SE $\simeq .02$.

(*continued*)

Table 5.4 (*continued*)
Assortiveness on More Specific and More General Abilities

Age-corrected assortiveness correlations
Age-adjusted spouse correlations for total Hawaiian
AEA and AJA samples[a]

Abilities	Corrected[b] AEA	AJA
Tests		
Vocabulary	0.22	0.23
Subtraction and multiplication[c]	0.06	−0.08
Card rotations	0.15	0.05
Hidden patterns[c]	0.22	0.00
Paper form board[c]	0.27	0.12
Number comparisons	0.14	0.09
Social perception	0.23	0.17
Progressive matrices	0.22	0.20
Intelligence probability g_f		
Spatial	0.14	0.03
Verbal	0.25	0.28
Perceptual speed and accuracy[c]	0.12	−0.02
Visual memory	0.04	0.19
First principal component (unrotated)	0.24	0.15
Size of sample	0.87	311

Source: From DeFries, Johnson, Kuse, McClearn, Polovina, Vandenberg, & Wilson (1979).

Note: This table has the special interest of a cross-ethnic comparison, showing appreciable stability.

[a] AEA and AJA are Americans of European and Japanese ancestry, respectively. Standard errors for AEA and AJA spouse correlations are .03 and .06.

[b] Corrected for test reliability.

[c] AEA and AJA spouse correlations significantly ($p \leq .05$) different.

main origin—are definitely lower than for intelligence. Before concluding this from Tables 5.5, 5.6, and 5.7 one should note that the results usually cited have been with prefactorial scales, and probably because of the mixtures of primary traits therein (e.g., in the Bernreuter [Hoffeditz, 1934: $r = .13$], the Allport [Crook, 1937: $r = −.07$], and the Whitman [Schiller, 1932: $r = .01$]) scales have yielded, in about half the studies, the nonsignificant results one might expect. The same seems to hold for MMPI pathology scales (Gottesman, 1965, found $r = .02$; see also Vandenberg, 1972). When purer source traits are measured—as in Thurstone's inventory (Schooley, 1936: $r = .30$), the 16 P.F. (Cattell & Nesselroade, 1967: mean $r = .22$), etc. —significant positive correlations appear on virtually all traits, as shown in

TABLE 5.5
Husband–Wife Assortiveness (r_{fm}) on Primary Personality Traits

Personality primaries	Barton and Cattell (1972) ($N = 171$)	Cattell and Nesselroade (1967) ($N = 112$) (S)[a]	Cattell and Nesselroade (1967) ($N = 33$) (U)	Chalmers (1980) ($N = 202$) (U)	Heather Cattell (1980) ($N = 127$) (U)	Karson (1980) ($N = 150$) (U)
A Affectia	.30	.16	(−.50)[b]	.18	.08	(−.04)
C Ego strength	.13	.32	(.05)	.19	.00	(.06)
E Dominance	.35	.13	(.31)	.00	.27	(.13)
F Surgency	.38	.23	(−.40)	.14	.32	(.03)
G Superego	.32	.33	(.19)	.13	.03	(.07)
H Parmia	.12	.23	(.12)	.00	.14	(.08)
I Premsia	.28	.15	(−.13)	.24	.06	(.11)
L Protension	.29	.18	(−.33)	.01	.08	(.11)
M Autia	.33	.22	(−.01)	.13	.25	(.18)
N Shrewdness	.13	.18	(.27)	.03	.06	(.17)
O Guilt proneness	.20	.11	(.36)	.02	.03	(.09)
Q_1 Radicalism	.26	.27	(.34)	.17	.14	(.18)
Q_2 Self-sufficiency	.27	.15	(−.32)	.21	.03	(.06)
Q_3 Self-sentiment	.25	.27	(−.02)	.07	.13	(.03)
Q_4 Ergic tension	.16	.16	(−.11)	.16	.03	(.07)
Mean	.27	.19	(−.01)	.11	.11	(.08)

[a] S = stable marriage; U = unstable marriage. Difference of stably and unstably married is significance (t) = 3.61 for A, 3.2 for F, 2.60 for L, 2.41 for Q_2, and 2.69 for mean of all traits.

[b] Values in parentheses are "clinical," not as others from general population.

[c] Mean r_{fm} for all primaries (unstable, parenthesized marriages [U] omitted) = **.17**. Attenuation-corrected assortive mating correlation for primaries (reliability .70) = **.24**.

TABLE 5.6
Assortive Mating Values for Second Stratum Personality Factors

	N	r_{fm}
QI. Exvia–invia		
16 P.F. (Cattell & Nesselroade, 1967)	102^a	.22
	37^a	$-.30$
16 P.F. (H. Cattell, 1980)	127	.15
Maudsley Scale		
(Kreitman, 1964)	79	.02
QII. Anxiety (or "neuroticism")		
16 P.F. (Cattell & Nesselroade, 1967)	102	.31
	37	.23
Maudsley Scale (Kreitman, 1964)	79	.36
Thurstone Scale (Schooley, 1936)	80	.30
Thurstone Scale (Willoughby, 1927)	100	.27
QIII. Cortertia		
16 P.F. (H. Cattell, 1980)	127	.09
(Cattell & Nesselroade, 1967)^b	102	.14
QIV. Independence		
16 P.F. (H. Cattell, 1980)	127	.33
(Cattell & Nesselroade, 1967)^b	102	.20
Mean for all second orders (Stably married:		.22
total sample 488)		
attenuation-corrected (reliability = .83)		.26

Source: From H. Cattell (1980): p. 32, Table 5.

[a] S = stably married; U = unstably (seeking counselling)

[b] QIII & QIV from Cattell & Nesselroade are compiled from same primaries as listed in Table 5.14.

Tables 5.5 and 5.6. Even with the comparatively new objective measures of dynamic traits (sentiments and ergic tensions—in the MAT, Woliver, 1979, given here in Table 5.7), the correlations remain largely positive. It is relevant also to note from Table 5.5 that *positive* assortiveness is characteristic of more stable marriages, and therefore is likely to be more characteristic of families enduring long enough to have children raised in them—as in our data here.

As Table 5.6 shows, the assortiveness extends to the broader second stratum general personality factors of exvia, anxiety, independence, and, probably, cortertia. And Table 5.7 shows that it seems to be of essentially the same level in dynamic traits.

The physical anthropologist is accustomed to examining assortive mating in man and animal by physical measures of well-nigh perfect measurement reliability. The psychologist who wishes, with his imperfectly reliable or factor-valid instruments, to get the best estimate of what the value really is, can, in the first place, correct for attenuation from unreliability by the

TABLE 5.7
Assortiveness of Mating in Unitary
Dynamic Traits[a]

Dynamic primaries (Ergs and sentiments)	r^b
Career interest	.11
Parental home attachment	.27[b]
Fear (security seeking) Erg	.17
Narcism, comfort	.35[b]
Superego	.03
Self-sentiment	.27[b]
Mating (sex erg)	.18
Pugnacity	.31[b]
Assertion	.16
Spouse	.01

[a] From Woliver (1979), N = 92 couples (newlyweds).
[b] It will be noted that though only four are statistically significant, all are positive, as with general personality primaries. Mean = .19; attenuation-corrected (reliability .75) assortive mating correlation for dynamic traits = .25.

dependability (test–retest) coefficients. The dependability may be taken (Cattell, Eber, & Tatsuoka, 1970) to be about .75 for the single scales of the primary trait list, about .83 for the second stratum factors (on the larger basis of about 40 items), and about .75 for the objective subtest combinations in the Motivation Analysis Test (MAT) (Cattell, Horn, & Sweney, 1964). On the basis of this correction for attenuation, we obtain from Table 5.3 a best estimate for true assortiveness for intelligence of .51 to .89, centering on .65; from Table 5.5 for primary personality factors, a value of .25; from Table 5.6 for second stratum personality factors, a value of .26; from Table 5.5, .24; and from Table 5.7 for primary dynamic source traits, of .25. The Table 5.3 mean value of .65 is, on circumstantial grounds, definitely too high. For example, Willson's and my data from a British, well-stratified, lower- and upper-middle class sample could well be reduced because of a wide, odd range, to compare with American and British samples today. A shrewdly considered better estimate in the intelligence area could well be .50 or .60. Even so, we surely have a real difference here between an assortiveness in intelligence of .50 to .60 and one in personality and dynamic traits of .24 to .26. Good reasons can be offered for this, namely, (a) in everyday interaction intelligence is both better focused and more unequivocally esteemed than are other traits; and (b) the social structure of education, socioeconomic class, and occupation already provides a preliminary step in selecting for intelligence those likely to meet each other.

Estimates of the true magnitude of assortive mating, *at the genetic (not the phenotypic level as above)* can be independently checked by the fit of our data to the MAVA model using different genetic assortive values, and we shall reevaluate them in Chapters 8, 9, and 10 on ability and personality data. It may be said here and now that in personality a value of .25 has been found, even as a *genetic* assortive *r*, to fit well, though lower values, around .13, generally fit better. Incidentally, Jensen (1972, p. 301) argues that an *r* of .25 best fits ulterior evidence, on fit to other models, but he there refers to intelligence; hence, if we followed Jensen, it would require us to pull down our intelligence *r* estimate to the level of other unitary source traits. For closer discussion of these values and their sources the reader is referred to, among others, Barton and Cattell (1973), Cattell and Nesselroade (1967, 1968), De-Fries *et al.* (1979), Jensen (1974), Vandenberg (1972), Woliver (1979), and Zonderman, Vandenberg, Spuhler, and Fain (1977).

2. How Assortiveness Affects Estimates of Genetic and Threptic Values

Assortiveness deserves further discussion because of its relevance both to calculations of genetic variance and to the total path coefficient calculations in which we shall try to put together the causal influences of different kinds of environmental variance with those from genetic sources.

A question we raised in the previous section was that of whether the resemblance of husband and wife depends more on their mutual intuitive appreciation, in selection of *genetic* qualities, or, of *acquired* traits. And, if the latter, does most of the acquisition of those similarities occur *before* marriage and the birth of children, that is, is it reached as a mutual selection effect, or does it come from a subsequent "growing together," so that later children would experience more parental harmony than the earlier born?

A brief digression into psychological theory is required here. Two major principles of personality selection in marital adjustment (Cattell, 1950) have been brought to experiment: that of *congeniality as similarity* and that of *completeness in marriage desirables*. The first states that people with similar abilities, personalities, and dynamic interests will be more stably congenial, and it has been borne out by Cattell and Nesselroade's (1967), Barton and Cattell's (1973), as well as other, findings of positive *r*'s in stably married couples and largely zero ones in marriages breaking up.

The second principle is both more complex and harder to prove. It states that a married couple, like any group, requires a certain *totality* of competent qualities successfully to "stand against the world." One would therefore expect evidence of supplementation, "completeness," or compensation in traits making an effective "team." For example, if one spouse is poor at making social contacts and the other is good at it, the family per se may successfully adjust socially. Since realizations of combined adjustments are in part conscious, this may lead to "bargaining" in "equity theory." To state it crudely, the unattractive rich man has the satisfaction of showing off a beautiful wife, the impoverished beauty has the recompense of financial security. And in personality qualities biographers tell us that the brilliant but often unbearable Bernard Shaw married a woman who was not particularly brilliant but who had the greater ego strength necessary to make marriage practicable. The first illuminating treatment in genetics of this departure from the first principle—that of simple homogamy—was given in Fisher's *Genetical Theory of Natural Selection* (1930), in which it is shown how different "desirables," for example, intelligence and emotional stability, come to be associated (granted lesser survival of the less fortunate combinations). There are several examples in the literature but a simple instance sometimes quoted (Loehlin *et al.*, 1975, p. 123) is the correlation of intelligence with lighter skin color among American blacks (since the latter is a "social desirable" and the former an aid to social success).

Proof of the completeness principle is complex (Barton & Cattell, 1973; Cattell & Nesselroade, 1967; Woliver, 1979) because in part it is masked by the effects of the similarity principle. In the studies I have cited it rests on showing that (*a*) marriages that last are higher on the pair's *sum* of desirables (intelligence, ego strength) than those that do not; and (*b*) significant correlations exist between *different* traits ("off-diagonal" *r*'s in the husband–wife trait correlation matrix), shown in Cattell and Nesselroade (1967), expressing an exchange.

At an operational, observable level the first of these two principles expresses itself as positive assortive correlations ("homogamy"). The statistical signs of the second and more complex principle are reduction of the positive r, correlation of the pair *sum* with marriage stability, and positive *r*'s (genetically), after several generations among adjustive desirables in *individuals*—the last from Fisher's (1930) argument. Our present introductory treatment forbids pursuit of the complexities in the latter. The genetic observations, however should not blind us to the fact that the effects of assortive mating need to be pursued not only in the genetic formulas, as is commonly done, but *also in the environmental aspects of genothreptics.*

As we proceed to use the assortive mating correlation as an extra piece of information helpful for solving for heritabilities—as already done in the limited MAVA design—it finally becomes necessary to know how much of that correlation is due to *purely genetic* similarity in the parents. The value of the father–mother genetic correlation, $r_{fm.g}$, affects the genetic correlation of parents and children, of sib with sib, and, especially, the relation of σ^2_{wg} to σ^2_{bg}, vital to certain solutions [Eq. (5.13) p. 138].

Of course, we always get, empirically, only the *observed, phenotypic correlation,* as in Tables 5.4 and 5.5, which we write r_0 or $r_{fm(g+t)}$ to remind us that the phenotypic correlation is based on both g and t elements. In the last decade some human geneticists have taken the view that the *genetic* as-

sortive r is negligible. Some psychologists agree, viewing the positive phenotypic assortive r's as due to spouses "growing more alike." Others, myself
included, consider the resemblance to be more due to mutual selection,
since it is very significantly present in newly married couples and there is *as
yet no evidence that it increases with length of marriage.* Furthermore, it is
as high for physical features like eye color, stature, chest depth, which do
not "grow more alike." The fact that it seems larger (at least for intelligence)
in middle-class, later-age marriages than in younger, less deliberated marriages also suggests selection. Further we shall argue here that it is as much a
selection on genetic as on threptic elements.

The equation by Crow and Felsenstein (1968), which appears to have
been accepted for some time by human genetics writers, namely,

$$r_{\text{fm.g}} = r_{\text{fm(g+t)}} \frac{\sigma_g^2}{\sigma_{(g+t)}^2} = r_{\text{fm(g+t)}} \cdot H_p, \tag{5.1}$$

where σ_g^2 is only additive genetic variance, makes the level of heritability
paramount in determining $r_{\text{fm.g}}$ and completely ignores what would be due to
the correlation that might exist independently in the threptic part. The assumption that the latter is zero seems to me improbable, and I present the
following alternative argument, depending on the assumption in (5.1) that
$r_{\text{fm.g}}$ and $r_{\text{fm.t}}$ are independent.

Let us *first* solve for $r_{\text{fm.g}}$ on the simpler basis of assuming no *genothreptic* covariance in the population generally. Then:

$$r_{\text{fm.g}} = \sigma_{g.c}^2 / \sigma_g^2, \tag{5.2}$$

where $\sigma_{g.c}^2$ is the common genetic variance of father and mother, and σ_g^2 is
the total genetic variance. Similarly the threptic correlation is

$$r_{\text{fm.t}} = \sigma_{t.c}^2 / \sigma_t^2, \tag{5.3}$$

where $\sigma_{t.c}^2$ is the common threptic variance of father and mother.

The actually observed, *phenotypic* assortive r will be

$$r_{\text{fm(g+t)}} = \frac{\sigma_{g.c}^2 + \sigma_{t.c}^2}{\sigma_g^2 + \sigma_t^2} = \frac{r_{\text{fm.g}}\sigma_g^2 + r_{\text{fm.t}}\sigma_t^2}{\sigma_g^2 + \sigma_t^2}, \tag{5.4}$$

assuming no correlation of genetic and threptic contributions.

The relation of the nature–nurture ratio to heritability is

$$\sigma_g^2 / \sigma_t^2 = H_p / 1 - H_p, \tag{5.5}$$

which yields $\sigma_g^2 = H_p\sigma_t^2 / (1 - H_p)$.

Substituting in (5.4) we have:

$$r_{\text{fm(g+t)}} = \frac{r_{\text{fm.g}}(H_p\sigma_t^2 / 1 - H_p) + r_{\text{fm.t}}\sigma_t^2}{H_p\sigma_t^2 / (1 - H_p) + \sigma_t^2}. \tag{5.6}$$

This simplifies to

$$r_{fm(g+t)} = H_p(r_{fm.g} - r_{fm.t}) + r_{fm.t}.$$ (5.7)

This can be used to get $r_{fm.t}$ in terms of $r_{fm.g}$, etc., thus

$$r_{fm.t} = \frac{H_p r_{fm.g} - r_{fm(g+t)}}{H_p - 1},$$ (5.8)

and $r_{fm.g}$ in terms of $r_{fm.t}$ and $r_{fm(g+t)}$, etc., thus

$$r_{fm.g} = \frac{r_{fm(g+t)} - r_{fm.t}}{H_p} + r_{fm.t}.$$ (5.9)

It can be seen, as would be expected on common statistical principles, that for a given empirical H_p and $r_{fm(g+t)}$ (phenotypic r) the sizes of $r_{fm.g}$ and $r_{fm.t}$ become related, though not simply so. One may note also that if H becomes zero $r_{fm.t} = r_{fm(g+t)}$, whereas if it becomes 1.0 then $r_{fm.g} = r_{fm(g+t)}$, which is as would be expected. Also, at $H = .5$, we see $r_{fm(g+t)}$ is the mean of $r_{fm.g}$ and $r_{fm.t}$. If $r_{fm.g}$ and $r_{fm.t}$ are equal we see further that they (separately) equal $r_{fm(g+t)}$ *regardless of the degree of heritability.*

However, a more striking argument develops from (5.7) if we can suppose

$$r_{fm.g} = r_{fm.t}.$$ (5.10)

Now in any selection on a trait by ordinary human observation the observer cannot tell the genetic part from the threptic part, so completely are they intermixed. As far as assortiveness through marital selection is concerned, therefore, its effect would be equal in the two domains, and the safest assumption is that $r_{fm.t}$ and $r_{fm.g}$ *are* equal. In this case the discovered phenotypic r gives us the required $r_{fm.g}$ directly. The only potential problem for this argument is that some of the genetic resemblance might arise, not from the same human psychological selection as operates on the σ_t^2, but from some homozygosity due to inbreeding in a class group, etc.

Support for this formulation comes later, empirically, from the finding that the values for $r_{fm.g}$ obtained independently of $r_{fm(g+t)}$ (i.e., of phenotypic observation), when either least-squares or maximum-likelihood methods are used (Chapters 8, 9, and 10), tend to fit a range from .10 to .25, much as do the phenotypic values. If the latter are attenuation-corrected, the former average slightly lower, suggesting possibly a slightly lower genetic than threptic assortiveness. The evidence for the last is too tenuous as yet, so for the present we shall stand by the probably equality of phenotypic and genotypic assortiveness.

The preceding introduction has the simplifying assumption of no correlation of genetic and threptic deviations, but when we pursue the analysis into the most realistic model, where genothreptic correlation *is* admitted, the derivation of $r_{fm.g}$ from the observed phenotypic correlation becomes more complex, as follows.

If $r_{gt} \neq 0$, then

$$r_{fm(g+t)} = \frac{r_{fm.g}\sigma_g^2 + r_{fm.t}\sigma_t^2 + 2r_{gt}(r_{fm.g}\sigma_g^2 r_{fm.t}\sigma_t^2)^{1/2}}{\sigma_g^2 + \sigma_t^2 + 2r_{gt}(\sigma_g^2\sigma_t^2)^{1/2}}. \tag{5.11}$$

Expressed as a basis for solution of $r_{fm.g}$, using the known heritability this becomes

$$r_{fm(g+t)} = \frac{r_{fm.g}[H/(1-H)]^{1/2} + r_{fm.t} + 2r_{gt}(r_{fm.g}r_{fm.t})^{1/2}}{[H/(1-H)]^{1/2} = 2r_{gt} + 1} \tag{5.12}$$

With the slightly different assumption that the correlation of the genetic deviation of the father with the threptic of the mother (and reciprocally) is the same as $r_{fm(g+t)}$ (since assortive mating is unaware of the difference) we can derive from the formula for correlation of sums. The sums are the $(g+t)$ value for each parent. This gives:

$$r_{fm(g+t)} = \frac{r_{fm.g}\sigma_g^2 + r_{fm.t}\sigma_t^2 + 2r_{fm(g+t)}\sigma_g\sigma_t}{\sigma_g^2 + \sigma_t^2 + 2r_{fm(g+t)}\sigma_g\sigma_t}, \tag{5.13a}$$

which becomes

$$r_{fm.g} = r_{fm(g+t)}\left[\frac{1}{H} + \left(\frac{1-H}{H}\right)^{1/2}\right] + r_{fm.t}\left(\frac{1-H}{H}\right) \tag{5.13b}$$

The algebraic awkwardness of Eq. (5.12) is no serious obstacle, but both Eq. (5.12) and Eq. (5.13b) require for solution in addition to the obtainable *phenotypic* assortive mating $r_{fm(g+t)}$ and H (for the population, i.e., H_p) the value $r_{fm.t}$. In some (limited MAVA) designs H_p would *depend* on $r_{fm.g}$ and an iterative procedure would be necessary, but the real obstacle to solution is that *when genothreptic correlation is assumed to be significant no solution [as in (5.1)] is possible without knowing also* $r_{fm.t}$. However, if either $r_{fm.g}$ (as, say, in cousin marriages) or $r_{fm.t}$, as when common upbringing can be quantified, is known, solution for the other is possible.

If the aim of obtaining $r_{fm.g}$ is to use it to get σ_{bg}^2 from a known σ_{wg}^2 by Eq. (5.14) given in what follows, and one wishes to use $r_{fm.g} = r_{fm(g+t)}$ only as a first approximation, then, as indicated in the last paragraph, one could use this value (starting with a .5 H_p assumption) in an iterative program to find the value for the best overall fit to the totality of empirical values (see Chapters 8, 9, and 10) available.

The importance of these approaches to finding the $r_{fm.g}$ value is that this value enables us to get solutions from MAVA where we are short of an equation (a concrete empirical family variance)—as in Chapters 8, 9, and 10—necessary to solve for the desired number of unknowns. To be precise it enables us to get σ_{bg}^2 when we cannot get it directly but only have σ_{wg}^2. This convenience is indispensable in the most limited MAVA and a useful check in the limited MAVA sets of equations. The possibility arises because the geneticist knows from purely chromosomal considerations that when there is completely random mating in a population $\sigma_{bg}^2 = \sigma_{wg}^2$, and that when there is assortive genetic mating σ_{bg}^2 is derivable as follows:

$$\sigma_{bg}^2 = \sigma_{wg}^2 \left(\frac{1 + r_{fm.g}}{1 - r_{fm.g}}\right) \tag{5.14}$$

If there are no resources for getting $r_{fm.g}$ as by (5.13) then one must fall back on the assumptions in (5.10) and use the phenotypic $r_{fm(g+t)}$ value (i.e., the observed father–mother correlation for that trait).[1] It should be noted that (5.14) is a formula for a population that has reached equilibrium by the continued practice of the given degree of assortive mating.

Genetically, assortive mating has effects both on the individual family and on population parameters. One must distinguish the effect of a one-generation change from random to assortive mating from a continued prevalence of assortiveness. In the latter, characteristic of most societies, the between-family genetic variance will be larger than the within-family variance, as show in Eq. (5.14). It will be seen that an assortive mating $r_{fm(g+t)}$ of .25 (Table 5.5) would lead us to suppose an $r_{fm.g}$ also of .25, for personality traits. This would make $\sigma_{bg}^2 = 1.67\sigma_{wg}^2$, while one as high as a .60 (which we have argued can occur for intelligence in stable, educated communities sensitive in marriage choice [Table 5.3]) would make $\sigma_{bg}^2 = 4.0\sigma_{wg}^2$.

In actual data, and where a "best fit" can be obtained, we have found that for personality this ratio is sometimes reached. But in our own particular data on intelligence, $r_{fm(g+t)}$ falls short of .60 and we are inclined to believe

---

[1] As mentioned earlier, certain behavior geneticists have hypothesized that the resemblance of married couples rests only on acquired, threptic parts. This was evidently Tennyson's belief, in his comment on an important marriage (in *Locksley Hall*):

> As the husband is the wife is: thou art
> mated with a clown.
> And the grossness of his nature will have
> weight to drag thee down.

Psychologists have yet given no answer to this. We do no know whether long-married couples grow more or less alike, nor do we know the rewards within people, or the circumstances and pressures within society, which operate to produce assortive mating.

If we turn to the broad evidence of the animal world it is obvious that all down the way from phyla through species "like mates with like" and, indeed, infertility results from certain interspecies sex activities. But experiments with breeds within mice show a slight tendency in the opposite direction to heterogamy, in the sense that mice seem attracted to those differing more from their parents. One must recognize also that it is probable that breeds of dogs would vanish but for the care of pedigree societies. As in so many psychological matters, however, generalizations from animal to human psychology can be misleading, and, since marriage is more than mating, it is not surprising that compatibilities seem to be sought involving a wide range of human traits. It seems to me gratuitous to assume that these resemblances depend only on the threptic portion of the individual's traits. Human intuitions go deeply into temperamental compatibilities, emotional sensitivities and tempos. Doubtless, when the results in Table 5.5 are followed up it will be found, however, that assortiveness is significantly different on different personality source traits. Meanwhile, the approaches suggested earlier, plus path analysis (Cloninger, in press; Cloninger, Rice, & Reich, 1978) will reduce our ignorance of the relative roles of genetic and threptic assortiveness in different traits. It may well be that assortiveness will prove particularly great in those largely acquired factor patterns we call sentiments (Cattell & Child, 1975).

that figure is more characteristic of the special cultural conditions described earlier, in the range of lower-middle and middle class, than of the population generally. However, where such conditions exist, and with a g_f heritability of about .65, the probability becomes very high that *genetic* assortiveness is significantly operative.

It is appropriate to notice both the individual family effects and the social causes and consequences of assortiveness. The origin in fastidiousness of human choice is surely likely to increase with a more advanced society. But the structure of society will also play a part, and the existence of classes and marriage within classes, will in some ways increase and in others decrease assortiveness, depending on some rather complex calculations. An educational system such as the American one, in which one person out of every three or four spends at least some time in college, constitutes a fairly powerful mechanism for getting assortiveness on intelligence and achievement-related traits. The effect of assortiveness within the family as we shall see, is—granted certain probable homogamy—to somewhat *reduce* genetic variance among the children. But the effect on society (depending on how long assortiveness has been at work) is a socially important one, namely [Eq. (5.14)] substantially to increase the *between-family variance,* and thus the total range in the population. (Because population variance is a sum of within and between variances, positive assortive mating more powerfully increases the between-family variance than it reduces the within-family value.)

Although social psychology is not our main theme here, this genetic effect is obviously one that simply cannot be ignored in any social-psychological calculations. As regards assortive *intelligence* mating there are social and economic effects on the needed supply of higher educational institutions, the growth of more technical occupations, the needs for special classes for the retarded, and the magnitude of general cultural invention. The central correlation in the assortive intelligence data would increase the sigma of IQ of the population relative to what would happen from random mating by roughly 1.15. If a fluid intelligence (largely innate) test of IQ found an IQ > 130 in a randomly mating population occurring in only 1.25% ($z = 2.24$) of the families, then with a sigma increased to 115, such an intelligence level would appear in 6.8% of the population ($z = 1.49$). The population mean would remain the same, and IQs below 70 would increase in the same ratio. One may well conjecture that the high level of creative performance of such interrelated families as the Darwins, Wedgewoods, Galtons, and Huxleys was partly due to more assortive mating in Victorian England, and of the Bachs in music to similar inbreeding in the guilds of musicians, segregating innate bases of musical talent. If the environmental variance and covariance of society were to remain the same, the heritability values would therefore rise with more assortive mating. This is only one example of the relativity of H values to racial and environmental conditions.

3. Using Purely Genetic, Mendelian Relations of Parents, Offspring, and Other Relatives as an Adjunct to Solutions

The preceding discussion on assortive mating is intended primarily as a necessary preamble to examining other avenues for obtaining heritability values than the MAVA offspring comparisons of Chapter 4, avenues that may serve as a prop to MAVA solutions when few constellations are available. In this chapter we propose to examine the parent–child avenue, as well as some methods that are really fragmentary parts of MAVA. As we noted, the lay person may be surprised that we have not made parent–child resemblances the *first* avenue to *H* values, but the reason is, basically, that there are complexities in defining the *threptic* part of the variances in this situation. In parent–child comparisons—in contrast to the case of, say, fraternal twins in a family—the threptic part is *a different kind of common background experience in the two people involved.* In the offspring it is that due to the *conditions in the present family*—partly from the parents, partly from sibs, partly from the outside world. In the parents most of the threptic part has already been built up from *conditions a generation earlier;* though somewhat, also, by the current behavior of the children. What one measures as a common threptic variance of parents and children, in σ_{bt}^2, is therefore a complex extract from several origins, with an inescapable duality.

Since we know on Mendelian principles how far a child will resemble the parent genetically, it might seem simple to find the *actual* correlational resemblance, and, from the difference of that and the genetic part, determine the role of the family (or other) environment as the threptic part. Methodologically we are saying that one is at an advantage in solving for threptic variance, as a second, supplementary part of the observable phenotypic variance, if one knows, on some firm Mendelian principle, what the common *genetic* variance is, instead of having to solve for it, as in MAVA.

Although ulterior knowledge of the common genetic variance first appears to us in the parent–offspring, intergenerational comparison and analysis of data, it *could* also have been used in the between-sib and between-half sib variance analysis within one generation. And we shall use it again when we come to the adoptive family fragment of MAVA methodology further on in this chapter. But in any case, to use that knowledge effectively we must, for this section and the next, step into some calculations in genetics proper, to see how dependably the purely genetic part *can* be determined in Mendelian principles.

It has been the plan of this book to cover heritability and personality–ability learning as they concern the psychologist, and to refer the reader to the many excellent books on genetics per se (Crow & Kimura, 1970; Cavalli-Sforza & Bodmer, 1971; Falconer, 1960; Jencks, 1972; Jensen, 1973 (appendices); Kempthorne, 1957; King, 1965; Mather & Jincks, 1971; Wright, 1968), for purely statistical genetic principles. However, in this chapter, in connection with having to use genetic quantities in our calculations, we must at least give a condensed rationale for the purely genetic calculations.

To begin with, let us recall that geneticists recognized two kinds of possible action of genes. On the one hand, genes—to be precise, the *gene values* from two alleles—can come together in a simple additive way. On the other, there are many instances in nature of interactive effects, which have been called *dominance* (D), *epistacy* (E), and *linkage* (L). The term *narrow heritability*, H_N, has been used for genothreptic outcomes based on simple additive gene action, and *broad heritability*, H_B, when the genetic term is due to the added interaction effects of D, E, and L. (Let the reader not confuse H_B with *between-family heritability*, written H_b.)

We shall therefore proceed in two steps. In this section we propose to study family genetic relations *in the simpler case of additive gene action* and the "narrow" heritability, H_N, derived from it. In the following section, we propose to look more closely at the complexities on the genetic side itself, defining more fully the effects that are called dominance, epistacy, and linkage, affecting H_B. The psychological reader who is not interested in the purely genetic arguments can skip these two sections and, accepting the verdict on the correlations to be expected—(5.25)–(5.34)—can continue with the next.

The reader will have noticed that in many behavior-genetics texts results are handled as correlations, not variances, and the same is often done in pure genetics. However, the reader should also be aware that *correlational* expressions and the alternative—prominent in the MAVA model—of using variances, are always equivalent. This has been shown at a simple level in the twin method, on p. 69, where we have gone in and out of the alternative representations by intraclass correlations and variances. To see more precisely, conceptually, what is happening, the variance method is almost certainly preferable, though one may wish to finish with correlation statements as a possible way of calculation. In this chapter we shall deal with the fragmentary MAVA approaches by experimenting only with parent –child constellations, the adoptive constellations, and the constellation of half sibs or sibs reared apart. Those working in these areas have often chosen to work with correlations. But it will be recognized that for every *variance* statement concerning a constellation, made in Tables 4.2, 4.6, and 4.8, there is a corresponding correlation. If the long expression on the right is called X (say for sibs reared apart), then the corresponding *correlation* (say for sibs reared apart) is $(\sigma_{GP}^2 - X)/\sigma_{GP}^2$. For X is their *difference* variance, so $(\sigma_{GP}^2 - X)$ is their *common* variance, which determines r. However, one fact must be kept in mind in this correlation–variance transformation, that is, that correlation treats the total population variance as the same—namely, 1.0—in the groups being compared and defines the constellation variance in question as a function of this. For example, in the twin method $(\sigma^2_{GP} - \sigma^2_{wt.t})/$ σ^2_{GP} is the fraction defined by r_{ITT}, and $[\sigma^2_{GP} - (\sigma^2_{wg} + \sigma^2_{wt.t})]/\sigma^2_{GP}$ that defined by r_{FTT}. If the variance of fraternals as a population is, say, greater than for identicals, the subsequent subtraction-derived σ^2_{wg} is in error. Some discrepancies of twin and MAVA heritabilities (H's) could be due to this oddity of σ^2_{GP} for the given group, as well as to some inconsistencies in results from rs with the adoptive-family method.

With this to prepare the reader to handle either variance *or* correlation statements of a model, let us return to the description and determination of correlations of relatives due to purely genetic mechanisms. In the additive case, where we suppose no dominance, etc., but do suppose polygenic action, the meiotic division of the sperm and ovum results in father and mother contributing with exact equality (the sex chromosome aside) to producing the genome of the offspring. The purely genetic correlation of one parent with one offspring ($r_{po.g}$) is therefore

$$r_{po.g} \quad \tfrac{1}{2}\sigma^2_g/\sigma^2_g = .5 \qquad (5.15)$$

as shown in Table 5.8, given in what follows.

The excess of any obtained r (a *phenotypic r, but corrected for attenuation*) over the purely genetic r's in Table 5.8 must be due (so long as we have random mating and additive gene action) to some additional greater correlation due to environment. Incidentally, we are already familiar with this kind of correlational calculation of H in the contrasting of identical and fraternal twin correlations to reach threptic variance and heritability, which calculation (from p. 69) we repeat for reference.

$$H_w = \frac{r_{ITT} - r_{FTT}}{1 - r_{FTT}} = \frac{\sigma^2_{wg}}{\sigma^2_{wg} + \sigma^2_{wt}}. \qquad (5.16)$$

Here we are not yet ready to consider the introduction of, and allowance for, threptic terms, because we first have the task of determining the correlations existing for purely genetic reasons. By reasoning similar to that just given for the meiotic division with parents, one can find what fraction of the total number of genes (polygenic traits) are shared in common by relatives at various distances. Table 5.8 gives the size of these shared variances and the correlations to which they lead. Note the subscript g is attached to all these r's to remind us we deal with purely genetic values.

In exploring the avenue that leads to heritabilities by contrasting the phenotypically obtained correlations with the genetically required ones, let us out of curiosity look first at older data, on physical traits. We will also look at some earlier intelligence test research. Appreciable data in this area was first provided by Karl Pearson and Lee (1903), who found, for example, father–son r's of .5 for height and eye color. (See also on the subject of assortiveness, Watkins, 1980.) The above, and Robson and Richards (1936) results on size and other physical measures are given in Table 5.9.

TABLE 5.8
Purely Genetic Correlations among Relatives on Simple Assumptions[a]

Relationship	Correlation symbol	Correlation magnitude	Derivation from variances
One parent–one offspring	$r_{\text{po.g}}$	.50	$\dfrac{\sigma_g^2}{2\sigma_g^2}$
Mid-parent–one offspring	$r_{\text{mpo.g}}$	.707	$\dfrac{\sigma_g^2}{\sqrt{2}\sigma_g^2}$
Mid-parent–mid-offspring	$r_{\text{mpmo.g}}$	.707	$\dfrac{\sigma_g^2}{\sqrt{2}\sigma_g^2}$
Grandparent–grandson	$r_{\text{gpgs.g}}$	.25	$\dfrac{\sigma_g^2}{4\sigma_g^2}$
Sib–sib	$r_{\text{ss.g}}$	.50	$\dfrac{\sigma_g^2}{2\sigma_g^2}$
Half sib–half sib	$r_{\text{hshs.s}}$	.25	$\dfrac{\sigma_g^2}{4\sigma_g^2}$
Cousin–cousin	$r_{\text{cc.g}}$	.125	$\dfrac{\sigma_g^2}{8\sigma_g^2}$
Uncle–niece[b]	$r_{\text{un.g}}$	.25	$\dfrac{\sigma_g^2}{4\sigma_g^2}$

[a] Additive gene action; no assortive mating.
[b] And Aunt–nephew. See Falconer (1960) p. 155, for derivation of the common variance in these cases. As he notes, "The covariance of any individual with the mean value of a number of relatives of the same sort is equal to its covariance with one of those relatives [p. 154]." The variance of the mean of the relatives is, however, reduced in the case of the mid-parent to $\sqrt{2}\sigma_g^2$, thus $r_{\text{mpo.g}} = .707$. This holds regardless of the number of offspring.

Erlenmeyer-Kimmling and Jarvik's classic review (1963) on intelligence findings is summarized here in Figure 8.3, p. 313. From there it will be seen that the parent–offspring $r_{\text{po.o}}$ (o for observed, phenotypic correlation or variance) stands almost exactly at .50—as does the $r_{\text{po.g}}$ given earlier—and that for siblings reared together a shade higher (.51) and reared apart a trifle lower. Fraternal twins together are close (.52) to sibs together and both are close to the genotypic, r_g. Jensen (1968)—see also Cattell (1971b)—gives .97 for identical twins reared together, .55 for fraternal twins, .52 for siblings together, and .49 for siblings apart. As for other relatives, he records a $r_{\text{po.o}}$ of .52, and gives .35 for grandparent–grandchild, .37 for uncle–nephew, and .30 for first cousins (apart). One notes signs of common family tradition in the last three correlations, in that they appreciably exceed the pure genetic values in Table 5.8, whereas the other phenotypic correlations are surprisingly close to the genetic values. Fuller data as far as intelligence relations are concerned is offered by H. E. Jones (1928), in an investigation not exceeded, after many years, for its accuracy (at any rate as far as traditional

TABLE 5.9
Resemblance of Relatives on Simple Physical Measures[a]

	Correlation	
	Parent–offspring	Sib–sib
Stature	.51	.53
Span of arms	.45	.54
Length of forearm	.42	.48
Birth weight		.50
Eye color	.50	

[a] From Pearson & Lee (1903) and Robson & Richards (1936).

intelligence tests are concerned). As Table 5.10 shows and as Jones himself noted, the values for the mid-parent and/or mid-child run systematically higher. But they fall slightly below the purely genetic expectation (Table 5.8) whereas the r's with one offspring fall slightly above. Somewhat less selected studies (for sample size and other conditions) in Table 5.11, however, show an r slightly below the *additive* purely genetic correlation. One thing that is obvious at this stage, and that has been noted by several reviewers, is that no significant difference has as yet demonstrated between the r's for intelligence and those for stature, weight, etc., which has correctly been taken to mean that intelligence is on the same footing of heritability as gross bodily features (whatever that heritability may be), as shown in Table 5.12.

While looking at abilities it is of interest to see what happens to the main primary abilities, where one may suspect environment in the family *might* play a larger role. DeFries *et al.* (1976) give mid-parent–mid-child correlations, which, as Table 5.8 reminds us, would be expected to give larger correlations than .5 on purely genetic grounds—namely, .707. Fortunately, DeFries *et al.* also correct for attenuation so that from values in Table 5.13 we can conclude that verbal and spatial primaries fall little below the purely ge-

TABLE 5.10
Phenotypic Parent–Offspring Correlations on Intelligence
(Crystallized Intelligence: Traditional Tests)[a]

	Son	Daughter	Mid-child	Mean of son and daughter
Father	.52 ± .04	.51 ± .04	.59 ± .04	.52
Mother	.54 ± .04	.56 ± .04	.65 ± .04	.55
Mid-parent	.59 ± .04	.59 ± .04	.69 ± .03	.59[b]

[a] From H. E. Jones (1928), p. 69, who notes that the multiple R of children with parents and with sibs, which Pearson indicated should be the same, should actually be larger for the former when assortive mating is present.
[b] On purely genetic grounds this would be higher (.707).

TABLE 5.11
Intelligence Correlations of One Parent with One Child in Ordinary Families[a]

Source	Test	No. of parents	Correlation
Burks (1928)	Stanford Binet	200	.46
Outhit (1933)	Army Alpha	102	.58
Leahy (1935)	Otis IQ	366	.51
Willoughby (1936)	Army Alpha	141	.35
Cattell & Willson (1938)	Cattell Scales II and III	202	.49[b]
Conrad & Jones (1940)	Army Alpha	441	.49
Higgins, Reed, & Reed (1962)	Group Tests	2,032	.44
DeFries et al. (1976)[c]	1st Component	739	(.61) .44
DeFries et al. (1976)	Matrices	739	(.51) .34
		Mean	**.46**

[a] Where son, daughter, father, mother have been separately given they are averaged here.

[b] Having a non-normal distribution, these researchers corrected in two ways, reaching mid-parent–mid-child rs of .54 and .78. The mean of .66 (p. 139) is here, for uniformity, transformed to that for one parent–one child.

[c] From DeFries et al.'s results we have not given the lower values for the smaller "Americans of Japanese descent" as probably culturally biased. The "1st Component" they report among primaries we have accepted as a good estimate of g_c. Since other rs are one parent–one child we have corrected their "mid" values (in parens) by Spearman-Brown to a one parent–one child value.

netic values, whereas perceptual speed and visual memory abilities fall decidedly below.

As we turn to parent–child similarity on primary and secondary personality factors the data suddenly become quite scarce, though one would have thought that conclusions here would be most valuable to clinicians. Table 5.14, however, presents data on reasonably good samples, in which the parents were measured on the 16 P.F. and children on the same source traits through the HSPQ and CPQ. The values are decidedly lower, in general, than would be expected on purely genetic grounds. Before one rushes to any

TABLE 5.12
Similarity of Family Relation rs on Intelligence and Stature[a]

	Intelligence	Stature
Father–son	.44	.29
Father–daughter	.38	.36
Mother–son	.50	.41
Mother–daughter	.67	.45
Sib–sib	.63	.55
Identical twins together	.93	.87
Identical twins apart	.97	.75

[a] These figures are taken, with kind permission of Dixon & Johnson (1979) from their larger list.

TABLE 5.13

Parent–Child Similarity on Primary Abilities (Calculated as
Mid-Child–Mid-Parent r)[a]

	Uncorrected	Corrected for attenuation
Verbal	.57 ± .04	.65 ± .04
Spatial	.57 ± .05	.61 ± .04
Perceptual speed	.41 ± .05	.46 ± .04
Visual memory	.32 ± .05	.44 ± 0.5

Source: From DeFries *et al.* (1976, p. 262).
[a] Sample size = 739.

TABLE 5.14

Parent–Child Similarity on Personality Source Traits in Q Data

	Dixon & Johnson (1979)	Loehlin, Horn, & Willerman (in press)	
	Mid-parent– mid-child ($N = 104$)	One child: With father ($N = 88$)	One child: With mother ($N = 90$)
Primaries			
A Affectia	.12	.02	.01
B Intelligence	.25*	—	—
C Ego strength	.23*	.16	−.01
E Dominance	.54**	.08	−.08
F Surgency	.26*	.15	.10
G Superego	.16	.22*	.09
H Parmia	.25	−.02	.11
I Premsia	.20	.10	−.01
L Protension	.36**	—	—
M Autia	.44**	—	—
N Shrewdness	.23**	—	—
O Guilt proneness	.33*	.40**	.04
Q_1 Radicalism	.30*	—	—
Q_2 Self-sufficiency	.23*	—	—
Q_3 Self-sentiment	.23*	−.10	.05
Q_4 Ergic auton tension	.15	.16	.09
Secondaries[a]			
QI Exvia (A, F, H, Q_2-)	.22	(Not calculated)	
QII Anxiety (C-, H-L, O, Q_3-Q_4)	.26		
$QIII$ Cortertia (A-, H-, I-, M-)	.25		
QIV Independence (E, F, H, L, M)	.37		

* $p < .05$.
** $p < .01$

[a] The secondaries are calculated here as a simple mean of the rs for the primaries in them (in parentheses). Dixon and Johnson's values from Eysenck's EPI are much lower, namely, .18 for exvia and .12 for neurotic anxiety.

conclusion that personality is generally less heritable, however, two caveats must be recognized:

1. Correction of the r's for attenuation would be considerable, first, because scales of only 8–12 items per factor have genetally been used, instead of combining scores from Forms A, B, C, etc., and, second, because as of this decade the research to produce progressively higher alignment of the scales of the 16 P.F., HSPQ, CPQ, etc., has not been completed, hence child and adult are being measured on slightly different source traits (see Chapter 9, p. 326).

2. One must *not conclude that an equality of the observed correlation of phenotypes with the expected correlation from sheer genetic causes proves high inheritance.* The inferences are more complex, as we shall see later.

4. How Dominance, Epistacy, and Linkage Affect Genetic Partitioning for Reaching Broad (H_B) and Narrow (H_N) Heritabilities

As pointed out on p. 264, the geneticist distinguishes between a "narrow heritability," which we write as H_N, and which deals only with an additive action among genes, and "broad heritability," H_B, which recognizes the effect of certain additional purely genetic influences, when they exist, that modify simple additive action.

The first of these modifiers is *dominance,* recognized since Mendel. When a gene with two alleles happens to assign to the offspring one allele from one parent and a different one from the other, the result is in this case *not* the mean of the double entry of the first and the second. The relations are most readily indicated by Figure 5.1.

In the simple additive case, the heterozygote Gg (one of each of the allelic genes) is midway in its effects between gg and GG. In the case of dominance, Gg is above the average of gg and GG effects, by an amount d, which is called the dominance deviation. Mendel hit on a case of virtually complete dominance (complete would be at 1.0 in Figure 5.1), but in most cases there is more or less dominance than the value $d = 1.0$, and, as Figure 5.1 shows, at (a) there can even be *overdominance,* in which Gg exceeds GG. More commonly there is partial dominance as at (b).

The practically oriented reader may ask at this point how we are able, as in Figure 5.1, to talk about values occurring in the *genotype,* as if we knew what happens before environment has come in to modify the former's expression into a phenotype. Whatever the dominance, epistasis, or other effects may be, it remains true that they can be seen and measured only through a phenotype. We have criticized elsewhere the habit of some genetics texts of talking about *the* phenotype. One can speak of *the* genotype, but

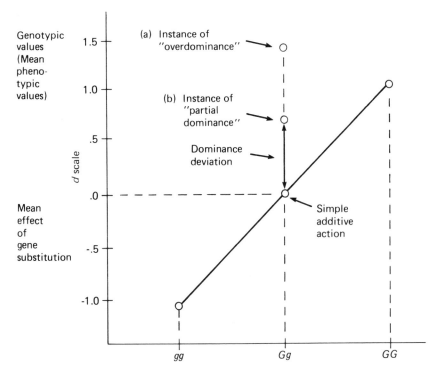

Figure 5.1. Dominance effect on phenotype.

We have already explained (p. 15) that *the* phenotype is a myth, though Johannsen's initial use to contrast with genotype was a roughly useful concept. There are as many phenotypes as there are environments. If, as in discussions here, *the* phenotype is defined as the average, abstract value across all developmental environments, it becomes a standard function (transformation) of the genotype. The genotype exists in the *genome,* which is some particular individual chromosomal entity, but can best be defined as a pattern which several genomes may have. A karyotype is the layout of chromosomal elements, not normally extending to definitions of the gene structures within each element, which fully defines the genome. As an abstraction defining the mean of phenotypes in many environments from many identical genomes, we speak of the genotype.

there are as many phenotypes as there are environments. Thus, when a geneticist talks of "phenotypic values" as his basis for the "genotypic values" in Figure 5.1, he had to be very particular in stating a defined environment (or random sample of environments).

Because we want to talk about genotypes, dominance, correlations of relatives on a purely genetic basis, etc., we have to get rid of threptic effects in the phenotypic value measures. This we can do by (a) holding environment defined and constant in all our measures, and considering the differences that we then measure (a, d, etc.) as genotypic (there being no interaction); (b) averaging phenotypes across a lot of environments; or (c) partialing out threptic variance by bringing in established heritabilities.

If we wish to keep in view the Mendelian genetic *mechanisms* for the genetic variances we are going to deal with, it becomes necessary to digress into the purely genetic arguments of the next few paragraphs. Figure 5.2 extends Figure 5.1 to the case of *two* gene loci, each with two alleles, and considers the four possible outcomes.

From knowledge of the possible genotypic structures for a trait, as in Figure 5.2, of the magnitude of dominance, etc., and of the frequencies in the population of the alleles, p_1 and q_1 for the first locus, and of p_2 and q_2 for the second, the variance can be calculated for the genotypic values of that trait in that population.

Let us consider first the problem of deciding how large the variance will be from purely additive action, that is, with no dominance effect. To more thoroughly explain what is meant by additive genetic variance, it is first necessary to introduce the concepts of allelic value and additive locus variance, σ_{AL}^2. Allelic value, denoted by α, is the average effect of the allele in its locus (averaged with respect to other alleles in effecting genotypic values). (This presupposes such assumptions as random mating.) Conceptually, it is the average effect of choosing an allele at the locus at random and replacing it with the allele of interest. The resulting *population* effect will depend not only on the size of this allelic value, but also on the relative frequency of the allele in the population (as a percentage, given by p or q).

The additive variance of the locus, referred to as the additive locus variance, σ_{AL}^2, therefore, is the sum of genic variances of the two genes: $\sigma_{AL}^2 = \sigma_{G_1}^2 + \sigma_{G_2}^2$. The final, total additive variance is the sum of such values over the total number of genes involved, written σ_A^2.

The magnitude of the dominance variance, σ_D^2, to be added to this will

		Male gametes			
		GH	Gh	gH	gh
	GH	GGHH	GGHh	GgHH	GgHh
Female	Gh	GGhH	GGhh	GghH	Gghh
gametes	gH	gGHH	gGHh	ggHH	ggHh
	gh	gGhH	gGhh	gghH	gghh

Figure 5.2. Two gene loci, each with two alleles: the beginning of polygenic complexities and frequencies.

    When meiotic division takes place the chromosome halves and we have the two element *gametes* (sperm and ova) as shown. Here 16 different types of fertilization can happen, but due to dominance only four kinds of phenotypes. If for a moment we suppose an entreme simplified personality inheritance with G = high intelligence, g = low intelligence, H = high surgency, h = low surgency, we would have from 16 fertilizations, 9 high intelligent and surgent, 3 high intelligent and desurgent, 3 unintelligent and surgent, and one unintelligent and desurgent. However, the frequencies in the total population as such (not this single pair of parents) would depend on the frequencies p_G and q_q for the intelligence alleles and p_H and p_h for the two surgency alleles (p and q are conventionally proportions out of unity, i.e., $p + q = 1.0$).

depend on the number of dominance loci (in the polygenic case) and the magnitude of the dominance deviations. We call σ_D^2 the dominance locus variance, and in the example (Figure 5.2) it would be

$$\sigma_D^2 = \sigma_{DL_1}^2 + \sigma_{DL_2}^2 \tag{5.17}$$

The second effect beyond additiveness is that due to *epistatic action* which we shall denote by C, and in variance as σ_C^2. (This is sometimes in purely genetic texts called "interaction," and represented as σ_I^2, which is confusing because that term applies also to true statistical interaction between genetic and threptic contributions [p. 66], and, incorrectly, to covariance of genetic and threptic influences.) Epistasis refers to the effect of one gene upon the action of another when they happen to occur in the same genome but on different loci. This is quite distinct from the interaction between parental alleles on the same gene locus that occurs in dominance effects. The action here is among nonhomologous parental genes, in which an allele of gene X modifies, when they come together in the same genome, the usual effect of an allele of gene Y. Such an effect is, of course, operationally detected by the genotypic effect when they are together being different from what one would have expected from their single action when alone in the genome, as illustrated in Figure 5.3. Commonly the effect is that one gene "masks" the effect of another, but it could also "catalyze" it. Research in this domain faces great complexities when *several* genes show mutual epistacy.

A third effect complicating simple additive action is *linkage*. The basic expectation in additive action among a good many genes is that in the reduction division (meiosis) of the parental zygotes (sperm and ova cells) the genes will split apart *randomly*, each gene being free to move independently of any other. However, in natural fact, the genes on one chromosome tend to move together and constitute a "linkage group." There are as many of these as chromosome halves after the reduction division. Except for what is called "crossing-over" (see King, 1965; Lerner, 1968; and others) linkage would be more prevalent than it actually is. As can be seen, it tends to modify a random additive action of the gene counts of the two parents.

From a geneticist's standpoint, therefore, the total genotypic (genetic) variance in a population can be partitioned as follows:

$$\sigma_G^2 = \sigma_A^2 + \sigma_D^2 + \sigma_C^2 + \sigma_L^2. \tag{5.18a}$$

A more advanced treatment of this partitioning is given by Kempthorne (1955) involving interaction terms, thus

$$\sigma_G^2 = \sigma_A^2 + \sigma_D^2 + \sigma_{C.AA}^2 + \sigma_{C.AD}^2 + \sigma_{C.DD}^2$$
$$+ \sigma_{C.AAA}^2 + \sigma_{C.AAD}^2 + \sigma_{C.ADD}^2 + \sigma_{C.DDD}^2 + \cdots, \tag{5.18b}$$

where $\sigma_{C.AA}^2$ is epistasis (interaction) of additive values of loci with one another, $\sigma_{C.AD}^2$ is epistasis of additive values of loci with dominance values of

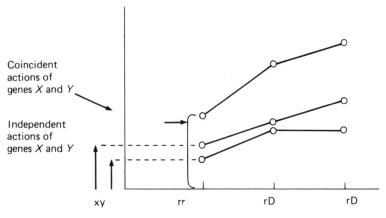

(a) Ordinary additive action

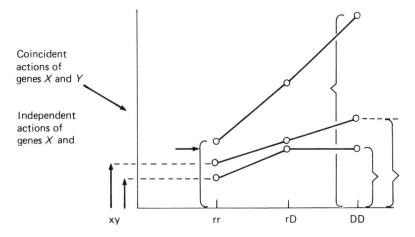

(b) Epistatic action (Interaction)

Figure 5.3. Epistatic action illustrated for certain alleles of genes X and Y.

other loci, $\sigma^2_{C.DD}$ is epistasis of dominance values of loci with one another, and $\sigma^2_{C.AAA}$, $\sigma^2_{C.AAD}$, $\sigma^2_{C.ADD}$, etc., are similarly three-way epistasis effects.

It will be observed that (5.18b) ignores the linkage term, σ^2_L, in (5.18a). Virtually all "practical" formulations do this because, though we may *conceptually* separate epistasis and linkage in the ideal formulations of (5.18a) and (5.19), they interact so complexly that no practicable operations can at present separate them.

Kempthorne (1955), who summarizes in a table the work on genetic correlations of relatives, by Pearson, Yule, Weinberg, Fisher, Wright, and Malécot, considers the Eq. (5.18b) to be the necessary basis for obtaining correlation of relatives in random-mating populations with an arbitrary number of unlinked loci and an arbitrary number of alleles at each locus.

In attempting, in this section, better to define the meaning of the inheritable part of a trait, so that broad, H_B, and narrow, H_N, heritabilities may be more closely evaluated, we are involved also in the ultimate aim of any genetic research: to determine the number and nature of gene loci. If we research a knowledge of what we may call the *genetic structure of the population* our power of inference about the combination with environmental effects is also increased. All approaches to analysis along the lines of the last two equations—leading eventually to hypotheses about number of genes, frequencies of allelomorphic forms (p and g), and magnitudes of dominance deviations, d's—at present proceed by obtaining the genetic variances, by achieving the separations from threptic variances that we have developed, one might then test the genetic structure hypotheses by goodness of least-squares fit to the data. One sees first how much of the genotypic variance is accounted for by additive variance, then by assumed dominance actions, and finally the residue in epistacy-plus-linkage. If we take σ_G^2 as the total determined genetic variance, this last step of is represented in (5.19). The parentheses around σ_C^2 and σ_L^2 indicate that, as Crow and Kimura (1970) point out, they are inseparably confounded:

$$(\sigma_C^2 + \sigma_L^2) = \sigma_G^2 - \sigma_A^2 - \sigma_D^2. \qquad (5.19)$$

An advanced treatment of handling covariances associated with epistasis and linkage, for various relatives, is provided by Cockerham (1954).

If, as psychologists, we want to think of (5.19) in the wider context of environment, and psychometric error, and in the familiar σ_g^2 with partitioning subscripts, we then have

$$\sigma_o^2 = [\sigma_{ga}^2 + \sigma_{gd}^2 + (\sigma_{gc}^2 + \sigma_{gl}^2) + \sigma_t^2] \\ + 2r_{gt}[\sigma_{ga}^2 + \sigma_{gd}^2 + (\sigma_{gc}^2 + \sigma_{gl}^2)]^{\frac{1}{2}}\sigma_t + \sigma_e^2, \qquad (5.20)$$

where σ_o^2 is total observed, phenotypic variance, gc is genetic epistatic component, gl linkage, e error of measurement, t threptic variance, and r_{gt} the r of total genetic with threptic variance. We perhaps need (5.20) to remind us of the approximation in the kind of calculation of heritability that subtracts an *a priori* genetic variance from an observed phenotypic variance to find the threptic residue. Mainly this occurs only in approaches in this chapter, when we use genetic knowledge to construct an "artificial" σ_g^2. When we solve independently for σ_{wg}^2 and σ_{bg}^2 in ordinary MAVA the σ_g^2 that we reach is genuinely the total expression in parens in (5.20), that is, it includes total genetic variance from all sources.

Although, in this partitioning, even present genetic methods cannot cope with separation of the epistacy and linkage terms, the effect of dominant alleles and their separation was handled (along with assortive mating) half a century ago by those giants in the field, Fisher (1930), Haldane (1965), and Wright (1922). These effects can be studied in particularly clear expositions later by Cockerman (1956), Falconer (1960), Kempthorne (1955), Mather (1949), Mather and Jinks (1971), and others, and, in the psychological field of intelligence, by Burt and Howard (1956), Jensen (1973), and Jencks (1972). As Falconer (p. 136) points out, if p and q are the population frequencies of two alleles, and d is the dominance deviation, the dominance variance from that source is

$$\sigma^2_{dg} = (2pqd)^2, \tag{5.21}$$

whence the sum of additive and dominance variance is

$$\sigma^2_{(a+d)g} = 2pq[a + d(q - p)]^2 + [2pqd]^2, \tag{5.22}$$

where a is the ordinary (additive) genotypic value.

Now it is possible by pursuing this line of action further to reach such tables as are given in Falconer, Jensen, Kempthorne, and others, giving coefficients for the extent to which additive, dominance, epistatic, and interactive genetic variances enter into the total genetic variance for *relatives of different kinds*. For, further consideration of what dominance means will show us that it can affect relations among some relatives and not others, as shown in Table 5.15. For example, dominance effects will not enter into the genetic similarity of a parent and offspring but will enter the common variance among sibs. The coefficients of the contributions from various sources are shown in Table 5.15. These weights, it must be kept in mind, do not state what fraction of the genetic variance will be due to, for example, dominance, but what fraction of the total population dominance locus variance will enter as a contribution to the common genetic variance of the relatives.

As regards the consideration in Table 5.15 of the action of epistatic and linkage effects, Falconer adds (from Cockerham, 1956) that the added covariance from linkage is contained in the interaction term, and that what has sometimes been ascribed to epistasis has, because of their present inseparability, often been partly linkage. Like dominance, $(\sigma^2_C + \sigma^2_L)$ can be separated theoretically as in (5.18a), but the final split is scarcely possible with human data and methods. As Falconer (1960, p. 139) points out an approach to C and L separation can be made in designs for animal experiment, but, fortunately, there is a consensus of geneticists that linkage is for human genetics a minor influence. Thus Cockerman (1956) concludes that parent–offspring correlations are unaffected by linkage, but that the effect of linkage is to raise somewhat the purely genetic correlation of sibs and half sibs, the increase being greater on gene effects from genes in closer linkage.

The effect of the realistically seen genetic components upon the correla-

TABLE 5.15

Mendelian Action Sources Underlying Broad and Narrow Heritability: Contributions from Additive, Dominance, and Epistatic Sources to Observed Genetic Covariances of Relatives[a]

Covariance of	Additive V_A	Dominant V_D	Interaction A × A $V_{I(A×A)}$	Interaction A × D	Interaction D × D $V_{I(D×D)}$	Interaction higher order $V_{I(A×A×A)}$
(1) Parent–child	$\frac{1}{2}$	0	$\frac{1}{4}$	0	0	$\frac{1}{8}$
(2) Mid-parent–child	$\frac{1}{2}$	0	$\frac{1}{4}$	0	0	$\frac{1}{8}$
(3) Half sibs	$\frac{1}{4}$	0	$\frac{1}{16}$	0	0	$\frac{1}{64}$
(4) Full sibs	$\frac{1}{2}$	$\frac{1}{4}$	$\frac{1}{4}$	$\frac{1}{8}$	$\frac{1}{16}$	$\frac{1}{8}$
(5) Fraternal twins	$\frac{1}{2}$	$\frac{1}{4}$	$\frac{1}{4}$	$\frac{1}{8}$	$\frac{1}{16}$	$\frac{1}{8}$
(6) Identical twins	1	1	1	1	1	1
(7) General	x	y	x^2	xy	y^2	x^3

[a] Extended from Falconer (1960), p. 157.

Interaction is epistacy and linkage.

Together, the sources sum to the *shared* variance of the relatives indicated, which means that we usually write as σ^2_{wg} is a function thereof. The unshared variance, which is complementary, gives σ^2_{bg} as a function. Thus in four cases in the table the unshared variance would be

	V_A	V_D	$V_{I(A×A)}$
(1) Parent–child	$\frac{1}{2}$	1	$\frac{3}{4}$
(4) Full sibs	$\frac{1}{2}$	$\frac{3}{4}$	$\frac{3}{4}$
(6) Identical twins	0	0	0
(3) Half sibs	$\frac{3}{4}$	1	$\frac{15}{16}$

From this table one can see whether the r of the relatives is decreased or left unchanged by the existence of dominance or epistasis. For example, (1), (3), and (4) will be reduced; (6) will be unchanged.

Let it be reiterated that we are dealing here purely with the genetic variance. Hence it is immaterial whether the relatives are raised together or apart. The functions state the function of the total population (genetic) variance from that source (A, D, or C) that enters the given relation. They do *not* state the absolute magnitude of that variance source. If in the given set of gene loci involved in determining a particular trait there happens to be few dominant alleles, then $\frac{3}{4}$ of that, as in the sib–sib case, would not be very large, and the correlation $r_{ss} = \sigma^2_{bg}/(\sigma^2_{wg} + \sigma^2_{bg})$ would fall relatively little above the .5 value for simple additive action. The literally (empirically) obtained σ^2_{wg} or σ^2_{bg} when the application of the twin or MAVA methods sets aside the threptic variance contributions, are thus *functions* of the values in the table and are written (f) σ^2_{wg} and (f) σ^2_{bg} above to indicate that they depend not only on the fractions, but on the degree of presence of dominance loci, epistatic interactions, *etc*.

tions as we first derived them on a simple additive assumption is not great. Burt and Howard (1956) comment on Wright's formula for epistatic (interaction) effect: "The effects of interaction are not likely to produce any great alteration in the theoretical value of the [parent–offspring] correlation, and for [intelligence] no concrete evidence is available to suggest that any such interaction exists, much less to compute its amount [p. 111]."

One final genetic mechanism we should briefly consider here is that of sex-linked genes. The X chromosome in humans (which is the carrier of female sex, Y being the corresponding male chromosome) would be expected to contain (besides the sex determiner) only a small fraction of what is carried by the rest, the autosomal chromosomes, and in fact Lehrke (1978) estimates that it contains only 5 or 6% of the gene content in a haploid (half) set of chromosomes (and presumably less for Y). Nevertheless, because of the many instances of significant sex difference (which could be genetic or cultural) in primary abilities, personality traits (see Table 5.16), and other traits, much attention has been given to possible sex-linked inheritance.

It will be seen that the differences are highly similar in Britain and the United States, and further sources show them to be similar on the HSPQ and on the 16 P.F. in Australia, Germany (Cattell, A.K.S., Krug, & Schumacher, 1980), and Japan (see 16 PF Handbook—Cattell, Eber, & Tatsuoka [1970]). We thus have here an approach to the research design in Table 5.21, except that instead of a race x culture factorial design we have a sex x culture design. What is substantially similar in all is that men are higher on E (dominance), H (parmia), L (protension), Q_1 (radicalism), and Q_3 (self-sentiment). Women are higher on A (affectia), I (premsia), O (guilt proneness), and Q_4 (ergic tension). All these differences are substantial and cross cultural, but even in the contrast of such tolerably similar cultures as Britain and America, the differences on C (ego strength), F (surgency), M (autia), N (shrewdness), and Q_2 (self-sufficiency) are either opposite or nonsignificant and usually both. It is noteworthy that on two factors that our later researches (Chapters 8 and 9) show to be most highly inherited, namely, surgency, F, and intelligence, B (or g_f), there is no sex difference, so *their* inheritance cannot be on the sex chromosome. On the other hand, the next highest primary factor in heritability—I, premsia–harria—has here as elsewhere the largest sex difference of all. This strongly suggests it would be profitable to look for loci for this temperament trait on the X–Y-chromosome system.

Obviously sex roles have been so similar in many cultures that much study of phenotypic regressions on environments (Chapter 6) would be necessary before reaching conclusions. All one can say is that if genetic differences are found they must almost certainly be on the X–Y-chromosomes, apart from any epistatic or linkage effects. A check by an independent approach is possible here from the physical linkage method, so much research having been done on medical–physiological differences, for example, on hemophilia and other disorders—apart from those sex differences well recognized by the general public!

Leads to sex-genetic traits can show up in behavior genetics not only in differences of means, but also in differences of parent–child correlations (Hogben, 1932) when the two are of same or opposite sex, and in differences of standard deviation. It has been recognized, at least since Havelock Ellis (1894) that though men and women average the same on intelligence, as in g_f tests, the standard deviation in the former is significantly larger. Lehrke (1978) has presented a thorough study of this phenomenon in terms of sex-linked genes.

TABLE 5.16
Sex Differences in Primary Personality Factors[a]

								Source trait								
	A	B	C	E	F	G	H	I	L	M	N	O	Q_1	Q_2	Q_3	Q_4
In the United States (Cattell, Eber, & Tatsuoka, 1970)																
Men																
Mean	18.49	13.44	32.96	24.28	26.18	28.21	28.50	17.83	16.70	22.15	21.93	19.91	20.13	20.49	23.54	22.21
S.D.	6.84	3.27	6.92	6.28	7.89	5.71	9.82	5.59	5.04	5.89	3.79	6.90	4.60	5.51	4.95	7.93
Women																
Mean	23.14	13.44	33.26	17.30	25.86	28.69	25.43	24.78	14.96	24.17	20.53	23.43	17.61	20.22	22.78	26.78
S.D.	6.12	3.27	7.87	6.54	7.75	5.01	9.90	4.58	5.04	5.67	3.98	7.30	4.32	5.10	4.92	8.30
Difference (positive if M higher)	-4.65	0.00	-.30	6.98	.32	-.48	3.07	-6.95	1.74	-2.02	1.40	-3.52	2.52	.27	.76	-4.57
T value	15.50	0.00	.91	24.93	.89	2.00	6.67	28.96	7.91	7.21	8.24	11.00	12.60	1.13	3.45	12.35
Significance of	<.001	NS	NS	<.001	NS	<.05	<.001	<.001	<.001	<.001	<.001	<.001	<.001	NS	<.001	<.001
In Britain (Saville, 1972)																
Male																
Mean	17.58	14.65	30.80	25.00	26.27	24.63	27.25	17.63	17.72	22.94	20.96	19.69	18.86	20.40	24.86	23.17
S.D.	5.85	3.47	6.99	7.30	9.23	5.90	10.30	5.76	4.95	5.63	4.80	7.79	4.97	5.76	6.00	9.20
Female																
Mean	21.72	14.16	28.07	19.09	26.42	24.79	23.48	25.01	15.89	21.75	22.34	25.35	16.49	19.68	23.13	28.78
S.D.	5.02	3.44	6.67	6.69	8.61	5.46	10.36	4.58	4.94	5.92	4.71	7.58	4.71	5.59	5.73	8.16
Difference in means (positive in M higher)	-4.14	0.49	2.73	5.91	-0.15	-0.16	3.77	-7.38	1.83	1.19	-1.38	-5.66	2.37	0.72	1.73	-5.01
T value	17.02	3.18	8.96	18.92	0.38	0.63	8.18	31.78	8.29	4.62	6.50	16.50	10.97	2.84	6.61	12.91
Significance of difference, *p*	<.001	<.01	<.001	<.001	NS	NS	<.001	<.001	<.001	<.001	<.001	<.001	<.001	<.01	<.001	<.001

[a] In both countries the data is based on (a) very large samples (2500 in each); (b) on use of both A and B forms of the test together; (c) on a nation-stratified (class, region) origin of the sample, quite unusually complete in Saville's British surveys; (d) on essentially the same age group, in the range of (or corrected to) 35–38 years. The values given are in raw scores.

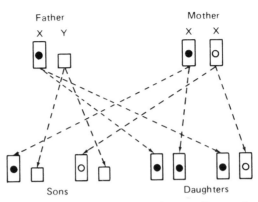

Figure 5.4. Expected relations from a sex-linked gene. Presented symbolically: Mo–So ≅ Fa–Da > Mo–Da > Fa–So ≅ Mo–Fa ≅ 0. From Stafford (1965). See also Lehrke in Osborn, Noble, & Weyl (1978).

Although no attack has yet been made on heritability of the personality primaries, several researchers have theorized about X- and Y-linked genes in the primary abilities, notably Stafford (1965) claiming evidence of male-linked spatial ability. Figure 5.4 by Stafford gives a useful presentation of the main effect expected in such transmission. As Stafford points out: "We would expect a zero correlation between fathers' and their sons' scores, but we would expect a significant correlation between fathers' scores and their daughters' scores since the father passes only his Y chromosome to his son, and passes his X chromosome, containing the gene determining the trait, to his daughter (see Fig. [5.4]). Since the son's X chromosome comes from his mother, the correlation between mothers' and their sons' scores should be significant and equal in magnitude to that found between the fathers' and their daughters' scores. The correlation of mothers' scores with their daughters' would be positive, although somewhat smaller in magnitude [p. 101]." Stafford summarizes this in symbols at the foot of Figure 5.4.

With this brief account of effects of epistacy, dominance, linkage, and sex linkage on the *genetic part* of the correlation and variances, as such, the reader must be left to pursue these issues further as he wishes in such texts as Kempthorne, King, Lerner, Li, McClearn and DeFries and others.

5. The Special Heritabilities Calculated by Parent–Offspring Methods with and without Genetic Partitioning

We are now approaching the parent–offspring research design, one of a series of designs that, seen in perspective, are properly regarded as pieces chopped out of the MAVA total design. They include use of the parent–offspring constellation, the adoptive family approach, the study of sibs reared apart, the constellation of ordinary sibs reared together, the study of half sibs, and, of course, the twin method already studied. It is an inescapable fact of the logic of statistical design that these "fragmentary" approaches cannot solve for the range of variances, covariances, and heritabilities that are possible when the full strategic design in the complete MAVA range is brought to bear. If one asks, therefore, why the fragmentary methods are used, the answer is largely that the individual researcher can rarely command the resources to cover the many constellations needed in MAVA. But occasionally, if one must be realistic on this theme, use of the fragmentary methods is due to some investigator's happy access to large numbers of one constellation, or even a fond fixation on a particular sub-method.

It will be seen from the analyses that follow in this chapter that these fragmentary methods are invariably incapable of delivering the full roster of within, H_w, between, H_b, and population, H_p, heritabilities *unless one adds an extra constellation or brings in more props in the way of assumptions than one needs to make in, at any rate, the full MAVA.* Furthermore, solutions for covariance are generally "out."

In the case of the parent–offspring constellation analysis, which, as we said, at first glance seems a "natural," the adjuncts or assumptions that have to be brought in are (*a*) an assumption about the purely genetic parent-child correlation, based on what we discussed earlier about genetic partitioning; (*b*) an estimated relation of σ^2_{wg} to σ^2_{bg}, based on an assumed assortive mating correlation; (*c*) either an assumption of equality of the common, family-shared threptic variance to that among ordinary sibs, or acceptance of a new term in that we deal here with a new type of family threptic variances. These we may write σ^2_{wt} and σ^2_{bt}, and note they are based on what we may call "intergenerational threptic" variance, or IGT, values: the σ^2_{bt} variance now shared by parent and child. As an alternative to some of these props the solution can be handled by going to data from one *extra* constellation taken as a supplement to the main constellation. These limitations and needs for props also apply to the other fragmentary approaches.

In regard to use of (*a*) we know that the purely genetic *r* of parent and offspring from *additive* genetic action is .5 (Table 5.8), but at present we usually have no idea how much dominance or epistasis exists for a given trait (see, however, Kempthorne's [1957] suggestion, given in what follows) and so do not know how much the parent–offspring or sib–sib correlation is reduced, by the fractions in Table 5.5. With a cautious eye on our conclusions we must therefore assume we are probably slightly in error through overestimating $r_{po.g}$ (the parent–offspring purely genetic correlation) as 0.5. There is also a question of what assortive mating will do to this correlation, to which we will attempt a tentative answer in Eqs. (5.27), (5.31), and (5.33).

On this problem for geneticists as such, Kempthorne (1957) comments, "It is remarkable that there has been essentially no further work on the general topic with the exception of Fisher's book (1930) for the nearly forty years up to the present time [p. 505]." Kempthorne himself has employed an approach of fitting the model to data as far as possible on an additive assumption, (see Equation 19) using a least-squares. He thus accounts for as much of the additive dominance variance as possible, the rest being epistacy.

It can readily be shown that the existence of dominance will tend *to increase* the general population variance, and to increase the common genetic variance shared by all ordinary within-generation relatives, that is, not parent–child, grandparent–child, etc. Thus the broad heritability, H_B, will in general tend to be slightly greater than the narrow heritability, H_N. For example, if we have a trait X with only additive action the common variance of sibs will be $\sigma^2_{bg(a)}$ and the between-family heritability will be [usually written as $\sigma^2_{bg(a)} = \sigma^2_{bg}$]:

$$H_{N.b} = \frac{\sigma^2_{bg(a)}}{\sigma^2_{bg(a)} + \sigma^2_{bt}} \left[\text{or, total population, } H_N = \frac{\sigma^2_{g(a)}}{\sigma^2_{g(a)} + \sigma^2_t} \right]. \quad (5.23)$$

But in a trait whose genetic variance involves dominance it becomes

$$H_{B.b} = \frac{\frac{1}{2}\sigma^2_{bg(a)} + \frac{3}{4}\sigma^2_{bg(d)}k}{\frac{1}{2}\sigma^2_{bg(a)} + \frac{3}{4}\sigma^2_{bg(d)}k + \sigma^2_{bt}} \quad (5.24)$$

$$\left[\text{or, total population, } H_B = \frac{\sigma^2_{g(a)} + \sigma^2_{g(d)}}{(\sigma^2_{g(a)} + \sigma^2_{g(d)}) + \sigma^2_t} \right],$$

where $\sigma^2_{bg(a)}$ is the role of additive variance; $\sigma^2_{bg(d)}$ is that of dominance variance; k is a statement of the number and power of dominance loci relative to additive gene loci in the totality of genes involved; and $\frac{1}{2}$ is the value if random mating obtains. It follows that H_B will in general be larger than H_N when dominance is present.

Our primary practical purpose here is to decide what purely genetic correlation to use in the parent–offspring (and some other) relations. There are four possibilities to consider, and strictly only the last is adequate in real data: (*a*) to assume additive gene action, A, and random mating; (*b*) to assume additive action and assortive mating; (*c*) to assume additionally the presence of dominance, D, epistasis, C, and linkage, L, with random mating; (*d*) to assume A + D + C + L with assortive mating.

However, the difficulties of assessing A, D, C, and L completely in any real case are, as we have seen, enormous. We shall therefore deal here with A, D, and C, considering linkage thrown in with epistacy in the residual which is C. It should also be noted that though we are interested in estimating the effects of dominance and epistasy on the purely genetic family correlations, and in distinguishing the narrow heritability (additive genetic) from the broad (with D and C action), *the actual genetic variances dealt to us by MAVA, or any method not requiring bolstering by preknowledge of genetic action, already contain whatever D and C action operates.* One normally gets an *H* value that is broad if, by the MAVA equations used, it needs to be broad. But the genetic effects on parent–offspring, etc., *genetic* correlations are worth understanding for scientific reasons *per se,* besides the practical reason that this $r_{po.g}$ when known, provides a prop to help solutions when we are short of constellations.

Four types here are (a) additive–random; (b) additive assortive; (c) additive–dominance random; and (d) additive dominance assortive. Considered both for parent–offspring genetic r's, as $r_{po.g}$, and for genetic r's between siblings, as $r_{ss.g}$, this will yield eight formulas (5.25–5.32), where A is additive action, D is any dominance entering, a is assortive mating, r is random mating, and G is the total population genetic variance symbol; that is $\sigma_G^2 = \sigma_{wg}^2 + \sigma_{bg}^2$. The simplest solution we have already set out in Table 5.8 for parent–offspring as (with $\sigma_G^2 = \sigma_A^2$):

$$r_{po.g.A_r} = \tfrac{1}{2}\sigma_G^2/\sigma_G^2. \tag{5.25}$$

In the same situation for sibs we have

$$r_{ss.g.A} = \tfrac{1}{2}\sigma_G^2/\sigma_G^2. \tag{5.26}$$

With assortive mating, more realistically added, these become

$$r_{po.g.A} = \tfrac{1}{2}\hat\sigma_G^2(1 + \hat r_{fm.g})/\hat\sigma_G^2, \tag{5.27}$$

$$r_{ss.g.A} = \tfrac{1}{2}\hat\sigma_G^2(1 + \hat r_{fm.g})/\hat\sigma_G^2. \tag{5.28}$$

All Gs are additive values here. The meaning of the caret on these estimates and others given in what follows is that these estimates are of the equilibrium values of the population parameters eventually reached through assortive mating. (Assortive mating initially changes population parameters but they eventually stabilize.)

As we meet the results with dominance we will need to be reminded when we are dealing strictly with the additive general variance which will be written σ_A^2. Then

$$r_{po.g.D_r} = \tfrac{1}{2}\sigma_A^2/\sigma_A^2 + \sigma_D^2, \tag{5.29}$$

which is the same as for A_r [in (5.25)] and means that the existence of dominance should reduce the parent–offspring correlation, here or in (5.31)— as can be seen by comparing the latter with (5.25) or (5.27):

$$r_{ss.g.D_r} = (\tfrac{1}{2}\sigma_A^2 + \tfrac{1}{4}\sigma_D^2)/\sigma_G^2. \tag{5.30}$$

The most likely real situation for most traits with both dominance and assortiveness is

$$r_{po.g.D_a} = \tfrac{1}{2}\hat\sigma_A^2(1 + \hat r_{fm.A})/\hat\sigma_G^2, \tag{5.31}$$

$$r_{ss.g.D_a} = [\tfrac{1}{2}\hat\sigma_A^2(1 + \hat r_{fm.A}) + \tfrac{1}{4}\hat\sigma_D^2(1 + \hat r_{fm.D})]/\hat\sigma_G^2, \tag{5.32}$$

where $r_{fm.A}$ is the correlation of the total additive genetic values and $r_{fm.D}$ is the correlation of total dominance values between fathers and mother. Note that Crow and Felsenstein omit the covariance term for dominance that we include at the end of (5.32). (Eq. [5.31] should thus be less than [5.25] and greater than [5.29].)

Beyond these eight combinations the following is offered for the closest approach to the full complexity when both dominance, D, and epistacy, C, are in action—with assortive mating:

$$r_{\text{po.g.DC}_a} = [\tfrac{1}{2}\hat{\sigma}^2_A(1 + \hat{r}_{\text{fm.g}}) + \tfrac{1}{4}\hat{\sigma}^2(1 + \hat{r}_{\text{fm.AL-AL}})]/\hat{\sigma}^2_G, \tag{5.33}$$

$$r_{\text{ss.g.DC}_a} = [\tfrac{1}{2}\hat{\sigma}^2_A(1 + \hat{r}_{\text{fm.g}}) + \tfrac{1}{4}\hat{\sigma}^2_D(1 + \hat{r}_{\text{fm.D}})$$
$$+ \tfrac{1}{4}\hat{\sigma}^2_{A-AL}(1 + \hat{r}_{\text{fm.AL-AL}}) + \tfrac{1}{8}\hat{\sigma}^2_{AL-DL}$$
$$(1 + \hat{r}_{\text{fm.AL-DL}}) + \tfrac{1}{16}\hat{\sigma}^2_{DL-DL}(1 + \hat{r}_{\text{fm.DL-DL}})]/\hat{\sigma}^2_G. \tag{5.34}$$

In Eqs. (5.33) and (5.34), σ^2_{AL-AL}, σ^2_{AL-DL}, and σ^2_{DL-DL} are, as explained previously, the components of epistatic variance due to additive genetic values among loci, additive genetic values and dominance values among loci, and dominance values among loci, respectively. Correspondingly, $r_{\text{fm.AL-AL}}$ is the correlation of interaction values of additive genetic action among loci for the fathers and for the mothers, $r_{\text{fm.AL-DL}}$ is the correlation of interaction values of additive genetic action and dominance among the loci for fathers and for mothers, and $r_{\text{fm.DL-DL}}$ is the correlation of interaction values of dominance among loci for fathers and mothers. A discussion of the underlying principles can be found in Malécot (1948).[2]

6. The Parent–Offspring Constellation as an Extended but "Fragmentary" MAVA Design

Dealing with heritability research by the parent–offspring constellation is a "fragment" of MAVA inasmuch as only one constellation, instead of

---

[2] The partitioning of the between family variance into the various subcomponents depends on the calculation of two quantities (Malécot, 1948): $\varnothing$ and $\varnothing^1$. If the resemblance between two relatives is due to their joint relation to a common ancestor, then $\varnothing$, for a single locus, is equal to the sum of the probabilities that each allele in the common ancestor is also in the relative. For example, in parent–offspring resemblance, the two genes in the father, considered as the common ancestor, would each have a probability of being in the child of $\tfrac{1}{2}$, so $\varnothing$ would be the sum of the separate probabilities: $\varnothing = \tfrac{1}{2} + \tfrac{1}{2} = 1$. Since the genes of the child from the mother are taken at random from the population, assuming random mating, $\varnothing^1 = 0$. That is, one would not expect the set of genes contributed by the mother to the child to be similar to the set of genes contributed by the father to the child. Now, since only half the number of genes is contributed by each parent, the portion of additive genetic variance the two relatives, father and child, have in common is $\dfrac{(\varnothing + \varnothing^1)}{2}$ which in this case is $\tfrac{1}{2}$. The portion of dominance variance the two relatives have in common is, for a single locus, the probability that both ancestors will have an identical locus which is $\varnothing \times \varnothing^1$. In a parent–child resemblance, $\varnothing \times \varnothing^1 = 0$, so there would be no common dominance variance. The subcomponents of epistasis can be found by multiplying the proportions of additive and dominance variance in common by the necessary number of times as shown in the last row of Table 5.15. For example, the proportion of epistatic AL × AL variance in common is $[(\varnothing + \varnothing^1)/2)]^2$ which in the parent–child case is $\tfrac{1}{2} \times \tfrac{1}{2} = \tfrac{1}{4}$. Similarly, the proportion of epistatic AL × DL variance shared in common is $[(\varnothing + \varnothing^1)/2)]$ $[\varnothing \times \varnothing^1] = 0$, and so forth. (See Kempthorne, 1955, for more detail.)

five or six, is used as a data source. It is extended from MAVA designs of Chapter 4, however, in that we operate primarily on the parent–offspring correlation, which we have not yet put into the framework of the MAVA constellations. The equations in Tables 4.6 and 4.8 made all their comparisons within one generation—the offspring–though they could be extended in the light of what follows here.

What we have *new* here is the correlation $r_{po(g+t)}$—the phenotypic parent–child correlation—and it might be *hoped* that the knowledge of the purely genetic correlation—the $r_{po.g}$ correlation—that we gained in the preceding section [in (5.27)] could be used to lever apart the threptic remainder of the variance. Incidentally we keep to Eq. (5.27)—assortive and additive —because we usually have insufficient knowledge to handle the assortive and dominant (5.31), and it suffices for present purposes to use the former.

We shall continue with the best simple assumption regarding the *genetic* assortive r, namely, (p. 136) that it is estimated with maximum likelihood by the observed (i.e., phenotypic correlation) as $r_{fm(g+t)}$. Before we employ *within-* and *between-*family terms as done in Chapter 4 with MAVA, we must pause to recognize that the two between generations terms should not be written the same because they are *not* identical. For although we can correctly proceed—using the additive assumption [(5.25) and (5.26)]—to suppose that the *genetic* variances are the same for offspring family pairs as for a family unit defined afresh as a parent and a child pair, accepting that $\sigma^2_{wg} = \sigma^2_{bg}$, we have to recognize that the within and between are both now different species. (In either case, however, σ^2_{bg} is larger when there is assortive mating.)

Whereas the common variance of sibs, which defines the size of σ^2_{bt}, derives from one environment—that of sibs in the same generation with the same parents and outer world—the common threptic variance of parent and child is probably reduced by their belonging to two different generations. The threptic results in the parent came from an environment of grandparents and the world a generation earlier. Those in the child come by one remove— so far as the parents pass on their values—and a different contemporary outer world. To keep this distinction in view let us write the parent–child common threptic component as resulting in $\sigma^2_{\overline{bt}}$ instead of σ^2_{bt}. (Incidentally this is one more instance of the value of separating "threptic" from "environmental." Parent and child may share the same family environment or atmosphere as much as sibs [except for role effects], but what happens to them threptically in σ^2_{bg} is still different.) Similarly, the within-family threptic variance $\sigma^2_{\overline{wt}}$ must be allowed to be different from σ^2_{wt}, especially because of role difference. Obviously if we use these values to get heritabilities, they also should not be exactly comparable with those from sibs, twins, sibs reared apart, etc., and we had best write them $H_{\overline{w}}$, $H_{\overline{b}}$, and $H_{\overline{p}}$.

The observed parent–child correlation, which we will subscript with an o [less clumsy than (g + t)] to distinguish from the purely genetic resemblances we have worked with in the last section, will be (ignoring covariances of g and t):

$$r_{po.o} = \frac{\sigma_{bg}^2 + \sigma_{bt}^2}{\sigma_{wg}^2 + \sigma_{bg}^2 + \sigma_{wt}^2 + \sigma_{bt}^2} \tag{5.35}$$

or

$$r_{po.o} = \frac{\sigma_{bg}^2 + \sigma_{bt}^2}{\sigma_{GP}^2}. \tag{5.36}$$

We may leave the bar off GP because a human general population sample is inevitably one of parents and children.

One may now note with some surprise that there is no way algebraically to bring in the ulterior knowledge that

$$r_{po.g.A} = \frac{\sigma_{bg}^2}{\sigma_{wg}^2 + \sigma_{bg}^2} = .5 \tag{5.37a}$$

or that

$$r_{po.g.A} = \frac{\frac{1}{2}\sigma_G^2(1 + r_{fm.g})}{\sigma_G^2} \tag{5.37b}$$

as auxiliaries to get a solution, without help from *another* constellation. (The correlation of sibs brings in only the *ordinary* σ_{bt}^2.)

If we ignore, as in most twin method work, that $\sigma_{wt.i}^2 \neq \sigma_{wt.s}^2$, a solution is possible by bringing in σ_{ITT}^2 as the auxiliary second constellation in the following:

$$1 - r_{ITT} = \sigma_{wt}^2/\sigma_{GP}^2. \tag{5.38}$$

Alternatively, we may use as an auxiliary source the data from unrelated children raised together:

$$r_{UT} = \sigma_{bt}^2/\sigma_{GP}^2 \tag{5.39}$$

or the parent–child σ_{bt}^2 in such circumstances. Yet another alternative might be to take children brought up entirely by grandparents.

If we can at a first approximation make assumptions that certain parent–child values in a family and between families are the same as among sibs, namely, that $\sigma_{\overline{wg}}^2 + \sigma_{\overline{wt}}^2 = \sigma_{wg}^2 + \sigma_{wt}^2$; that $\sigma_{\overline{bt}}^2 + \sigma_{\overline{wt}}^2 = \sigma_{bt}^2 + \sigma_{wt}^2$; and that $\sigma_{\overline{GP}}^2 = \sigma_{GP}^2$, then

$$\sigma_{bg}^2/\sigma_{GP}^2 = 1 - (\sigma_{bt}^2 + \sigma_{wg}^2 + \sigma_{wt}^2)/\sigma_{GP}^2 \tag{5.40}$$

will lead to

$$H_b = \sigma_{bg}^2/\sigma_{GP}^2 = r_{po.o} - r_{UT}/r_{po.o}, \tag{5.41}$$

this being in fact a solution for all four variance values whence all heritabilities can be calculated.

There remains the question of the relation of the intergenerational, parent–offspring environmental variances in similarity and difference, relative to those of two sibs. It is suggested in the next section that the difference of σ_{bt}^2 and $\sigma_{\overline{bt}}^2$ will yield values useful in the psychological study of the family.

Two caveats might be mentioned in looking at the parent–child design. First, if ways of contrasting $r_{po.g}$ and $r_{po.o}$ are used, the latter must be corrected for attenuation, since $r_{po.g}$ is an unmeasured estimate. This appears not to have been done in empirical comparisons so far made, for example, in Table 5.10. A second caveat here is to avoid the conclusion sometimes made that if the (corrected) $r_{po.o}$ value virtually equals $r_{po.g}$, as has been true of several physical measures and intelligence, then these traits must be *wholly* genetic. If $r_{po.g} = r_{po.o}$ we have

$$\frac{\sigma_{bg}^2}{\sigma_{wg}^2 + \sigma_{bg}^2} = \frac{\sigma_{bg}^2 + \sigma_{bt}^2}{\sigma_{wg}^2 + \sigma_{bg}^2 + \sigma_{wt}^2 + \sigma_{bt}^2} \;. \qquad (5.42)$$

For this equality it requires only—if we may simplify by dropping covariances that $\sigma_{bt}^2/(\sigma_{bt}^2 + \sigma_{wt}^2)$ reach an equal value to $\sigma_{bg}^2/(\sigma_{wg}^2 + \sigma_{bg}^2)$. From various trait studies (Chapters 8, 9, and 10), it is not an unusual result for the ratio of between-family threptic variance to total threptic variance to reach such an equality of value. In short, an excess of $r_{po.o}$ over $r_{po.g}$ merely indicates *a large between-family to within-family (and total)* threptic variance ratio. The whole problem of estimation of heritability by regression of offspring on parent is further discussed by Kempthorne and Tandon (1953).

The value of σ_{bt}^2 is worth finding, quite apart from attempting to get H by the parent–offspring route, for it leads us as psychologists to an evaluation of the parent influence on the child relative to the largely sib–peer group value in σ_{bt}^2. Conclusions in this area require simply an evaluation of σ_{bt}^2 and σ_{wt}^2 in relation to σ_{bt}^2 and σ_{wt}^2. The difference in magnitude, if significant, can be reached, with ordinary family data, by comparing the correlations (or variances) within (r_{ST}) and between (r_{BNF}) offspring families (see σ_{ST}^2 and σ_{BNF}^2 in Table 4.2) with those just discussed, $r_{po.o}$ and $r_{po.g}$, since the genetic contribution is the same in the relation of sib–sib and parent–offspring, assuming only additive genetic variance.

7. Other Single- or Two-Constellation and Across-Generations Designs: Adoptive Families, Sibs Apart

The addition of data beyond the MAVA design by intergenerational observations has the virtue of a truly independent source of evidence, though results may not be entirely comparable. Most of these approaches, however, though mutually independent designs, are simply expansions of particular fragments of the MAVA design. Two of these that have been most popular are the study of biological offspring raised apart—σ_{sA}^2, Eqs. (6) and (14) in MAVA Table 4.8—and of unrelated children raised together in an adopting home—σ_{UT}^2, as in Eqs. (6) and (15) in Table 4.8.

If indeed, we consider all possible fragmentations of MAVA, we see that three of the four possible constellation subsets in particular have given rise to special subdesigns—the twin method, the sib apart approach, and the *adoptive-family method*. The fourth, essentially half-sib or cousin comparisons, has so far received relatively little attention.

It should be clear, by the end of this chapter, after studying the "fragmentary approaches" that, as would be algebraically expected, no one of them is capable of defining as many unknowns as the full MAVA method. For example, solutions for the genothreptic covariance terms are generally missing. Each has to be propped up by more assumptions or auxiliaries, and to be content with more ambiguous values, than the full dress MAVA method. But when one considers the demands on organization of collection of data in the latter, and the complexities of the analytical solutions, it is understandable that behavior geneticists under pressures of limited funding should follow the somewhat less efficient and fruitful, but still valuable, single- or two-constellation designs.

As mentioned earlier, the adoptive family has the attraction of seemingly offering a very direct comparison of the results of biological and adoptive family environmental effects, permitting conclusions rather easily in the applied social field, for example, on social class effects. However, the design, on closer scrutiny, runs into complications, and is plagued by possible distortion from selection in adopting and adopted samples, and of placement correlation effects inherent in the practice of social work agencies.

If we consider first the analysis only at the offspring level (omitting parent–child correlations) as occurs when σ_{UT}^2 data are gathered as part of MAVA, one heritability value, H_b, can be obtained readily, provided we bring in, as a second constellation, sibs reared apart, thus:

$$H_b = \frac{\sigma_{bg}^2}{\sigma_{bg}^2 + \sigma_{bt}^2} = \frac{r_{SA}}{r_{SA} + r_{UT}}. \tag{5.43}$$

We can get beyond this to H_w and H_p if we make the usual assumption in family genetics that, with random mating, $\sigma_{wg}^2 = \sigma_{bg}^2$[or $\sigma_{bg/k}^2$ if assortiveness exists, yielding k by Eq. (5.14)]. Then σ_{wt}^2 can be obtained as

$$\sigma_{wt}^2 = \sigma_{GP}^2 - \sigma_{wg}^2 - \sigma_{bg}^2 - \sigma_{bt}^2 \tag{5.44}$$

and H_w and H_p can follow from the resulting complete knowledge of these four terms. Note this still uses a model ignoring genothreptic covariance.

Let us, secondly, approach this same adoptive-family experimental research design from the standpoint of parent correlations, as we did earlier for ordinary families. In so doing we shall meet again the fact that σ_{bt}^2 does not equal σ_{bt}^2—the shared variance among offspring—which fact is, on the one hand, an obstacle to solution, and, on the other, an avenue to possibly finer analysis of within-family threptic influences. It also involves, in the study of the "placement effect," an introduction to some subtleties that will also need to be faced in some other constellations.

The adoptive-family approach has subdesigns in which both children studied are adopted, and in which one child is adopted and the other is the bio-child (if we may so shorten "biological" for present discussion) of the adopting parent. We shall also suggest, as having some advantages, a further design, in which an adopted child is brought into a twin family.

Debate sometimes arises as to whether the placement effect—the similarity, before adoption, of the child to the adopting parents—is based on observation of the child or of its bio-mother, and as to whether it applies both to genetic and threptic similarities of biological and adopting parents. As to the first question, social worker evidence suggests that apart from superficial (e.g., coloring) features, the placement cannot rest on any knowledge of qualities of the newborn, and must therefore rest on knowledge of the bio-mother (the father being often not known). As to the second question, our argument is that when selection is made for resemblance, either simply of the two mothers, or, in assortive mating, with husband and wife, the "judge" is quite unable to know which part of the trait compared is genetic and which threptic, and therefore *makes a selection that is equally based on genetic and threptic parts.* The proportions will of course differ with the population heritability of the trait. In Figure 5.5, given in what follows, the placement effect is represented by a correlation $r_{m_1 m_2.o}$ between the biological and adopting mothers. If the child of a relative were adopted, $r_{m_1 m_2.o}$ would have a larger genetic fraction, but, as stated, researchers have largely avoided such cases. We have not complicated Figure 5.5 by adding the adopting father, but if necessary it could use a mid-parent value for the adopting family rather than the mother only.

Let us examine Figure 5.5 by first explaining what the variance terms are, and then proceeding to the resultant correlations, and, finally, to the desired solutions for heritabilities. Since there is evidence that the mean and sigma of adoptees is not significantly different from the general population, at least for intelligence, and we may initially assume the same for adopting families, the total variance for all classes of persons involved will be assumed to be the same, namely,

$$\sigma_{GP}^2 = \sigma_{wg}^2 + \sigma_{wt}^2 + \sigma_{bg}^2 + \sigma_{bt}^2. \tag{5.45}$$

The ordinary family variance relations can be assumed to hold in the adopting family, so that the common variances of adopting mother and her own bio-child $= \sigma_{bg}^2 + \sigma_{bt}^2$, but that of adopting mother and adopted child $= \sigma_{bt}^2$. A quite special common variance applies to the two mothers because of placement, symbolized by $(\sigma_{g.n}^2 + \sigma_{t.n}^2)$, n being for placement. If we *knew* heritability beforehand we could estimate $\sigma_{g.n}^2$ from socially observed phenotypic σ_n^2 by $\sigma_{g.n}^2 = H\sigma_n^2$. Meanwhile the correlation between them can be written

$$r_{m_1m_2.o} = (\sigma_{g.n}^2 + \sigma_{t.n}^2)/\sigma_{GP}^2. \tag{5.46}$$

The main point to note is that the resemblance of the biomother of the adopted child to her child (adopted out) is not simply σ_{bg}^2, through belonging to the same biological family, but includes some threptic resemblance too, because of placement. This arises through the mediation of the adopting mother, m_2, because within the common threptic variance for the child based on the family atmosphere will be included the common variance with m_1. It would be unrealistic to suppose that this total common threptic variance is passed on to the children without reduction, because the mother is only one influence, along with the sib peers, etc., determining σ_{bt}^2. Let the discount be k, so that

$$\sigma_{bt}^2 = \sigma_{bt}^{2\prime} + k\sigma_{t.n}^2, \tag{5.47}$$

where $\sigma_{bt}^{2\prime}$, is that part of σ_{bt}^2 (assuming σ_{bt}^2 as a whole is normal in such families) not involving $k\sigma_{t.n}^2$. The shared variance of the bio-mother, m_1, and her adopted child, c_1, will be $\sigma_{bg}^2 + k\sigma_{t.n}^2$ or, in correlation form

$$r_{m_1c_1} = (\sigma_{bg}^2 + k\sigma_{t.n}^2)/\sigma_{GP}^2. \tag{5.48}$$

The correlation of the adopting mother, m_2, with her adopted child, c_1, will also differ from that due to σ_{bt}^2 in (5.46) and become

$$r_{m_2c_1} = (\sigma_{bt}^2 + \sigma_{g.n}^2)/\sigma_{GP}^2. \tag{5.49}$$

The *genetic* placement similarity of m_1 and m_2, namely, $\sigma_{g.n}^2$, will also be shared by m_2 and c_1.

The correlation of adopting mother with her own child will remain unchanged at

$$r_{m_2c_2} = (\sigma_{bg}^2 + \sigma_{bt}^2)/\sigma_{GP}^2 \tag{5.50}$$

But, if the experimenter should include the correlation of the bio-mother, m_1, with the bio-child of the adopting mother, c_2, (which has data on four people permits), it would not be zero, as if it were a correlation of two people at random, but would be

$$r_{m_1c_2} = (k\sigma_{t.n}^2 + \sigma_{g.n}^2)/\sigma_{GP}^2 \tag{5.51}$$

since the bio-child of the adopting mother includes in his or her genetic resemblance to child c_1's bio-mother the $\sigma_{g.n}^2$ component.

The last of the six possible observable phenotypic correlations is that between the bio-child and adopted child of the adopting mother, m_2, as follows:

$$r_{c_1c_2} = (\sigma_{bt}^2 + \sigma_{g.n}^2)/\sigma_{GP}^2. \tag{5.52}$$

As in (5.49) the $\sigma^2_{t.n}$ does not enter the numerator because, as indicated in (5.47), the purely threptic contribution to common variance of c_1 and c_2 from $k\sigma^2_{t.n}$ is already absorbed in the common family atmosphere in the σ^2_{bt} and σ^2_{bt} normal to all families.

Since we know σ^2_{GP} and can use all six equations—(5.46), (5.48), (5.49), (5.50), (5.51), and (5.52)—a least-squares or maximum-likelihood solution can be found for σ^2_{bg}, σ^2_{bt}, σ^2_{ft}, $\sigma^2_{g.n}$, $\sigma^2_{t.n}$, and k. From the first two, the between-family heritability, H_b, can be obtained as usual. However, as Eq. (5.45) reminds us, we still lack separate values for σ^2_{wg} and σ^2_{wt}. The discoverable k value is of psychological interest as an expression of the role of the personality of the mother in the family atmosphere.

If we turn to that further adoptive design in which two *adopted* offspring (the second c_3 not shown in Figure 5.5) are measured and compared, we find that if placement is considered significant we do not have the simple expression

$$r_{c_1c_3} = \sigma^2_{bt}/\sigma^2_{GP} \qquad (5.53)$$

but

$$r_{c_1c_3} = (\sigma^2_{bt} + r_{gm_1m_3}\sigma_{g.n.m_1}\sigma_{g.n.m_3})/\sigma^2_{GP}, \qquad (5.54)$$

where the covariance of the genetic variances that each bio-mother shares with the adopted mother is involved. Here a new unknown enters: the mutual genetic correlation of the two mothers of two adoptees in the same family. This value may be trivial, but if not, the other design for adoptive-family research seems preferable.

It is instructive to look at the adoptive-family research model also by *path coefficients*. The latter are more fully described in Chapter 7, p. 251, and in Cattell (1978a). They are an attempt to fit causal sequence contributions, by arrows, to a frame of correlations in such a way that the magnitudes of variance contributions assigned to the arrows (path coefficients) fit the observed correlation magnitudes. "Cause" is sometimes an elusive concept (see discussion of methodology and epistemology in Cattell, 1966). If a parent and child have gene block X in common, the parent's genes "cause" the child's genes, yet we can say better that the genes cause the genotype values of *both*. And as a "determiner" we could consider even a reverse arrow from child to parent, since the relation of gene to genotype is timeless.

Figure 5.5 presents the problem of making inferences from adoptive family data, in the typical situation of having measures on both mothers and both the child of the biological mother and child of the adopting mother. To present a first illustration of path coefficients (more fully examined in Chapter 7) the diagram presents both the path coefficients and the actual observable correlations that they are called upon to fit. Path coefficients (broken lines) are simply arrows for precisely presenting hypotheses about causal action, in close relation to correlations found between cause and effect. An initial statement of equations relating p's and r's, to understand Figure 5.5 is given in Footnote 3.[3]

The path coefficient p_1 represents the effect of the mother's personality and status in causing placement choice of the adopting mother's phenotype. The adopting mother's effect on her own child is represented by p_2 for environmental effects and p_3 for genetic contribution; p_4 is her environmental effect on the adopted child; p_5 and p_6 are the mutual environmental influences of sib and foster sib, and p_7 the genetic effect of the bio-mother on her own child. These path coefficients are set out in Figure 5.5 and the analyses below it.

As typical of this situation it will be noted that we get some correlations which do not correspond to those direct causal actions to which we have mostly been accustomed. Thus a significant correlation could be expected, as $r_{m_1c_{2.0}}$, between the biological mother of sib c_1 and the child of the adopting mother, c_2. And the correlation of the biological mother with her own child—$r_{m_1c_{1.0}}$—will be due not only to the genetic correlation $r_{m_1c_1.g}$, but—though to a much lesser degree—to the placement similarity to the foster mother "echoing" itself on through the foster mother's environmental influence on the adopted child.

A path coefficient, like a correlation, is a statement about a variance contribution, with the difference that the latter is a purely statistical statement, susceptible to the usual "amtrivalence" of three possible causes of correlation, whereas the path coefficient stands by one defined direction of action. The chief difficulties in use of path coefficients are that (a) a given set of empirical correlations can sometimes be accounted for by different networks of path coefficients; (b) there is especial difficulty with two way action, as in p_5 and p_6 in Figure 5.5; and (c) tests of goodness of fit to correlations obtained by different hypothesized causal networks are not yet highly developed. We shall not proceed here into the refinement of inferences from correlations to path variance contributions necessary in the adoptive family approach, but enough has been indicated for serious researchers by that method to do so. In the last resort it will be found that in the adoptive design, as with the ordinary parent–child family analysis, to proceed to H_w and H_p solutions requires some adjunct information or assumption *outside* the data from the one constellation used.

A first approach to solution by supplementation is (reasonably) to assume only additive gene action in the parent offspring r (which simply means

---

[3] The rule in path coefficients, when variables are all in standard score, equal variances, is that $r_{p_1p_x}$ in a chain $p_1p_2 \ldots p_x$ equals $(p_1 \times p_2 x \ldots x\ p_x)$. If there is a convergence of two paths on a single consequent as in Figure 5.5 where p_4 and p_5 both contribute to the threptic part of the phenotype of the adopted child then the contribution to the latter is $(p_1^2 + p_2^2 + 2rp_1p_2)^{\frac{1}{2}}$.

If there is a bifurcation, as there is at m_2 in Figure 5.5, with p_1 contributing to p_2 and p_4 then p_2 and p_4 would square and add to unity (the variance of the total "cause") provided p_1 were the total contribution to m_2. The fit to the actually obtained correlation, that is, whether the empirical $r_{m_1c_{10}}$ in Figures 5.5 actually comes to .53, is a test of the hypothesis about paths.

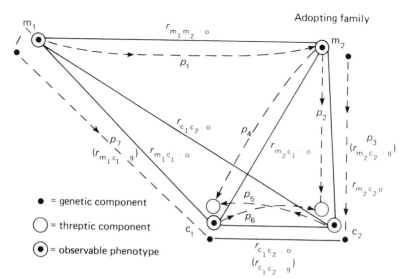

Figure 5.5. The model of influences—as p's and r's—in adoptive families. The basis for inferring H from adoptive-family data. m_1 = biological mother; m_2 = adopting mother; c_1 = adopting mother's child (adopted); c_2 = her own child.

Correlations
Known (observable)

$a = r_{m_1 m_2 . o}$ of phenotypes of mothers ($r_{m_1 c_1 . g}$) genetic r of $m_1 c_1 g = .5$
$c = r_{m_1 c_1 . o}$ of bio-mother and child "away" ($r_{m_2 c_2 . g}$) genetic r of $m_2 c_2 g = .5$
$d = r_{c_1 c_2 . o}$ of foster sibs
$e = r_{m_2 c_2 . o}$ of adopting mother and own child
$f = r_{m_2 c_1 . o}$ of adopting mother and adopted child
$h = r_{m_1 c_2 o}$ of bio-mother and child of adopted mother
Required to find $r_{c_1 c_2 . g}$

Inferrable (The non-observable r's in the figure are in parens)

Variance Components
Full variance of all persons $= \sigma^2_{wg} + \sigma^2_{bg} + \sigma^2_{bt} + \sigma^2_{wt} = \sigma^2_{GP}$
Common variance of m_1 and $c_1 = \sigma^2_{bg}$ i.e. genetic only.
Common variance of m_2 and $c_2 = \sigma^2_{bg} + \sigma^2_{bt}$
Common variance of m_1 and $m_2 = \sigma^2_{g.n} + \sigma^2_{t.n}$. This is the resemblance due to placement workers choosing foster parents somewhat like bio-parents. It could be (as phenotypic) both genetic and threptic.
Common variance of m_2 and $c_1 = \sigma^2_{bt}$

Note σ^2_{bt} in above is not the same as that for sibs, that is, from means of sibs, but due to common environmental variance of parent and child which is a different quantity. The same is true of all variances in which the *family* is the *parent–child pair*, genetically and threptically. It so happens that σ^2_{wg} and σ^2_{bg} are the same as for sibs, but, strictly, σ^2_{wt} and σ^2_{bt} are different. From variance components to correlations, giving a, c, d, *etc.*, values for further calculations.

$$r_{m_1 m_2 . o} = \frac{\sigma^2_{bg} + \sigma^2_{pt}}{G^2_P} = a$$

$$r_{m_1 c_1 . o} = \frac{\sigma^2_{bg}}{\sigma^2_{GP}} = c$$

$$r_{c_1 c_2 . o} = \frac{\sigma^2_{bt}}{\sigma^2_{GP}} = d.$$

no epistacy, in this case), and, knowing the assortive mating $r_{fm.g}$ derive σ^2_{wg} from the σ^2_{bg} reached. This permits all four variances to be known (neglecting genothreptic covariance) and H_w and H_p to be calculated as shown earlier.

As a second approach, we can go to twins to get $\sigma^2_{wt.i}$, and, if we are willing to call this σ^2_{wt}, again get all four variances, knowing, as we do σ^2_{bg} and σ^2_{bt}. Since there are advantages in keeping the reference population uniform, that is, in not going outside the given population for the twin data, this suggests the third adoptive design proposed earlier in which twin families with an adopted third child are sought.

Since it is impossible in this space to proceed exhaustively through all the "one constellation" fragmentary MAVA designs, we shall leave half sibs to be handled by the reader on the principles illustrated here (except to mention that half sibs too will require an adjunct, auxiliary group for solutions for H_w, H_b, and H_p).

There remains the analysis for sibs reared apart, by using sib–sib or parent–sib correlations. If there is placement effect, the correlation of bioparent with "adopted out" child is modified as in (5.47) and with adopting parent as in (5.48). Without significant placement effect, the r with the bioparent gives immediately $\sigma^2_{bg}/\sigma^2_{GP}$ and with the adopting parent $\sigma^2_{bt}/\sigma^2_{GP}$, yielding immediately

$$H_{\bar{b}} = \sigma^2_{bg}/(\sigma^2_{bg} + \sigma^2_{bt}). \qquad (5.55)$$

For the σ^2_{bt}—common environment among sibs as used in MAVA—one must go to the correlation of the sib with his foster sib, and, substituting this for σ^2_{bt}, obtain an H_b strictly comparable to other H_b's. In any case, for H_w and H_p solutions one must again, from an adjunct source, get σ^2_{wg} or σ^2_{wt}.

It has sometimes been suggested that it would be a neat, advantageous design to use the *same* families as a basis for data for both a combination of sibs-apart and unrelated children (adoptees) raised together. This design would not really bring gains. One can proceed to get σ^2_{wg} as a function of σ^2_{bg} in either case, and to do so together is no easier. One can still proceed toward getting σ^2_{wt} only by bringing in identical twins reared together, and assuming $\sigma^2_{wt.i} = \sigma^2_{wt.s}$.

Another "composite" design that has sometimes been suggested is that of using sibs raised apart with twin pairs, providing a check by a double assessment of a resemblance term in the adoptive design. But, in both this and the "composite" in the last paragraph the assumption that $\sigma^2_{wt.i} = \sigma^2_{wt.s}$ is one that a careful investigator will do much to avoid.

The combinations of two-constellation designs that, with one or two assumptions, will lead to solutions for H_w, H_b, and H_p are fairly numerous. On considering the nine or so constellations in the full MAVA (identicals, together and apart, fraternals or sibs, together and apart, unrelated together, half sibs, together and apart, and on father's and mother's side, and the general population) one sees there are theoretically more than 30 of such possible designs using pairs of constellations, but some may not be easy to get and all require one or more simplifying assumptions. The particular assumptions that make these designs inadequate compared with a full MAVA design are (a) that $\sigma^2_{wt.i} = \sigma^2_{wt.s}$; (b) that the parental genetic assortive mating r is known and therefore k to calculate $\sigma^2_{wg} = k\sigma^2_{bg}$; and (c) that genothreptic covariance terms can be omitted, which in at least half the traits we have studed (Chapters 8, 9, and 10) unfortunately proves false.

Next to the twin method, the adoptive-family design has probably been most used among the fragmentary methods. (Sibs apart and half sibs, well authenticated, seem to have been relatively hard to get.) As to the empirical results, Scarr and Weinberg (1977) and Scarr, Scarf, and Weinberg (1980) have concluded that the larger part of adoptees consists of illegitimate children, but (that) no significant difference can be found between them and the general population on mean and sigma of IQ. As mentioned earlier a clinical or social psychologist may justifiably be suspicious that, though such differences are not significant on intelligence, in *personality* traits the illegitimate children might show genetic differences (depending on customs in the culture). This hypothesis is well supported by the evidence of Loehlin, Horn, and Willerman (in press) who find their sample of adopted children differing significantly from their legitimate birth counterparts (adopting families) by higher surgency, F, ($p < .05$), higher dominance, E, ($p < .05$), and lower guilt proneness, $O(-)$, ($p < .05$). (There is a suggestion of lower premsia, I, at borderline significance.) It needs no special psychological experience with the meaning of the 16 P.F. primaries to recognize (see criterion relations in Cattell, Eber, & Tatsuoka, 1970) that these traits would be a powerful combination for defying convention.

Meanwhile, in the intelligence domain primarily, fairly substantial data on the correlations and variances commonly considered in the adoptive family design exists in systematic work of Horn, Loehlin, and Willerman (1979), Scarr and Weinberg (1977), Scarr, Scarf, and Weinberg (1980), Scarr, Webber, Weinberg, and Wittig (in press), Teasdale (1979), Rosenthal (1972), and others. However, except for Rosenthal and others on *psychiatric syndrome* development in adoptees, which is a powerful adjunct approach to the genealogical method, virtually all of these findings confine themselves to abilities (see Jencks *et al.,* 1972). The magnitudes of heritabilities reached are consistent with those from other designs, though Scarr has recently added some lower values. What little has been done by the adoption design on continuous *personality* variables, as contrasted with *psychiatric syndromes,* is hard to interpret, psychologically. For it has rarely been done, except for the work of Loehlin and Nichols (1976), with scales for primary and secondary personality source traits thoroughly replicated in factor analysis, and checked across ages and cultures. But on intelligence the adoptive parent–adopted child correlations as empirically obtained are illustrated by the main studies in Table 5.17. Corrected for attenuation the mean of .19 in Table 5.17 would rise to .22. The best estimate may lie between this and Jencks's (1972) corrected value (which lacked the last two studies in Table 5.17) of .28. In most of these studies the biological parent correlations were between .23 and .42 (one parent), and in the case of unwed mothers .32. Honzik (1957) found a value of .48, and Skodak and Skeels (1949) found .38, but these have been omitted from the Table 5.17 means for the good reasons given. The averaging of adequate studies existing today gives an intelligence correlation of foster parent and adopted child that falls well below the .50–.58 obtained for a parent with his own child in his own family. By some complex relations and assumptions best seen in the original, Jencks (1972) has recognized as we do here that information simply on the biological (ST) and adoptive (UA) parent correlations is not enough for a definitive solution and he concludes a "guessed" heritability (i.e. with special assumptions) somewhere between .29 and .76. (This involves accepting the purely genetic parent–child correlation at the level he assumes.)

In the personality area, and with defined primaries or secondary traits available, evidence is extremely thin, though the extensive study being undertaken by Loehlin, Horn, Nichols, Willerman, and others may soon fill the gap. Some results yielded by Loehlin *et al.* (in press) are given here in Table 5.18. Because of the possible differential effects of mothers and fathers on threptic variance the results (courtesy of the authors) are set out in detail. The general indication from these more scientifically substantial results is clearly that the resemblances are small to negligible, with an interesting exception, which developmental psychologists might well note. Daughters acquire significant similarity to the adopting fathers on exvia–invia (ratings), inhibitory control (Q data), and anxiety proneness (Q data). The values in Tables 5.17 and 5.18 should be compared with those for the true parents on abilities in Table 5.13 and personality traits in 5.14 (p. 147).

It is of interest in terms of family dynamics generally, and in dividing the

TABLE 5.17
Correlations of Intelligence Scores of Children and Adopting Parents

Source	With father's IQ	With mother's IQ
Burks (1928)	.07 ($N = 178$)	.17 ($N = 280$)
Freeman (1928)	.37 ($N = 180$)	.28 ($N = 255$)
Leahy (1935)	.19 ($N = 178$)	.24 ($N = 186$)
Honzik (1957)	Essentially 0	Essentially 0
Willerman, Horn & Loehlin (1977)	.12 ($N = 405$)	.15 ($N = 401$)
	.14 ($N = 662$)	.19 ($N = 459$)
Mean[a] (uncorrected) .19; corrected .22		

[a] This averaging omits Honzik (1957) as being on a poor sample, and the much-quoted Skeels and Dye (1939) (see also Skodak & Skeels, 1949) is not on the list because of the unfortunately major weaknesses brought out by McNemar (1940).

TABLE 5.18

Resemblances of Children and Adopting Parents on Personality[a]

| | | By child tests | | | | By child ratings[b] | | | |
| | | Father | | Mother | | Father | | Mother | |
		Son	Daughter	Son	Daughter	Son	Daughter	Son	Daughter
QI									
Exvia–invia	A	−.02	.04	−.07	.04	−.06	.13	−.03	.06
primaries	F	.12	.07	.14	.12	.05	.17*	−.03	.18*
	H	.10	.06	.05	.03	.09	.17*	.02	.09
$QVIII$									
Inhibitory con-	G	−.04	.23*	.11	.10	.00	.01	.05	.04
trol primaries	Q_3	−.08	.23*	.19*	−.07	.07	.13	.20*	.06
QII									
Anxiety	C	.06	.11	.00	.07	−.01	.08	.12	−.04
primaries	$Q(-)$	.04	.18	.12	.26*	−.02	.04	.09	−.02
	$Q_4(-)$	.17	.14	−.07	−.06	.09	.04	.08	.02
$QIII$									
Cortertia	E	−.05	−.02	−.02	−.07	.12	.14	.14*	.23*
primaries	I	.05	.13	.25*	−.00		(I not rated)		
N (pairs)		111	94	111	93	223	188	225	188

* $p < .05$.
** $p < .01$.

[a] Courtesy of Loehlin, Horn, & Willerman (in press). These authors factored the primaries and came out with the usual second orders (four of them), which we have titled as usual (Cattell, 1973a). The object was to see if secondaries behaved uniformly in their parts, and there does appear to be some such tendency.

[b] Instead of being tested on the HSPQ factors that match the adult 16 P.F. primaries, the children were rated by observers on those primaries.

origins of within family threptic agreement ($\sigma^2_{GP.t} - \sigma^2_{wt}$), to compare Table 5.18 with 5.19 which gives resemblance among sibs in adoptive families.

Here the interesting indication for family developmental studies is the greater same-sex imitation, and the suggestion that sibs have a greater environmental effect—almost all toward convergence—than parents. The common influence toward (a) surgency or desurgency; (b) high or low self-sentiment, and (c) high or low guilt proneness, is particularly striking.

In due course data such as these will need to be analyzed in the light of the model in Figure 5.5 and Eq. (5.46)–(5.54), but it must be recognized that at present the data we have in this field is not exact enough, in terms of replication as such for the application of a refined model. One is moved to make several suggestions.

First, psychometric measures should be both longer and more factor pure. The uneven results in the adoptive studies almost certainly arise from psychologists' trying to measure a personality factor in one-tenth of the time they have long recognized to be necessary for an ability factor (intelligence) of the same generality. The WAIS and the 16 P.F. each take most of an hour, but the latter is expected to measure 16 factors, the former only one. Parallel equivalent forms—six per factor—have been constructed for the 16 P.F., and if all were used, as recommended, each broad personality factor would at least reach close to one-half the time given to fluid or crystallized intelligence, with a commensurate rise in reliability and validity, to give us a much firmer grasp of the true value of (age, corrected) intrafamilial correlations.

Second, we need to know more about the effect of the very real social selection of *means* that goes on in unwed mother populations and families

TABLE 5.19
Resemblances in Personality among Sibs in Adopting Families[a]

Personality secondary	Personality primary	Male–male pairings	Female–female pairings	Male–female pairings
QI Exvia–invia	A	.12	.00	−.11
	F	.26	.33	−.02
	H	.13	.05	−.08
QVIII Inhibitory control	G	.04	.23	−.04
	Q3	.68**	.26	−.05
QII Anxiety	C	.02	.21	.13
	O	.21	.43*	.03
	Q4	.21	.01	.08
QIII Cortertia	E	.38	−.06	−.01
	I	−.11	−.03	−.02
Degrees of freedom		18	16	88

    * $p < .05$.
    ** $p < .01$.

[a] Courtesy of Loehlin, Horn, & Willerman (in press).

TABLE 5.20
Placement Selection Correlation of Traits of True and Adoptive Parents

	Leahy (1955)	Burks (1928)	Leahy (1935)	Skodak & Skeels (1949)
Education levels of fathers			.31 (N = 124)	
Occupations of fathers	.09 (N = 89)	−.02 (N = 80)	.08 (N = 104)	
Education levels of mothers	.25 (N = 94)		.29 (N = 836)	.27 (N = 100)
IQ of child at placement and education level of adopting mother	.34 (N = 93)			

interested in adopting children, as well as to determine the placement *correlations*. Table 5.20 shows that these are real enough and positive. There are therefore several separate findings:

1. There is a significant—and appreciable—correlation between the educational level of the adopting parents and that of the original (biological) parents. The value in Table 5.20 would be about .28. Scarr and Weinberg (1976) indicate about the same. Teasdale (1979) using SES (social status) rather than years of education found, in three large samples, .15, .26, and .15.
2. There is also a correlation of about the same size between the IQ of a child at time of placement and the educational level of the adopting family.
3. With some indirect evidence assumed, Jencks estimates that the usual correlation, in placement, of the child's and the adopting father's IQ is close to .13, attenuation-corrected to .16. Other data (see Table 5.17) raises this, corrected, to .22.
4. Less relevant, but of social-psychological interest, is Teasdale's finding (N = 11,094) of rs of the *later life* SES of the adopted child, reaching .20 with adopting father and .23 with biological midparent.
5. Adopting families are not only above average in mean, in education, as indicated earlier, but also significantly restricted in range (with respect to both education and IQ) relative to the full general population.

There are questions not only of more accurate allowances, but also of methods and analytical goals. Both here in the adoptive family approach, and in the previous section on the approach through parent–child (rather than sib–sib) resemblance, we have stressed that there are three main kinds of (or components in) the within-family threptic similarities.

First, earlier, we stressed that within-family *difference* (variance) springs partly from effects within the family and partly from the different "adventures" of the sibs outside, that is, σ^2_{wt} is not all within-family *influences* though it is within-family *effects*.

Second, we have seen good reason to recognize that what is yielded by parent–child analysis and sib–sib analysis as "family atmosphere" are different but overlapping values. Adoptive family studies contribute to determining this separation. So also does the analysis by path coefficients, as described earlier, originally introduced by by Wright (1934) and now developed by Fulker, Eaves, Rao, Morton, Cloninger, and others. It is interesting to note that Wright as long ago as 1934, on data of 1928, actually evaluated a split of the threptic variance of the foster *home* (foster sibs and foster parents) as such from that due to the foster *parents* as such, in the ratio of 17% to 3%, the percentages being of the total variance from all sources (illustrated in Tables 5.18 and 5.19, and elsewhere). Incidentally, the student should not assume from our estimating that these are truly low (but significant) correlations of adopted children with foster parents and foster siblings that environment is not really important in the traits concerned. There are many other environmental influences—even *within* the family—than parents and sibs. And if as the last paragraph suggests, we do not yet know how much the correlations with foster parents and sibs spring from a common atmosphere and how much from independent personalities, it could be, with little overlap that together these foster relatives contribute appreciably. For example, if the sources in Tables 5.18 and 5.19 were independent the correlation with environment would be $\sqrt{.10^2 + .15^2} = .18$.

With these complexities to take into account it can be seen why, despite the attractive features that make the adoption design popular, scientific weight should not be put unduly upon its present conclusions—at least not when considered in isolation, as some of its enthusiasts would have us do. One should in future look to necessary modification of those conclusions in the light of the more comprehensive evidence from MAVA and the analysis which shows that this partial MAVA model cannot stand on its own feet. At the same time we must recognize that other constellation designs, for example, that of sibs reared apart (or twins reared apart), are almost certainly not free from need for allowances for some of the same placement effects, etc., that have been brought out in studying adoption (since "apart" commonly implies adoption!). However, the sibs reared apart constellation is, in any case, less popular, since most investigators have found, as we did (Chapters 8, 9, and 10) that is the most difficult of all (except for identicals apart) for which to get acceptable cases (reared apart from birth).

Finally, we may note that the adoptive design has been as popular in syndrome study as in biometric, continuous data. And here it has contributed some of the clearest evidence on heritability of schizophrenia, manic-depressive disorder, neurosis, and psychopathic personality. The studies have compared, in adopting homes, the incidence of *diagnosed cases*, for biological and for adopted children, and have compared rates for children raised with and away from schizophrenic parents (see Kety, Rosenthal, Wender, & Schulsinger, 1971; Rosenthal, Wender, Kety, Welner, & Schulsinger, 1971). For example, in one study, 50 adopted children with normal parents yielded no schizophrenics, whereas 47 with one parent (mother) schizophrenic yielded 5 schizophrenics in the adoptive home. For evaluation of this methodology the reader is referred to among others, McClearn and DeFries (1973, p. 276) and Kety *et al.* (1971).

8. The Macroscopic Approach through Racial and Cultural ANOVA Designs

In this and the previous chapter we have looked at the main analyses that can be made through family relationships—the convarkin methods—and the student or researcher should have no difficulty in extending the methods and equations to more unusual kin relationships either "sideways," across members of one geneneration, as in MAVA, or "up-and-down" to ancestors and offspring.

Regardless of whether the comparisons they make are horizontal or vertical, *all* the designs so far studied center on the family as the unit of social variance, as it necessarily is of genetic variance. In this section we shall ask more speculatively whether it is possible to take other units than the family, such as different orphanages, boarding schools, races, social classes, or other groups differing in definable ways in culture, in genetics, or in both. Some of these, for example, schools, would be even susceptible to manipulative experiment, as in determining how much variance can be educationally produced by special training in accomplishments or personality in groups essentially genetically alike, that is, taken randomly from the same genetic pool. This last design deserves consideration despite our earlier observation that important aspects of personality can rarely be experimentally manipulated. Consequently, *existing* differences analyzed *in situ* for different educational, religious, national, and class groups must remain here, as elsewhere, our mainstay.

The student's familiarity with an array of investigative methods will by now have led to the perception that the essence of all behavior-genetic methodology is to find or produce situations where genetic and threptic means or variances *appear in different combinations and amounts* and yet permit equations that lead to separate assessment. The last is most readily done if we have the same genetic make-up in different learning situations, and the same learning influences applied to groups of known genetic difference, or if we have a situation permitting separate manipulative control, or even just assessment, of each.

Our designs up to this point have been bounded by the impossibility in human genetics of manipulation of major environments or matings, and by our having confined ourselves to families as the measurable units. Naturally, the human behavior geneticist glances a little enviously at the animal experimenter, who uses manipulative control both in breeding and in producing and defining major environments. Tryon's classical experiment (1940) showing that with learning experience standardized, different breedings of rats led to strains with highly different performances, was a first, rather startling, example. But the annals of animal behavior genetics now record a great variety of such clearcut findings. Actually, with humans, even were manipulation of mating a possible, the greater length of generations would alone have been a sufficient obstacle to the development of such approaches.

A daydream of the human geneticist would be to have twins reared apart, not only in sufficient numbers, but assigned in a balanced design to national cultures A and B, to different lengths and kinds of schooling, to different religious values in upbringing, etc. Essentially the same design, but with some blurring of the genetic quality, can, however, be brought to practicality by taking a homogeneous racial group—a restricted gene pool—and comparing its performances under different cultural upbringings. Possibly a stratified sample of sufficient size could be taken from the main racial gene pool and examined in different cultural settings. This would give us half the answer—namely, the variance from cultural influences. Conversely, *different* racial groups could be taken in the same culture, as mentioned earlier. Theoretically, at lest, we could therefore reach a comparison of variances possible from cultural ranges and from racial, genetic ranges, and obtain the equivalent of heritabilities for various performances, as set by present cultural and racial variations. The design would amount essentially and initially to a two-way ANOVA FACTORIAL design as indicated in Table 5.21, with the implication of some world population of all racial and cultural combinations. Thus R_1 might represent the Negro race, with the C's as diverse cultures within Africa, Haiti, the United States, etc. In turn, C_1 might represent the culture within the United States, and the column contain Negro, white,

TABLE 5.21
Genothreptic Analysis by Race and Culture in an ANOVA
FACTORIAL Design

		Cultural Environment					
		C_1	C_2	C_3	C_4	C_5	C_6
Racial (or	R_1						
physical-	R_2						
anthropol-	R_3						
ological gene	R_4						
pool defined)	R_5						
sample							

Hispanic, Chinese, and other groups such as have been measured on intelligence performance as dependent variable by Jensen (1971, 1972, 1973), Knapp (1960), Loehlin *et al.* (1975), McGurk (1967), Rodd (1958), and in Cattell (1971), Cattell, Eber, and Tatsuoka (1970), Shuey (1966), Vernon (1969) and others in the United States, and Eysenck (1971) and Lynn (1977, 1979) in Britain.

Cultural anthropologists have so far made virtually no attempt (except for rare instances like Jordheim & Olsen, 1963) at analyses such as that in Table 5.21. One can ascribe this obvious lack of interest in analysis of performances into genetic and cultural contributions partly to absence of the psychometric training necessary to handle the dependent variables, partly to the sheer magnitude and expense of data gathering, and partly, we must frankly recognize, to the last generation's embarrassed hiding from the term *race* (see Baker, 1974), for which they substituted the vague term *ethnic*, used at one extreme as a synonym for race, at the other for a culture, and in between for a particular raciocultural combination. It is perhaps not surprising that cultural anthropologists, as distinct from physical anthropologists, have been shy of the concept of race, even apart from any political prejudices. For there are complex technical difficulties in separating patterns. However, as shown by the Cattell, Bolz, and Korth (1973) experiment with breeds of dogs, by the analysis of data yielding national cultural groupings (Cattell & Brennan, in press; Cattell, Graham, & Woliver, 1979), and by the systematic work of Sokal and Sneath (1963), such computer programs as Taxonome *can* provide *objective* sortings of collections of individual organism patterns into definable "species" groups, when these exist. The real problem for the ANOVA type of design described here is that history has left a tangle of combinations of races and cultures, for each of which the term *ethnic* is properly applied.

Nevertheless, although the ideal set-up for Table 5.21 may be extremely difficult to arrange, investigators of sufficient determination and ingenuity are likely (probably over the criticisms of certain cultural anthropologists) to approach it, as Eysenck (1971) and Lynn (1979) have done in analyses of school achievement, etc., of both Anglo-Saxon mixtures and Negroes within essentially the same British culture. Undoubtedly, statistical, psychologically meaningful interaction effects will be found to be substantial; nonetheless, meaningful comparison of means effects from the two main sources is possible, as argued in the researches of Jensen (1971, 1972, 1973), Humphreys (1973), Humphreys and Taber (1973), Loehlin, *et al.* (1975), Swan (in press, 1981), Osborne (1980), Osborne, Noble and Weyl (1978), Cattell and Brennan (in press), Lumsden and Wilson (1981), and others.

When relative levels and variances for racial mixtures of various kinds are estimated, there will remain the problem for the geneticist of linking results in the various traits to the gene frequencies of various genes. It is essentially the same as the problem of tracing from the determined heritabilities, specific to given cultural and racial variances in the groups studied, respectively the particular environmental elements and the particular gene frequencies responsible for the σ_t^2 and σ_g^2 values found. A beginning has been made with discovery of particular gene frequencies (e.g., for blood groups, and sickle cell anemia) for particular races.

The ordinary assumption has been made here that the dependent variable is the average value for persons in a given raciocultural cell in Table 5.21 on a particular common source trait—intelligence, extraversion, sex erg tension, etc. The discovery of source trait factors for scoring the *syntality* profiles of nations (Cattell, Breul, & Hartman, 1952; Cattell, Graham, & Woliver, 1979; Cattell, Woliver, & Graham, 1980; Rummel, 1969, 1972) opens up the further possibility of asking to what extent the variance of total national scores on one of these cultural dimensions can be broken down on the one hand into genetic, population variances and on the other into the environmental variances (resources, climate, international events, and pressures) that also determine these syntality levels. The examination of such genetic and environmental contributions in better than journalistic terms, and with explicit scientific intentions, models, and methods, is recent and rough. But the writings of Loehlin *et al.* (1975), Eysenck (1971, 1973), Lynn (1977), Jensen (1971, 1973), Osborne (1980), Harrison (1961), Osborne, Noble, and Weyl (1978), Scarr (1980), Baker (1974), Kuttner (1968), Mather (1964), Vandenberg (1965a), Roberts and Harrison (1959), Cattell and Brennan (in press), and others promise conceptual and methodological advances.

9. Summary

1. In looking for possible independent sources of evidence by convarkin methods, one steps beyond the within-generation multiple comparisons of kin in MAVA, to the "vertical" comparison across generations, most commonly by parent–offspring relations. These vertical approaches are practically always associated with designs centering on a single "fragmentary MAVA" constellation, for example, sibs reared apart and adoptive families. To the average person, studying the parent–child relation would seem *the* method for finding out about heredity, but in fact it is less powerful than MAVA because it loses the leverage of the other constellations and because it introduces new unknowns, mainly σ_{bt}^2 defined by the *common threptic variance of parent and child* rather than the σ_{bt}^2 defined by the common family threptic experience of child and child.

2. As we study parent–child relations we encounter a new need, namely, to determine the degree of genetic assortive mating in the parents. Actually setting a value for this unknown is already required in some single generation MAVA solutions but *only* when we are short of sufficient equa-

tion constellations, whereas in parent–offspring analysis its recognition and assessment is unavoidable. Assortive mating has been found to run *positive*, as to phenotype, in all areas investigated—in physical, personality, and ability traits. It is partly rooted in general social homogamy, by class and ethnic group, and therefore is subject to historical change.

3. The effect of assortive mating operates, as far as genetic research is concerned, only through that part which represents *correlation of the genotypes*. Granted that we can estimate this the genetic effects can be clearly handled in genetic formulae. Social and personality psychologists have not made such definite headway in handling effects of the threptic part of assortiveness. Psychologically, assortiveness derives from the principles called *congeniality* and *completeness* in marriage, and these actually result also in significant values for what we call "off-diagonal" correlations; so the effects become complex.

4. The main effect of positive assortive mating of father and mother as regards genotype—$r_{fm.g}$—is to increase the total population variance in any trait, through increasing the between-family variance, σ^2_{bg}. Its effect on the within-family offspring variance, σ^2_{wg}, depends on several other factors, but is generally a slight reduction. The overall effect of assortive mating in increasing the general population genetic variance can have important social consequences. In the case of intelligence, it is responsible for greater supplies of both very low and very high intelligence, the consequences of which, however, it is not our present task to pursue.

5. Some geneticists have argued that the degree of genotypic assortive mating (compared to phenotypic) is negligible, and since the value is not determinable exactly by methods used to this point, the issue must remain in doubt. However, I have presented arguments on a statistical basis, with certain moderate assumptions, that the most probable value of $r_{fm.g}$ is the same as the observable, phenotypic $r_{fm.o}$. Later results on a wide range of traits (Chapters 8, 9, and 10) by other methods (least-squares fit, etc.) indeed point to $r_{fm.g}$ values being apparently unbiased inferences from the observed phenotypic (attenuated corrected) $r_{fm.o}$ values. Phenotypic values center on about .25 (ranging from 0 to .50), but in unsatisfactory and broken marriages values center on about +.08. Nevertheless, in inferring heritability from parent–child correlations, which is affected according to a definite formula by the assortive $r_{fm.g}$, the imprecise value of the latter introduces a weakness in such present approaches as need to assume an $r_{fm.g}$ value.

6. It would be a useful lever, especially in the "fragmentary" MAVA designs, if one could know what the correlation between the relatives in various constellations is from genetic factors alone. A table of such correlations for major relative pairs is given, on the simple assumption of purely additive polygenic action. The extent to which effects from Mendelian *dominance*, *epistacy*, and *linkage* can modify the additive correlation is discussed and set down. However, since these genetic correlations are difficult to deter-

mine with precision, being complexly affected by dominance, assortive mating, etc., the utility of this genetic knowledge as a lever for solving for phenotypic heritability relations (H_w, etc.) when only one or two constellations are available is not great. Nevertheless, the examination of MAVA-based determinations of genetic variances, in terms of various purely genetic models should eventually lead to knowledge of the roles of dominance, epistacy, etc., action in human genetics.

7. The fact that in parent–child, adoptive-family, and other fragmentary MAVA designs one solves for $\sigma^2_{\bar{b}t}$—the extent of common threptic variance of parent and child—rather than σ^2_{bt}, depending on threptic variance common to sibs, is not a defect, but must alert the behavior geneticist to the fact that H_b, as $\sigma^2_{bg}/(\sigma^2_{bg} + \sigma^2_{\bar{b}t})$ is likely to be different from H_b as $\sigma^2_{bg}/(\sigma^2_{bg} + \sigma^2_{bt})$. (Later it will be seen that a σ^2_{bg} and σ^2_{bg} can also be different when Mendelian dominance action occurs.) The same holds for H_p.

The difference of σ^2_{bt} and $\sigma^2_{\bar{b}t}$ however, can be turned to advantage in (a) finding out more about family atmosphere structures by calculations using σ^2_{bt} and $\sigma^2_{\bar{b}t}$ as basic variables; and (b) getting evidence on the existence of Mendelian dominance by comparing σ^2_{bg} and $\sigma^2_{\bar{b}g}$. As to (a), the parent gets a good fraction of his threptic variance from the action of the grandparents and their generation and the common threptic variance with the children, evaluated in σ^2_{bt}, would be less than $\sigma^2_{\bar{b}t}$ in an epoch where less is passed on by parents and more of the common sib atmosphere comes from outside forces and peers. The unique possibilities of quantitative findings from genothreptics seem not yet to have been used by analysts of the family and its social trends.

8. Several research designs that are sufficiently attractive to have been extensively pursued on their own—the twin method, the adoptive-family approach, sibs reared apart, and comparisons of sibs and half sibs—are properly seen in perspective as special fragments or sectional studies within the MAVA design. The study of such constellations in isolation seems to be partly a matter of research endowment (full MAVA is a major expense), partly of predilection and opportunity, but probably arises also through failure to realize that, mathematically, the MAVA method offers "compound interest" on data-gathering time invested. Only with four or more constellations can complete independent solutions be obtained for all unknowns. The twin, the adoptive family, the sibs apart, etc. designs on their own all require a second, adjunct constellation, and/or assumptions of a genetic, assortive, and other natures that one would prefer not to have to make.

9. The most fully pursued of these sectional approaches—other than the twin method—has been the adoptive-family method. This appeals through addressing itself directly to a socially important practice and the applied sociopolitical question of how far traits and attitudes might be changed by attention to the class-cultural differences of family backgrounds and collective "boarding out."

Technically, it has theoretically possible weaknesses of placement effects and of biased selection from the general population in the adopted children and the adopting homes. As far as intelligence is concerned, there is already data showing that selection of adoptees is not significant (though it appears to be so for *personality*), but that the level of adopting families is above average and the range subnormal. The placement effect (placement with relatives being ruled out), due to social agencies choosing adopting parents with some similarity to the biological parents, is real and significant. The SES and education of the true parents has a low significant positive correlation with those of the adopting parents, as does the preadoption intelligence of the (late-adopted) child with the intelligence of the adopting parent. Nevertheless, the correlation of the offspring's intelligence is higher with the biological than the adopting parent.

Although the actual heritabilities obtained from the adoptive and other "special section" approaches agree quite well, at a rough level, with those from MAVA, there is actually a complexity about the adoptive-family relationships that calls for a more refined treatment of the statistical analysis. Such a treatment is offered here in terms of variance analysis, correlations, and path coefficients. The "sibs reared apart" approach has been neglected because of difficulties (as with twins reared apart) in getting sufficient samples, and it is not the ideal adjunct constellation it might seem for effective comparison with the adopting constellation. To go beyond the H_b solution to H_w and H_p, the twin constellation is the most efficient adjunct, though it involves the assumption $\sigma^2_{wt.t} = \sigma^2_{wt.s}$.

10. Since the essential experimental situation required for separating and evaluating genetic and threptic components is examining empirical values known to combine these sources in different proportions, the convarkin designs can theoretically be pursued with social and genetic units other than the family. In principle, one could turn to threptic variances from cultural differences of classes, nations, religions, and whole "civilizations" (Toynbee, 1947) and genetically to differences where recognizable races or relatively inbred populations define different gene pools. Theoretically, an ANOVA type of design, or factorial design, could be set up in a two-way plan of different races and different cultures. History has not presented many instances suitable for such a set of race-by-culture combinations. *Ethnic* groups, presenting only *particular* combinations of race and culture are more prevalent than the needed full array of racial and cultural combinations. Nevertheless, a beginning has been made by psychologists with methods indicating significant components in behavior associated with different gene pools. It is suggested that scores on the dimensions of national syntality, now replicated and measurable, could be used as dependent variables for analysis, as traits of personality are used for individuals.

6

Models of Interaction of Learning and Genetic Processes

1. Relating Individual Personality Development to the Ambient Culture by the Econetic Matrices

As a special emphasis of this book we have stated that behavior genetics would be treated as potentially as great a contribution to learning as to genetics. It has been the custom in behavior-genetic writings to extract the gold of genetic mechanisms and throw away the dross of environmental interference. The learning theorist, in contrast, does not throw away the genetic findings—he or she generally simply does not recognize that they are there. Yet as the shape of oceans delimits the land, and the map of land defines the shape of the oceans, so a proper grasp of genetics tells us more precisely the shape of what we have to explain by learning, and a sound learning theory is aware of all its interactions with genetic potentials and motivation. So long as reflexological learning theory concerned itself with rat experiments of a few weeks, or with the learning of strings of words in a laboratory session, these relations could be overlooked, but with the emergence of *structured learning theory* (Cattell, 1980a)—which is concerned with personality, with the rise of dynamic structures, under *the impact of the real world*—this sterile narcissism in both human genetics and human learning theory becomes academically obsolete.

The main concepts, methods, and functional equations in human behavior genetics as such should now be sufficiently clear to the reader so as to permit us to explore the learning frontier without losing ourselves. That frontier connects with learning theory, clinical psychology, education, social psychology, sociology, and cultural anthropology.

Structured learning theory extends classical learning theory in two directions, by studying; (a) the change in *level* (frequency, strength) of a trait performance though action of existing structures; and (b) the mode of formation of the known, factorially discovered trait *structures*.

Structured learning theory has broken new ground in attacking the latter problem and has also opened up new ways of conceiving of and calculating the former, notably through consideration of the role of *existing structures along with reward patterns* (Cattell, 1980a).

In studying both (a) and (b) our goal is to consider, and form clear *models for, the interaction of learning and genetic maturation,* and in both, as indicated, we must consider the roles of trait *structures, temporary states, and sequential processes,* in the temperament, ability, and dynamic trait areas. Let us begin by stating the behavioral equation for any response act or performance at a given moment, that is, a single performance, not a learning gain as such. The reader may wish to see it built up by simpler steps (in Cattell, 1963a, 1979b) but most succinctly it is

$$a_{hijk} = \sum_{x=1}^{x=n} p_{hk} e_{hj} s_{hk} T_{xi}. \tag{6.1}$$

Here T_x is any trait x out of a total of n traits, and its subscript i indicates a score for a particular individual i. The subscript h is a focal stimulus; j is a particular *kind* of response to that stimulus, measured in magnitude by a. Incidentally, T is either an ordinary *trait* or a *state proneness or liability* to get into a certain state in a given situation, so *both trait and state levels* enter into the behavioral equation. In Eq. (6.1), k is the ambient (total) situation in which the focal stimulus, h, is encountered, and it *modulates* the state liability, T_x, of the individual to produce the observed state level. Thus S_{xki}, the state level of i on emotion x in situation k is calculable by $S_{xki} = s_{hk}T_{xi}$. The behavior indices, p and e, represent, respectively, the effect of any given trait on *perception* of h and k and on the *dynamic intensity* or *execution power* expressed in the response j to stimulus h.

It will simplify presentation of the next analysis if instead of taking all n of the person's traits, as in (6.1), we take just one trait, x, and consider it to have a factor loading (weight) virtually equal to unity, so that we can forget p, e, and s for the present argument. Then

$$a_{hijk} = T_{xi}. \tag{6.2}$$

Turning to genetics, let us now recognize that every trait has a genetic part, T_{xgi}, and a threptic part, T_{xti}, so that the following holds:

$$T_{xi} = T_{xgi} + T_{xti}. \tag{6.3}$$

Recognizing that these two parts, written as deviations from the population mean (i.e., as t_{xgi} and t_{xti}) may be correlated, we wrote [originally as substantive Eqs. (3.12) and (4.1)]

$$\sigma_x^2 = \sigma_{xg}^2 + \sigma_{xt}^2 + 2r_{xgt}\sigma_{xg}\sigma_{xt}. \tag{6.4}$$

We also recognized, as a more *probable* model, that there would be interaction of t_g and t_t so that

$$t_{xi} = t_{xgi} + kt_{xgi}t_{xti} + t_{xti}. \tag{6.5}$$

However, admitting that with correlation *and* interaction the variance expression becomes too complex [Eq. (3.13)] to permit dependable experimental resolution at this juncture in the development of genothreptic research, we have proceeded with covariance only.

Keeping to the simpler model in (6.3) let us now succinctly represent the fact that we regard the genetic part of a trait, T_{xg}, (the geneticist's "genetic value") as deriving by some function f_g from a particular genome (set of genes in the individual's chromosomes), G_{xi}, and the threptic part T_{xt} by a learning function f_t from the environment, E_{xi}, the individual has experienced up to that time, thus

$$T_{xgi} = f_g(G_{xi}) \tag{6.6}$$

and

$$T_{xti} = f_t(E_{xi}). \tag{6.7}$$

The function f_g we have looked at, with as much closeness as is here appropriate, in Chapter 5. It concerns Mendelian and later laws that can be read by the student in greater detail in several good genetics texts, some of which were cited in Chapter 5. The f_t, however, is not yet available in learning texts and so in this chapter we shall pay particular attention to the relation of discovered threptic gain and variance per se to observed environmental impact and variance.

It might be thought that this relation has been abundantly dealt with in texts on learning theory, but as the more detailed discussion given in what follows will show, this is not the case. If appreciable learning occurs in, say, school mathematics, between the ages of 10 and 14, it has been customary in learning theory to treat this as learning gain and relate it to, say, different teaching methods in the provided environment. Actually, it is partly a threptic gain from teaching and partly a genetic gain from intelligence and primary ability maturations. Any "learning law" based on the assumption that the gain measure is purely one of learning is as erroneous in calculation as would be a law of gravity based on dropping steel balls onto a powerful magnet.

Granted that we have recognized first the difference between pheno-typic and threptic, and then the difference between threptic and environ-mental, we are in position to investigate the relations of the "teaching," by environment, to the threptic gain which results, and thus finally, holistically, to put the functions (effects) of the genome and the environment together.

Unfortunately, attempts to handle the learning of personality traits and abilities in the life environment by clear models are very recent (Cattell, 1977a,b; Cattell, 1980a; Cattell & Child, 1975) and so not yet subjected to sufficient criticism, experiment, and developmental refinement. They in-volve the concepts of a *cultural press matrix, adjustment process analysis, path learning analysis,* etc. (Cattell, 1979a, p. 264.) These concepts consist-ently handle interaction of a multidimensional personality with a multidi-mensional environment by matrix representation. In this space one can only hope to indicate enough features of structured learning to give some insight into the present application of the theory; the reader must turn elsewhere to understand just how the econetic model can handle the threptic–environ-mental learning interactions for personality. In particular this involves a model for the life situations that constitute the context in which the vari-ances we have studied here are born.

In encompassing the life environment, the theoretical model begins with a *cultural press* matric, **C**, which covers the importance of such features as families, religion, bank accounts, SES, national identification within the given culture. (This is a sociologist's list with general psychological impor-tance added.) Upon this **C** matrix there operates an *individual position* matrix, transforming the score (importance) of the element as such, in the general culture as such, to an importance or strength of impact, adjusting to the personal situation (connections and status) of that particular individual within the society. Next there operates a *psychological effect* matrix based on general learning laws (so far for the *average* individual) governing person-ality change under certain impacts. (This last is enlarged upon in the next section of the chapter as *path learning analysis.*) Finally, as in all learning, there must be an addition to prediction from the motivational state of the individual. Since the dynamic structures are multiple this last must also be in matrix form.

The full econetic model thus results in five matrices in successive multi-plication: (*a*) The cultural press matrix, **C**; (*b*) the individual position matrix, **P**, yielding the individual impact matrix, **I**, by $\mathbf{P} \times \mathbf{C} = \mathbf{I}$; (*c*) **E**, an effect matrix; (*d*) the **G** matrix of normally generated psychological threptic changes; and (*e*) the dynamic matrix, **D**, for the individual, which individu-alizes the *commonly* generated changes in response to the individual's per-sonal dynamic make-up. When these are put together, the result is the gen-eral cultural environmental (or econetic) matrix equation (Cattell, 1979b, p. 270), which quantitatively relates the cultural pattern to the threptic person-ality developments of a given individual.

The new structured learning laws thus go beyond reflexological learning laws (which are adequate to describe a specific conditioning, especially in an animal or laboratory situation). They need to be invoked especially when it is a question of dealing with learning as personality dimension change in the real world environment. However, this model tells us nothing directly about the relative roles of heredity and environment on the acquisitions thus defined. It is solely concerned with learning laws relating phenotypic change to life experiences in a given culture. That is to say, it deals with the same measures and variances in phenotypic traits as have been the material of our procedures for splitting into genetic and threptic parts. The usual derivation of learning laws from reflexological measures as such does not recognize this division. In short, in order to relate experimentally obtained measures of H_w, H_b, and H_p (for particular source traits) to ranges of cultural elements we need structured learning theory, and its expression in the econetic matrices. That somewhat extensive branch of personality–learning theory must be left to be read elsewhere (Cattell, 1979b, 1980a). Meanwhile, as a first step in understanding genetic–environmental interaction, we need nevertheless to use the essence of the methods and concepts that have been discussed to find how the threptic gain is related to environmental features. It will be found that this gives us increased control in checking the magnitude of genetic contributions.

2. Discovering Learning Sources for a Particular Trait by PLA Experimental Design

A vital but as yet quite unresearched step toward fuller understanding of genetic–environmental interaction is the comparison of heritabilities—H values—obtained in the same culture with different (a) genetic groups; (b) ages; and sometimes (c) also at different cultural epochs. The second might be done by cross-sectional studies at, say, 5-year intervals from 5 to 75 years of age. Research designs of this kind with MAVA equations we call comparative MAVA research. Incidentally, for comprehensiveness, we shall include under comparative MAVA research with this aim the supplementation of such study also with fragmentary parts of MAVA, as in twin or adoptive constellations, and we shall include cross-cultural as well as cross-age comparisons.

Let us consider such comparisons first in relation to age. Our model must naturally consider two ongoing processes, as shown, for example, in Figure 6.1, one being the inherent maturation curve and one the effect of the learning impact of a succession of environmental experiences. The actual data that we shall always have to take as our starting point is the joint effect of these two curves, as shown in Figure 6.1 in the continuous line. As usual,

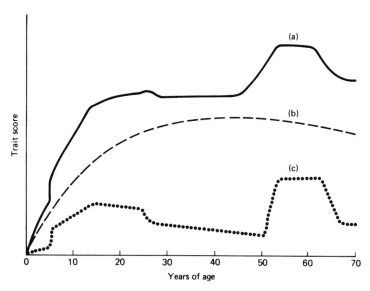

Figure 6.1 Basis of analysis of phenotypic into genetic and threptic process curves, in summation. (a) ——— phenotypic, observed curve; (b) ----- genetic, volutional curve; (c) ..... threptic
curve representing changing environmental impacts.

in the best model we must not assume that the curve comes from the simple
addition of the two; rather, it admits interaction, though in Figure 6.1 the
latter is kept small.

Certain such broader principles must be discussed before we get to grips
with the problem of "splitting the developmental curve" for any given trait.
First, we must recognize that both contributory curves can go down as well
as up. For "maturation" we must substitute the broader concept of *volution,*
which described the effects of the inner biological clock in *maturation and
involution.* (It is the latter which stops athletic champions at 30.) Second, we
must recognize that learning and time can both do and undo learned habits,
so it would not follow that the threptic level and variance would always increase with time any more than the genetic. On the other hand, we must also
firmly recognize that what is genetic is not identical with what can be perceived at birth. It is mathematically best to consider both the genetic and the
threptic curves as based on cumulative action, which, however, can also
have negative terms, in one case based on forgetting and unlearning and in
the other on decline in inner (e.g., hormonal) physiological influences.

We shall find that the breakdown of the observed developmental curve
into genetic and threptic parts, which is the aim of the developmental behavior-genetics approach in this chapter, is difficult to achieve by any truly general, relatively mechanical, analytical procedure. As in much of science, we
have to proceed by reconnaissances, leading to the tentative testing of spe-

cial, likely, hypotheses. These hypotheses can be developed for both of the contributory curves, but we shall begin with environment and the threptic curve.

Starting with the concept of a cultural press matrix, introduced in the preceding section, let us recognize the fact that for the typical individual, the position matrix (Cattell, 1979b, p. 264) will so change with time that the resulting product, the impact matrix acting upon that individual, I, will consistently change ($P \times C = I$). Table 6.1 sets out to describe the changes in the average person's cultural impact matrix that occur with age. The numbers in Table 6.1 represent rough hypotheses as to the periods and intensities of impact of the main cultural institution on the average person.

To touch on practical instances we may speculate that, in all probability, for ego strength, surgency, and superego (C, F, and G, in the symbols for primaries) the biggest impact of environment could be expected to arise from parental home influences and to be over by 20 years of age. A trait like radicalism—conservatism ($Q1$) would probably have an H value changing the most between 10 and 20, from the peer group cultural pressures, whereas sentiments (acquired dynamic traits) having to do with economics ("the bank") could well acquire their most substantial threptic components in middle age.

Although the cultural impact matrix will suffice for certain calculations we may have to make relating general population threptic variances to age and cultural difference, we need to shift at this point to a more refined development which deals with the *cultural paths* followed by *particular individuals,* instead of the gross connection of cultural features with the average personality. The two individual-oriented models developed for this use are *adjustment process analysis, APA,* and *path learning analysis, PLA* (Cattell, 1980a, pp. 294, 307), of which the second is most relevant here.

In PLA we analyze the individual's learning experiences into a number of *experiential paths,* for example, going to school, getting and holding a job, running a home with a spouse, and bringing up children. Typically a person is simultaneously following several paths. The traditional scientific design of the "controlled experiment" is quite impossible, since all else cannot be held constant while a given amount of experience of a particular path is "administered." The solution is a multivariate, nonmanipulative experimental design, embodied in matrix analysis terms in path learning analysis, as represented in Figure 6.2. It is a "macroscopic" approach to personality and ability learning, as it actually occurs in life, and affecting known unitary source traits. (See examples in Barton & Cattell, 1972, 1975; Cattell, Kawash, & De Young, 1972.) The paths will coincide, incidentally, with the paths between subgoals in the dynamic lattice (Cattell & Child, 1975).

The model supposes that traversing any one such path for a given unit period of time (or a single transit, if repetitions are to be our basis for counting) typically produces just so much change in each of the several personal-

TABLE 6.1

Learning Experience Sources Responsible for Threptic Components[a]

Age	Parental home	Sibs	School and college	Peer group	Sweetheart, spouse	Religion	Occupation	Own family and children	Socio-economic problems "the Bank"	Hobbies, art, sports	Reading and TV	Patriotic, political activities	Travel	Health and self-sentiment	Culture pattern of society (customs, law agencies)
0	10	7													
5	7	7	1	2											
10	4	8	6	8		1					4				
15	3	6	10	10	2	4	5				6		5		1
20	2	1	5	7	8	8	6	4	4	6	8	6	6	5	6
25	1		1	4	9	6	6	6	5	8	7	6	7	6	7
30	1				8	5	6	8	6	8	6	6	8	6	8
35	1				7	4	6	9	7	6	5	7	7	5	7
40					5	3	6	8	8	5	4	7	6	5	6
45					4	2	6	8	8	5	4	6	5	4	5
50					3	1	6	4	9	5	4	6	4	4	5
55					2	2	5	3	8	5	5	6	3	4	5
60					2	2	5	3	6	5	6	6	3	5	5
65					1	3	4	3	5	5	6	5	5	6	5
70					1	4	1	3	5	5	6	4	5	7	5
75					1	4		3	4	5	7	4	2	8	4
80					1	4		3	4	5	7	4	1	9	2

[a] This table presents hypothetical values for the "press" of interest in various institutions. They are presented here as a basis for quantitative experiment and to illustrate the possibility of relating cumulative environmental teaching impacts to threptic variance values found by MAVA for particular traits at particular ages.

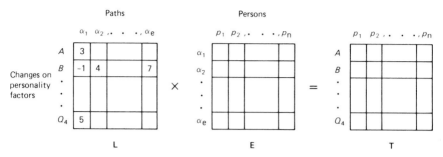

Figure 6.2 Trait change related to environment by path learning analysis. L is learning law matrix; E is experience matrix; T is trait change matrix. In L, A, B . . . Q_4 represent path potency vectors, regarding path effects on traits. In E, the columns represent the frequencies (and/or intensities) with which people are exposed to the paths.

ity source traits a person possesses. Thus a vector of unit *change* values, covering all traits, for the average person, with *one* unit of experience of such a path, can be written, as in column α_1 in the learning law matrix in Figure 6.2. This is a gross, *empirical* statement of what the average, final source trait changes are in a typical experience of this path. It does not negate the operation of reflexological laws, in Skinner's CRI and CRII, or other learning paradigms (Cattell, 1980, p. 188). Indeed, as is shown elsewhere (Cattell, 1980a, p. 304), it is possible to break down the learning law matrix, L, into further matrices of dynamic reward, repetition, and other determiners of the empiricaly evaluated end result in L.

What we would get as data in a personality learning experiment is the trait score change matrix T in Figure 6.2, for *n* people, and the record of path transits for those people. The learning law matrix is obtainable by the following calculation.[1]

$$L = TE'(EE')^{-1}. \qquad (6.8)$$

(Whether to use a right or left inverse will depend on the ratio of persons to paths.)

Thus we have successfully separated the effects of the several simultaneously experienced paths, and, what is more important for our present purpose, we have, reciprocally, located and assigned quantitative values to the several path experiences that account for the change in a given source trait. In Figure 6.2, that is given to us by a row, such as for trait B in matrix L.

However, in this change measure and its environmental quantitative sources, we are still confounding learning with genetic volution, for the path

---

[1] This is no place for intensive examination of the PLA calculation as such. In this simplest form two approximations need, however, to be recognized: (*a*) learning is taken as a linear function of the number or duration of transits; and (*b*) learning gain in L is initially *not* treated as in part a function of the absolute level of the trait. If it were, a further step in the analysis would have to be added.

transits (unlike miniature laboratory experiments) have gone on long enough for volution to occur and get included. The problem of separation we leave for a later section, but discovery of the change relations is an indispensable beginning.

3. Genothreptic Splitting of Developmental Curves: Four Methods within and across Cultures

Just as it has seemed to the amateur observer that the most natural approach to understanding genetics is to begin with the parent–child relation, so it has often seemed to him that separation of influences might be gained simply by watching development through the life span. Unfortunately, despite the general good sense of these approaches, their execution leads through thickets of increasing methodological complexity.

As we have said, a first falsifying simplification resides in the naive view that what is innate is present at birth and that what appears thereafter must be credited to environment. The emergence of sex interests at puberty, the reduction of hearing range with age, and the emergence of such highly heritable disorders as Huntington's chorea (a Mendelian dominant) only after about 45, are instances that negate this mode of thinking. Thus we are not on any firmer ground in supposing that genetic effects are limited to, say, the first 21 years of life and that curves thereafter are determined by environment. Volutional effects last through life. However, we shall argue for some simplicity in their action, in that the maturational part of the preadult life curve has a high probability of taking a certain form.

No entirely satisfactory approach to discovering genetic–threptic contributions and interactions by study of the life curve can yet be designated, but we shall consider four procedures that might prove useful to researchers with tactical skills, and with definite hypotheses to be tested.

Method 1: Alignment of Genetic Component with a Biological Growth

Substantial mathematical–biological studies of growth curves exist which offer technical possibilities. In the first place, if we know observed trait developmental curves that closely approach some form of Eq. (6.9), there are grounds for hypothesizing that we are dealing with a largely genetic curve. Of course, in psychological data the curve may be inverted, since the maturation may be that of an inhibiting agent. Also, unfortunately there has been less study of the characteristic involutional than of the maturational part of volution.

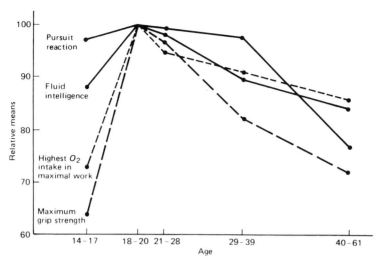

Figure 6.3 Conformity of psychological and physiological volutional process curves (Horn & Cattell, 1966a; Robinson, 1938; Miles, 1942; Burle *et al.*, 1953).

If the assumption could be made that environmental influences as frequently reduce as augment a trait, then the "bumpy" observed curve, due to their impact, coming and going at different life periods, could be a guide for reaching a "best fitting" volutional growth curve which would be taken as the genetic component. When this assumption does not seem reasonable we may turn to the likelihood that the psychological trait has some associated biological, physiological "markers," presumably little subject to environmental influences, and which follow closely the typical growth curve in Eq. (6.9). Many maturation curves are undoubtedly tied to the typical growth period of the species and in humans "flatten out" somewhere between the age of 16 and 21. They may, additionally, show a much slower involution, beginning almost from the prime of 21–22, as in hearing, fluid intelligence, metabolic rate, rate of repair of tissue. Thus it has been found, for example, that the developmental curve for fluid intelligence follows closely that for several physiological growth indices, as shown in Figure 6.3, whereas crystallized intelligence, g_c, does not (see Figure 6.7). This form can be seen in other psychological traits (Figures 6.4 and 6.5) similarly the difference of g_f and g_c curves in Figure 6.7 showing g_f to have a more volutional form, will be seen to support the evidence of Chapter 8 that fluid intelligence is the more inheritable factor.

The (b) curve in Figure 6.1 follows (up to 40) a logarithmic curve[2] of growth well known in other areas, namely, that of:

$$G_t = G_f(1 - \bar{e}^{rt}), \tag{6.9}$$

[2] Alternatively this could be written $\text{Log}_e(G_f - G_t)/G_t = rt$.

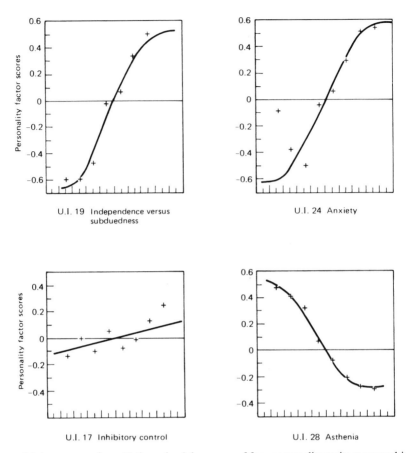

Figure 6.4 Age curves, from 10 through adolescence, of four personality traits measured in *T*-data. By maximum-likelihood analysis, the total population heritabilities for these are .21 for U.I. 17, .50 for U.I. 19, .52 for U.I. 24, and .34 for U.I. 28. (From Cattell, 1978b).

where G_t is the growth level at any time t from birth, e is exponential e, G_f is the final growth level approached, and r is a rate of drawing upon growth resources. It is possible to think of G_f as the full, finally expressed level contained in the genes and r as some rate (inverted) of enzymatic action in leading to that expression. A survey of the whole subject of natural growth in plants, animals, bacterial colonies, human populations, etc., would doubtless be very rewarding in showing more of the nature of this curve, for example, when it is continued into involution, and in revealing any possible natural growth curves of a radically different nature. But for the present let us call it the *basic volutional growth curve* and consider the possibility that if sufficient segments of it can be found in any curve of total (genetic + threptic) change we can hope for a technique of fitting the ideal genetic curve to the central tendency of those portions (as a physical anthropologist fills in

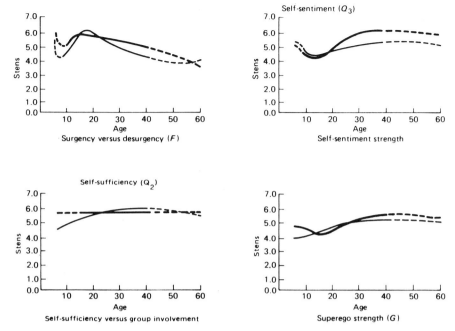

Figure 6.5 Life span age curves for four personality traits measured in Q-data. The general population heritability, H_p, later found for these primaries (Chapter 9) averages at .65 for F, .59 for Q_3, .37 for Q_2, and .12 for G, that is, the two upper scales have high H_p's; the two lower scales have low H_p's. The thick line indicates male subjects; the thin line indicates female subjects. (From Cattell, 1973a).

the missing parts of fossil skulls) in a way to permit by subtracting it from the irregular form, of the observed curve, an evaluation of the threptic curve additions and subtractions.

If we consider the presently known age curves for primary and secondary personality factors, what do we find? In objective test measures (T data), for some 10 of which curves are now known (Cattell, 1978b; Cattell & Schuerger, 1978), it is possible, since we now have both developmental curves and heritabilities, that certain generalizations can begin to emerge at an inductive level. It is noticeable that source traits of decidedly low heritability, such as U.I. 17 (control by upbringing) and U.I. 28 (asthenic overeducation) do not show the pattern discussed, whereas U.I. 19, independent temperament (which also has a substantial sex difference), U.I. 24 proneness to anxiety, and U.I. 25 realism, which have substantial heredity, do. (On the other hand, U.I. 20—unless inverted—and U.I. 23 do not fit this heritability generalization too well.)

In Q data we already have (Cattell, Eber, & Tatsuoka, 1970) data carried further into the full life span than is the case with T data, as shown in Figure 6.5. With adequate space for discussion we should probably move to the conclusion here that the coincidence of "growth form" curves with higher heritability is not as clear as with T data traits, and that some secondary qualifying principle is at work. Incidentally, the early dip in F, H, and Q_3 can be discounted, being a "dotted line" addition on data affected by what is probably an artifact. The more highly heritable factors (Chapter 9) A, F, I, and Q_3 do show the characteristic sharp growth curve to about 20, (considering I inverted, that is, measured as harria, not premsia) but Q_3 is anomalous, rising to the mid-thirties. Trait D (excitability) also shows a typical biological growth curve, but we have no evidence on its heritability.

The low heritability source traits, G (superego) and O (guilt proneness) have curves quite different from the typical growth curve, but E (dominance) and Q_4 (autonomic ergic tension) of slightly higher heritability behave anomalously, at least in the case of E. The *secondary* Q data factors (QIV-U.I. 19 and QIII-U.I. 22, cortertia) fit the higher heritability found (agreeing with U.I. 19 in T data), but invia–exvia (QI) and anxiety (QII) show a peculiar falling-off well before the age of 20, which perhaps indicates a strong environmental impact in later adolescence.

Along with examination of (*a*) resemblance to the biological growth curve and (*b*) possible association with physiological measures such as rate of myelination and axon growth of brain cells in the case of fluid intelligence (see Hendrickson & Hendrickson, 1980), we may perhaps accept at a probability level that (as mentioned earlier) (*c*) the more completely genetically determined trait will show a *smoother* curve, because of being unaffected by cultural impacts coming and going with changing age situations.

*Method 2: Separation by Subtracting Estimates of
Environment Impact from Annual Change*

This method begins at the opposite pole from Method 1, by getting an estimate of threptic change and subtracting it from the observed annual change. There follows in Method 3 what is essentially a variant on this, in which calculation is made from two (or more) groups in 1 year (with different experiences) instead of one group in 2 years. It will be remembered that in path learning analysis (PLA), estimates of changes on a trait for a group of people are obtained from the frequency with which they have had particular environmental experiences. Each component in the ongoing environmental experience is referred to as a *path*, represented by a series of learnings which characteristically occur to persons following that path. Each person is simultaneously experiencing several paths. In a manner analogous to multiple regression analysis, PLA amounts to finding the weights of the frequencies of a set of environmental experiences (paths) in order to estimate the total changes in a trait over time. Stated in matrix algebra, a row vector of changes in a trait over a designated period of time for a group of people, C_{1n}, (where n equals the number of people) is said to be a function of a row vector of weights corresponding to the various environmental paths, L_{1p}, (where p

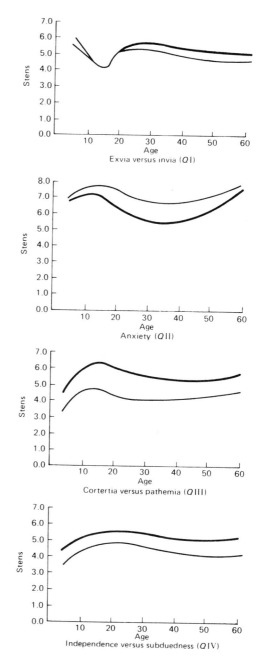

Figure 6.6 Age curves on secondary personality factors measured in *Q*-data (seemed order 16 P.F.). Age trends for secondaries I, II, III, and IV. The thick line indicates male subjects; the thin line indicates female subjects. (From Cattell, 1973a).

equals the number of environmental paths) post-multiplied by (applied to) a matrix, E_{pn}, containing the frequencies (or intensities) with which each person has had a particular environmental experience. Initially these weights are established through individual difference measures, but since the means of any group are simply derived from the latter we can similarly use the between-group weights also to estimate the shift of the *mean* of a *group* under the given impacts. A row of weights from the L matrix in Figure 6.2 for *trait* x we will call $\mathbf{1}_x$ and a column of quantified environmental experiences from the matrix E for the *year* y (being also a vector from Table 6.1, where it is a row) we will call $\mathbf{e}_y$.

We are going to assume that research has progressed to the point where the learning values in $\mathbf{1}_x$ cover so complete a range of the environmental experiences for the general population that the estimation of the observed change, in a year, from the measurable observables in $\mathbf{e}_y$ is virtually complete. (Incidentally, as in a multiple regression equation, weights will take account of any intercorrelations among the strengths of the environmental impact experiences.)

Now the observed change on trait x during a year y, namely, c_{xy} will consist of both threptic change and genetic change. (Incidentally, throughout this exposition we shall assume we are dealing with true measures (i.e., the equations will not be complicated by experimental error terms) and no g–t interaction. With no necessary connection of genetic make-up with path exposures we can simply sum the two contributions, thus:

$$c_{0.x(y)} = c_{t.x(y)} + c_{g.x(y)}. \qquad (6.10)$$

The genetic change, $c_{g.xy}$ has no direct casual connection with the environmental influence, but since genetic and threptic change are both tied to *time*, the coefficients in the vector $\mathbf{1}_x$ will owe part of their weight to the prediction of the included genetic change, so that

$$\mathbf{1}_x \mathbf{e}_y = c_{t.x(y)} + c_{g.x(y)}. \qquad (6.11)$$

If we are to separate the threptic from the genetic change in that year we would likely begin with a somewhat different l matrix to predict from $\mathbf{e}_y$ the *threptic* change only. This we shall call the pure learning or threptic effect matrix or vector, and write it $\mathbf{1}_{xe}$. The equations from which we may seek to derive this are for the first year:

$$\mathbf{1}_{xe} \mathbf{e}_y = c_{t.x(y)}, \qquad (6.12a)$$

$$\mathbf{1}_{xo} \mathbf{e}_y = c_{t.x(y)} + c_{g.x(y)} = c_{0.x(y)}, \qquad (6.12b)$$

$c_{0.x(y)}$ being the observed actual change and $\mathbf{1}_{xo}$ (o for "overall") the best predictor of this total (not threptic only) change. This assumes, as henceforth, that the regression weights in $\mathbf{1}_{xe}$ and $\mathbf{1}_{xo}$ remain the same for subjects of adjacent ages.

For the second, ensuing year we have:

$$\mathbf{l}_{xe}\mathbf{e}_{(y+1)} = \mathbf{c}_{t.x(y+1)} \tag{6.12c}$$

$$\mathbf{l}_{xo}\mathbf{e}_{(y+1)} = \mathbf{c}_{t.x(y+1)} + \mathbf{c}_{g.x(y)} = \mathbf{c}_{o.x(y+1)} \tag{6.12d}$$

Two assumptions should be noted here: (1) as stated above, the estimated change for the year—$(\mathbf{c}_{t.x(y)} + \mathbf{c}_{g.x(y)})$—is considered equal to the total observed change, $\mathbf{c}_{o.xy}$; and (2) The genetic curve is assumed to be smooth and, indeed, here linear, so that $\mathbf{c}_{g.x(y+1)}$ which would normally go into (6.12d) can still be written $\mathbf{c}_{g.xy}$. (Assumption of a linear genetic change is not essential to this solution: any assumed fixed relation of $\mathbf{c}_{g.x(y+1)}$ to $\mathbf{c}_{g.x(y)}$ would do.)

Taking (6.12b) from (6.12d) (second parts, that is, the total change in the first year from the second) we have:

$$\mathbf{c}_{o.x(y+1)} - \mathbf{c}_{o.x(y)} = \mathbf{c}_{t.x(y+1)} - \mathbf{c}_{t.x(y)}, \tag{6.13a}$$

whence

$$\mathbf{c}_{t.x(y+1)} = \mathbf{c}_{t.x(y)} + \mathbf{c}_{o.x(y+1)} - \mathbf{c}_{o.x(y)}. \tag{6.13b}$$

Substituting $\mathbf{l}_{xe}\mathbf{e}_{(y+1)}$ for $\mathbf{c}_{t.x(y+1)}$, shown equal in (6.12c), we have:

$$\mathbf{l}_{xe}\mathbf{e}_{(y+1)} = \mathbf{c}_{t.x(y)} + \mathbf{c}_{o.x(y+1)} - \mathbf{c}_{o.x(y)}. \tag{6.13c}$$

Subtracting (6.12a) we have:

$$\mathbf{l}_{xe}(\mathbf{e}_{(y+1)} - \mathbf{e}_y) = \mathbf{c}_{o.x(y+1)} - \mathbf{c}_{o.x(y)}. \tag{6.13d}$$

If we wrote the next step as if it were an ordinary division we should proceed to:

$$\mathbf{l}_{xe} = \mathbf{c}_{o.x(y+1)} - \mathbf{c}_{o.x(y)}/\mathbf{e}_{(y+1)} - \mathbf{e}_y. \tag{6.13e}$$

(Note the denominator is a matrix subtraction, yielding $\mathbf{e}_{(y+1)-y}$, a *difference* vector which we can write $\mathbf{e}_d$ for short.) However in matrix algebra the calculation from (6.13e) is naturally made by multiplying both sides by the right generalized inverse of $\mathbf{e}$, that is , $\mathbf{e}^{-1}$, reaching:

$$\mathbf{l}_{xe} = (\mathbf{c}_{o.x(y+1)} - \mathbf{c}_{o.x(y)})\,\mathbf{e}_d^{-1}. \tag{6.13f}$$

Since $\mathbf{e}_d^{-1}$ represents at this point a vector, not a square matrix, the solution is not unique. That is to say by inverting it (or any rectangular, non-square matrix) by the generalized inverse (which needs to be a right-handed one: $\mathbf{e}_d^{-1} = (\mathbf{e}_d^1\mathbf{e}_d)^{-1}\mathbf{e}_d^1$ would yield an $\mathbf{l}_{xe}$ that functions mathematically in giving the $\mathbf{c}$ answer, but which mathematically is *not* unique and therefore not psychologically unique and useful. Now, $\mathbf{l}_{xe}$ is supposed to be a psychological statement defining what each of several different environmental encounters ("paths") does to the threptic part of a trait x, and is useful psychologically only if unique so either an additional mathematical assumption must be made[3] or we must add to the experimental design.

Thus although we have illustrated, for initial simplicity, by using a single year's increment we now recognize that the solution is mathematically more satisfactory if we make a square matrix E (in order to obtain a true and unique inverse). The experiment must now be modified so that its data comes from as many different periods (years or fractions of years) as we have environmental paths to deal with. That is to say, we must take our experimental group through, say, 4 years, with different intensities of experience in each (represented by vectors e_y, $e_{(y+1)}$, $e_{(y+2)}$, $e_{(y+3)}$ in the E matrix) recording four increments $(c_{0.x(y+1)} - c_{0.x(y)}; c_{0.x(y+2)} - c_{0.x(y+1)},$ etc.) in a vector c, and solving by:

$$\mathbf{l}_{xe} = \mathbf{c}_{1y} \mathbf{E}'_{yp} (\mathbf{E}_{py} \mathbf{E}'_{yp})^{-1} \qquad (6.14)$$

(Incidentally $\mathbf{l}_{x.e}$ can also be a square matrix if we care to take measures on additional traits to x at the same time, that is, the number of traits will be made to equal the number of environmental paths. We may also note parenthetically that the group and individual treatments are such that one could handle the experiment by standing with individual PLA analyses and calculating therefrom the averages for the groups studied.)

This mode (Method 2) of derivation of threptic gain predictors is simplest if, as just stated, we can suppose that maturation over these years is taking place evenly—in a linear fashion in this case, at least over this short span of years. It will be noted also that the E matrix is one of *difference* of experiences from one year to the next and that the derived L matrix, in consequence estimates the effect of a difference in frequency of experience upon a difference in annual (or other period) gain. It seems psychologically reasonable however to assume that this represents also the effect of an *absolute* frequency of experience in a given time upon an *absolute* gain. (This can

---

[3] As indicated, we have had to use a generalized inverse in the above equation (6.13f) because E is not, in the general case, a square matrix and therefore has no true inverse. For the reader who wishes to look further at the possibilities it can be pointed out that various kinds of generalized inverses could be used, each conforming to a different set of properties. If we choose to use a Moore-Penrose inverse (Penrose, 1955), E^+, (sometimes referred to as a pseudoinverse), then it will be unique (i.e., only one right Moore-Penrose inverse for E exists). The solution for $\mathbf{l}_{xe}$ as in Eq. (6.13f) will therefore be unique for a solution using this Moore-Penrose inverse but other solutions would exist when other kinds of generalized inverses are used based on different sets of properties. Each element (weight) in $\mathbf{l}_{xe}$, therefore, is not psychologically meaningful *per se* unless in some unobvious and extremely abstruse manner it is meaningful relative to the set of properties upon which the particular generalized inverse employed is based. The vector as a whole, however, *is* meaningful, in the sense that it does estimate the changes that have occurred in a group of people as a partial result of particular environmental experiences. A more detailed description of the psychological use of PLA may be obtained from Cattell and Child (1975, pp. 59–64), and more information about generalized inverses may be obtained from Green (1976, pp. 323–350).

be assumed even if **E** is not square, provided the Moore-Penrose (see Footnote 3) or other satisfactory generalized inverse is used.)

Thus the *threptic* gain (or change) in given year (or period) can be estimated, knowing $\mathbf{l}_{xe}$ and $\mathbf{e}_y$ (the experience rates for that period) by:

$$\mathbf{c}_{t.x(y)} = \mathbf{l}_{xe}\mathbf{e}_{d(y+1)}, \tag{6.15a}$$

and similarly for the second year

$$\mathbf{c}_{t.x(y+1)} = \mathbf{l}_{xe}\mathbf{e}_{d(y+1)}. \tag{6.15b}$$

If it is possible to include so many environmental predictors in $\mathbf{l}_{xe}$ that the threptic increment can be substantially predicted then we may obtain for each period an estimate of the genetic increment by subtracting from the observed increment, as follows:

$$\mathbf{c}_{g.x(y)} = \mathbf{c}_{o.xy} - \mathbf{c}_{t.x(y)}, \tag{6.15c}$$

and similarly for the following year

$$\mathbf{c}_{g.x(y+1)} = \mathbf{c}_{o.x(y+1)} - \mathbf{c}_{t.x(y+1)}. \tag{6.15d}$$

After having obtained the threptic and genetic change scores, $\mathbf{c}_t$ and $\mathbf{c}_g$, through a series of time periods, say y to $(y+k)$, then the cumulative threptic change curve over the total period of time is going to be, for the end of each period, the change of that period plus all previous changes as in Eq. (6.15e). Likewise, the cumulative genetic change for any time period is going to be summed similarly, as in Eq. (6.15f).

$$\mathbf{c}_t(\text{cum at } y + k) = \sum_{i=y}^{y+k} \mathbf{c}_{t(i)}, \tag{6.15e}$$

$$\mathbf{c}_g(\text{cum at } y + k) = \sum_{i=y}^{y+k} \mathbf{c}_{g(i)}. \tag{6.15f}$$

If we begin with the first period of life, say the first year, then the cumulative curve is also a curve of the absolute magnitude of that particular component, threptic or genetic, over the time periods, since we begin at a zero magnitude on the trait. It would be possible also by this mode of analysis to extrapolate a curve into the future.

Method 3: Comparison of Equigenetic Groups Differing in Environmental Experiences

A further, new, experimental approach, but using the same general form of calculation, differs in its data basis and assumptions. It proceeds by taking *several* groups (4 if we work on the same scale as above) at the *same* age, but undergoing *different* environmental experiences. The assumption is that if they are of the same age and from the same gene pool (but obtained by a random selection of such groups) their maturational (volitional, in the more general sense) gain scores will be, within sampling error among groups, the same.

Dropping the y subscript for a particular year, since all are the same, and calling the observed gains in the four groups $c_{0.x(1)}$, $c_{0.x(2)}$, $c_{0.x(3)}$ and $c_{0.x(4)}$, we have, analogously to the (6.11) and (6.12) equations:

$$\mathbf{l}_{xo}\mathbf{e}_1 = c_{0.x(1)} = c_{g.x(1)} + c_{t.x(1)} \qquad (6.16a)$$

$$\mathbf{l}_{xe}\mathbf{e}_1 = c_{t.x(1)} \qquad (6.16b)$$

$$\mathbf{l}_{xo}\mathbf{e}_2 = c_{0.x(2)} = c_{g.x(1)} + c_{t.x(2)} \qquad (6.16c)$$

$$\mathbf{l}_{xe}\mathbf{e}_2 = c_{t.x(2)}, \qquad (6.16d)$$

where there are groups 1, 2, etc. instead of years of y_1, $(y + 1)$, etc. Operationally $\mathbf{l}_{xe}$ is here somewhat differently defined from Method 2. It defines weights for estimating differences in observed learning changes (in each trait) between groups.

With enough groups to equal the number of paths we would then have similarly a solution proceeding beyond the two group solution:

$$\mathbf{l}_{xe} = \frac{c_{0.x(2)} - c_{0.x(1)}}{\mathbf{e}_2 - \mathbf{e}_1} \qquad (6.16e)$$

to:

$$\mathbf{l}_{xe} = c_{xe}\mathbf{E}'(\mathbf{EE}')^{-1} \qquad (6.16f)$$

At this point we deal with differences among groups in c score and differences in E experiences.[4] These formulas could just as easily be applied to individuals as groups, but groups, as means, offer reduction of experimental, circumstantial error. It should not be difficult in practice to find same age groups in different E's (e.g., 19-year-olds of equal IQ, some in jobs and some in college, some married and some unmarried).

An alternate analysis that suggests itself for both Method 2 and Method 3 is to include genetic change also as an "experience" in the L and E matrices. The value could be simply a fraction of a year, over which the change

---

[4] The author is indebted to John Campbell for raising the question of whether Method 3 works adequately with more than two groups because the number of possible differences (changes) is combinatorial and not simply equal to the number of groups. He suggests the modification:

$$\mathbf{L}_{fp}\mathbf{E}_{pc} = \mathbf{T}_{fc}$$

where f is the number of traits, p of paths, and c the number of *combinations* of groups (two at a time). T_{fc} is then the matrix of differences on each factor, and $\mathbf{E}_{pc}$ is the matrix of differences of experiential frequency between the groups for each path. Whence

$$\mathbf{L}_{fp} = \mathbf{T}_{fc}\mathbf{E}'_{cp}(\mathbf{E}_{pc}\mathbf{E}'_{cp})^{-1}$$

scores are taken. In using the answer obtained one faces the difference of scale between genetic (g) and path experience (e) values, but this problem has already existed in differences of scale among e values anyway. What happens is that the values in **L** adjust themselves to the scale sizes in **E**. In this case one need not deal first with change in **E** related to change in **C** but would get without assumption

$$\mathbf{L}_{xo} = \mathbf{C}_{xo}\mathbf{E}_{xo}^{-1} \qquad (6.16g)$$

where o is observed, as before and a weight for g enters the path change matrix, $\mathbf{L}_{xo}$ (which is here both learning and maturational in its terms) and literal frequency values enter in $\mathbf{E}_{xo}$. Method 3 gives the proportions of threptic and genetic contribution in a change score and thus gives us curve plots in which the absolute level of each is not known [unless one grossly applies (6.16g) to a **C** which is absolute levels after crudely summing life experiences to that date], whereas Method 2 gives absolute levels.

Only by some further assumption can an absolute level be inserted. For example, if we assume that the genetic curve must be zero at birth then the summed $\mathbf{C}_g$'s to the age n will give the absolute value at n. An alternate assumption would be that the population *variances* (or standard deviations) of g and t have the same ratio as their *absolute* levels of contribution. If we know the hertitability, H_p, at age n we could then fix the absolute g and t curve levels at that point.

Methods 2 and 3 have one assumption that is hard to meet at our present stage of substantive ignorance, namely, that we can claim *knowledge of sufficient values in the* **L** *matrix to predict most of the variance in the trait change score*. That is to say, we make the assumption already in Eq. (6.10) that $c_{t.x(y)} + c_{g.x(y)} = c_{o.x(y)}$. A correction to bring them in line may be desirable (as when we find a z in zR^2 to make the latter equal to 1) though it would not actually eliminate error of estimate.

Method 4: The Use of Epogenic and Ecogenic Developmental Plots

As any psychometrist will recognize the sheer obtaining of a phenotypic development curve for a particular trait is not to be assumed itself to be as simple and well defined as may first appear. Let us therefore pause to summarize, by Table 6.7, the concepts more or less independently developed here by Baltes (1968), Cattell (1969, 1970), Nesselroade (1967), and Schaie and Strother (1968) concerning building curves respectively on cursive, cross-sectional and other data. For exact work one also needs to consider how to get comparable raw scores for a given trait across a wide age range, as that trait changes its factor pattern. (See equipotent and isopodic methods in Cattell, 1970a.)

TABLE 6.2
Separating Endogenous, Ecogenic (Prevailing), and Epogenic (Epochal) Components of a Change Curve

(a) *Possible combinations of observations*

	Same age at testing		Different age at testing	
	Same birthday	Different birthday	Same birthday	Different birthday
Same year of testing	No series	Impossible	Impossible	SC
Different year of testing	Impossible	FCE	SL and CL	EE

Note: Only one category permits a further subdivision into same subjects or different subjects (from the same age group), namely, SL and CL.

(b) *Resulting series*

Different persons tested

Calendar year at birth	Age at testing						
	10	20	30	40	50	60	
1910	1920	1930	1940	1950	1960	1970	SL
1900	1910	1920	1930	1940	1950	1960	
1890	1900	1910	1920	1930	1940	1950	
1880	1890	1900	1910	1920	1930	1940	FE
1870	1880	1890	1900	1910	1920	1930	SC
1860	1870	1880	1890	1900	1910	1920	
1850	1860	1870	1880	1890	1900	1910	
	FCE						

Same persons tested

	Age at testing						
	10	20	30	40	50	60	
1910	1920	1930	1940	1950	1960	1970	CL_1
1900	1910	1920	1930	1940	1950	1960	
1890	1900	1910	1920	1930	1940	1950	CL_2

Suggested Designations of Six Major Experimental Series

SL = *Simple Longitudinal* series: same birth year, different subjects, different ages, different testing dates.

CL = *Cursive (or Cohort) Longitudinal* series: same birth year, same subjects, different ages, different testing dates. Two sub-series, $CL_{(1)}$ and $CL_{(2)}$ are put in here because one may test *all* of the cohort at every point, as proposed, designated $CL_{(1)}$; or test all at age 10, divide into five groups, and retest each at a different decade, to avoid retesting effects (practice), as in $CL_{(2)}$.

SC = *Simple (Fixed-date) Cross-Sectional* series: different birth years, different ages, same testing date.

FCE = *Fixed Age Changing Epoch Cross-Sectional* series: different birth years, same age, different testing data.

FE = *Fixed Epoch* series: different birth years, different age at testing, different testing date, but with life span centered on the same calendar year (epoch).

CCL = *Combined Cursive Longitudinal* series: same as CL above, except that for a planned collation of results for several different age groups in the same epoch.

Table 6.2 *(continued)*

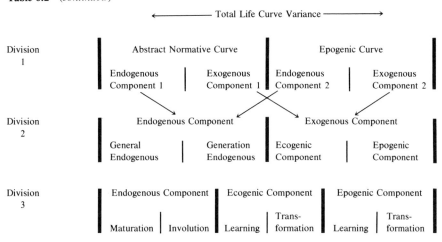

Main possible conceptual divisions in analyzing the total life course curve

Granted attention to both of the preceding requirements for getting a score at a given age point, the *ecogenic* life span curve is defined as the typical plot of a trait, representing the *combination* of normal, average genetic source and learning from environmental situations in a given culture. Down the ages poets and doctors have described its psychological and physical characteristics (e.g., Shakespeare's seven ages of man). The *epogenic* curve, on the other hand, depicts the average curve across people in a single, particular historical epoch. Since the genetic curve will remain pretty constant, (granted selection is trivial and slow) the difference of the epogenic and ecogenic curves will describe the environmental, threptic portion due to a particular epoch. It will *not* immediately separate out the normal endogenous (genetic, volutional) component from the normal exogenous (environmental, threptic) component, because they are *together* the substance of the *normal,* ecogenic curve. But approaches given earlier in this chapter could be applied for such separation. In any case a major alteration of the environment without any change of trait curve would argue for a relatively genetic trait. For example, as Weyl (1974) and others have pointed out, despite a period of severe starvation in a certain province of Holland during World War II, the later development of intelligence there did not differ from that of other provinces. Thus longitudinal evidence supports evidence from entirely different approaches as to the relatively high genetic determination of IQ. On the other hand, because of the improbability of any major shift in genetic make-up of a large population in, say, half a century, an appreciable change in mean, or in the total epogenic curve for a given trait, makes a substantial environmental component likely. For example, in Q data there are indications (as yet unpublished) of shifts of norms, suggesting that superego, G, and possible introversion, $QI(-)$ have declined between 1960 and 1975. This would be compatible with the Chapter 9 evidence that QI has appreciable environmental determination and G a considerable amount ($H = .18$).

4. Factorial Separation of Curves by P-Technique: The Case of Intelligence

It has happened in the case of intelligence, and it may happen elsewhere if our eidolon model (p. 234) is correct, that what for some time has been regarded as a single factor will prove to be a pattern of superimposition of two factorial influences of similar pattern—one genetic, one threptic. Even apart from this particular model the possibility of separating genetic and threptic sources as distinct factors *per se* has been raised by Royce (1957), Thompson (1967), and myself (1971b), though I have pointed out the hopelessness, because of the enormous number of variables required, of factorially isolating single genes, even of wide pleiotrophy. Vandenberg (1965b) and Loehlin and Vandenberg (1968) seem the only investigators so far to have given a practical demonstration on experimental data of this approach to separating purely genetic patterns by factoring. They factored the same variables as within-pair *differences,* first of identical and then of fraternal twins. The argument is sound that the former should yield only genetic and the latter both genetic and threptic factor patterns. But until this pioneer study is repeated with improvements and good personality factor markers, the empirical demonstration of a clear instance of one of our three kinds of genothreptic combination (p. 233) can scarcely be considered accomplished.

What may be considered our fifth design for a developmental approach calls for some *preliminary* recognition, by such an approach as Vandenberg's and Loehlin's, or by Hs determined by MAVA, of variables marking factors that, respectively, might well be concerned with largely genetic or largely threptic factors. Since both types of factor structure are known to show short time fluctuation (with what relative magnitude we do not know), the P-technique proposed here would take its measures as averages over occasions within each of a succession of, say, 3-month intervals. Also, since we are interested in *common* factors, the scores would be averages across a group of people. Third, since the experimenter cannot wait for, perhaps, 100 three-month points (a 25-year wait), observations would have to be cross-sectional rather than cursive, as in Figures 6.4, 6.5, and 6.6.

It will be recognized that in this design, we should not be splitting a single curve into genetic and threptic parts, but making an initial guess (from H values as earlier) that some variables will mark largely genetic factors and others largely threptic. What we could hope to get from P-technique would be support or nonsupport for this original hypothesis that certain variables mark largely genetic and others largely threptic factors. The check would come in terms of their developmental courses and response to environmental institutions.

The *P*-technique approach (but also other longitudinal approaches permitting distinct factor plots) could also throw light on eidolon and other relations of genetic and threptic factors. For example, the investment theory of crystallized intelligence says that g_c is a product of a largely innate g_f and its cumulative investment in the more complex learning of life and school. Since these two large and readily identifiable factors, g_f and g_c, have long been well isolated factorially and their longitudinal courses well plotted (Baltes & Schaie, 1976; Horn, 1972a, 1975, 1978; Horn & McDonald, 1980), they can well be studied as a trial of this approach. However, we cannot actually take our life curves from *P*-technique but, instead, assuming their factorial natures to be well defined, we will look at curves reached by cross-sectional studies as in Figures 6.4, 6.5, and 6.6.

As to their general nature, fluid intelligence, g_f, as measured by present devices (culture-fair intelligence tests) has more genetic than environmental origin, whereas g_c is significantly more environmental, but of neither can a pure genetic or environmental origin be asserted. Nevertheless, we take the position that except for physical brain damage, special physiological deprivation, and sheer lack of ordinary exercise, g_f is a pattern of abilities acquired by maturation. The volutional curve as shown by culture-fair tests is approximately as shown in Figures 6.3 and 6.7.

Now the *investment theory* (Cattell, 1971b) of the rise of the crystallized intelligence, g_c, structure is that it arises by the investment of g_f in complex learning material, as in school, and that its unitariness arises from the unitariness of g_f and the uniformity of the environmental exposure to socially demanded skills, acting in conjunction. Granted a fairly constant school learning situation for all individuals, the level of complex judgmental skills (e.g., in synonyms, mathematics) that an individual can acquire in a given year will depend on his mental age in g_f plus the effects of interest, etc.

The correlation among, say, 20-year-olds, between scores on g_f and g_c will depend on the uniformity of education (including common interest in education) and the length of time the various individuals have had in school to invest their g_f capacities. From 12 to 18 years this *r* reaches and remains at about .5, and if g_f were wholly hereditary, and if the g_c "achievement" had no component from other personality factors with hereditary origins, we should therefore expect an *H* of .25 for g_c. Actually *H* for g_f is lower and *H* for g_c is probably higher, through multiple determiners not considered here.

Before setting out, in Figure 6.7, the developmental curves that the investment theory would lead us to expect, let us recognize that the empirical testing of the theory by existing alleged tests of g_c is not really possible. The uniformity of a content of complex skills ends with high school. As people go into various careers their investment of g_f veers from the common school curriculum into judgmental skills as diverse as those in handling automobile engines, law, social work, etc. Designers of traditional tests of g_c, such as the WAIS and WISC, have not responded to this challenge but have re-treated to the relatively safe ground of continuing to use the content of the last age at which people experienced a common content, namely, the end of high school. But this is not truly safe, for that content of skills suffers age attrition by memory loss. (Studies in Britain show decline even in vocabulary in young women confined to domestic life.) Nothing short of the Herculean task of developing equivalents for, say, 40-year-olds, in different occupations, classes, and countries, could tell us what happens to g_c through adult life, and that task has not been undertaken. Nonetheless, let us pursue the theory and see what happens as far as can be checked.

It is rewarding to pursue the notion of investment of a genetic endowment in environmental experience in the greater detail and the special contact with psychological experimental realities which the case of fluid and crystallized intelligence permits.

The psychological realities which we shall incorporate in the model, as drawn in Figure 6.7, are as follows:

1. That the curve of fluid intelligence reaches an early biological maximum around 15–16 years of age (see Figure 6.3) and falls steadily thereafter, such that in the average man or woman it stands at about a 10–12 year mental age (g_f units) by 70 years of age.

2. That in each year, y, an increment occurs to crystallized intelligence which is equal to the level of g_f in that year, namely, $g_{f,y}$ multiplied by a value m which represents *engramming capacity,* that is, the capacity to commit a perceived idea to memory. This m can be considered part of a more general power of forming engrams, especially involved in the *imprinting* phenomenon, and which covers both cognitive and emotional effects of experience. That power seems to decline from birth. We suggest there is biological evidence, in, for example, the rate of tissue regrowth, that this power declines as an exponential function of age. This natural value we will call q. Then engramming capacity at year y would be

$$m_y = q^{-y}. \tag{6.17}$$

(We could put some constant before q but it would merely be a scaling change. Also we have considered an alternative decrement rate such that $m_y = q.y^{-1/2}$, which proved less satisfactory.) If we now combine m_y (Eq. 6.17) with the capacity to handle and entertain complex ideas (which is the fluid intelligence level $g_{(y)}$) in an area of general experience in year y we would obtain the increment of g_c in the year y which can be designated ($l = learning$, increment)

$$l_{g_{c,y}} = m_y g_{f,y} . \tag{6.18}$$

Substituting (6.17)

$$l_{g_{c,y}} = q^{-y} g_{f,y} \tag{6.19}$$

And when g_c is accumulated to a year x

$$\sum_{y=1}^{x} l_{g_{c.y}} = \sum_{y=1}^{x} q^{-y} g_{f.y}. \tag{6.20}$$

   3. The level of g_c in any year is the cumulative value of increments over all years to that point. However, we now must suppose a third influence, namely, an attrition of whatever is stored in memory, which may be further considered as either an actual loss of engrams or a decline in power of recall. This for initial simplicity, we shall consider in one sense constant over age. That is to say it causes the loss of a *fixed percentage of the total* [reached in Eq. (6.20)] *at the given year.* [Eqs. (6.20) plus (6.21) state what occurred in the piscatology professor who forgot the name of a fish whenever he learned the name of a new student.] The process will be designated as having a percentage rate k, and we may note that in further developments we propose to give this value trials in the model as a linear function of time ($k_y = yk$). As it stands (a timeless percent) the decrement, d_x, up to year x would be

$$d_x = \sum_{y=1}^{x} k l_{g_{c.y}} = k \sum_{y=1}^{x} l_{g_{c.y}} \tag{6.21}$$

using x as the year to which individual years, designated y have accumulated, where $g_{c.y}$ is the investment (c for learning) gain in year y as in Eq. (6.18), and which we may now substitute as

$$d_x = k \sum_{y=1}^{x} m_y g_{f.y} = k \sum_{y=1}^{x} q^{-y} g_{f.y}. \tag{6.22}$$

The balance of increment and loss is the actual level of crystallized intelligence at year x, as follows:

$$g_{c.x} = \sum_{y=1}^{x} q^{-x} g_{f.x}(1 - k) \tag{6.23}$$

Alternatively to this conclusion to (3) we can experiment with a tendency for k to *increase* linearly with age:

$$g_{c.x} = \sum_{y=1}^{x} q^{-x} g_{f.x}(1 - kx). \tag{6.24}$$

   In Figure 6.7 the plot is given, however, for (6.23), using the g_f values in the given curve, together with values for k (.10), and q (1.06), which best give what seems the currently most favored (Baltes & Schaie, 1974; Horn, 1976, 1978) values for the resulting curve of g_c.

   The currently most accepted curve for traditional intelligence test scores in Figure 6.7 is not that of Jones and Conrad (1933), C. G. Miles (1934), and others, which dips somewhat as g_f does (Cattell, 1971a, p. 159), but rather the curve for which Baltes *et al.* have argued, from *cursive* (not cross-sectional) measures, as showing virtually no drop after the plateau reached at 20 years (interrupted line). The curve for g_c constructed on the

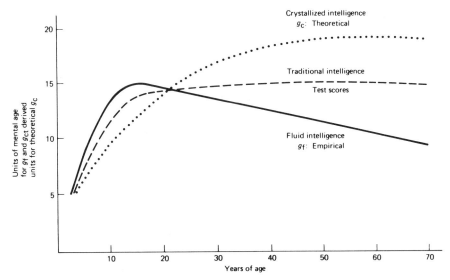

Figure 6.7 Derivation of phenotypic from genotypic maturational process illustrated by investment theory of fluid into crystallized intelligence. Age: 5, 15, 25, 35, 45, 55, 65, 75; g_f values: 9, 15, 14, 13, 12, 11, 10, 9; g_c values: 6.7, 13.0, 16.2, 17.9, 18.8, 19.3, 19.5, 19.6; g_{ct} traditional intelligence test curve follows general argument of Baltes and Schaie (1974).

other hand on our theoretical model and with the constants we have chosen, obviously suggests continued gain to 40 or 50 years. This curve is one that politicians, who claim to direct complex matters best in their sixties and seventies, would like to believe in! The fact that traditional test scores are just about the mean, throughout the entire age range, of g_f, and of what g_c might be expected to be if we could measure it with a good theoretical foundation seems to support my own contention (1971b) that tests like WAIS and WISC are mixtures of g_f and g_c factors. As indicated, the diversity of people's domains of learning after high school presents problems for meaningfully measuring the true age curve for g_c that have never yet been faced, so Figure 6.7 must remain, in part, theory. (Note that this theory requires a late drop in g_c.)

This theory of g_f and g_c, which illustrates also the eidolon theory of rise of observable unitary traits, leads to inferences that may be checked in other ways. First, if we can measure the trait and its genetic part separately, we should expect a population correlation between them, changing with age (g_c and g_f rise to an r of .5–.6 in late high school which apparently falls thereafter). Second, we should expect a relation between H values and the population age cross section at which they are determined. In this connection a brief panoramic glance at human performances is helpful. As noted by Lehman (1936) and others, "peak performance" as seen in everyday life is reached at different times in different areas. Gallois and Newton illustrate the well-supported fact that in mathematical performance the peak is close

—perhaps by a 6-year lag—to that of fluid intelligence, as these geniuses were most productive at around 20–24 years. In this respect, mathematics contrasts with, say, biology or history, because it is contentless, involving pure abstract reasoning which is the essence of g_f, whereas history and, say, Darwin's perception of evolution, require "wisdom." Some of the world's best "wisdom" writing has been done by authors in their seventies. Plutarch, for example, wrote the fascinating observations on human nature in his *Lives of the Noble Grecians and Romans* while in his seventies or eighties. In contrast, Newton was said by some contemporaries not to have understood in later life some of the methods and equations he developed when he was 21. (He had the good sense to turn to practical improvements in monetary matters, as Master of the Mint!) As indicated earlier, the age-trend question as regards abilities has been excellently developed in the last decades in the researches and mutual debates of Horn (1976, 1978) and Baltes and Schaie (1974) and recently advanced further by Cattell and Brennan (in press).

This amount of delving into the particular case of intelligence is appropriate because it gave us one of the few empirically firm and measurably precise bases for generating ideas in the next section, otherwise relatively speculative. It is also apposite to Section 7 of this chapter inasmuch as it shows for the "eidolon" model there described *why* the factoring result is sometimes one factor and sometimes two, and illustrates the causes of changing correlation between them in the latter case.

5. The Use of Comparative MAVA in a Developmental Setting

In the models—such as that of PLA—considered so far we have recognized the greater exposure to certain environmental influences occurring at certain ages, and the effects that this will have on heritabilities calculated for different age groups. But we have taken little note—except in the case of crystallized intelligence—of interaction effects which cause the gain to be different when the same amount of learning experience occurs at two different ages or levels of the volitional curve.

The literature of developmental psychology is now replete with instances of such interaction effects, some of them well quantitatively documented, principally under the rubrics of "critical periods" and "imprinting." In school, for example, children who begin to read at 4–5 years and others at perhaps 7, may be indistinguishable in level at 9 years of age, and in this case the reason is largely a cognitive one, as in our analysis of g_c progress. Personality theory, while not ignoring the cognitive, is in this matter more concerned with the greater retention of earlier *emotional* conditioning and cathexis, or of learning at the critical period when some drive, for example, the sex erg at adolescence, shows a maturational spurt.

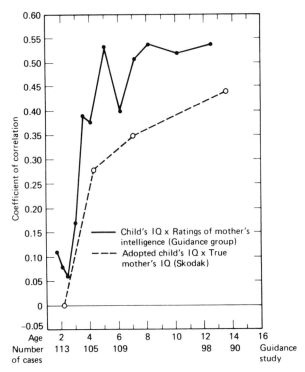

Figure 6.8 Possible indication of greater imprinting effect in earlier years. The correlation, at different ages, between true mother–offspring ability for children reared by biological parents (guidance groups) and for children reared by adoptive parents (Skodak's group). Experimental data of this type (from Honzik, 1957, p. 220) on Skodak's material could assist methodologically in getting at magnitude of "imprinting" effect—the m value in Figure 6.7.

It is obviously not possible within the compass of this book to examine and weigh all the evidence in developmental psychology on such periods. In any case the presently available developmental researches practically never contain independent evidence on the genetic curve, but only show different learning rates at certain age periods. The complexities of analysis are great and evidence conflicting. For example, though schizophrenia and manic-depressive disorders show susceptibility to manifestation characteristically at adolescence and early middle age, respectively, Gershon (1976) found no resemblance of relatives in age of onset of the latter (within the range indicated). One of the main reasons for complexity of analysis of heritabilities calculated at different ages is the role of cumulative effects—integrations of effects over time. Eaves and Eysenck (1976) have examined differences in anxiety ("neurosis" scale) which seem to show accruing environmental effects. We discuss elsewhere whether H values in general would be expected to get greater or less with age, and recognize that in psychological measures we as yet have no firm evidence. Figure 6.8 *suggests* that in the intellectual domain the influence of environment (or interaction with given maturation) is greater earlier. Much more solid data exists on physical measures and Rao, MacLean, Morton, and Yee (1975) show clearly that in weight and height, the age-corrected values show an increasing correlation, with the passage of time, for parent and child, but a decreasing one for sib and sib. McAskie and Clarke (1976) find similarly. We need pause only to note here that the causal analysis admits of several alternatives.

The approach to such problems which follows most naturally from the methods of analysis into abstract, contributory variance and variances used here that we have been alluding to as *comparative MAVA*. Comparative MAVA belongs in the convarkin class of methods and compares variances, covariances, and heritabilities as obtained by applying MAVA to samples of populations from (*a*) different racial–cultural groups, (*b*) different sexes, and, especially (*c*) different ages.

Let us suppose now that we are concentrating on comparing variances and heritabilities obtained with the same trait measure on populations at a series of age intervals, say, at 5, 10, 15,. . ., 60 years of age. Our question concerns what the relations of the variances and heritabilities would be expected to be to such underlying threptic curves as we have just examined in intelligence, or other traits. In passing we may note that it happens that the beginnings of such an approach have appeared in that "fragmentary MAVA" that has been keenly developed in the "adoption method." There the question has naturally soon been asked whether greater length of adoption produces, as it theoretically surely should do, a greater development of threptic but not genetic resemblance with time. The results in Figure 6.8 have been interpreted as showing that some decline in resemblance to the foster mother, relative to the bio-mother, occurs with increasing age. This may show (in accord with the usual early age imprinting theory) that the foster home environmental effect can be more powerful in the early years, but falls behind relative to the steady march of maturation-based differences. (Some other hypotheses might also fit.)

The advance from an empirically obtained set of genetic and threptic variances, etc., across a series of population ages, to the inferences that follow about genetic–environmental interaction is not simple. In general, perhaps, it is safe to depend on some constancy of the *coefficient of variation,* so that a demonstrated increase of variance means an increase in mean, and vice versa. Thus, in intelligence, if the *absolute* contribution of g_c to a given performance, relative to the contribution of g_f, increases markedly with age, as Figure 6.7 suggests, we should expect the variance contribution also to increase. In studies treating intelligence as a single g factor ($g_f + g_c$), therefore, our prediction would be that H would diminish with age, though both ingredients, g_f and g_c (as σ_g^2 and σ_t^2), would increase in variance, at any rate up to 15 years.

Some contribution to this question could be made, incidentally, by comparing the ages of peak performance in various areas where opportunities for learning change little with age. Records, for example, in Olympic athletic events, and in tennis, skiing, boxing, chess, mathematical contributions, scientific research, political leadership, and authorship, show various lags beyond 21 years in peak performance. Boxers, as representative of the physical skills, gain sufficient threptic additions of motor and tactical skill for their peak to fall in the late twenties, since in the twenties these skill additions come faster than the volutional decline. The common generalization that intellectual powers peak much later than physical powers is, as we have seen, oversimplified. Powers in the content subjects, which rest on cumulative investment of capacity, rise steadily and fall quite late relative to some slight steady decline in the abstract subjects, like mathematics. In the *content* subjects, therefore, we should expect more threptic variance later than earlier.

The expectation of variance contribution from the genetic source is fairly straightforward, but that from environment we shall need to discuss at greater length. Our initial discussion has been simplified by assuming learning from environmental experience to be a steadily continuing "opportunity of learning" with cumulative effects. But in real life the traffic is very uneven. In the next section we shall consider breaking this total effect down into effects of separate cultural elements and separate times. But in the simplest conditions of noninteraction, noncorrelation, and an environmental variance about equal in effect to the genetic variance at biological maturity, say 21, we should perhaps expect an early childhood low heritability, due to maturational differences not having had opportunity to stabilize, followed by a high H as the genetic variance increases rapidly to maturity, followed by a slow lifetime reduction in H as the cumulative effect of environmental differences produces a predominance of the threptic over a no longer increasing genetic variance.

If there is interaction—say, as a *product* of g and t as in the investment theory of intelligence—an increase of the absolute level of *g* will multiply the increasing *t* variance to produce an increased total observed variance. If H is calculated in the gross way (see Chapter 7, p. 269), there will be an increase or decrease in H depending on the size of the change in the mean of t relative to that in the sigma of g. If there is not interaction, but only covariance, as in our main MAVA model, an increase in mean t without change of variance will leave H unchanged. However, we have accepted the probability that a change in the mean level of environmental effect is in general likely to be accompanied by some change in its variance. Most increases of *mean* in psychological data tend to be accompanied by an increase of *sigma*, except when the population is pushed toward a ceiling.

Regarding interaction effects we must, furthermore, keep in mind the clinical and animal experiment observation that an *earlier*, and/or *more appropriately biologically timed* environmental experience may be far more powerful than the same impact at other times. The weight in the genetic–environment interaction will come out as a weight affecting both r_{gt} and t itself. It would seem that—particularly in that massive emotional learning (as in the APA model, Cattell & Child, 1975) which is of most concern to *personality* psychologists—the following facts have to be integrated.

1. Environmental engrammings are more powerful and lasting in early life. This may not be a simple linear function of inverted age score. Particular periods, for example, that of suckling and, in humans, of toilet training, may show the effect as some specially weighted function of the age score.
2. The cumulative effect of lengthy life experience may be expected to increase the *absolute* contribution of environment with increasing age, but to reduce threptic population variances. (One recognizes that the simple phenotypic parent–child correlation increases with age [McAskie and Clarke, 1976].)
3. The consequences of the individual's own decision (especially as in a Freudian "repetition compulsion") build up a selection of encounters, often proceeding steadily in some direction. Thus the environmental samples accumulated by individuals from the total environment are far from random and do not permit us a model in which the environment is a passive set of situations to be randomly encountered. (Our econetic model [p. 192], incidentally, already approaches a more discriminating treatment of the last effect.)

To base a theory of age change of *H* merely on, say, this assumed constancy of the *coefficient of variation* (σ/M) in threptic accumulations, is perhaps untenable, and we propose a more detailed model which formulates from the preceding discussion four influences as follows:

1. *The Imprinting Principle.* A given environmental experience has a greater influence at an earlier than a later age.
2. *The Equity Principle.* This principle supposes a cultural population of environmental elements that can be encountered between birth and old age. Incidentally it will not affect our main argument whether we consider each element to have only *positive* influence on a trait, scoring thus 1 or 0, or (as is more probable) either an augmentative *or* reductive effect, scoring $+1$ and -1. By the equity principle we mean that the older person is treated by environment with greater equity than is the younger one. In spite of the repetition compulsion luck tends to even out. Statistically we are saying that the younger person has a smaller sample, namely n_1, than the older person, n_2, so that the variance of the *mean* effects, if sigma is the variance of the elements, is $\Sigma^{n_1}d^2/n_1$ in the young and $\Sigma^{n_2}d^2/n_2$ in the old. The absolute threptic variance in the old would thus be less, as the variance of means is less with larger samples. Consequently individual threptic differences would fall and *H* would rise.
3. *The Stochastic Process Principle.* A psychologist will recognize at once that the principle in (2), though true as a main effect, shall require some

correction for what we have alluded to as a repetition principle. For thereby the equity of environmental experiences, as argued, becomes modified. The intelligent lad who works hard has a much higher chance of encountering a college experience than one who does not. The lad who happens to get into bad peer company may go widely astray. The dictum, "It's the first step that counts," summarizes this principle. A stochastic Markoff process is one in which events at, say, Stage 3 can be in part predicted from those in Stages 1 and 2. The statistics of stochastic process must therefore be invoked to handle this effect. It is necessary to separate here the effect of one *experience* (a threptic trait) upon choice of the next, and the effect of *genetic* trait factors upon choice, which will be separately considered in (4).

A special subdevelopment within the stochastic model is required by the fact that the culture provides several "self-contained" types of life channels, for example, different occupations, which narrow the likelihood of meeting certain experiences. Clergymen have a subnormal probability of getting into a prison gang fight. The existence of such partly segregated, institutionalized channels will, like the stochastic process in general, increase the magnitude of threptic variance produced by age from that expected by the main model.

4. *The Principle of Genetically Derived Threptic Variance.* Biometrics seems not quite to have decided what to do with genothreptic interaction and covariance terms when it comes to stating heritability. We shall deal with this statistically in Chapter 7. Meanwhile we have recognized, in the eidolon developmental model, and concretely in the evidence on developmental curves for fluid and crystallized intelligence, that both interaction and covariance can be appreciable. What covariance does to heritability values hinges on a subjective conceptual question of how we prefer to define heritability (Chapter 7, p. 269). Handling interaction similarly sometimes depends on how we define a trait. Is crystallized intelligence, g_c, a new trait or is it the threptic variance associated with fluid intelligence? If the latter, (as would be in cases where so definite a new factor emergent as g_c does not happen to appear) then threptic variance is a function of genetic variance plus or times experience. If, by the equity principle, experience evens up the $g \times t$ product, once again the heritability will increase with age.

The main inference we therefore offer for testing by experimenters with comparative MAVA or any other method determining variances and heritabilities of groups at different ages is that at least after adolescence heritability should increase with age through a reduction of threptic variance, despite increase of threptic mean, if effects of (1), (2), and (4) outweigh (3).

The folklore belief that "John grows more like his father with age" could be dismissed with the comment that similarities cannot be recognized easily except when they reach the same age context. But, on the other hand, it may well represent a reality of the ecogenic curve. There are also, naturally, epogenic effects—between 1920 and 1960 John was generally taller than his father for nutritional reasons. But, in variance terms relative to the given generation, the hypothesis we present is that similarity due to environmental influences should increase, while genetic variance remains essentially constant. The finding of Rao and others of declining H with age for intelligence could well be due to poor adult intelligence tests.

6. Learning Calculations Refined: The Contributions of Particular Environmental Elements to Pure Threptic Measures

We have tried to point out that behavior genetics can contribute almost as much to learning theory as to genetics, and we shall illustrate this further in the present section. In discussing econetics and path learning analysis we have learned that change during life in a particular trait can be analyzed into weighted effects from each of the many common life path encounters that people in a given culture experience more or less.

However, it was insisted that in life experiences of ordinary (not laboratory) learning the *change* score is not merely *learning*. Indeed, the student of learning theory seems to need fairly often to be reminded that the great majority of learning experiments make an error that would be fatal to their scientific value if they were not saved, like a poor boxer, by the bell of time. Only their brevity saves most bivariate, manipulative learning experiments from the charge that what they call learning gain must be considered, as discussed earlier in this chapter, as really a *mixture* of learning gain and genetic maturation.[5] The 1-week laboratory experiment is actually safe from the accusation; but many studies that use its model unchanged for real-life learning ("macroscopic") experiment—often involving primary personality and ability factors, which change only slowly—seriously confound volutional change with the learning gain. This distorts any extracted law that may attempt to relate the gain to, say, measures of certain prolonged clinical conditioning procedures, as with children in therapy. Thus educational experiments (Cronbach, 1980) with different teaching methods on small groups, not genetically equated (e.g., an experiment concerning a course in arithmetic or 8–10-year-olds) would show a learning gain almost certainly confounded with ability maturation. At the opposite end of the life span, a measure of learning of botanical names for 80–85-year-olds would be confounded with involution of memory capacity.

Incidentally, though our concern here is with different genetic growth rates (a) being confounded with, and, (b) differentially affecting, learning at different ages, it will be recognized that throughout this chapter we are also

---

[5] This is unquestionable in repeated measures designs, where gains of individuals are related to some independent variable, but could occur also in matched groups where the matching does not explicitly extend to known genetic characters.

concerned with the more obvious fact that interactions affect the calculations of learning by individuals of different genotype. The immediate learning laws and the constants in them must be different for the same experiences falling on different genotypes. Though the defects in conditioning laws through total neglect of genetics by reflexology may not be great enough to justify Hirsch's (1967, p. 421) designation "The 50-year fiasco that was behaviorism" yet in principle, classical reflexological learning theory is far behind structured learning theory (Cattell, 1980), in preparing to handle these problems.

At any rate in terms of developmental interaction we must see what we can do to avoid confounding learning and maturation. Let us consider some methods that promise separation of the threptic, purely learned gain. In principle, the four methods of curve splitting and the longitudinal (*P*-technique) factoring given earlier apply equally to data on the *mean* of a group and on a *single score* for an individual. But mainly we have thought of life span analysis for typical cultural groups, because of the larger error (irrelevant unknowns) in individual measures, not to mention the unlikelihood of an experimenter of today starting a 60-year-long individual *P*-technique experiment!

The further methods to be discussed in this section depart from those already discussed in making one of their primary objectives the hitherto scarcely touched problem in behavior genetics of asking how much of a measured trait is, *in the case of a particular individual,* due to heredity and how much to environmental experiences. The two curve splitting methods yet to be discussed, and their attack on individual evaluation, require us to utilize again the concepts in PLA (path learning analysis) (p. 190). The central principle, it will be remembered, addresses itself to the reality that the individual in life is subject, as regards learning gain in *any* trait, to *several* influences acting simultaneously. For example, the increase during late adolescence of ego strength comes partly from maturation, partly from experience of successfully handling a job, partly, in some, from learning effective adjustment compromises in marriage, and partly from other environmental experiences.

As we reach discussion of the learning component in development we realize that it is not enough simply to ascertain the *total* threptic variance contribution. In the interests of further scientific advance—notably in estimating the pure learning gain in curve-splitting—we want to know also *what each of the several environmental elements contributes.* Naturally the quantitative evaluation of what each element "weighs" will take into account the correlations among these environmental elements in their impinging on the individual.

In early attempts to get at these generators of the threptic component psychologists have actually been accustomed to use as the predicted criterion, the learning gain due to the experiences—what is really *learning gain plus a fairly small and constant genetic gain.* But we have stated as our ultimate aim in structured learning theory (Cattell, 1980a) the transformation of this biased estimate of trait–environment association to one of correlation of the *given* environmental element with *the gain purely in the threptic component of the trait, only.* (This can also be viewed as finding the portion of the curve of increase in the acquired threptic component of that trait that can be "sliced off" as due to the experience of that *one* particular environmental feature.)

Assuming linear regression of that gain on the amount of exposure to the learning element the first possibility that naturally occurs to the psychometrist is to employ partial correlation. Let us designate the correlation (across individuals) found between score on the amount of teaching by (or measured environmental experience of) an element x, and the resulting trait gain or level (gain from zero) as $r_{e_x(g+t)}$. Here e_x is the amount of experience of the specific environmental element, and g and t as usual are genetic and threptic gains simply summed to the total observed gain for trait y. We shall assume that experiment has already provided the ratio of g to t variances, from the research on the heritability of y, so that we know $r_{g_y(g+t)_y}$ from

$$r^2_{g_y(g+t)_y} = \frac{\sigma^2_{g.y}}{\sigma^2_{g.y} + \sigma^2_{t.y}}. \tag{6.25}$$

It will be noted that this involves an assumption that the genetic–threptic ratio (or heritability) for the variance in absolute levels, at a given moment, also holds for the gain magnitudes, which is approximate because neither the learning nor the volutional curves follow simple linear equations. A way out would be to apply MAVA to *change* scores at that age, instead of to the usual absolutes. In connection with handling change scores I would, incidentally, repeat the warning I have substantiated earlier elsewhere (Cattell, 1966; see also Nesselroade, 1967) against tampering with change scores by partialing out initial or final levels from gain between the repeat measures.

The third correlation, $r_{e_x.g_y}$, that one would like to know, for any partialing out operation we can fortunately assume to be zero. For although genetic and threptic deviation values *may* get correlated, as in the covariance terms in MAVA equations, such covariance is often shown to be absent in maximum-likelihood fit (Chapters 9 and 10), and most of any correlation that may exist does not arise from the single cause of a genetic characteristic generating an environmental characteristic as would be involved here. We are saying that a person's genetic endowment in trait y is not likely to be much related to the experience he or she is about to have on path x. For example, as school children sometimes painfully realize, there is no providential correlation of, say, the individual's genetic component in spatial ability with the amount of teaching he or she gets in geometry. On the other hand, if the source trait in question were intelligence and the environmental element were parental salary (or SES) we know that a low but significant $r_{e_x g_y}$ relation (about + 0.2) does exist. Yet generally we adopt the simplifying assumption that the correlation $r_{e_x g_y}$ is zero (6.26a), because it is zero at least in many known empirical cases. (Nevertheless, desire to avoid this approximation has moved us to introduce in what follows an alternative, independent approach, by comparative MAVA.) Meanwhile, we proceed on this basis to the partial r, as follows:

$$r_{e_x t_y} = r_{e_x(g+t)_y g_y} = \frac{r_{e_x(g+t)_y}}{\sqrt{1 - r^2_{g_y(g+t)_y}}},$$ (6.26a)

which can alternatively be stated as

$$r_{e_x t_y} = \frac{r_{e_x(g+t)_y}}{r_{t_y(g+t)_y}}.$$ (6.26b)

In either case we know the numerator from PLA (or sometimes less sophisticated) experiment and the denominator from a predetermined heritability, namely by:

$$r^2_{t_y(g+t)_y} = 1 - H.$$ (6.27)

It will be understood that we are considering the calculation here for a single environmental element x in its effect on trait y. Our hope is to reach a multiple R with such known weights, and so many elements—x_1, x_2, x_3, etc.— that $(1 - R^2)$ will become very small, and permit us to estimate most of the threptic score from knowing scores on the environmental element exposures.

A second, independent approach to the estimating a person's g_y score through his t_y subtracted from the observed y rests on comparative MAVA which uses the statistical principle (McNemar, 1962, p. 399) that if we have several groups, which can be considered random samples from a population, and measures therein on two traits, a and b, then the correlation r_{ab}, *within* the population (and the central value within the groups) will be the same as the correlation $r_{\sigma_a^2 \sigma_b^2}$ of the variances across the groups. Thus to determine r_{ab} we must be able to calculate the threptic and genetic variances for each group. The essentially normal distribution obtained by Cattell, Breul, and Hartman (1952) and Cattell, Graham, and Woliver (1979) for variances of cultural measures on up to 120 countries, suggest such groups are random samples from a larger population, justifying hope of usefulness of Eq. (6.28). However, there is no doubt whatever that type groupings exist among national cultures (Cattell, 1950a; Cattell, Breul, & Hartman, 1952; Cattell & Brennan, in press), and although this does not deny normal distribution on single variables (Cattell, Coulter, & Tsujioka, 1966), one might do better to use other than national groups for this approach. The equation:

$$r_{e_x t_y} = r_{\sigma^2 e_x \sigma^2 t_y}$$ (6.28)

on this principle certainly opens the way to obtaining the relation (as a linear regression) of environmental experience measures to threptic gain as such, cleared of volutional contamination. But the data required for it—namely, variance values on community environmental variables (years of education, amount of sport participation, etc.) and population psychological "dependent" variables—across an adequate sample (100?) of societies belonging to the same genus—may prevent its use until social and genetic research is better endowed.

It is illuminating in methodological terms to reflect upon the fact that many reflexologists, disparaging and neglecting attention to heredity, have pointed to the variability of H from culture to culture as proof of its uselessness. As we have just seen in comparative MAVA, that variability is itself the gateway to new realms of knowledge.

Supposing now that $r_{e_x t_y}$ values have been obtained by one of the preceding two methods for a sufficient number of x's (i.e., elements of the cultural environment), we shall find that several valuable new avenues of research in social and clinical psychology are opened up.

First, in learning theory (broadened and advanced to structured learning theory) the regression plot (from $r_{e_x t_y}$) of the amount of environmental experience of type x on the real gain—t_y, not $(t + g)_y$—in threptic addition to a given trait offers a better basis for understanding the learning process involved. Of course, being born of a calculated regression value, the "learning curve" *has* to be linear, but a nonlinear curve is available as a succession of linear solutions at various levels. Personality and social psychologists have here a means to relate actual gains in personality and ability traits to quantified levels of influences in the social environment, without the maturational contamination which presently invalidates virtually all of their findings. From such results one would be able to build up matrices parallel to those in PLA (Figure 6.1)—now with the improvement, however, that the learning law matrix, L, would contain values (for the general population) that are more accurate, by being freed of error from intrusion of maturation.

A second advance lies in being able to discover how much of the threptic change in a trait has been accounted for by the range of environmental elements and path experiences already considered, and how much still requires exploration of yet unknown or unevaluated influences in the environment. Of course, the influences, x_1, x_2, x_3, etc. as they became known as significant predictors will naturally be found to be to some extent mutually correlated. For example, the salary of parents and the child's number of years of education would be expected to be positively correlated, and the number of illnesses and the amount of participation in sports might be negatively correlated. These correlations are readily ascertainable from social data and permit transforming the $r_{e_x t_y}$ values to beta weights, thus

$$R^2_{y(x_1, x_2, \ldots, x_n)} = b_{x_1} r_{e_{x_1} t_y} + b_{x_2} r_{e_{x_2} t_y} + \cdots + b_{x_n} r_{e_{x_n} t_y}. \tag{6.29}$$

Since R^2 here is a statement of the fraction of the threptic variance of y that is accounted for by known environmental elements, the gap between R^2 and unity indicates the extent of the challenge to the social psychologist to explore further.

The third and last gain to which we will draw attention is a surprising one, but one that, as research progresses, could be of great *practical* value to clinics and schools. If we know, from biographical records, the values of x_1, x_2, etc. for *a given individual,* we can estimate, by R, the amount of trait y measure *for this individual* that is threptic in origin. Of course, to be practically useful, R would have to have been raised, by organized sociogenetic research, to a respectably large figure. Now, up to this point, most methods and concepts in behavior genetics have had to do with *population values, as averages.* Trait heritability values, if applied to individual members of a group, will merely put the estimates (by regression) of their individual genetic endowments in the same order as their phenotypic measures. With the usual H of about .5 there will be considerable error of individual estimates. If now, however, through biographical knowledge of the individual's past environment, and of the standard regression values and weights in Eq. (6.29) we could achieve an R to make a reasonably good estimate of the individual's threptic component, the magnitude of the individual's genetic component in the trait follows immediately by subtraction.

Hitherto it has been only in the case of a few discrete syndromes, whose gene basis has been exposed by genealogical, pedigree methods (McKusick, 1968), that the clinician or educator could be reasonably sure about something in the genetic part of an individual's make-up of which he or she should be taking due account. But here, if R is large enough, and the individual's past environment is known, it is possible (*a*) to speak not only in general terms of, say, a 65/35 population N value for intelligence, but of the ratio of genetic to threptic components in the *intelligence make-up of a given individual;* and (*b*) to say what fractions of that threptic score came from various distinct sources.

A new possibility that deserves brief mention in relation to relating personality to culture is that resulting from the factoring of national cultures, and the definition of measurable dimensions that could then be related to personality mean scores of the actual populations. An introduction to this syntality analysis has already been given in Chapter 5. Where correlations are obtained, as in the findings of Cattell, Woliver, and Graham, 1980, the task would still remain of deciding whether the culture traits produced the personality traits, or vice versa, and, if the latter, how much is due to the genetic components. Incidentally, as has been pointed out, the analysis of variance design suggested in Table 5.20 treated each culture as a single total pattern and examined, for a given trait as dependent variable, the relative variance across national populations due to racial–genetic variances and culture variances. Over the past 25 years, however, programmatic factor analytic research on national characteristics have been made, covering over 100 countries as entries, and some 80 social, economic, medical, and political variables (Cattell, 1949; Cattell, Bruel, & Hartman, 1952; Cattell, Graham, & Woliver, 1979; Rummel, 1972). At least eight major dimensions of

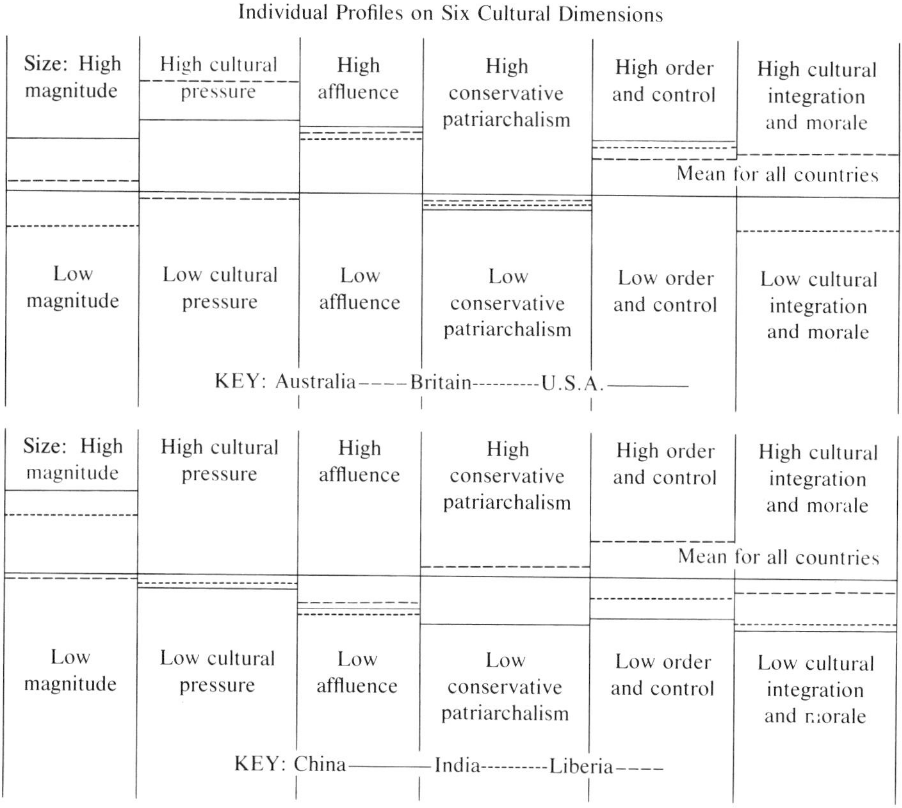

Individual Profiles on Six Cultural Dimensions

Size: High magnitude	High cultural pressure	High affluence	High conservative patriarchalism	High order and control	High cultural integration and morale
					Mean for all countries
Low magnitude	Low cultural pressure	Low affluence	Low conservative patriarchalism	Low order and control	Low cultural integration and morale

KEY: Australia----Britain----------U.S.A.————

Size: High magnitude	High cultural pressure	High affluence	High conservative patriarchalism	High order and control	High cultural integration and morale
					Mean for all countries
Low magnitude	Low cultural pressure	Low affluence	Low conservative patriarchalism	Low order and control	Low cultural integration and morale

KEY: China————India---------Liberia----

Figure 6.9 Illustration of syntality profiles of nations, with typological groupings.

syntality, upon measures of which a profile can be constructed for any given nation, have proved stably replicated from study to study and across at least 50 years of history. Among these checked patterns—their characteristics are listed elsewhere (Cattell, Woliver, & Graham, 1980)—are the following:

Population and general size effects
Cultural pressure
Enlightened affluence
Conservative patriarchal solidarity
Morale level
Vigorous, adapted development

To give some illustrative substance to these patterns, the scores on two "types" of country are given in Figure 6.9. The ray of promise in this approach resides in the finding (Cattell, Woliver, & Graham, 1980) of significant correlations between national syntality dimensions and dimensions of personality in the population. It has been known for some time that national

228 Models of Interaction of Learning and Genetic Processes

populations differ significantly on intelligence and personality scores—probably more on the latter than the former. (See cross-cultural studies on particular behaviors, and Lynn, 1977, 1979; Cattell, Eber, & Tatsuoka, 1970; and Meredith, 1965, on factored source traits.) However, it is only recently that significant relations have appeared between *population personality means,* and *characteristics of the whole culture,* for example, of low score on the C factor of emotional stability, and frequency of riots. A factoring of countries racially virtually identical and those differing both in race and culture and analogous to the Loehlin and Vandenberg's (1968) Vandenberg's (1959, 1968) factoring of identical and fraternal twin groups is a possible approach to the next step of separating genetic and cultural factors.

A more detailed account of concepts and methods in this section and the next can be found in my article "Unravelling maturational and learning developments" (Chapter 6 in Nesselroade & Reese, 1973).

7. The Rise of Unitary Trait Patterns by Genothreptic Interaction

The developmental approach in this chapter has mainly asked how trait scores may be expected to change through joint action of volitional and environmental forces. An equally important question concerns how the *unitary structure* of the traits themselves, as shown by factor analytic results, arises by such interaction.

The traits to which we refer, and which are genothreptically analyzed in the data of Chapters 8, 9, and 10, are now the common coin of clinical, educational, and personality psychology—for example, the traits of ego strength, intelligence, superego development, and exvia–invia—as well as of motivational psychology—for example, the ergs of sex, fear, gregariousness, etc. These concepts are the products of multivariate experimental research over the last half century. The recognition that a statistical factor, granted certain conditions of experiment, is a unitary influence, determiner, or cause was propounded by Spearman, Thurstone, and myself in the first half of this century against considerable opposition from Burt and Thomson, as well as many psychologists who were less familiar with the mathematics involved. It seemed to me and to my coworkers that concrete illustrations might win over, more quickly than logic, those who believed that factors were "mere mathematical abstractions" or "reductive conveniences." With this didactic purpose, two examples were worked out, less restricted than Thurstone's box, showing that physical "behaviors" of objects yield the factors which physicists and chemists would call "causes," namely, mass, temperature, volume, elasticity, etc. (Cattell & Dickman, 1962; Cattell & Sullivan, 1962).

The "certain conditions" clause in factor research design is, however, vitally important, since perhaps only one factor analysis in five reaches them. The conditions are the following: (*a*) proof of unique rotational resolution by simple structure or confactor principles; (*b*) replication of patterns across samples; and (*c*) checks on functional unity through the addition of *P*- and *dR* factorings to ordinary *P*- technique. The conception of factor as a determining cause has made considerable progress as reviewed in connection with path coefficients by Cattell (1978b) Cloninger (1981) and more generalized causal models by Bentler (1976).

Now in the present context of seeking out distinct environmental and genetic influences, it is not surprising that the notion has been entertained that a trait pattern could arise from a single large gene—an idea which Royce (1957), Thompson (1967), and others have propounded. Geneticists noted early that a single gene could be pleiotropic. That is to say, it could simultaneously affect *several* bodily features. Given that subsequent observation has revealed that practically all genes are pleiotropic, the adjective as a distinguisher of the character of certain genes becomes superfluous. In the pedigree approach in behavior genetics, instances of such genes occur (e.g., in Huntington's chorea and phenylketonuric mental defect) but, except with incomplete dominance, they have basically an all-or-nothing character. Such a genetic basis cannot be used to explain the continuous variation from polygenic action that we see in such unitary traits as intelligence, surgency, dominance, premsia.

Conceivably the genetic basis of some traits could rest on as few as 6 to 12 loci (Hurst proposed 9 for intelligence), the various combinations of which would produce a rather fine, but still ultimately, in the *genotype,* steplike, set of increments. That form of distribution would then be smoothed over by the numerous and varied touches of a diverse environment to give the appearance of the continuous, usually normal curve that we in fact find. Thus it is possible that relatively small numbers in polygenic action could cause the distributions that we see. Nonetheless, other considerations suggest that larger numbers of genes are generally involved in virtually all the normal unitary traits with which psychology deals.

By accepting polygenic action, however we are confronted with some difficulties—compared to the "one big gene" theory—in accounting for the appearance of a *unitary* genetic trait. Let us consider a trait such as spatial ability, the phenotypic unitary character of which as a primary is well established and which will serve as an example of a trait of high genetic determination. Presumably a dozen or so genes, each pleiotropic, add together in favoring bits of the neural structure underlying spatial ability. To produce an obvious unity in spatial ability only, however, it must be that that they do *not* overlap and add together, systematically, on their *other* pleiotropic aspects, and we must suppose that one such side effect favors, say, a slight thickening of the skull, another some speed of eye movement, another a motor skill in fingers, and so on. For if they added up on other features they might produce a genetic trait of quite a different nature, highly correlated with, or indistinguishable factorially, from, spatial ability. Conceivably there are instances where some manifestations phenomenally quite different will be found to factor into one and the same genetic trait due to some degree of summation of "secondary," pleiotropic effects. For example, spatial ability might go with straight hair, a variable not generally included in educational test batteries! Thus more wide-ranging research in the future may produce perhaps slight loadings of some odd secondary variables on our psychologically familiar factors, when those factors happen to be largely of a genetic nature. Incidentally, when we have polygenic action, the unity of some trait such as a spatial factor could be naturally explained by Godfrey Thomsons' (1951) sampling theory as he applied it to intelligence, that is, by various spatial performances calling on overlapping subsets of the some total polygenic neural resource.

Although we shall not pursue this concept of a gene set defining a genetic unitary trait to its ultimate fastnesses of complication, we have to recognize that there *are* conceptual complications. We need to note that there is a dual level of evidence of structure here: On the one hand, we have a unity shown by correlating phenotypic behaviors, on the other some neural dimension from overlapping action of a set of genes. Genetically there is no real unity, only the chance overlapping or systematic common "crossing-over" (linkage) of genes, producing a certain common behavioral effect resting on neural (or hormonal) structure. As suggested earlier, it must happen that the genes overlap less repeatedly in other bodily features, so that no marked, easily observed continuity of range is produced elsewhere. The brain region (or other physiological organ) in which appreciable range is thus produced then becomes one pre-requisite condition for the appearance of a whole set of psychologically similar, correlated, *behavioral* performances. The unity does not (by this model) necessarily exist in an organic sense that the genes fall on the same chromosome. At most, there might be some initial unity through some linkage of genes in "crossing over." The range of a unitary capacity occurs at the psychological, behavioral level because of additive action occuring in some neural effect that these genes happen to share.

On the threptic side, however, we are much more familiar with environmental conditions that would definitely produce unitary trait structures. As far as acquired dynamic traits—sentiments—are concerned, three main mechanisms have been described for their development (Cattell, 1980a):

1. Common learning schedules where a social institution of stereotyped form simultaneously teaches a whole set of attitude and skill variables, bringing to bear an equal intensity and frequency of reward to all of them. This would readily create a common factor in individual differences. It might, in some cases, produce also a *functional* unity as observable in *dR* and *P*-techniques. A sentiment to sports produced through a sports club would be such a common learning factor, in which say, drinking at the bar, not really logically connected with the game of golf, becomes part of the ac-

quired pattern. Several such unitary dynamic traits have been replicated by research (Cattell, 1979a, 1980a)

2. *An inherent growth agency,* a cognitive tool or principle that links (in "classical conditioning," i.e., coexcitation) a widening circle of response tendencies, such that the possession of one comes in future, through cognitive linkage, to be accompanied by all. A Piagetian insight occurring at a certain point to some children and not others could similarly be the agent for growth of a differentiating unitary ability. In dynamic traits a sentiment to liberalism, for example, might form as a unitary factor in the emotional field, based on some emotional experience of breaking a conservative mold.

3. *Budding,* by which a preexisting powerful sentiment may require for its effective functioning the formation of a new sentiment subsidiating to it. For example, a professional chemist may find he needs to read German journals and thus become motivated to pick up an array of skills and knowledge subsidiating to his sentiment to chemistry. Factor analysis in a group of similar people would then pick up a unitary, integrated sentiment to German, which, though a distinct factor, would show some correlation with the older factor.

The question now arises as to what we think we are doing when we work out a nature–nurture ratio or *H* value for one of the well-known unitary personality and ability structures defined just as it issues from R-technique research. The puzzle is that actual behavior-genetic findings on *H* very very rarely permit support for the conclusion that a trait is wholly genetic or wholly threptic, as the preceding discussion of unitary influences might initially seem to require. Consequently, additional possibilities to account for the observed factors and the observed *H*s must be entertained, and we shall in fact consider *four* models in order to cover both concepts already discussed and some new concepts. These models exhaustively cover possibilities as follows:

1. The pattern is purely genetic in origin, created as a continuous trait by genic summation as described earlier.
2. The unity is entirely threptic, produced by coordinated learning processes as has been described for sentiment structures.
3. The unity is *apparently* that of a factor but is on closer analysis only a correlation *cluster* produced by the overlap of moderately different patterns belonging one to a genetic and one to a learned factor origin. Where their significant variances sum they produce a *cluster* of highly correlating variables.
4. The at first unlikely seeming phenomenon exists that has been called the "box-and-lid" or "eidolon" model, in which threptic and genetic components have the *same* factor pattern on variables. In Model 3, more skilled factor analysis would split the cluster, but this case, even more sophisticated factor methods might generally fail to pull the two apart and yield more than a single factor. The difference be-

tween Models 3 and 4 is brought out, and readily seen in Figure 6.10, given in what follows.

Let us call Models 3 and 4 respectively *amalgamated* and *combined* models (since the phenotypic, observed, trait is a mixture of contributions from genetic and environmental contributions) and Models 1 and 2 *divisive* models (as in Cattell, 1973b, p. 133). The behavior geneticist's search for H values, if he firmly believes that the differences of his centrally ranging discovered values from .1 to .9 are significant but that 1.0 and 0 are never found, implies that only the amalgamated model is true. On the other hand, I have argued—on slender sample evidence it is true (see Figure 9.1)—that H values point to a platykurtic and perhaps ultimately bimodal distribution of H values for the dozen or so source traits yet studied. That is to say, it is hypothesized that with *correction,* especially for influences and errors that we know reduce H, an appreciable fraction of source traits would be found to be of the divisive kind in types 1 and 2. Thus, fluid intelligence, g_f, surgency, F, and temperamental independence, U.I. 19, could, with some correction, approach an H of 1.0, while superego, G, inhibitory control, U.I. 17, and asthenia, U.I. 28 approach an H of 0.

The probability of these traits having Hs actually of 1.0 and 0 is best technically considered when we come to corrections for this and other sources of distortion in Chapter 8. But the basic point can be made that in the estimation of any factor score there is appreciable pollution from the summed specifics from the various subtests for the wanted factor and that the average H of a random collection of these alien specifics is likely to be toward the middle. Thus the high true Hs would be pulled down and the low raised. This effect of invalidity could be quite substantial, on the order of making an H of 1.0 become .75 and an H of 0 become .25. On some theories fluid intelligence could be an example of a unitariness based wholly on a genetic contribution (except for physical brain trauma) perhaps identifiable with the fixed number of cortical cells given at birth. On the other hand, the superego factor, G, could be a unitary structure arising wholly from the coordinated conditionings of childhood, and, later, from those of religious institutions.

The already mentioned ingenious methodological contribution of Vandenberg (1965b) and Loehlin and Vandenberg (1968), which involved factoring a number of variables measured as *difference* scores between identical twins in pairs and fraternal twins in pairs, could assist in solving the divisive versus amalgamated problem. For the factors from identicals could only be pure threptic ones (i.e., as in Figure 6.10(a)2) whereas the fraternals could yield any of the other three classes. If their pioneer study were followed up with data that cover the personality sphere, and with sophisticated factoring methods, some real grasp of the genothreptic nature of certain structures could well be gained.

Discussion of the amalgamated action will be facilitated by an examina-

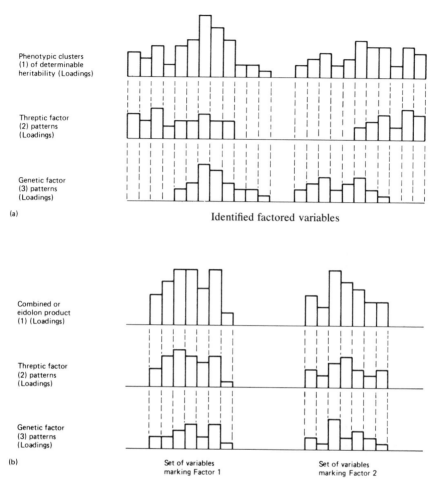

Figure 6.10 Two models of possible genothreptic interaction in producing observed unitary phenotypic structures: (*a*) amalgamated model (in divisive basis): factor overlap of distinct genetic and threptic patterns; (*b*) the box-and-lid or eidolon model of overlap of factor patterns. The top row is in each case the sum of the two below.

tion of Figure 6.10. The profiles are each an ordinary factor loading matrix, graphed and laid on its side (relative to the usual matrix), thus having a series of various variables along the base, and factor loading sizes represented by the vertical coordinate. In Figure 6.10a we suppose in Row 2 two examples of a genetic factor contributing to certain variables some of which happen to be rather strongly affected by environmental influences also, as shown by their loadings on some threptic factors in Row 3. For example, an array of variables in the area of spatial perception might be affected genetically and some of them might also come in for extensive training in a culture using much mechanical gear. Row 1 represents the addition of Rows 2 and 3 and, as suggested earlier, two outstanding *clusters* might appear, which, if examined with only average skill in factoring (especially with defects in deciding on number of factors) would appear as two factors only (Row 1) instead of the actual four comprised by Rows 2 and 3. With interaction rather than simple addition of genetic and threptic components these two clusters could even more easily be taken for true factors. The N (nature–nurture) value (and therefore the H) found would be a ratio of the two "divisive factor" variances. A psychological instance that might fit this would be an overlap of spatial ability (possibly almost entirely genetic) with mechanical training (wholly threptic), producing a cluster which might get called mechanical aptitude, with a middling H value.

Model 4—see Figure 6.10b—supposes that for reasons inherent in psychological development the genetic pattern that is initially maturing (Figure 6.10(b)3) acquires threptic additions (b(2)) which almost exactly replicate its own pattern. Thus in Figure 6.10b, unlike Figure 6.10a, the loading patterns in Rows 2 and 3 are in consequence closely similar. In this case even sophisticated factor methods would have difficulty in splitting the two factors appearing in Row 1 each into its "box" and "lid," respectively in Rows 2 and 3. For this model in which (2) is an image of (3), we have used the label *eidolon*, from the Greek word for image.

The question that critics will naturally raise is "Why should such a remarkable similarity arise?" Part of the answer is that if we look broadly across other scientific domains of natural observation we see analogous instances, which may give clues to what the inherent causes are. For example, in letterpress printing only the raised parts take up ink and produce an image. Or again, the pattern of streams in semidesert country, that would themselves be invisible from 40,000 ft, becomes visible through the band of vegetation they create around them.

Speculation on mechanisms for eidolon production in psychology could take us far afield. At one extreme, one can suppose that the natural assertiveness (esteem-seeking) in human interaction causes those who are genetically gifted in a trait pattern to develop a paraphernalia of means to express it still further. The individual with high motor dexterity turns to sport, dancing, and athletics more than the born clumsy individual; the person with a good ear for music turns to flute and violin; and the oversexed male acquires the skills of a Don Juan. At another extreme of explanation, one may point to indications that the hunting and agricultural phases of human evolution produce some genetic characteristics favorable to each, that is, evolution shaped genetics to culture. Thus (3) imitates (2). And turning to historical times, perhaps Darlington (1969) is right in arguing that over 20 or 30 centuries of intermarriage within craft families, certain acquired skills required by society, for example, those of tinker, tailor, builder, or musician, have produced genic accumulations fitted to the occupation, thus yielding some alignment of genetic endowment patterns with particular institutional learning patterns arranged by society.

Fortunately we already have in the illustration given earlier of fluid and crystallized intelligence a fairly intensively investigated case of what is undoubtedly the eidolon model, with the genetic pattern coming first. Here we find in g_f an appreciably genetic ability factor, loading most highly any form of *complex relation and correlate eduction*, but only trivially those cognitive skills that are simple and more dependent on rote memory. It is perhaps no accident that for centuries schools have concentrated on teaching the more abstract subjects, as in the medieval quadrivium, the learning in which demanded precisely those complex relation-perceiving capacities evident in g_f. Meanwhile, at least until recently, it was assumed that *simpler* matters were learned by the way, in the family and everyday life, though with Jensen's advocacy of teaching more of his "Type II abilities" in school this may change. (Type I and Type II correspond to what Spearman called high and low g-saturated skills.) But as long as schools attend especially to mathematics, classics, syntax, and science, this concentration on the more complex judgmental skills involved will mean that the effects of years of *schooling* experienced and of years of maturation of fluid intelligence *mental age* will load much the same pattern of observable variables.

Since this question will be more concretely encountered in Chapter 8, on ability inheritance, it suffices here to note that when we encounter the box-and-lid (eidolon) model we have the alternative of recording some sort of middling H ratio for the whole, composite, summed factor, or of discovering separate H values, approaching 1.0 and 0, for the two more analytically separable factor traits. However, in many apparent unitary traits that one suspects really fit the eidolon model the box and lid are so "stuck" that we are likely not to be able to assess their separate heritabilities, and thus behavior genetic research may continue for some time to get a greater frequency of middling H values than should correctly be found. And although it may often be psychologically convenient or "simplifying" to treat them as one, to do so eventually lands us in inconsistencies and clumsy predictions, as happens in saying that traditional intelligence tests measure "intelligence."

The whole of the present discussion has strictly applied to source traits isolated as primaries. As Chapter 10 shows there are still other possibilities of conceiving genetic and threptic components when secondaries (second-order factors) in personality and ability structure are studied.

8. Summary

1. This chapter approaches the problem of genetic and environmental interaction with special emphasis on developmental concepts and methods. To understand interaction of maturation and learning it is important to try to

match the precision of the geneticist's handling of genetic mechanisms with an equally exact, quantitative taxonomy of cultural, environmental structure. An approach to this has been provided by the *econetic model*, with its matrix handling of *cultural elements, individual position,* and *environmental impacts*. An illustration of drawing up an impact matrix apportioning cultural presses at different ages, for the average individual, is given.

2. Controlled learning experiments are possible in the laboratory, but to handle at the macroscopic level the important personality changes due to life situations it is necessary to introduce multivariate experimental methods and cultural experience measures, for which the design of *path learning analysis* (PLA) was invented. This manages to assign change on a particular trait (essentially by a regression relationship) to length or intensity of exposure to each of a variety of standard life situations.

3. The goals of methods and concepts here developed are, first, to split developmental curves into genetic and threptic components, second, to make learning laws more accurate by relating amount of situational learning experience to magnitude of *pure threptic gain*, uncontaminated by associated genetic increments, and third, to attempt quantative division of genetic and threptic components in the trait score of a single individual.

4. Six methods, some practicable only on a basis of more extended findings than we now possess, are proposed for splitting developmental curves. The first proposes to recognize pure maturation curves by their mathematical form and their attachment to measurable physiological maturations and to subtract them from best fitting observed curves, containing threptic deviations. In T data factor age curves so far known, some follow, up to 19 years, a typical organic maturation exponential growth curve and others do not. There is a tendency, at present of uncertain significance, for traits with such a growth curve to show higher heritability (Chapter 10) than others. In Q data the relation is at present less clear.

5. The second method aims to separate components by the inverse process of estimating threptic gain independently and subtracting from the observed curve to get the genetic components in the curve. It takes the same people over two adjacent years, and, assuming a steady relation of the genetic gain of the first and second years, demonstrates that the estimation by the PLA method of the *gross* gain (g + t) can be recalculated to give the threptic (t) gain only. In practice it is limited by (*a*) our incomplete research on values in the learning vector, l, and (*b*) our incomplete knowledge of the strength of the actual environmental elements, operating in each of the two years.

6. The third method is formally the same as the second except that it takes two random samples from the same racial-mixture pool, exposed to two different environments at the same age. It thus has less control than taking the same group through the changing environments of two years.

The second and third methods yield initially only a breakdown of the observed increments (change score) into threptic and genetic components, leaving the absolute levels of the two curves unknown. However, a second step can lead to the absolute levels of the threptic and genetic curves.

7. Life-long developmental curves have also to be considered in the perspective of world history. This has been psychometrically recognized earlier than in the present genetic context in the concepts of an *ecogenic* curve—the curve of life change in the *average* ecological environment of man—and an *epogenic* curve peculiar to a particular historical *epoch* (Cattell, 1970b). Comparison of cross-sectional and cursive psychometric analyses has been suggested for separating these. Although this does not amount to achieving the required separation of genetic and threptic curves, it has interest in separating a threptic portion that is epochal. There are suggestions that some of the 16 P.F. factors—superego and exvia—have altered levels in the cultural changes of the last 20 years.

8. A fifth form of life curve analysis is to utilize *P*-technique. Because of the brevity of the experimenter's research life this has to be group *P*-technique and carried out by the cross-sectional method (10-, 20-, 30-, etc. year-olds all measured in the same calendar year). Its contribution rests on the possibility that some factors are wholly genetic and some wholly threptic and that the *P*-technique factor patterns and their age plots will either discover or confirm hypotheses of this nature.

9. The sixth method, *comparative MAVA*, has a broader use than the present, age analysis alone, for it yields comparisons of magnitudes of genetic and threptic variances, not only across ages, but also across cultures, and across educational methods, etc. The present use across ages yields (*a*) if we assume constancy of the coefficient of variation, σ^2/M, an inference concerning the changes in *mean level of the threptic and genetic components from year to year* and; (*b*) the correlations of magnitudes of environmental influences with associated magnitudes of purely threptic changes such as were approached by an independent method in (5).

10. Besides the two methods just mentioned for obtaining the learning law—as a correlation, connecting an environmental experience with a trait growth measure free of maturational (broadly, volitional) contamination—a third solution by partial correlation is suggested.

These solutions to the problem of relating environmental experiences to the purely threptic change component offer mutual checks and the hope—when research has supplied l and e vector values—of several research determinations of genothreptic relations not previously possible. These include the splitting of developmental curves into their genetic and threptic components; the determination of imprinting potency at different biological ages; and also the determination of the magnitude of *the genetic part of a trait in a single individual.*

The usual heritability values are of general interest to the scientist, and offer real guidance to the therapist and vocational counsellor, but the applied psychologist needs to know the genetic limits in a particular case. If a biographical profile, based on a systematic treatment of environmental elements as in the econetic model (Cattell, 1979a), can be produced for a particular client, the threptic part of that client's observed trait measure can be estimated from the regression values in the L matrix derived from PLA. His or her genetic component is then obtainable by subtraction. (This is, in fact, what the clinician is already doing in a rough way in attempting to determine whether the client is a process schizophrenic or a person with little schizophrenic genetic predisposition who has encountered abnormally stressful situations.)

11. The relation of learning to genetics includes not only the genesis of *levels*, but also of *structural forms*. The multivariate experimental researcher in personality has in the last 75 years gained extensive knowledge of the common trait unitary patterns in abilities, temperament, and dynamic traits, which classical reflexological learning has not explained. Structured learning theory and behavior genetics now offer explanations of how these unitary structures come about. It is concluded that there are four sources of generation.

> (*a*) From a genetic source producing unity, either as a single gene, or, much more commonly, as the overlap effect, on a neural structure or hormonal chemical output, of several pleiotropic genes, overlapping primarily in effect on the given target. This target organ or physiological ingredient then contributes, in degrees depending on its maturational development, to the whole range of behavioral variables that are discovered in the psychological unitary factor.
>
> (*b*) A unity of purely threptic–environmental origin. Three mechanisms producing such unity have been extensively discussed elsewhere (Cattell, 1980a): (i) common learning reward schedules for the variables concerned, arising from the impact of a single institution; (ii) the development of single agency ("a discovery," as in Piaget, or a rewarding general concept) which spreads reward to a whole set of behaviors; and (iii) budding from a preexisting sentiment, as a necessary auxiliary set of subsidiating behaviors.
>
> (*c*) An *apparent* unity appearing as a salient *correlation cluster,* which is actually not a unitary factor, but an overlap of cooperative factors loading the same variables. The overlapping factors could be both genetic and environmental in origin. With sufficient psychometric attention to separation, this would resolve into cases of true unities as (*a*) and (*b*).
>
> (*d*) The development of a special kind of unity, called the eidolon or "box-and-lid" phenomenon, which can be logically and statistically regarded as either a single factor or two factors. The

eidolon theory states that there is sone inherent sociopsychological cause (or causes) for the variables that have a prominent unity from a genetic cause also to acquire in common a threptic increment. Analogies are fairly common in nature, and appear, in a different setting, in evolution, where patterns of learning, in each generation, cluster around an evolving organ. The reasons for the eidolon phenomenon are open to psychological research, but meanwhile we hypothesize that with special experimental and statistical care, in certain instances, an apparently single factor could be split into genetic and threptic factors, loading almost exactly the same variables, as happened in splitting intelligence into g_f and g_c.

The first two origins we call "divisive" models and if successfully pursued they should yield heritability values for the underlying factors of, respectively, 1.0 and 0.0; the second and third, which we call the "amalgamated" models, should give intermediate range values—at least *initially* for source (d).

That 1.0 and 0.0 values for H have virtually never been found does not negate the possible reality of the first two origins; for in existing data there are suggestions of platykurtic distributions of Hs. The nature of error, especially in factor score estimation from subtests, would produce high and low values in place of true 1.0 and 0 values. Experimental error also favors reduction of average H values, making 1.0 values less likely. It is suggested that the presently obtained H values in three cases—fluid intelligence, surgency, and temperamental independence (U.I. 19)—when corrected, would be very close to 1.0 and compatible with their being essentially purely genetic unity patterns. At the other end of the distribution (Figure 9.1) the low values for superego, radicalism, and asthenia are compatible with their being virtually pure environmental patterns.

12. Partly as an exploration of the models considered here, partly for its intrinsic interest to psychology, we discuss the relation of fluid intelligence, g_f, and crystallized intelligence, g_c, in light of the *investment theory* and with regard to the factual findings in Chapter 8. The model supposes that fluid intelligence operates each year on the more complex skill acquisitions which the school particularly, and later life technical work, demand. The factorial unity of g_c is therefore of the eidolon type, in which the acquisitions encouraged by environment are those in which g_f can have the greatest influence. This eidolon structure explains, incidentally, why it took 50 years after Spearman's discovery of g to split it into g_f and g_c.

The model here calculates the derivation of the g_c level partly by continuing g_f investment and partly by secondary influences changing with age. After about 18 years of age an age decline is assumed both in g_f and in the capacity to engram what the application of g_f immediately produces. What it produces is cumulatively contributory to g_c, so that on this basis alone g_c

would increase in life indefinitely, though with deceleration through the volutional decline of g_f and the age decline of engramming capacity. But a third influence enters in a linear memory attrition of stored skills. The upshot, with parameters as tried here, is a g_c curve which goes on increasing well beyond 18 years, whereas g_f declines after 18. Since no adequate measures of g_c exist after high school no empirical check can yet be made. Traditional tests like WAIS, WISC, and various Binet translations are almost certainly ill-defined mixtures of g_f and g_c, and the age curves obtained for them fall between g_f and g_c as shown here.

13. The case of intelligence throws light also on the general discussion here of how heritability values would be expected to change with the age of the population on which they are determined. Three principles affecting age change in H are considered: (a) imprinting, by which earlier (or appropriately timed) experiences produce greater threptic variance than later ones; (b) equity, by which an even balance of favorable and unfavorable life experiences is more likely to be achieved in an older than a younger population; and (c) stochastic process, by which the selection of later experiences is affected by earlier choices. In this last we can include the effect of heredity in causing selection of environments. Only (b) is easily handled in a mathematical model. With random environmental encounters it follows that the population variance of the total environmental contribution to individuals' scores decreases with age: This (b) result would be unaffected by (a), but could be reversed by (c). Probably the most reasonable hypothesis at present is that H should be higher in older populations due to relative reduction of σ_t^2 despite a rising mean in threptic contribution.

7

Evaluating Interactions: Path Coefficients and Diverse Heritabilities

1. Interaction and Covariance of Genetic and Environmental Influences

In setting out the convarkin designs (twin, MAVA, etc. as specialized biometric developments), it was pointed out that genic and environmental forces could be regarded simply as creating (*a*) two variances in a trait, genetic and threptic, or (*b*) as also adding variance from interaction and from covariation. At this stage of research support of the attempt to handle interaction might be impracticable, but covariation reaches, in my own judgment (but not of all behavior geneticists), a magnitude not to be overlooked, and definitely investigatable. Accordingly, and certainly to be on the safe side, covariation but not interaction has from the beginning been incorporated in the MAVA models and their solutions.

What we are saying regarding this covariance *within* the family is that a child who initially deviates in his or her genetic make-up from the mean of his or her sibs may encounter a treatment by others, or run into environments, that likewise deviate from the mean of those experienced by his or her sibs. As to *between*-family covariance all children in a family that deviates in its average genetic make-up from that of the population may suffer or enjoy environmental circumstances different from the average of the population. As we saw on p. 93 this produces a within-family covariance term of

$$2r_{\text{wgwt}}\sigma_{\text{wg}}\sigma_{\text{wt}}$$

and a between-family covariance (if within- and between-family variances are to add to the population variance) of

$$4r_{\text{bgbt}}\sigma_{\text{bg}}\sigma_{\text{bt}}.$$

In this chapter we propose to examine possible causes and consequences of covariance, and also to look at varieties of correlation that might conceivably extend beyond those we included in the immediate MAVA equations. This will take us into discussions of causal rather than merely statistical connections, and to the consideration of what have been called *path coefficient* methods. These conceptual advances will prepare us to do credit to the findings of a covariance nature in Chapters 8, 9, and 10.

Parenthetically, let the reader be reminded of the distinction we made in Chapter 3, p. 65 between the two possible interpretations of a *statistical interaction,* as appearing in ANOVA. First it can mean that the dependent variable is contributed to by some product or other interaction of the two variables, as in w.gt in Eq. (3.11). Second, it can mean that the correlation scattergram of the two variables expresses so much correlation that any *random* selection of points among them for entries to the ANOVA cells is likely to contain more than the expected number of instances high on one that are also high on the other.

It is psychologically reasonable to expect that this might sometimes be *true* interaction [not covariance masquerading as interaction but in the true sense of w.gt in Eq. (3.11).] Unfortunately the consequence of pursuing such a model is to meet a forbidding degree of complication as shown in Eq. (3.13), and in a book for the general psychologist we do not propose to develop its complexities. After all, the general factorial behavior-specification equation has worked well in practice, both as to prediction and conceptualization, despite our ignoring the analogous situation there, where something more than additive action might be reasonably hypothesized between certain factors. A psychologist who would like to pursue the interactive possibility further, however, might start with some general equation of the form

$$a = k_1 g^m + k_2 g^n t^o + k_3 t^p \tag{7.1}$$

and see what data bases are required to solve for k_1, k_2, k_3, m, n, o, and p, from concrete variances based on $(a_1 - a_2)^2$, etc.

With this glance at possible interaction treatments we propose in this chapter to work only on the covariance problem, considering it quite broadly, however, to occur as both linear and nonlinear correlational covariance. As with any scientific problem we can attack either (*a*) by looking for inductive generalizations given by data, with a minimum of preconceived analytical plans, or (b) entering with theories begetting deduced hypotheses possibly leading at once to checking inferences by complex data analytic procedures. In view of the fact that, for some not too obvious reason, several conservative behavior geneticists doubt the very existence of genothreptic covariance, it seems more appropriate to emphasize the first approach, and to begin by asking about the existence and descriptive nature of the phenomena of covariance.

2. The Logically Possible, Empirically Researchable Forms of Genothreptic Covariance

Let us begin with the relations among offspring—twins, sibs, half-sibs —which have been the basis of the main MAVA sets of equations and solutions we have given earlier. There are four kinds of deviations that are measured and four corresponding variances. The individual deviates from the family genetic mean, by d_{wg}, yielding the σ^2_{wg} values, and from the family threptic mean by d_{wt} yielding σ^2_{wt}. At the same time each family mean deviates from the grand population genetic by d_{bg} and the grand threptic mean by d_{bt}. If we wish to keep count of special possibilities we can separate the deviations peculiar to identical and fraternal twins, as we do in Figure 7.1. The object of Figure 7.1 is to remind us of all correlations that are mathematically and logically possible, as combinations, namely six, among four things, but among which we settle (the solid lines) on those that are naturally and empirically possible. The mathematically possible combinations of 4 remain 6 but become 7 if we (a) split the fourth referent, not, as shown, but (b) drop the impossible $r_{wg.wt}''$, $r_{wg.bg}$, r_{wtbt}, $r_{wt'bt}$, and $r_{wt''bt}$.

Such correlations, as between genetic and threptic deviations within the family, and others set out without parentheses in Figure 7.1, have already been discussed and accepted as, psychologically, possibly significant, and have therefore been embodied as unknowns in the MAVA equations. Now in ANOVA, with subgroups in cells chosen to meet its assumptions, any correlations of a member's deviation from his or her group mean with that mean's deviation from the grand population mean is impossible. The splitting of concrete group variances into four abstract variances does not in itself make it any more possible. But the family is not a randomly sampled group of individuals, and if one considers, not just linear correlations but also curvilinear, then one rather surprisingly finds that there are good psychological reasons for expecting an individual's deviation from his or her group mean to correlate with the group mean's deviation from the grand mean.

For example, a curvilinear r_{wgbg}—or η_{wgbg} if one wishes—in parentheses in Figure 6.1 could arise from the fact that genetically more assortive

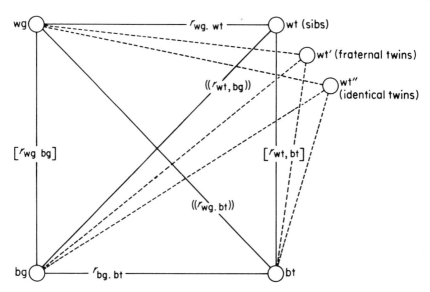

Figure 7.1 Logically and naturally possible covariations of offspring measures in the genothreptic realm.

matings will tend to yield greater between-family variance and somewhat smaller within-family hereditary variance in the children. Secondly since personality differences of the parents, as such, are part of the common family environment of their children, a child more deviant from his or her family hereditary mean (and therefore having parents more different) is likely to be brought up in a family in which the environmental effects of having more genetically discordant parents are greater. Thus a curvilinear r_{wgbt} (η) possibly becomes significant. Let us note also that there are possible situations leading to a significant r_{wtbg}. This would be illustrated by a family with a high manic-depressive genetic endowment, compared to one that is low, tending to create an environment more uneven for different sibs, depending on their being at a crucial age when the parental instability happens to show itself.

The relations of within environment with between environment, $r_{\text{wt.bt}}$, and of within heredity with between heredity, $r_{\text{wg.bg}}$, which could be dubbed "same medium correlations," would not, mathematically, be able to yield *any* linear correlation, for the usual reason in ANOVA. They could, however, yield relations of the kind shown in Figure 7.2. In Figure 7.2a, for example, the families that deviate positively in environment could systematically contain individuals who deviate more environmentally in either direction than do families below normal. For instance, a family high in interest and capacity for giving educational aid to its children might discriminate among them more, according to capacities, and adjust the individual education given, whereas in a family with low educational interests an indiscriminating low level of educational stimulation might uniformly prevail.

In environmental influences:

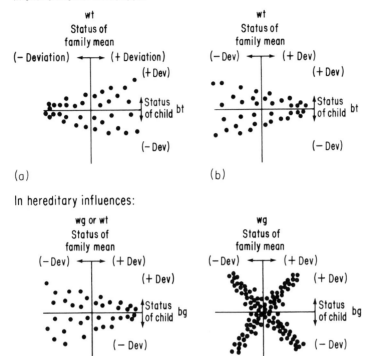

(a)

(b)

In hereditary influences:

(c)

(d)

Figure 7.2 Forms of significant curvilinear correlation of genetic and threptic components. Some major examples of forms of association of within- and between-family deviations (and, therefore, variances) which do not create correlations. Scattergrams illustrate hypothetical relations produced in variances.

The opposite direction of such an association, in a different trait area, could be illustrated by the tendency of parents with greater and more enlightened affection for their children to bestow it more evenly, whereas careless favoritism might prevail in parents less conscientiously interested in their children (see Figure 7.7b).

An instance of another pattern of relation of within deviation to between deviation, this time in heredity (Figure 7.7c), could result if matings are less assortive at lower than at higher levels. (The data, Table 2 in Higgins, Reed, & Reed [1962], and from Cattell & Willson [1938], suggest this may be true in the case of intelligence.) Then greater within-family hereditary variance would prevail at lower hereditary family levels.

Effects of the kind in Figures 7.2a, b, and c could yield significant curvilinear correlations of absolute parent and child scores, statistically expressible in eta (η) coefficients; and in Figure 7.2d could result in a correlation coefficient never before described which might be formulated as double eta.

However, none would yield a linear covariance term. If they were evenly balanced curvilinear correlations they would neither add to nor subtract from the observed concrete variance, (i.e. taken as the sum of the genetic and threptic variances). Only if they were not balanced would they yield some rough simulation of a linear correlation. Thus we are justified in omitting any terms for etas from MAVA, and staying with the main four linear correlations only—r_{wgwt}, r_{bgbt}, r_{wtbg}, and r_{wgbt}. However, to remind us of the curvilinear presence the last two are in double parentheses in Figure 7.1 because they represent both linear and curvilinear relations, the linear relations arising in adoptive families (Table 4.8, p. 115). It will be realized that we have no way at present of evaluating possible etas because the entries needed for calculation are *abstract* deviation scores that we cannot directly obtain. However, if progress is made, as indicated in the last chapter, in assigning threptic and genetic values to *individuals,* determination of such relations would be simple.

Certain variants of these four basic correlations, denoted by primes in Table 4.8, need to be considered now that we have space to do so. They are entertained in the fully extended MAVA design in Table 4.8, whereas we were generally not able to afford them as separate values in more condensed models. They are

r'_{wgbt} The correlation of within-family genetic deviation with between-family threptic deviation that could occur as a placement effect when for example the brighter children in a family that is breaking up are placed in a "brighter" adopting family.

r_{bgwt} The correlation of between-family genetic deviation with within-family threptic deviation that could occur in children adopted into a family and sibs raised apart. Equation (7).

r'_{bgbt} The main placement correlation (supplementary to r'_{wgbt}) in Table 4.8, from any attempt to match the genetic make-up of the adopted child to the threptic standard (biologically and educationally determined) of the adopting family.

The first is almost certainly quite trivial and though perhaps real has never been definitely demonstrated. The last we have accepted (Chapter 5) as a necessary unknown requiring regard for selection effects in adoptive-family studies. The second, we have argued, might reasonably be eliminated by considering that it is actually the same as r_{wgwt}. That is to say, the linear coefficient describing the effect of genetic endowment upon environmental treatment in the family remains the same regardless of genetic range so that the covariance increases only through the genetic range in these adopting families becoming higher. Thus in Chapter 5 we weighed the assumptions either that the correlation or the regression coefficient can remain unchanged, but that σ^2_{wg} changes to a within-adopted-family genetic variance of $(\sigma^2_{wg} + \sigma^2_{bg})$. Thus if the former seems psychologically more acceptable, $r_{wgwt}\sigma_{wg}\sigma_{wt}$ becomes $r_{wgwt}(\sigma^2_{wg} + \sigma^2_{bg})^{1/2} \cdot \sigma_{wt}$.

3. Causal Influences in Genothreptic Correlations

In approaching explanations and hypothetical origins for the various genothreptic correlations we must consider the same three scientific, causal, possibilities as in *any* correlation. In this setting they are: (*a*) that a genetic deviation brings about an environmental deviation, (*b*) that an environmental deviation brings about a genetic association, and (*c*) that some third influence produces both.

Since heredity and environment are supplementary and exclusive when we consider the isolated individual, the third possibility *may* seem illogical. But, of course, we are here planning to step outside the immediate genetic and environmental influences and embrace also secondary causes. As we shall see later in this chapter, some investigators have preferred to get to causal path models by first setting up a reticulum (network) of path coefficients and then seeing how well it fits observed correlations. We feel it a better research approach first to look at *observed genothreptic correlations,* to make an analysis of possible causes in each case, and *then* to turn to explicit path coefficient models. Accordingly we take these three possible directions of causal action and ask how significant within- and between-family correlations would arise from each type of action.

1. Genetic Causation of Threptic Correlations with Genetic Components

Parenthetically, one must beware of confusing any true genetic effects of environment with the fallacious concept sometimes appearing in such statements as "the same environment is more stimulating of intelligence in intelligent children because they see more in it." But if the psychologist so writing means to abolish any fixed term for an objective environment he is heading toward scientific madness! The specification equation defining a perception recognizes, of course (Cattell, 1963a) by the different trait scores or weights applied, that every individual's perception is different and that every individual's learning from a perception of what is objectively the same situation is different.

Thus our model of an *objective* environment nevertheless fully recognizes that the threptic gain is a function of interaction of a pre-existing trait with environment. Since the trait is partly genetic, it is thus possible for a correlation to arise between genetic and threptic magnitudes. This we have seen empirically in the last chapter in the roughly .5 correlation of g_c and g_f, where g_c can be considered the threptic part of intelligence. (This correlation was over a set of individuals, but Figure 6.7 shows an overtime correlation for the mean individual, at least over the 0–20-age part of the curve.)

The effects of original genetic endowment in leading to correlations with environmental action are, however, more numerous and subtle than that just discussed, and we shall proceed to analyze them as follows:

(a) WITHIN-FAMILY GENOTHREPTIC CORRELATIONS

(i) *Associations by* Choosing *Environment*. Here the genetic endowment causes the individual to move into a certain kind of environment, as when a schizothyme temperament causes a person to take up such occupations as forestry or research physics, where experience in social situations is avoided and reduced.

(ii) *Associations by* Shaping *Environment*. The genetic endowment results in environmental correlation because it changes environment (as birds come to experience a nest environment denied to animals). Countless examples can be cited. A fairly subtle instance is the finding by Cattell, Blewett, and Beloff (1955) of the apparent tendency of a high I factor (tenderminded) person to create an environment in which esthetic development is cherished ($r_{wgwt} = .10$ to $.50$).

John Stuart Mill, in his autobiography (1874) after citing life experiences, states: "Though our characters are formed by circumstances, our own desires and efforts do much to shape those circumstances." Psychology knows two primary senses of environment: (a) as it is, and (b) as the individual perceives it (for which personality factors in a behavioral equation can be weighted to define the perception (Cattell, 1979a, 1981). There are two other interactions with environment which make a fourfold operational field with these, namely, environment as it acts on the person, for example, in an illness and environment as it is manipulated by the person. The correlations we are considering under this heading are the latter as it affects (a) and (b).

(iii) *Associations by* Producing Reactions *from Environment*. From the social and the physical environment a genetic endowment may provoke characteristic environmental treatment. The example found by Cattell, Blewett, and Beloff (1955) of negative r_{wgwt} for dominance, indicating a popular tendency to "slap down" the more dominant, shows that it operates fairly strongly within the family. Here must be included also the tendency of other people than the subject to select their associates according to preferred characteristics.

(b) BETWEEN-FAMILY GENOTHREPTIC CORRELATIONS

All three of the preceding principles obviously operate here also. An example which shows all three in action is the upward social mobility of families endowed with high intelligence. One would assume that correlations from (iii) would be smaller here than within families, because, inasmuch as *invitation* to advance is involved, society does not know each family as well as a family knows its members. On the other hand the effects in (i) and (ii) should be larger, because there might be less error in the mean performance of a group than in the evaluation of a single individual. Additionally we have specific to between-family effects:

(iv) *Associations by Differential Survival.* If a certain genetic endowment favors survival by differences in birth rate or death rate within a certain social environment than r_{bgbt} correlations will ensue.

2. Environmental Causation of Hereditary Associations (Converse of the Preceding)

(a) WITHIN FAMILY

In general, environmental correlation within the family would be zero, since by hypothesis the child is already born! However, differential death rates might operate: for example, an overprotective attitude toward the child might favor his or her survival if he or she is of weak constitution, bringing about the conjunction of overprotective threptic effects with defective genetic constitution.

(b) BETWEEN FAMILIES

All four mechanisms discussed under 1 (b) can act here, but in the reverse direction—environment as cause. Environment, stimulating migration, could produce correlations, for example, sociologists report lower capacity in migrants to a city in times of agricultural distress, whereas the opposite may hold at other times. Second, environment can shape heredity, as when persons in a more neurosis-producing environment might avoid neurosis-prone mates. Third, it can act by selection as when educated families have fewer children. And fourth, it can produce differential survival, as in (iv). One would expect correlations from Section 2 to be trivial as compared with those in Sections 1 or 3.

3. Common Causes Operating on Both Heredity and Environment

(a) WITHIN FAMILY

Only in the exceptional instance of something that produces mutation (e.g., radiation) simultaneously affecting heredity and health environment could there be anything but a general zero correlation from any effect immediately and directly affecting both. However, just as we have observed that some within-family variances and correlations, as measured, actually arise from between-family, outer-world influences, so here some influences in the next section could reflect within-family correlations. For example, the social institution of promoting the bright would produce some correlation of genetic and threptic elements also *within* families, in the process of doing the same between families.

(b) BETWEEN FAMILY

Many influences can simultaneously change environment and select for heredity between families. In a more refined analysis one could make claims for all four mechanisms noted earlier. There can be deliberate selection by society, as when intelligent children are sent to college or delinquents to jail. Diseases may select for both environment and heredity, for instance, malaria in some tropical areas for debility and sickle cell anemia, and tuberculosis which selects by being associated with lower social status and, at the same time, with leptosomatic-schizothyme constitution (Cattell, 1950). Slum dwellers in time might become by genetic selection less subject to tuberculosis than well-nourished families. Such various influences as social placement agencies or eugenic ideals may produce similar correlations; and as Fisher pointed out (1930), the practice in the middle class of dysgenic, competitive-restrictive, family attitudes on reproduction (environment) may cause competitive social value attitudes to become correlated with higher intelligence, and other genetic qualities. Altogether one might expect fairly substantial correlations of between family genetic and threptic deviations from this cause (as well as mutual correlations in gene incidences, as Fisher indicated).

Almost every genothreptic correlation observed will evidently result from several influences, and one seeks for some reductive procedure such as factor analysis that might group these influences more simply to give better guidance for path coefficient analysis. Due to the fewness of the particular correlations that can be correlated a factor approach, however, is not practicable and the researcher today must launch more speculatively upon path coefficient hypotheses, as in the following section.

4. Path Coefficient Hypotheses

In studies on personality the group using multivariate experiment has suggested and explored three methods—one old, two with newer forms—of reaching *causal* (see definition in Cattell, 1966) conclusions from correlational evidence:

1. *Finding, by simple structure, or confactor methods, unique factorial resolution.* When unique resolution by a general principle is reached it can logically be argued that the direction of action is such that changes in the factor strengths are the cause of changes in variable scores (Cattell, 1978b). It has just been mentioned that this new approach may be impracticable here.

2. *Introducing time sequence into experimental measurements.* This is very old in *manipulative experiments,* where the experimenter changes the independent variable first (in time) to see what happens, with dependable frequency to one or more subsequently measured variables. This principle has recently taken a new form in *lead and lag correlation* (Cattell, 1966) particularly in *P-technique* (Birkett & Cattell, 1978; Cattell & Cross, 1952) and sequential *dR*-technique. In these designs, *without* manipulation, the same evidence is obtained concerning invariable sequence as was formally largely derived from ordinary manipulative experiment. An example, of lead and lag *P*-technique would be that if a certain behavior is thought to be a consequence of anxiety, the correlation of the anxiety factor score with that behavior score should become higher when the correlations are staggered (anxiety at time t with the behavior variable scores at $t + 1$ or $t + 2$) than when simultaneous, or with reverse staggering. In the present setting, if we had sequential measures on parents and on children, or on sibs, (say at 3 month intervals) we could, in principle, by experimenting with staggered correlations, answer such questions as "Does a decrease in the threptic component of, say, dominance, in the parent contribute to an increase in the threptic component of dominance (or super ego, or whatever) in the child, or vice versa?

3. *Use of path coefficients.* This is an old approach (Wright, 1934) which has grown to popularity in the last decade or so (Cavalli-Sforza & Feldman, 1973; Cloninger, Rice, & Reich, 1978; Jaspars & deLeeuw, 1980; Jencks, 1972; Rao & Morton, 1974; Rao, Morton, & Yee, 1974; Rao, Morton, & Cloninger, 1980). The principle has already been briefly described here. One lays down a network of hypothesized causal actions, but unlike, say, some clinical and social theories that do the same, one *quantifies,* in the theory, the amount of variance contribution of *each* cause to its consequence. One then asks if the empirically obtained correlations fit these values.

In the diagrams (Figures 7.3–7.6) a direction of hypothesized causal action is represented by an arrow, with a path coefficient value on it which represents the amount of variance it contributes to the succeeding measured entity. The relation of path coefficients, p's, to observed correlations, r's, is simple—identity—in the case of a single step with unit variance in each variable. Then both the path coefficient, p, and the correlation, r, describe a variance contribution, as the *coefficient squared.*

In a chain of, say, four variables, in which each is only *part* of the contribution to the next, that is, where there are also side tributary causes if we bring each variable's variance back up to unity, the contribution of the first to the fourth element is given by

$$p_{14} = p_{12}\, p_{23}\, p_{34}. \tag{7.2}$$

Where two variables, a and b, are the sole determiners of c, then

$$p_{ac}^2 + p_{bc}^2 = 1. \tag{7.3}$$

Philosophical discussion of the meaning of cause and effect is apt to concentrate on one effect with one cause; but the natural scientist has to recognize that an effect generally has several causes (and any one causal influence several effects). And how much effect there will be depends on the strength of the single causes in the simplest case as their sum. Thus Figure 7.3b is a simplified example where the many contributing causes are cut to two, but generally there would be $a^2 + b^2 + c^2 + \cdots + z^2 = 1.0$. Furthermore, 7.3b) assumes g_m and g_f are completely independent but sometimes they are correlated and then

$$p_{ac}^2 + p_{bc}^2 + 2r_{ab}\, p_{ac}\, p_{bc} = 1. \qquad (7.4)$$

If instead of convergence we have divergence so that a single cause, a, spends itself contributing to two variables, b and c, which derive their magnitudes from no other source, then we can relate the correlation of a and b to the path coefficients thus

$$r_{ab} = p_{ab} \cdot p_{ac}. \qquad (7.5)$$

Anyone familiar with factor analysis will see that these rules can be derived from quite simple factor analytic models. For example, Figure 7.3b is an example of getting the correlation of two variables, g_{s1} and g_{s2} as the "inner product" of two rows of a factor matrix, where they both load on factor g_m by amounts p_{gms1} and p_{gms2} and on factor g_f by amounts p_{gfs1} and p_{gfs2}. Thus the value $r_{g_{s1}g_{s2}}$ would be $p_{gms1} \cdot p_{gms2} + p_{gfs1}\ p_{gfs2}$. (Actually $.5 \times .5 + .5 \times .5 = .5$; Table 5.8) A simple introduction to and illustration of path coefficients in a factor framework and in relation to psychological instances is given in Cattell (1978a, p. 427). It shows the relation of causal analysis by this third approach to the first approach (unique factor rotation) given earlier.

Path coefficient hypotheses can be presented for both *within*-generation relations, for example, hypothesizing the courses of similarity of mono- and dizygotic twins, sibs, etc., and for "genealogical," intergeneration correlations *across* two or three generations. Figure 7.3 takes equations (1) and (3) in the MAVA Table 4.2 and re-expresses them in path coefficients, for comparison.[1]

In Figure 7.3 the thick lines are genetic determiners, and these have values fixed by Mendelian laws. The symbols g and t have their meanings; while m, f, t1, t2, s1, s2, pc and ps refer to mother, father, twin 1 twin 2, sib

---

[1] Cloninger (1980) aims at a distinction in path analyses between what he calls "natural causes" and "associations" in which the former actually "determines in part the variance of its dependent variable," while the latter "influences only their joint distribution without contributing to their variances." This seems a distinction between on the one hand direct action, here represented by p's, and on the other, prior circumstances, such as have caused the parents to be mated assortively and which we represent as r's. The latter is essentially a common factor operating on two dependent variables.

(a) Twins (identical)

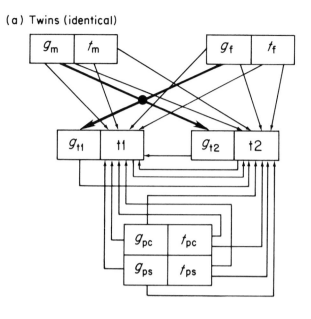

(b) Siblings (genetic only)

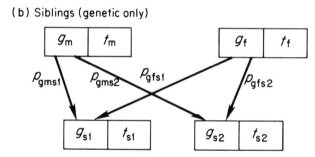

Figure 7.3 Path coefficients for offspring; (a) twin and (b) sibling families.

1, sib 2, the outside population genetics and culture *common* in impact, and that *specific* to the experience for each child. In Figure 7.3a the single genetic source for twins (represented as a circular junction of the two parents, contributions) means that the twins do not differ, whereas the sibs (Figure 7.3b) each get two unique contributions.

The real complexity of the threptic part is at once evident; since the threptic part of each twin receives no fewer than ten "arrowheads." (These could be repeated for sibs in Figure 7.3b.) It is supposed, of course, that *both* the genetic and threptic part of each of the three "significant other" individuals and the general population act on the threptic part of the off-spring. To run them together as a *single phenotypic agent* is possible, but would lose some of the analytic objectives. (No complexity from g and t *interaction,* however, is here assumed.) Of the ten impacts on the threptic part of a trait, six are common and four (g_{s1}, t_{s1}, g_{ps} and t_{ps}) are different, for sibs. For twins, seven are common and three are different. This is because, in the interaction of twins, $g_{t1} = g_{t2}$, whereas g_{s1} and g_{s2} are different. This is one reason why $\sigma^2_{wt.i}$ is likely to be smaller than $\sigma^2_{wg.f}$ or $\sigma^2_{wg.s}$. In saying six or seven are equal we assume the t_m and t_f values remain the same for the two different offspring, though in a more refined analysis later we admit that situations and recipients alter what is received from the same, e.g. parental, source. This difference, if significant, would be empirically perceived, however (if the correlations defined the magnitudes of the variance contributions completely), since it would be observed that p_{tmt1} and p_{tmt2}, for example, are of different magnitudes though they spring from the same source.

Because path coefficient analysis has become popularly recognized in the last decade or so as leading to the ultimate desired causal interpretation of the correlations of the variance relations empirically observed, we must not rush to the conclusion that its complexities are anywhere near being understood. Figures 7.3, 7.4, 7.5, and 7.6 are offered to the reader to illustrate the variety of emphases and forms of causal analysis that different investigators have brought into this new field. Anyone can draw a causal network, but a statistical check against correlations is another matter! In no case has the offered model been completely solved and checked, but another twenty years may see adequate checking methods developed. And in any case, these path coefficient models are presented to stimulate reflection on the nature of the interactions, and as a foundation for theory, based on clear diagrammatic statement.

One reason why the path coefficient approach has difficulty in getting a unique solution is that each diagram is in reality a cut out from a larger map. Of course, we have to stop somewhere and draw a boundary. Thus in Figure 7.3 we start with the parental make-up *as given,* and the input from the larger ambient culture is also stated as it is received, not as it is generated. And just as the rise and fall of the Nile cannot be fully predicted from observations on a stretch from Cairo to Alexandria, so the ancestral momentum is not fully given in the descriptions of the parents.

Figure 7.4 like 7.3 offers my proposed path diagram in connection with MAVA analysis, with initially a comparatively simple relation of path coefficients (p's) to correlations (r's), whereas Figure 7.5 is by Jaspers and Figure 7.6 expresses a combined attack by Rao *et al.* (1980), who, incidentally, have begun to develop the necessary r to p checking methods. Whereas Figure 7.3 introduces path coefficients (p's) in relation to the familiar example of twins and sibs, Figure 7.4 takes *any* child in relation to the parents and society and follows the paths more comprehensively. It does so, for example, in regard to interaction between parents and back-influences of child on the threptic trait fraction of the parents. As any parent knows, over the course of years these can be appreciable, and it was through the second of these that King Lear lost his mind.

In Figure 7.4 the phenotype is introduced as an intermediate term, acting as a whole, instead of in two parts, threptic and genetic (as in Figure 7.3), on the assumption that the whole may have more properties than the parts. Thus P_m, P_f, and P_o are the *phenotypes* of mother, father, and offspring. However, in each case P itself is derived as some function purely of the threptic and genetic components here shown above the P's, and contributing to them.

As in Figure 7.3, the threptic causation is complex and the genetic simple, except that we allow an input g_x into the offspring's genetic makeup for some effects such as mutations or reversions to ancestral alleles that are not fully accounted for by observed, existing parental genetic variances. The phenotypes of the parents act mutually on each other's threptic parts

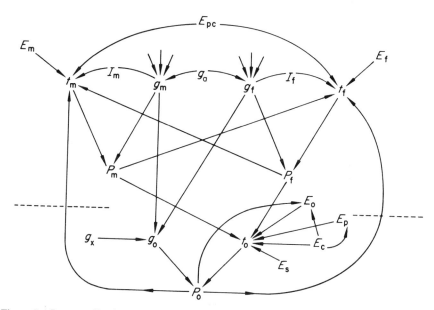

Figure 7.4 Parent–offspring relations analyzed by path coefficients. E_m and E_f = noncommon environments to which mother and father, respectively, were exposed; E_{pc} = common generational environment to which mother and father were exposed; t_m and t_f = resulting threptic components in M and F, respectively; g_m and g_f = genetic components in M and F, respectively; P_m and P_f = the phenotypes of M and F. Since these are environments for the child they have (E_M) and (E_F) attached; g_o and t_o = genetic and threptic components in the child; P_o = the phenotype of the offspring; g_x = aspects of the child's genome not predictable from the parent genomes, through epistacy, mutation, etc.; E_c = the larger common environment of society outside the family; E_p = the peer environment; E_o = the environment offered by the other offspring; E_s = the historical individual environment specific to the given offspring.

(arrows *upwards* from P_m and P_f) and, of course, on the threptic part of the offspring's traits. The latter is affected, however, by several other inputs, namely, E_o, the environment created by the other offspring, E_p, the peer group, E_c the common outside cultural environment (which incidentally impinges similarly on the other offspring and the peer groups) and finally E_s, the quite *special*, unique environmental (biographical history) experiences of the given offspring. Other breakdowns might be suggested here, but these seem best for present measurement resources.

Two lines at the top of Figure 7.4 *have two* arrow heads because they represent two way influences of common influences. Thus g_a includes possible homogamous (inbreeding, assortive mating) that is, shared variance of g_m and g_f, as g_a. The three upper arrows on the g's are meant to represent additive, dominant, and epistatic fractions. The horizontal broken line cuts the lower part, depicting the determination of the offspring's character from the upper which concerns the parents'. We have finally to reckon with the feedback across this frontier in which P_o, the phenotype of the child affects the threptic components of the parents. In this diagram we have not labeled the arrows as p's, with particular path coefficient magnitudes or with particular associated r values, as is done later. Even path coefficients, incidentally, do not consitute a single type of causation, as the barbed and nonbarbed lines indicate, and Cloninger, for one, has attempted a distinction by two terms to cover these.[2] Figure 7.5 maintains our distinction between environment, as such, as E, and the threptic part of a trait, t_i, recognizing, of course, that any g or t can be an environment for someone else, as in deriving the E's for, say, the peer group.

The path coefficient map in Figure 7.4, though not as complex as some that we shall glance at later, is fully realistic in that it accepts (a) the existence (Epc²ga) of assortive mating at both genetic levels, $r_{fm.g}$, and threptic levels, $r_{fm.t}$; (b) the possibility correlation, across the population, of genetic and threptic deviations; (c) proceeding beyond the simple additive model for Mendelian action, by allowing an intrusion from g_x, an extra of genetic endowment; and (d) the living reality of a *recursive* model (See discussion of recursive phenomena in Buss, 1974), as illustrated in p-s from $P_o t_o E_o t_o t_o$ and P_o to t_m and t_f.

---

[2] These sources in the MAVA method are restricted to two products—a within-child-family threptic variance and a between-child-family threptic variance. But the actual sources in Figure 7.4 are (a) mother's personality effect on child, (b) father's personality effect on child, (c) sibs' personality effect on child, (d) child peer group's (the school and its children) effect on child, (e) position of child's family in society in its effect on child, and (f) culture pattern's effect. The last, as a whole, will be the same for all measured in one culture and hence no contributor to variance. However, as shown in Chapter 6, and Section 5 here, we can ultimately go to analyzed *elements* in the culture, which may differ in action from person to person and age group to age group, so that ultimately (f) is a whole spectrum of possible environmental element sources in itself.

Recursive or "feedback" models are decidedly more awkward to handle when one produces the equivalent statistical and algebraic statements to a path coefficient map. They belong to systems theory (Cattell, 1980a, p. 422). But their fidelity to nature cannot be ignored. Particularly, in the present case, we cannot deny that the threptic development of the parent is often as powerfully affected by the phenotype of the child as the child's threptic development by the phenotype of the parent or another child. By the feedback loop which is shown in Figure 7.4 by the P_o-t_f line on the right the behavior of, say, a "difficult" child affects the threptic part of the father's make-up, which in turn affects his total phenotype, which effects (a) the mother; (b) the other offspring (path omitted); and finally (c) the child himself, in terms of his threptic component.

5. Further Path Coefficient Models in the Parent and Offspring Generation

In this section we shall put forward models by different investigators to show that there can be some variety of hypotheses. Indeed, there are, in the present creative stage of path coefficient use, differences of emphasis and handling which can almost be called matters of style, and which spring also from the type of professional background and the psychological or genetic concepts considered important. The main difference at present seems to be that which arises between the models of psychologists, which embody essentially the psychological concepts in the MAVA method (Fulker, Jaspers, Loehlin, Eaves, Cattell, Cloninger, Jensen, Jencks) and those of geneticists (Rao, Morton, Yee, Jinks, Wright). The latter, though in the same family as regards the broad basis of biometrical genetics, tend to give less emphasis to the refinements of environmental sources to which the psychologist is sensitive.

In Figure 7.5 we reproduce without modification, and therefore without bringing the notation and the diagram forms to those used throughout this book, Jaspers and de Leeuw's (1980) path coefficient models for the five of the six main offspring constellations used in the MAVA method: ITT (MZT in Figure 7.4), ITA (MZA), ST (FST), SA (FSA), and UT (URT). FTT, fraternal twins together, is not included, being presumed the same as ST (FST).

The Greek alphabet symbols in Figure 7.5 are correlations: α is what we have been writing as r_{gt}; β is r_{gt} in the special case of sibs reared apart, as in our *placement correlation;* η is any correlation that *might* exist between the genetic endowments of unrelated reared together (placement again); π is any similarity in the environments of twins or sibs reared apart (placement); and μ the correlational similarity of the genetic make-up of sibs, which we know in additive Mendelian action to be .5. Jasper's diagrams are only different

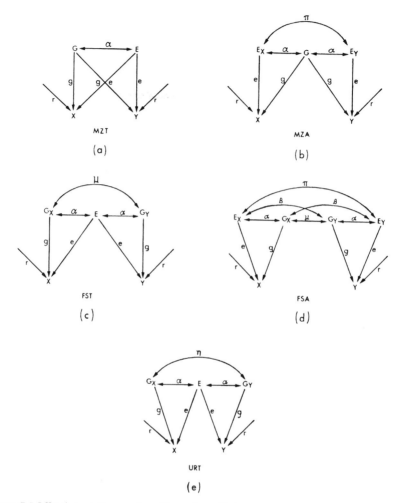

Figure 7.5 Offspring relations analyzed by path coefficients. (From Jaspers & de Leeuw, 1980.)

from those we would write from the MAVA model in two respects: (a) in allowing an error term, entering these diagrams as r (ultimately, this is not different, we handle the error in allowances in the concrete empirical variances—see Chapter 8); (b) in putting environment into a single term, E, where we have the environmental exposure divided into that within the family, appearing as variance, σ^2_{wt}, and that between, as σ^2_{bt}.

As pointed out earlier, any path coefficient model necessarily leads to a set of observable covariances and variances that can be deduced from it. Unfortunately, as we shall see, the converse is by no means always possible, when we want to get to the specific values for the path coefficients that will reproduce uniquely the given *t*'s, Jaspers and de Leeuw proceed from these path coefficients to the inferences:

$$\text{COV MZT} = g^2 + e^2 + 2\alpha ge, \tag{7.6a}$$

$$\text{COV MZA} = g^2 + \pi e^2 + 2\alpha ge, \tag{7.6b}$$

$$\text{COV FST} = \mu g^2 + e^2 + 2\alpha ge, \tag{7.6c}$$

$$\text{COV FSA} = \mu g^2 + \pi e^2 + 2\alpha ge, \tag{7.6d}$$

$$\text{COV URT} = \eta g^2 + e^2 + 2\alpha ge, \tag{7.6e}$$

$$\text{VAR} = g^2 + e^2 + 2\alpha ge + r^2. \tag{7.6f}$$

Their analysis proceeds with a linear model as follows. The model presented in Figure 7.5 has eight parameters and only six observables. Consequently the parameters are not identified, and we must rewrite the model in terms of six identifiable parameters. We also want these parameters not to be negative, and we want them to vary independently. Thus we cannot choose ηg^2, μg^2, and g^2 as parameters, because they are connected by the restrictions $\eta g^2 \leqq \mu g^2 \leqq g^2$. Consequently, we choose ηg^2, $(\mu - \eta)g^2$, and $(1 - \mu)g^2$. For the environmental and covariance components we proceed in the same way. This gives the following parametrization of (7.6), in which **c** is the vector with the five different covariances and the common variance.

$$\mathbf{c} = \mathbf{A}\theta. \tag{7.7}$$

The 6×6 nonsingular design matrix **A** is given by

	θ_1	θ_2	θ_3	θ_4	θ_5	θ_6
MZA						
MZT	1	1	1	1	1	0
MZA	1	1	0	1	1	0
FST	0	1	1	1	1	0
FSA	0	1	0	0	1	0
URT	0	0	1	1	1	0
VAR	1	1	1	1	1	1

$$(7.8)$$

The new parameters are obviously identifiable (because **A** is nonsingular). They are related to the path diagram parameters in the following way:

$$\theta_1 = (1 - \mu)g^2, \tag{7.9a}$$

$$\theta_2 = (\mu - \eta)g^2, \tag{7.9b}$$

$$\theta_3 = (1 - \pi)e^2, \tag{7.9c}$$

$$\theta_4 = 2(\alpha - \beta)ge, \tag{7.9d}$$

$$\theta_5 = \eta g^2 + \pi e^2 + 2\beta ge, \tag{7.9e}$$

$$\theta_6 = r^2. \tag{7.9f}$$

If we analyze intraclass correlations, we use only the 5×5 submatrix of **A**.

Let us now look at a more complex model for offspring, in relation to *parents* also, by Rao *et al.* (1980). Here again the symbols are more specialized to a particular theory than those we have used, and for those now discussed we therefore supply a new key. The reader wishing to get more than an overall sense of the model can go to the original.

In essence this is a putting together of things examined separately in Figures 7.3, 7.4, and 7.5, in that, like Figures 7.3b and 7.5 it takes two sibs and like 7.4 it takes the parent–child relation, now expanded to two sibs. As regards notation, P_f and P_m mean the same as in my Figure 7.4, and P_1 and P_2 are the phenotypes of the two offspring. Instead of g for genetic and t for threptic Rao *et al.* use G and C, and instead of allowing the path coefficients uniformly to drop down from g and t components to the constituted phenotype they do this for the offspring but not the parents. The objective appears to be to have interactions of the components more readily visible in the center of the diagram.

These interactions differ from those in my Figure 7.4 in that only the parents' *threptic* components act directly on the child's threptic components, the genetic component acting only indirectly with the phenotype as an intermediary. To me, and probably other psychologists, this model seems less satisfactory. First, without leaning unduly on psychoanalysis, we can surely recognize (as in the U and I components of the Motivation Analysis

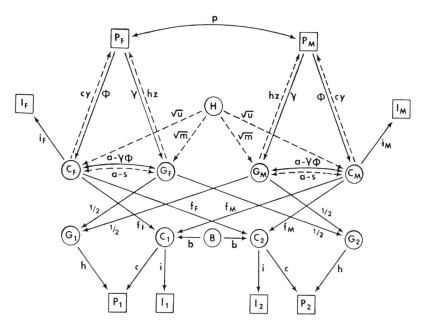

Figure 7.6 Path coefficients in parent–offspring map. (From Rao, Morton, and Cloninger, 1980.)

Test) that a genetic endowment may, through repression, get no phenotypic expression of a kind that can be measured. Yet associates of the individual may nevertheless react to the pattern. For example, when the pugnacity erg in the MAT has a low I and high U component even children can sense the unexpressed irascibility. And again, in terms of variance contribution how does C_m in Figure 7.6, though receiving a variance contribution from P_m get directly affected by variance from G_m, since C_m is by definition environmentally generated variance? If the continuous lines are accepted purely as correlations, then sense can be made of that part, but we would argue that a directly caused and sensed effect of a genetic component in person X upon the threptic part of Y is possible only as set up in Figure 7.4, and omitted from Figure 7.6.

The reader who wishes to pursue these questions in depth must at this point be referred to the original article containing Figure 7.6 and others in press by Cloninger. We would only add, in completion of the description, that what we handled as an assortiveness correlation of parental mating is here represented in the H value and that the correlation p shows what we called "mixed homogamy." That is to say the assortiveness values for the genetic and the threptic parts are here considered different, and for those writers similarity in the latter belongs to special concepts of social and phenotypic homogamy. The I symbols represent a part of the environmental (presumably threptic) make-up of parent and of child that does not go into the family interaction. That is to say one does suppose that the whole of the child environment value C_1 or C_2, or the father's environment value, C_f, goes into the corresponding phenotypic and other contributions. Actually this part matches, in different terms, my model in Figure 7.4.

6. Construing and Contextual Dimensions in Environmental Action

In the preceding chapter we have struggled to make some advances in learning theory by relating learning experience (with it repetitions and rewards) to the true threptic gain, uncontaminated by underlying volitional change. (This is unimportant, as we have recognized, to the *laboratory* reflexologist and *animal learning* experimenter, but vitally important to the psychologist, for example, the clinician and the teacher, dealing with long term learning, especially in personality, in everyday life.)

It will be recognized that if path coefficient hypotheses could be tested,[3]

---

[3] The desirables in a model, which we have led up to in the contrasting of Figure 7.4 with some others, seem to me to be recognition: (*a*) of assortive mating, as applying to *both* genetic and threptic elements in the spouses, (*b*) of the distinction between environmental and threptic

and threptic variance contributions tied down to particular sources, a new avenue would be opened up to studying learning laws. Despite some disagreements in models in Figures 7.3, 7.4, and 7.6, natural to a newly developing area, we may hope that methods of checking models against correlations, as in the fertile contributions now being made by Rao, Cloninger, Rice, Morton, and others (See Asher, 1980 on causal modeling), will lead to definite checks on values. The uncertainty, incidentally, does not lie in the behavior geneticists' own area—that of path coefficient values for *genetic* links among various relations—but in the *environmental* interchanges, and here it is the task of the psychologist, in the new, *structured learning* developments, to provide the most likely models.

So far we have offered path learning analysis, in which the experiences are defined as courses of action in relation to particular objects and situations (e.g., taking a college course, getting married). Econetic theory (Cattell, 1980, p. 310) presents the concept of further analysis of a path learning vector into a *potency of determiners* matrix P_d, by steps we need not follow here.

Living with parents and sibs is a *path* (though a long one!) by our PLA definition, and among the determiners in that complex path are the personality spectra (vectors) of parents and sibs. Considering this, an inadequacy becomes evident in path coefficient diagrams at their present stage, in that they commonly deal with a *single* trait in the main subject and the *same* trait in the relatives. This is correct as far as genetic elements are concerned, but constitutes an oversight as far as threptic parts are to be calculated. The genetic connection of, say, the offspring's intelligence is with the parents' intelligence, but the threptic part may well be affected by their ego strength, super ego strength, surgency, etc. (See especially Johnson, 1981) The path coefficient diagrams have tended to be "flat," that is, concerned with only one trait as dependent and independent variable, whereas in fact they should be multidimensional—in the *threptic* part only.

While we cannot detour here into the complete inventory of parts in the *potency of determiner* matrix, P_d, we do well to structure the part which behavior geneticists put into path coefficient figures, namely, the personality of parents and sibs. An "effect of important others" equation can be set down in the same form as the perception by others, in *trait view theory* (Cattell, 1973, 1979a, 1981), since the acts of others follow from their perceptions. It will therefore have both a *construing* part (the second in Eq. [7.10]) due to the personality of the parent, and a *contextual* part (the first in Eq. [7.10]) representing the effect of the *other* traits in the child on the treatment he re-

---

terms, (c) of several—as many as six or more—influences with path coefficients to the threptic variances, for example, genetic and threptic components of parents, sibs, peers, and of culture common and biographically specific contributions, (d) of feedback of child phenotype upon parental threptic variance (e) of significant genothreptic correlation, and (f) of the effect of one trait in the parent upon a different trait in the child, as described in the text.

ceives for that particular trait. (For example, if he is surgent and intelligent his surgency will be more encouraged than if he is surgent and unintelligent.)

$$Tx_{tike} = (b_{xol}T_{li} \cdots + b_{xon}T_{ni}) + (b_{xkl}T_{le} + \cdots + b_{xkn}T_{ne}) \quad (7.10)$$

where Tx_{tike} is the effect on trait Tx_t in its threptic part, in individual i, in situation k through "important other" e the Te's are the n traits of this impinging person e in the environment and the b's the life situation k loadings for these traits in their effects on i's (threptic) trait T in that situation. The behavior indices b_{xol} etc. are the effect on the average associate, o, of the subject's other traits as well as T_x on the treatment of his trait T_x. Similar equations would operate for other impinging individuals, cultural groups, (also dimensionizable) and physical circumstances, and the contribution from the sum of these equations for acting elements would approach as a limit the path coefficient found for that total experience. For future reference let us call Eq. (7.10) the *threptic–effect* or t-e equation.

The empirical evidence for this cross-trait effect of parent on child, that is, of the need for the full specification equation as in 7.10, might be said to go back to the autobiographies of John Stuart Mill and many others, and, in quantitative terms, to the work of Symonds (1931) and all who followed. Symonds, for example, found children of high dominance parents to be more introverted, interested in school, reliable, and responsible (compared to the opposite parentage, and G. Watson (1934) found adult students with dominating parents to be overconscientious, anxious, given to quarreling with associates, and prone to depression (actually, desurgent). Johnson (1981), found child intelligence related to personality factors C, G, etc., in parents. If part of the effect in Cattell and Nesselroade's (1967) results on married couples is learning rather than selection then we see such effects as high super ego (G) husbands developing high ego strength (C) in wives (r = .29, $p < .01$). More direct, unambiguous evidence comes from H. Cattell's (1980) primary and secondary personality factor measures on family members showing, for example, a highly significant relation ($p < .01$, $N = 231$) of high affectia ($A+$) and parmia ($H+$) in the father with strong super ego (G) in the son. The supposition that such "off-diagonal" correlations in the parent-child-bounded matrix must be from threptic action, needs however, to be cautiously examined because, if *second* order factors were inherited as such a correlation of, say, surgency (F) in the child with group affiliation (Q_2-) in the parent would be due simply to F and Q_2- being inherited together. But second orders do not seem strongly inherited (except perhaps QIV) and the correlation of A and H with G, above, falls outside any second order pattern.

Granted that the correlation of the threptic scores *per se* can be obtained (by methods just discussed) with the impinging phenotype traits of the parents, sibs, peers, etc. (i.e., that the b values can be found in the threptic-*effect* equation), several interesting possibilities arise for psychology. In the first place the multiple R squared should equal or approach the path coefficient variance contribution from that source hypothesized in the path coefficient map. Secondly some order may be found in changes of b weights with various situations (since b's are situation-bound) with different types of relations or parents, sibs, etc., permitting an objective taxonomy of such situations and family atmospheres. This relation of the threptic–environmental equations to path coefficient values is an enquiry for the future. Meanwhile we are facing the methodological challenge that a single, unqiue, network of correlations can in general be approximately met by several alternative sets of path coefficient diagramatic hypotheses, for deciding among which more refined statistical methods need to be found.

7. Heritablilities, Broad and Narrow: The Genotypic Structures Fitting Genetic Variances

Now that we have, here and in Chapter 6, surveyed much of the interaction of heredity and environment, we are in a position to see the need for a more sophisticated set of heritability definitions than the less analytical, gross H or N (nature–nurture ratio) with which so much genetics research is content to conclude. In the first place, without casting the least doubt on the ultimate importance and convenience of the traditional H, for much discussion and theory, we can yet see that the several diverse variances and covariances from which it is harvested are often in themselves more important than as a mere means to an H value. In particular the actual magnitudes of various *threptic* variances for psychologically important traits need to be recognized by social psychologists and clinicians. These values are a research avenue—indeed often the *only* avenue—toward information and theoretical solutions that social and clinical psychologists have struggled toward in vain in their neglect of genothreptic research. And the learning theorist as such can find important principles about *personality and ability learning in the life situation* by plotting relations of measured environmental "teaching" features to discovered *threptic* variances, as taken up above and again in the next section.

One might indeed point out today that behavior-genetic research on continuous variables has probably yielded less to the geneticist than it has to the social learning field. The reason is that the inference from genetic variances to Mendelian, chromosomal structures is extremely difficult. Indeed, at the moment, the genealogical, pedigree, syndromal method (p. 21), now exceeding a yield of 2000 disorders, functions, malfunctions, and physical features with known Mendelian action (McKusick, 1968), has given the *geneticist* something far more definite than have biometric methods.

Continuous variables imply polygenic underlying action. A generation ago some powerful developments by Malécot (1948), Mather (1949, Haldane (1919, 1932, 1938), Fisher (1918, 1930, 1936a, 1936b), Wright (1922, 1934, 1968), and others in *population genetics* pointed to possibilities of taking the discovered genetic variances for various degrees of inbreeding, for example, first cousins and second cousins, and reasoning therefrom to estimates of the number of genes, dominance effects, epistasis, etc., that genetics as such wishes to discover. Valuable inferences can be drawn particularly when there are data on inbreedings continued in certain patterns, for example, offspring of pairs from first cousin marriages, two twins marrying two twins or sibs, and "back crosses" (see McClearn & DeFries, 1973, and earlier literature cited earlier.) In animal research, with manipulative breeding, quite substantial progress has been made along these lines of studying greater and lesser homogamy (Broadhurst, 1957, 1967, 1977; Fulker, 1966; Royce 1955, 1966; Royce & Covington, 1960) and its effect on behavioral variance. A good introduction to population-genetic analyses of data in relation to gene frequency, etc., is available in Falconer (1960) and in McClearn and DeFries (1973).

A problem in all such genetic research, and particularly of humans, is that the *same* external features (to keep simpler physical data for the moment) may have been evolved from *different* genes in different, isolatedly evolving subvarieties and races. For example, height among the Gaelic Scots may in part rest on different genes than among Zulus. In the theory in the preceding chapter, our first model: That a single structural factor in personality, for example, fluid intelligence, could appear from overlapping genes (each pleiotropic) illustrates this point. It is the unity of an intermediate producer—the organ—not the unity of a particular set of genes that determines the behavioral unity. In this instance we theorize that the total cell count size of the cortex is determined by many genes, each of which may have various other consequences besides the contribution to cortical development. *Different* subsets of genes from the population gene pool could then result in the same level of cortical development. The unitary nature of g_f, as a behavior-inferred entity, then arises from the fact that a considerable diversity of complex relation-perceiving behaviors simultaneously have their limits fixed by cortical size, but this latter could be fixed for individuals by different polygenic sets.

We must leave to future genius the methods of determining the operative genes from the variances which our methods now supply; but we can nibble at the problem by recognizing possible instances, for example, spatial ability, depressive tendency, schizophrenia, of contributions specifically at least from the X and Y chromosomes. And we can consider alleged instances of dominance and recessiveness that modify the simple additive gene action.

Up to this point, in discussing the interaction of heredity and environment, we have stayed with the simpler assumption of the genetic variance being additively determined so that we could conclude, with random mating, that $\sigma_{wg}^2 = \sigma_{bg}^2$ and, with assortive mating, that one relates to the other as in Eq. (5.14) on page 139. According to whether the simpler genetic mechanism, or a Mendelian action with dominance and epistacy is assumed, geneticists have referred to a *narrow* or a *broad* heritability, and used different symbols for them, which we must now clarify.

In Chapter 5, a summary was given of purely genetic concepts, notably of additive gene action, dominance, the evaluation of partial dominance, and overdominance, and the nature of epistacy. The reader may refresh his concepts in this field in Falconer (1960), Kempthorne (1957), Crow and Kimura (1970), McClearn and DeFries (1973) and others. What we wish to focus on here, however, is the fact that the common *genetic* variance for two members of a type of relative—for example, parent–child, sib–sib—will have different proportions of these genic action origins, and that the genetic variances, of, say, sibs, when there is *not* dominance in the acting genes and when there is, will be different. Tables 5.8 and 5.14 give some details on this.

The effects of dominance and epistacy are to increase the genetic variance in offspring relative to that which would exist if the genes in action were acting purely additively, while assortive mating, on the other hand, under usual conditions of population hetero-homozygosity will reduce it. Burt and Howard (1956) have examined what will happen to heritability calculations for intelligence, with and without dominance, etc. (Their theory is clear and correct, though their data have justifiably been questioned.)

As indicated earlier, there are two aspects to these genetic models. First, by comparison of relations among the common genetic variances found for various pairs of relatives we can begin to move toward some understanding of the genome structures of various traits. Second, we are alerted to the existence of possible differences of heritability coefficients worked out on different bases. Regarding the latter an interesting distinction was introduced by Lush (1940) in regard to the difference already mentioned between the *broad heritability* of a trait (which is what is obtained when dominance, epistacy, and other effects are involved) and *narrow heritability*, when the genetic part of the phenotypic variance is taken in a situation where it is produced only by additive gene action.

8. The Six Heritability Coefficients and Their Interpretations

Calculating heritabilities, as pointed out earlier, is not the be-all-and-end-all of behavior-genetics research, but to obtain H's for important personality and ability source traits, and to interpret them correctly is scientifically important. (As regards the first part of the last sentence we refer to concepts beyond heritability such as the inferences on Mendelian structure, just mentioned, the opening up of laws of structured learning theory in relation to the threptic gains isolated in comparisons of cultures, and examination of the family and cultural structures that produce significant genothreptic correlations.)

There are basically six heritability coefficients concerning a given trait, but, at least as history, we may briefly follow some textbooks in indicating other variants which inventiveness has suggested. They usually point to:

$$\text{Falconer's } h^2 = 2(r_{\text{ITT}} - r_{\text{FTT}}), \tag{7.11a}$$

$$\text{Jensen's } h^2 = \frac{r_{\text{ITT}} - r_{\text{FTT}}}{1 - r_{\text{ss.g}}}, \tag{7.11b}$$

where $r_{\text{ss.g}}$ (a nonobservable) is the correlation of sibs raised together due to genetic similarity only. Going back we encounter also Nichols' formula[4]:

$$HR = \frac{2(r_{\text{ITT}} - r_{\text{FTT}})}{r_{\text{FTT}}}, \tag{7.11c}$$

and, finally, of course, Holzinger's basic H which is used in essentially the same sense as H here, but derived only from twins. (As pointed out elsewhere, we avoid h^2 as a *symbol* of heritability because psychologists are long accustomed to h^2 as the symbol for test commonality.) As Mittler (1971) points out, Jensen's is the same as Falconer's, if one makes the assumption that $r_{\text{ss.g}} = .50$, that is, that genetic variance is additive and we are then in the domain of a *narrow* heritability.

In Chapters 3 and 4, [notably Eqs. (3.18) and (3.21) on p. 69], we have shown how the equations for H (h^2 in the preceding) values can be stated equivalently in *correlations* (preferably intraclass[5] correlations), on the one hand, and in concrete *variances* on the other. At that stage our formulas—(3.18) and (3.21)—dealt only with the genetic variance *within the* (offspring) *family,* and the earlier (7.11) formulas of Holzinger, Jensen, Nichols, and others are aimed to get the population heritability from twin (id-frat) data only, which we have had to consider, in evaluating the twin method, as an unsatisfactory procedure. Such data *can* lead to between-family heritability values, but only with some assumptions which the MAVA method can avoid. For example, in the Falconer and Jensen formulas one assumes (*a*)

---

[4] Nichols' formula becomes, if we follow his implicit dropping of genothreptic covariance terms, $HR = \sigma_{\text{wg}}^2/(\sigma_{\text{wt}}^2 + \sigma_{\text{bt}}^2)$. It is thus a nature–nurture ration or N value not a heritability. It is, however, a hybrid N, neither $\sigma_{\text{wg}}^2/\sigma_{\text{wt}}^2$ nor $\sigma_{\text{bg}}^2/\sigma_{\text{bt}}^2$.

[5] Let us remind the reader of the earlier statement that the intraclass correlation is a correlation with *double entries,* that is, in the case of fraternal twins each twin goes in both columns and there are twice as many cases as pairs. The intraclass can therefore readily be stated as a variance ratio, as given in the formula in Chapter 3. The ordinary correlation can be regarded as a (slightly) biased estimate of the intraclass correlation, but is frequently used in practice in the formulas.

narrow, additive genetic action only (i.e., that $r_{ss.g}$ or in our consistent notation $r_{s_1 s_2.g} = .50$; (b) no assortive mating, which accounts for the 2 in Falconer's formulas and the $1 - r_{ss.g}$ in Jensen's, since the between-family genetic variance is then simply assumed to be equal to the within (σ_{wg}^2) and the total population genetic variance thus has to be $2\sigma_{wg}^2$; and (c) it is assumed that twin families give a nonselected, unbiased estimate of the variance of the usual population, typically about 98% nontwin.

Completing our earlier reference to equivalence of correlational and variance formulations we note that the correctional formulas in (7.11a) and (7.11b) can be seen as follows:

$$r_{ITT} = \frac{\sigma_p^2 - \sigma_{wt}^2}{\sigma_p^2} \quad (p = \text{total population}) \quad (7.12a)$$

and

$$r_{FTT} = \frac{\sigma_p^2 - (\sigma_{wt}^2 + \sigma_{wg}^2)}{\sigma_n^2}, \quad (7.12b)$$

where p designates the phenotype total population variance.

Hence Falconer's h^2 becomes, as a total population heritability,

$$H_p = \frac{2\sigma_{wg}^2}{\sigma_p^2} = \frac{\sigma_{wg}^2 + \sigma_{bg}^2}{\sigma_{wg}^2 + \sigma_{bg}^2 + \sigma_{wt}^2 + \sigma_{bt}^2} \quad (7.13)$$

(or with covariances, as in MAVA, if desired). That is, it represents total genetic variance over genetic plus threptic variance. However, these "totals," if obtained from the twin correlations in Falconer's (7.11a) formula, are partly spurious in ways just indicated. This is an argument against the use of these heritability formulas when derived from twin correlations.

The six most basic forms of the heritability coefficient are set out in Table 7.1. They are all forms of *broad heritability*, because the genetic variances as they emerge from such a method as MAVA already take account of whatever dominance and epistatic action is at work. The narrow heritability is an abstraction of interest to the geneticist as such. But the genetic variance that the psychologist as a practitioner and personality theorist has to deal with is that finally given as a result of various genetic machineries.

The six values we finally deal with are thus the story of three heritabilities, each in two forms.[6] The three are (a) H_w, the heritability within the family, as a fraction of the observed variance among children in one family;

___

[6] If broad and narrow heritability were included there could be 12 H's! The reason we do not suggest proceeding to three subscripts, for example H_{bgB} is that in psychologists' empirical determination of H's they are given a σ_{wg}^2 (or σ_{bg}^2 or σ_{pg}^2) which represents the total genetic component as it actually appears—broad or narrow in origin. It is this that one has to deal with in psychological transactions and labeling it B or N does not alter it. On the other hand it is of interest to the pure geneticist to take one result which is narrow and another which is broad, in order to get at the role of dominance, epistacy, etc., in the Mendelian understanding of the genotype.

(b) H_b, the heritability analyzing the observed variance among families in a society, taking the mean of the offspring to define each family level; and (c) H_p, the population heritability, which is the breakdown we should obtain if we took as our starting point the phenotypic variance of a sufficient sample of the total population (in most studies the child population) at random. Incidentally, the word *child* in parentheses reminds us that heritabilities are always relative to an age group and a generation. For later, comparative work, the narrower the age group the better, since, as Chapter 6 argues, H values should change appreciably with age and epoch.

When we come to actual personality and ability results in the next three chapters it will become apparent that differences among H_w, H_b, and H_p values can be psychologically illuminating. For since the genetic parts of the equations have fixed relations the differences of H's provide a way to appreciate the comparative variabilities of within- and between-family *environments* regarding influences on particular traits. The extent of assortive mating is also involved since with high genetic assortiveness H_b will tend to be bigger.

These three divide each into two because there are two ways of calculating *any* heritability, which we call *gross* (g) and *net* (n) and represent by subscripts as in Table 7.1. The difference is that the gross includes the covariance ($r_{gt}\sigma_g\sigma_t$) in the denominator, that is, it takes the total genetic plus threptic variance *as it literally is*. (It would do the same with covariance in the numerator if it ever existed.) H_g and H_n can differ considerably if r is moderately large, and since in most instances r has proved to be negative, H_g is more often larger than H_n than smaller.

The majority practice has been to report by the net method, H_n, but (though we also commonly do that here) it is not easy to find a completely convincing argument for doing so. The reasons could be that most authorities in animal biometric methods have established a tradition, through not

TABLE 7.1
The Six Forms of the Heritability Coefficient[a]

	Broad heritability			
	Gross, H_g		Net, H_n	
Within family	$H_{w(g)} = \dfrac{\sigma^2_{wg}}{\sigma^2_{wg} + \sigma^2_{wt} + 2r_{wgwt}\sigma_{wg}\sigma_{wt}}$.		$H_{w(n)} = \dfrac{\sigma^2_{wg}}{\sigma^2_{wg} + \sigma^2_{wt}}$	
Between family	$H_{b(g)} = \dfrac{\sigma^2_{bg}}{\sigma^2_{bg} + \sigma^2_{bt} + 2r_{bgbt}\sigma_{bg}\sigma_{bt}}$.		$H_{b(n)} = \dfrac{\sigma^2_{bg}}{\sigma^2_{bg} + \sigma^2_{bt}}$	
Total population	$H_{p(g)} = \dfrac{\sigma^2_{wg} + \sigma^2_{bg}}{\sigma^2_{wg} + \sigma^2_{wt} + 2r_{wgwt}\sigma_{wg}\sigma_{wt} + \sigma^2_{bg} + \sigma^2_{bt} + 2r_{bgbt}\sigma_{bg}\sigma_{bt}}$.		$H_{p(n)} = \dfrac{\sigma^2_{wg} + \sigma^2_{bg}}{\sigma^2_{wg} + \sigma^2_{wt} + \sigma^2_{bg} + \sigma^2_{bt}}$	

[a] The narrow heritabilities would be similar except that the g values would be derived from observed (extracted) g values by subtracting estimates for dominance and epistasis.

having to deal much with significant genothreptic correlations, and, when they enter human genetics, seem reluctant to entertain the idea that r_{gt} can differ from zero. Yet in fact psychologists can think, as we have shown earlier in this chapter, of a dozen cultural and family dynamics reasons for a significant r_{gt}. An argument can perhaps be made that σ_g^2 and σ_t^2 reflect basic characteristics of a society and its situations, whereas r_{gt} is a relatively accidental "state" of the society at the time. Certainly σ_g^2 is a value little likely to change in, say, a century. But a change in σ_t^2 and a change in r_{gt} are surely more on the same footing as changing cultural features. For example, with the Education Acts in Britain and the associated developments in schools, between say, 1830 and 1930, σ_t^2 for educational achievement almost certainly reduced considerably. But the correlation of intelligence with length of schooling probably changed, through scholarship schemes, in greater degree, toward the end of that period.

A possible compromise is logically definable between H_g and H_n, namely, to consider that the covariance in $r_{gt}\sigma_g\sigma_t$ should be proportionately divided between its contributors. We know σ_g^2 and σ_t^2 separately, and if the ratio σ_g^2/σ_t^2 is m then the compromise heritability, which we can call H_c, would be:

$$H_c = \frac{\sigma_g^2 + 2mr_{gt}\sigma_g\sigma_t}{\sigma_g^2 + \sigma_t^2 + 2r_{gt}\sigma_g\sigma_t}. \qquad (7.14)$$

The result of this laudable compromise is, however, as a moment's algebra will show, that $H_c = H_n$! (This would not be so, however, if $2r_{gt}\sigma_g\sigma_t$ were simply cut in half.) This is perhaps another, though still weak, argument for using H_n. However, the fact remains that if for some practical calculations and social policies we wish to know what fraction of the literal, *observed* total phenotypic variance is genetic, then H_g is the answer. Perhaps the main conclusions are that we should (*a*) be clear which H we are talking about, and, when in doubt, insist on the subscripts $H_{w(g)}$, $H_{w(n)}$, $H_{b(n)}$, etc.; and (*b*) recognize that in *most cases* the $H_{(n)}$—which is generally used—will be an understatement of the $H_{(g)}$ heritability, because of the prevalence of negative genothreptic r's. The case of intelligence, in the next chapter, is an example, in point.

A caveat that is at least equally important is to remember that H's from the twin method that are based on the assumption that $\sigma_{wt.i}^2 = \sigma_{wt.s}^2$, that is, that within-family environment for twins and sibs is the same, will *over*estimate heritability for ordinary families, and by an appreciable fraction, which we now have some data to calculate (pp. 305, 350, 366).

9. Summary

1. The mutual relations of genetic and environmental forces are covered, broadly, by interaction and covariance. In this chapter we give due

consideration to both, but recognize that quantitative analysis of interaction depends on further development of the MAVA model and its solutions.

2. Research on covariances and variances needs to be approached by designs beginning "from both ends": (a) By setting up hypotheses about causal action of various sources of genetic and threptic variance, in models of path coefficients, as in Figures 7.3, 7.4, and 7.6, and inferring, to meet empirical checks, the correlations and variances to be expected; and (b) by carefully observing what replicating empirical results are emerging, especially in genothreptic correlations, and logically analyzing therefrom a taxonomy of possible influences for each, as in Figures 7.1 and 7.2.

3. Pursuit of the latter can begin with the six logically conceivable correlations among the four abstract variances—σ_{wg}^2, σ_{wt}^2, σ_{bg}^2, and σ_{bt}^2—or with the intrafamilial correlations of father, mother, son, and daughter. Beginning with the former we find that more correlations could actually be significant than if one thought in an habitual ANOVA framework, not recognizing that the various types of families are not equal variance random subgroups. Psychological meaning is given to four instances of curvilinear correlations, and all correlations are examined in terms of (a) a genetic deviation causing a threptic deviation; (b) a threptic deviation leading to a genetic selection and deviation; and (c) some common cause affecting genetic and threptic deviation simultaneously.

4. Among linear correlations that are possibly significant, but often not expected to be are $r_{wt.bg}$ applying to sibs reared apart in adopting families and $r_{wg.bt}$ applying to "placement" effects. But beyond these a series of *curvilinear* correlations, η's, could possibly be quite significant. Fortunately, though the latter are worth knowing (but difficult to track down!) they do not upset the MAVA equations, and the unusual "placement" correlations are already handled by being entered as unknowns therein.

5. In passing from observed correlations and established common variance fractions to causal maps one has to cope with the usual three possibilities of interpreting a correlation as stated in (3) above. In this decade behavior geneticists have turned for solutions to Wright's relation of hypothetical *path coefficients* to observed correlations. However, one must not overlook other approaches particularly the establishment of causality by (a) the principle that a simple structure factor is a cause of variation in its dependent variables, and higher order factors are causes of the variations on lower orders and (b) The possible of discerning causal action by either manipulating the independent variable or observing temporal sequences. The latter could be introduced in behavior genetics by the developmental analysis of Chapter 6, by determining changing variances with time differences, and by lead-and-lag correlations.

6. As indicated in 2 and 3 hypothetical models of causal action can be precisely set out by path coefficients, the variance-contributing magnitudes of which can be translated into expected correlations among various elements in a chain or lattice. A path coefficient map, based on our earlier anal-

ysis, is presented, suggesting the most probable contributions to the child's phenotype from his or her parents' genetic and threptic trait endowments, and from four other kinds of environment elements impinging on the child. It is pointed out that any realistic path coefficient model must accept recursive (feedback) paths, for example, the child's effects, during his upbringing, on the threptic personality portion of the parent's personality. The recursive requirement at present offers difficulties in checking path coefficient models against correlations.

7. Geneticists have created more complex path coefficient models than Figures 7.3 and 7.8 referred to₁ in (6) above., a good example of which is illustrated here, for the whole parent-offspring constellation. There are some differences between psychologists and geneticists, however, in what are considered the best breakdowns into conceptual elements. Geneticists have used more molar concepts as far as environment is concerned, whereas psychologists cannot get full satisfaction for their theories from such lumping together of all the non-genetic. And the geneticist, if he splits environment, as in Figure 7.6 above, is likely to do so into child environment and parent environment. To the psychologist the latter is better taken care of as it appears in the measurable parental phenotype, for otherwise it is transmitted "through a glass, darkly."

8. Some major methodological problems have still to be solved in using causal path coefficients and systems theory. For whereas correlational networks can be derived (with difficulties in recursive models) from hypothesized causal networks the reverse inference requires new statistical developments. Many alternative theoretical path maps may fit tolerably one empirical, unique set of correlations. Computer programs are becoming available to give correlational network derivations, and check the fit, but this remains a trial and error game of examining as many path coefficient hypotheses as one cares to think up.

9. This chapter, like Chapter 6, aims to bring to the learning theorist and the developmental psychologist the hitherto little appreciated gains in research goals that behavior genetics can bring to these branches of science. The most basic gain is the separation, in long term personality learning of threptic from volutional components. This leads to a linear, matrix grasp of the action of environment in terms of the personality trait scores of the significant others in the child's life, and of the dimensions of cultural groups. The threptic–effect (t-e) equation which results for estimating change in the offspring from parental, sib, peer, and culture vectors is identical in form with that used in trait view theory (Cattell, 1973, 1979) containing, therefore, a *contextual* vector and a *construing* vector, though these now describe action on rather than perception of the subject as in the origin spectrad model (Cattell 1981).

This model encompasses the experimental observations now richly accumulated that the development of trait X in a child correlates not only with

X but also with Y, Z, etc. in the parent (Johnson, 1981; H. Cattell, 1981). The geneticist's path coefficient treatment, which correctly assigns the *genetic* contribution in X wholly to the parental X, has tended, incorrectly, to do the same in the threptic part. The one dimensional parent-child threptic path coefficient representation has therefore to become multidimensional.

10. Genetic and threptic variances, and their interaction in covariances eventually get brought together in genetics and psychology in conveniently condensed statements as heritability coefficients (H's subsume and imply nature–nurture ratios, N's). However, knowledge of the magnitude of genetic and threptic variances and covariances, in and of themselves, is important for several purposes. The contribution of knowledge of magnitudes of the *threptic variance* portion to personality learning principles in the life situation has already been stressed in (9). Attention is finally drawn here, therefore, to the contribution of available knowledge of genetic variances, at various degrees of relationship, to the determination, by Mendelian algebra, of genotypic, chromosomal structures. Genetics is facing a difficult task in doing this for continuous, polygenic traits, but another decade may see useful resolutions.

11. In this latter connection the difference is between the genetic concept of a *narrow* inheritance, in which we deal only with the additive part of the gene action, and *broad* inheritance. In the latter we recognize effects from genetic dominance and epistacy, which increase, for example, the variability of offspring from the parental values. These concepts are important in Mendelian interpretation of the finally obtained, empirical genetic variances from MAVA and other forms of biometric genetics, and affect the expected purely genetic variance in common among different types of relations.

12. Broad and narrow aside, there remain basically six heritability coefficients consisting of *gross* and *net* ratios, H_g and H_n, for each of three constellation relations: within-family, $H_{w.g}$ and $H_{w.n}$; between-family, $H_{b.g}$ and $H_{b.n}$; and across the general population, $H_{p.g}$ and $H_{p.n}$. Some historically earlier proposed coefficients are instances of these. They have been given in the correlational forms that can be seen as equivalent to the basic variance statements here, and have involved sometimes unstated special assumptions. Earlier results were also often expressed as N, the nature–nurture ratio, which is related to H by $N = H/(1 - H)$. As between *gross* and *net* the net heritability equalling $\sigma_g^2/(\sigma_g^2 + \sigma_t^2)$ seems to have been most reported in studies so far. Whether it is larger or smaller than the gross— which is $\sigma_g^2/(\sigma_g^2 + \sigma_t^2 + 2r_{gt}\sigma_g\sigma_t)$—depends on the sign of the genothreptic correlation. If one thinks to compromise by a hybrid of H_g and H_n, in which the genetic fraction of the genothreptic covariance is added to the genetic term in the numerator, he finds he has returned to H_n. H_n can be defended as a value unaffected by covariance, but as a statement of the fraction of *the actual, observed, population variance assignable to pure* genetic variance H_g is more correct. The argument for H_n is that the covariance is perhaps a

more ephemeral phenomenon, and harder to determine accurately; but the important warning to the human behavior geneticist is to label these two H's differently and remember their properties.

13. At the twin stage of research a number of alternative formulae for heritabilities—H's, h^2's, etc.—were suggested by various writers one of which proves actually to be a nature–nuture ratio, N (normally $H = N/1 + N$). Additionally to these sources of confusion the student needs to be alerted to the fact that today's surveys and summaries of heritabilities are haunted by the differences of values resulting from different assumptions in twin and MAVA methods. The former studies—*far* more numerous in the literature—report on the basis of the assumption that $\sigma^2_{wt.i} = \sigma^2_{wt.s}$ whereas MAVA recognizes what has now been reasonably proved, that environmental differences between sibs (even when properly corrected for age differences) are significantly larger than for identical twins. The result of the equality assumption is that twin study H_w values run significantly higher than those based on recognition of the larger threptic variance in sibs.

14. The three values H_w, H_b, and H_p are not expected to be the same for any trait, though they should have appreciable resemblance, with H_p falling between H_w and H_b. Actually, their *differences* give valuable information; for since the genetic component alters relatively little from within to between these indices bring out mainly the relative importance of within-and between-family environmental influences in shaping the trait in question.

8

The Inheritance of Abilities: Some Psychometric Requirements

1. Three Approaches toward a Strategic Choice of Psychological Variables for Genothreptic Research

Abilities are the oldest traits under behavior-genetic study, and have the distinction also of being involved in the most notoriously polemical and politically determined debates! Our aim in this chapter is to avoid the latter, while improving upon some of the unsophisticated methods and concepts in science itself that have left the subject so grievously open to misinterpretations.

The present chapter is, in any case, the appropriate point at which to focus psychometric and statistical issues that up until now were left implicit. They must become explicit and clear as we now come to handle real data on abilities and personality traits. The psychometric issues include (a) the definition and measurement of the unitary traits used; (b) the effects of error of sampling and measurement; (c) the effects of trait fluctuation; (d) the manner of correction of variances for reliability and validity; and (e) the standard errors of the obtained abstract variances themselves.

The logical first question, which we will study in this section, is "With what kinds of traits should behavior geneticists concern themselves in a well-planned strategy?" They have, in fact, concerned themselves with an almost random array of specific behavior traits, for example, reaction time, sensitivity to tastes, speed of arm movement, visual acuity, memory for letters, and attitudes to food. As in other fields, the problem soon arises that there are far more behaviors than investigators, so that the idiosyncratic object of one man's interests rarely coincides with and becomes subject to check by researches of another. Perhaps there are three principles to give us guidance in choice: (*a*) knowledge of the larger and fewer structures in which abilities and personality traits group themselves; (*b*) indications in evolution or elsewhere that a trait may have a peculiarly high heritability, or a contention that it has none at all; and (*c*) recognition that knowledge of the degree of heritability will have some unusual practical importance, for example, in educational, occupational, or clinical psychology.

The impact of the last is abundantly evident in the researchers' choice of intelligence (Burt, 1969), spelling ability (Thurstone, Thurstone, & Strandskov, 1955), and verbal ability (Vernon, 1965, 1969) in educational psychology, of color vision, sense of pitch, mechanical aptitude, etc., by vocational researchers, and of schizophrenia, manic-depressive psychosis, etc., by psychiatric geneticists, such as Kallmann (1950) and Rosenthal, (1970) not to mention the medical disorders listed by McKusick (1968).

The second director of choice, namely, a suspicion of unusually high heritability, or evolutionary growth or a polemic that trait X has no heritability *at all,* is much less evident in existing research examples. It seems perhaps to have guided Kretschmer (1929) to relate forms of psychosis to body build. And early students of identical twins reared apart (Burks, 1928; Rosanoff *et al.,* 1941) were impressed by high heritability, but often to behaviors that are quaint or surprising rather than important (Bouchard, 1980; R. Guttman, 1970), for example, a special kind of laugh, liability to unusual headaches, way of folding hands, choosing similar kinds of names for offspring, and style of walk. The pursuit of traits related to blood groups (Cattell, Young, & Hundleby, 1964; Swan, 1980b), fingerprints (Swan, 1980a), and eye color belongs here inasmuch as these studies start with highly inheritable physical traits. The evolutionary findings of paleontology also give leads. From an evolutionary point of view, since one of the greatest differences between man and man-apes is the opposed thumb and the high manual dexterity that goes with it, one might expect a trait of manual dexterity to show appreciable inheritance. Similarly the indication of evolutionary development of altogether larger cortical areas for language might suggest concentrating on verbal ability.

These varied sources have all led to some "finds," but as work becomes more systematized more studies are likely to begin, in the human behavioral area, with the best defined and widely researched factorial source traits. The primary abilities of Thurstone (1938) extended by Hakstian and Cattell (1974, 1978) and primary personality traits in the replicated studies surveyed by Hundleby, Pawlik, and Cattell (1965) offer examples of these. These source traits are in any case likely to link on to and subsume the concepts from the sources (*a*) and (*b*) given earlier. For example, factor analyses of bodily dexterities show little evidence of a *general* dexterity, but a very clear and extensive factor of *manual* dexterity, such as evolution would suggest if human performance levels tend to be most "spread out" on what is most rapidly developing. The same applies to a verbal ability factor, independent of any particular language. And if the theory of "divisive" traits in the last chapter is correct, namely, that *one* source of unitariness in correlation of performances is a single genetic action of many overlapping genes, and another is the effect of a single, common conditioning schedule, then more definite and useful findings will result from using measures of unitary source traits. By contrast, taking a host of particular, narrow behaviors chosen almost at random, each likely to be a complex mixture of source traits, some genetic, some threptic, is the road to scientific futility. Consequently the concentration on psychometrics in the next sections will finish by orientation especially to factorial source trait measurement.

2. The Triadic Theory of Ability and Its Genetic Implications

From this point let it be understood that when we refer to unitary structures, both here in ability and in the subsequent chapters on personality, we are not speaking of subjectively conceived entities, but of factor-analytically discovered and replicated patterns. Nevertheless, this does not mean an indiscriminate attention to alleged "concepts" in all published factor-analytic studies. This is no place to spread into factor-analytic technicalities, available elsewhere (Cattell, 1978; Gorsuch, 1974; Harman, 1976; Rummel, 1970). But it must be said that the surveys of Kameoka and Sine (1981), Cattell (1974), DeYoung (1972), Vaughan (1973), and others show that vital conditions have been omitted in many published studies, for example, in push-button computer factorings that present merely principal components, in use of guessed numbers of factors, in factors incompletely rotated to simple structure, or not carefully replicated across several studies. Moreover, as mentioned earlier, in support of functional unitariness, the field needs research demonstrations that go beyond *R*-technique (individual difference) evidence, into evidence of unity of growth, by *dR*-technique (Nesselroade & Reese, 1973), of organic unity of fluctuation, in *P*-technique (Birkett & Cattell, 1978; Horn, 1972a), and of uniform change in manipulative multivariate experiment (Cattell, Kawash, & DeYoung, 1972). These are the needed hallmarks of a unitary structure.

There has been some tendency in the trait field, because of many personality theorists' inability or unwillingness to go to the roots of the differences of research findings (e.g., of Thurstone and Guilford in the ability field) to accept Brown's factors, Jones' factors, and Smith's factors, as equally correct. This may be tact but it is not science. There are unitary traits present in a given racial–cultural population, which are real. That is to say they no more belong merely to one person's ideology than the decision on the number and nature of the elements in the Earth's crust. Where programmatic research has been conducted, in the last 40 years it has presented psychologists with 30 or more unitary trait concepts in ability and personality that they are at liberty to accept or reject on technical bases. Most are now measurable with known validity, replicated a dozen times or more, and indexed for operational identification (replacing a jungle of subjective labels).

As pattern matching shows, some of them *do* give good support to older clinical concepts, for example, Freud's ego strength as C factor and Jung's extraversion-introversion, as QI. They bring such clinical entities into measurable forms, with specification equations for effective predictive of real-life performances—school achievement, recovery from delinquency, occupational success (by salary), likelihood of drug addiction, etc. But Freud's ego strength, indexed as factor C, his super ego, as G, Jung's extraversion as QI, and so on now stand and operate within a *larger company* of traits than could be tied down by the clinical method. The knowledge of criterion associations, growth curves, sex differences, etc., that has accumulated around these factors is now considerable, and the most conspicuous need in connection with them has now become research on their heritability.

The theories of *ability* structure considered in behavior-genetics research—where there has been any explicit statement—appear to have been four: (*a*) Spearman's theory of a single general intelligence factor (in Jensen [1972, 1977] and implicitly in the bulk of twin data based on traditional tests like the WAIS, WISC, Binet); (*b*) Guilford's theory of very numerous regional abilities, (1967); (*c*) Thurstone's discovery of primary abilities (1938); and (*d*) the theory of fluid, g_f, and crystallized, g_c, intelligence, that is, accepting *two* general intelligences. All four of these allow for primary abilities —verbal, spatial, inductive, etc.—outside the intelligence factor, though in Guilford's case their relationship is different. Indeed, in the last case, based on arbitrary orthogonality, we would be dealing, as Horn (1979) has cogently shown, with an artificially distorted representation of the natural structure. However, we need not for the present purposes pursue that difference, but can think of it as the kind of distortion that a Mercator projection map produces relative to the round globe. For example, if genetic results appear based on Guilford's "divergent thinking" we shall come pretty near to consistency by considering them as the equivalent of findings on the fluency factor of Spearman (1904) and Thurstone (1938).

The evidence for two general factors of cognitive discrimination capacity (Cattell, 1963) has now been out for 18 years, and, especially through the programmatic work of Horn and his collaborators (Horn, 1977, 1979, 1980; Horn & Donaldson, 1980; Dixon & Johnson, 1979; Vandenburg, 1967), and others, we are now provided with pretty sound knowledge on how their natures are distinguished. They differ in age curve plots (Baltes & Nesselroade, 1973; Horn & Cattell, 1966b; Horn & Donaldson, 1976; Horn, Donaldson, & Ergstron, 1981), in cross-cultural transmissibility of tests used to measure them, in magnitude of standard deviation of true IQ's, in response to brain injury, in correlation with social status, and, if recent evidence can be counted (Cattell, Graham, & Schuerger, 1981) in heritability.

In a broad search for higher-order factors, across carefully identified primary abilities (Hakstian & Cattell, 1974; Horn & Cattell, 1966a; see Horn, 1978, for a review), factor analysis shows these two—fluid and crystallized general intelligence—factors stand clearly along with four or more other factors, for example, fluency, perceptual speed, at the third *general* factor level, as shown in Table 8.1 (see also Cattell, 1971, p. 106; Horn, 1980).

The term "general factor" here carries the usual continuation of Spearman's meaning of g and leads to representation by a g symbol, though today the term *broad* would be, factor analytically, a better term (Cattell, 1978a) since, granted a sufficient wide range of primaries, a *truly* general factor is practically unknown, for no broad factor subtends more than a good fraction of them.

However, although this is the literal factor analytic outcome the final *triadic theory of abilities* (Cattell, 1971) argues that visualization, g_v, a general motor-kinesthetic ability factor, g_k, and the auditory factor recently found by Horn and Stankov (1979) form *structurally* an intermediate rank, as shown in Figure 8.1. This brain-regional set of *provincial powers* contributes in a role between the true *general capacities,* fluid intelligence, g_f, crystallized intelligence, g_c, general (perceptual) speed, g_{ps}, memory, g_m, and retrieval rate (fluency), g_r, in the top stratum, and the true primary abilities or *agencies,* the a's, in the bottom factor stratum, immediately above variables.

These middle structures have been classified and called *provincial*[1]

---

[1] This is a new term (Cattell, 1971) but seems most aptly to designate the concept, for these powers stand to the general capacities as a national *province* (or state in the United States) stands to the central national administration. The p's handle the genetic and acquired capabilities of a tolerably defined *sensory* interpretive input or motor output area. And developmentally they both contribute to and receive aid from the general capacities.

Since the best substantial instances are a collection of skills in the visual and auditory areas (and similar factors might be found around the olfactory skills of cooks, and the taste memories of wine-tasters) one might be inclined to call this class of factors "crystallized sensory discrimination powers." However, the various analyses of bodily dexterities and kinesthetic sensitivities, as in a tight-rope walker or a trapeze artist, point also to capacities centered on the cerebellum, and the well-explored motor areas of the brain (notably that of hand dexterity). What these psychological provincial powers have in common, therefore, is a definite degree of brain localization, in the form of a sensory or motor center, incorporated with a growing association area

TABLE 8.1
Twenty Primary Abilities Examined for Higher Stratum Factors[a]

	Second-stratum factor[b]						
Primary ability	Gc (1)	Gf (2)	Gv (3)	Gps (4)	Gm (5)	Gr (6)	h^2
Verbal ability (V)	**59**	−02	−06	18	05	06	52
Numerical ability (N)	−05	**45**	01	29	**32**	−02	60
Spatial ability (S)	07	**68**	**37**	03	02	04	54
Speed of closure (Cs)	02	02	**28**	**89**	−33	−04	63
Perceptual speed and accuracy (P)	−25	**40**	**27**	**37**	−01	00	40
Inductive reasoning (I)	00	**42**	22	23	13	02	40
Flexibility of closure (Cf)	02	23	**27**	26	00	11	24
Associative memory (Ma)	03	06	−02	00	**66**	−03	46
Mechanical ability (Mk)	**57**	17	**37**	−03	−02	−09	40
Span memory (Ms)	−13	05	00	**31**	11	22	22
Meaningful memory (Mm)	12	03	01	06	**38**	06	25
Spelling (Sp)	10	−13	03	**63**	05	−04	42
Auditory ability (AA)	17	00	24	23	−03	11	18
Esthetic judgment (E)	09	01	06	−08	04	**32**	13
Spontaneous flexibility (Fs)	13	24	23	09	07	**27**	29
Ideational fluency (Fi)	01	28	01	04	00	**78**	70
Word fluency (W)	07	09	02	**53**	02	06	40
Originality (O)	19	02	15	−01	−16	**41**	23
Aiming (A)	00	−06	**44**	02	17	−01	26
Representational drawing (RD)	−04	−07	**42**	07	17	03	27

Source: A. B. Hakstian and R. B. Cattell (1977, p. 662).

Note: Decimal points have been omitted. Salient factor pattern coefficients, or loadings used to interpret particular factor, appear in boldface. Gc = Crystallized Intelligence, Gf = Fluid Intelligence, Gv = Visualization Capacity, Gps = General Perceptual Speed, Gm = General Memory Capacity, and Gr = General Retrieval Capacity.

[a] Since there has been some tendency to equate crystallized intelligence, g_c, with "verbal or scholastic intelligence" (Vernon, 1965, 1969) it should be pointed out that crystallized intelligence arises from investing g_f in *any* cultural area. This is well illustrated here by mechanical ability being just as much such an acquired intelligence as verbal ability.

[b] Numbers in parentheses are factor numbers.

powers, *p*'s, because each operates within the domain of a special sense organ or motor capacity. Presumably they represent the resolving power in the associations formed around a particular sensory input. Horn and Stankov's (1979) work on auditory ability brings out well that these are typically not just a sensory acuity but a capacity *to perceive and resolve relations* in the given sensory area. Any agency is affected both by one or more *p*'s and one or more *g*'s, and there are also two-way connections between *g*'s and *p*'s. For our present theory— and it is little more—supposes that between *g*'s and *p*'s there is action not just in one direction, but that the individual's

around it. Agencies, *a*'s, are smaller than this, and borrow from several sensory areas. The provincials are, in a neurological sense, "provincial" relative to the total cortex (or to the mid brain unity, in noncognitive matters), so the symbol *p*, for provincial powers, has seemed the best designation of the cognitive ability factors at this factor stratum.

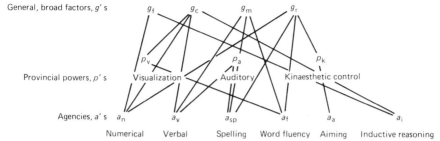

Figure 8.1 Model of the triadic theory of ability structure. The diagram would become a tangle if all directions of influence were drawn as path coefficients. A sufficient sample is drawn to indicate, as in the text, the agencies draw on both p's and g's, and that some degree of recursive action occurs between g and p's. This last prevents a ready separation of g and p's by simple factor analysis.

level on, say, visualization and auditory resolution may add into his or her overall general relation-perceiving capacity in, say, g_c and g_f.

The triadic theory of overall ability structure also has physiological and other implications. It supposes that the general capacities are parameters of the total cortex, such as total number of neurons, cholinesterase supply, etc; and that p's are associated with the well-known sensory input areas for filtering tasks and with equally definite motor areas (different in kind but factor analytically on the same level). Meanwhile the well-known, established primary skills or agencies, a's, are hypothesized to being as widespread neurological paths, but to finish as more local and narrow in area. This last fits the evidence of Lashley (1963) and others since, that acquired skills are initially spread over many neural tracts, but later become localized.

The triadic model thus offers the behavior geneticist a basis for a tolerably definite set of hypotheses about heritabilities. Already we have expounded, p. 211, an investment theory, for the relation of fluid and crystallized intelligence, which, incidentally, we believe also has more general application, namely, to certain *personality traits* at the second order. Initially this implies nothing more than the familiar structured learning equation (Cattell, 1980a) showing again as a function of reward plus weights on a *combination of genetic and learned existing structures*. But it carries within it also a possibility of explanation by particular mechanism of the rise of "box-and-lid" structures. The triadic theory would also lead to the hypothesis that there should be many *agencies,* of varied form, stamped from the dies of the varied cultural institutions to which we are exposed, and that these consequently should have low heritability values. Conceivably *some* of the g's, such as perceptual speed, might also prove very susceptible to environment, for example, in terms of hormone or vitamin insufficiencies, but generally they should be more heritable.

From the standpoint of neurological indications we might therefore expect several of both the g's (broad capacities) and the p's (provincial powers) to yield, in behavior-genetic analysis, the eidolon type of structure. That is to say some particular relatively unitary neural p structure, such as the visual occipital area, could generate in its immediate association area a host of visual skills correlating with the boundaries of the genetic endowment. As we have pointed out, in such a case we may get, according to the technical finesse employed, a quantitative split of genetic and threptic components within a single factor, seemingly applicable to a single factor, or, two factors of highly similar loading pattern, one wholly genetic resting on the neurological endowment, and one wholly threptic, developed by the years of exposure to learning.

Although we have hypothesized that agencies in general will be of low heritability, they could in some cases have a similar relation to neurological structures. The speech area of Broca, which could be substantially connected with Thurstone's verbal factor, stands at an interface of motor control of the tongue (and hand) and the auditory area. Here the two million years in which man made one of his three great separations from the man-apes, by acquiring speech, have doubtless left their mark in a speech-apt neurological area *as such*. But in the build up of the verbal ability agency, a_v, it seems that there is much scope for establishing learning connections also between the major two or more adjacent provincial power zones aided by some *general* cortical action too, as suggested by the path coefficients (as yet unquantified) in Figure 8.1. In this case the high g saturation of V (A_v) certainly suggests appreciable involvement of a genetically limited total cortical resource for each person, and the quite high heritability for V supports this explanation (Table 3.3, p. 76).

Several concepts here will be illustrated further in relation to g_c and g_f in the last section of this chapter. As stated earlier, our aim in this introduction has been to put the kernel of up-to-date concepts about ability structure in the nutshell of a single chapter section. Among those familiar with the area the simplifications will perhaps be forgiven. Among those unfamiliar or committed to a different emphasis we would suggest, first, a resort to the data sources (Cattell, 1971; Horn, 1972; Horn & Donaldson, 1976; Spearman, 1923; Thurstone, 1938) and then a return to look at the model.

The model emerging from studies like those in Table 8.1, as drawn schematically in Figure 8.1, would definitely direct future genetic research differently from many ability views previously available, and, we believe, with more promise of clear findings than in the past. For example, researchers have been content for the most part to base genetic research on older intelligence tests such as the Binet, WAIS, and WISC which we now recognize to be mixtures of g_f and g_c, plus various primaries special to each test. They attempt vainly to stay on center, for g_c, but it is really impossible to adjust neatly an ever-changing and Protean cultural spread. The uselessness of trying to compare IQ scores for, say, a middle-aged engineer and a housewife (say 10 years older) by such traditional tests has surely been made clear, and only an elaborate construction of a dozen different tests of WAIS design, each carefully statistically constructed to an explicit existing cultural zone and age could justify genetic or any other conclusions about intelligence with due reference to a representative selection of humanity over age, cultural, and geographical areas. The question is simpler and the research means are available (in the three culture-fair intelligence scales, the Raven matrices, etc.), on the other hand, as far as g_f is concerned.

If we were to ask what theory has generally been behind the existing fairly numerous genetic investigations of "heritability of IQ" we should probably get the answer that the intelligence measured is Spearman's g. As one who knew Spearman well, I would aver that, with the advances in the 60 years since the g theory was propounded, Spearman himself would not have backed that theory today. The crystallized intelligence factor, g_c, which tests like the WAIS in America, and various similar tests in Britain, Germany, etc., essentially aim to test, is, we must repeat, so Protean in pattern, across classes, ages, and cultures, that from a scientific point of view it is undefined by such tests. Until far more precise definitions have been worked out for standard cultural content, age changes in factor composition, etc., g_c measures are as uncertain in meaning and comparability as say, a count of windows to determine the size of houses.

Fortunately, at least within a common school culture, the known 20 ability primaries (Hakstian & Cattell, 1978; Thurstone, 1938) have proved relatively definite and stable, and these, rather than random arbitrary collections of skills should surely be the concepts concerning which questions of heritability, environment variance, age of impact, or environment, etc., should be asked. Moreover, one may hope that the determination of H's for validly measured broad factors—g's, such as fluid intelligence, speed, and retrieval capacity—and good measures of provincial powers—such as auditory and visual—will have greater tendency to link up with physiological and other independent evidence, thus helping to clarify their ultimate structural nature.

3. Correcting for Error in Measurements and Samples: General Considerations

After the question of *what* to measure comes the question of *how* to measure it. It may seem sufficient to answer the latter by referring the reader to texts on psychometrics. But, though brief reference to the latter is seldom wasted in *any* psychological endeavor, we do have here some issues relatively specific to behavior genetics. In what follows we shall assume that the question of what to measure has been answered by choosing unitary source traits (unitary across R-, P-, and dR- techniques, with experimental manipulations) and that the validity of our test measures is a *concept validity*— defined by correlation with the pure factor. We shall also suppose that since the pattern of expression of a factor changes with age, etc., the isopodic and equipotent principles (Cattell, 1970b) for producing comparable scores across age, sex, and subculture differences are employed in getting the scores. Finally we shall assume that we use the stratified uncorrelated determiners (SUD) model to represent what factor analysis finds. SUD means that second-order factors *add to* independent primary "stubs" to produce the observed primary correlations (Cattell, 1978a, p. 213). The question of what we have and what we are measuring specifically in these contributing second-order factors is taken up in more detail in Chapter 10, pp. 377–382.

On this basis let us turn to psychometric issues in handling the test measures as such. In considering the variance values in twin and MAVA methods typically based on *two* offspring (such as we have considered in Eqs. (4.5) and (4.6), initially) a question is often raised about the effect of dealing with families of different sizes. This really contains a statistical question and a psychological question. From a purely statistical point of view the estimate of the same interoffspring variance can be made from families of any size, provided we use the correct degrees of freedom. For families of two, from the common mean of each, the degrees of freedom are one, for those of three this becomes two, and so on, that is, it is $(n\text{-}1)$ (n for family size and N for number of families). Thus the calculation of within-family variance can take the following form:

$$\sigma_{\mathrm{w}}^2 = \frac{\Sigma^{2N} d^2}{N(n-1)}, \qquad (8.1a)$$

which for sib or twin pairs

$$\sigma_{\mathrm{w}}^2 = \frac{\Sigma^{2N} d^2}{N}, \qquad (8.1b)$$

or, with more complete formality, and to include families of different sizes becomes

$$\sigma_{\mathrm{w}}^2 = \frac{\Sigma_{j=1}^{N} \Sigma_{i=1}^{n} d_{ij}^2}{\Sigma_{j=1}^{N} (n_j - 1)}, \qquad (8.1c)$$

where j is the family and i the individual, and where d is each sib's deviation from the mean of the sibs in the case of each of the N sibships (families).

In Chapter 3, Sections 2 and 5, we have explained in a preliminary way how concrete variances are calculated and used in obtaining heritabilities. Now that we are about to handle actual data it is appropriate to go further to consider practical conveniences in those calculations.

Since in practice it is easier with two offspring simply to take their difference, $X_{i2} - X_{i1}$, the useful calculating formula is

$$\sigma_{\mathrm{w}}^2 = \frac{\Sigma_{j=1}^{N} (X_{j2} - X_{j1})_j^2}{2N(n-1)}. \qquad (8.1d)$$

Like (8.1a) this applies to two sibs, so the denominator can be written $2N$. For getting the variance of the *family means* around the grand population mean we have

$$\sigma_b^2 = \frac{\Sigma^N D^2}{(N-1)}, \tag{8.2a}$$

where D is deviation of the family mean from the grand mean. Here the more detailed statement is

$$\sigma_b^2 = \frac{\Sigma_{j=1}^N(\bar{X}_j - \bar{X})^2}{(N-1)}, \tag{8.2b}$$

where σ_b^2 will be called henceforth the *between-family variance* [such as we have considered *in* Eqs. (4.5) and (4.6), initially]. Here $\bar{X}_j$ is the mean of the sibship and $\bar{X}$ of all sibs.

It often happens in data gathering aimed at getting sibs of, say, 12 through 18 years of age that families of three, four, and even more offspring will be brought into the net, and it will then seem wasteful to throw away and cut to twos only, merely for uniformity of calculation. The psychological issue then enters of whether the environmental forces are in fact differently distributed in larger and smaller families. They almost certainly are, for example, through some diminution of parental influence in the larger family. For this reason behavior geneticists should eventually set out to determine threptic values for families of different sizes. (Our rejection of the disturbing effect of $r_{g_1t_2}$ in MAVA, pointed out by Loehlin [1965b] would be still better justified in larger families, and there is much to be said for working with families of three or four rather than two, also as being more representative of world population generally.) Both family size and birth order have been found related to personality and ability. Ideally, therefore, the relation of σ_{wt}^2 to the (unchanging) σ_{wg}^2, that is, should be investigated separately for each family size.

Meanwhile, for the sake of a broader data basis, most psychologists will probably want to throw the available two, three, four, etc. families together. The formulas for calculating variances from mixed size families are not complex.[2]

Statistical error is encompassed by measurement and sampling error. Let us consider measurement first. An obvious source of measurement error, in the broader sense, which has been taken care of this and the next two chapters, but which has been sadly neglected in perhaps half of all pub-

---

[2] One could estimate variances separately for two, three, and four size families and average with weights for the sizes of samples of these that are combined or enter the computer directly with (after Rao) $\sigma_w^2 = (\Sigma^{2k_2} d_2^2 + \Sigma^{3k_3} d_3^2 + \Sigma^{4k_4} d_4^2)(k_2 + 2k_3 + 3k_4)$ where k_2, k_3, and k_4 are the numbers of families respectively with two, three, and four children, and d is as in (8.1a).

Correspondingly for between-family variance estimation: $\sigma^2 = \lfloor(k_2-1)\sigma_{b_2}^2 + (k_3-1)\sigma_{b_3}^2 + (k_4-1)\sigma_{b_4}^2\rfloor/[(k_2-1) + (k_3-1) + (k_4-1)]$.

The former in full formality would, incidentally, better be set out as:

$$\sigma_w^2 = [\Sigma_{j=1}^N \Sigma_i^{nj}(X_{ij} - \bar{X}_j)^2]/[\Sigma_{j=1}^N(n_j - 1)].$$

lished studies in personality, rendering convergence of results unlikely, is that of the age differences of sibs, and those of parents. If a trait has an age development curve—and most do, as shown in Figures 6.4, 6.5, and 6.6 (p. 198)—individual differences must be corrected for age. This is a very obvious need when we compare variance of brothers with that of twins, but it is needed also in *between-family* variances, and some error exists in several published studies from failure to watch this, and also through the mean age of, say, the adoptive family constellation being different from that of, say, the identical twins with which it is compared.

Fortunately, age curves of reasonable accuracy are now known for primary personality factors in Q-data (Cattell, 1973a; Cattell, Eber, & Tatsuoka, 1970) and in T-data (Cattell & Schuerger, 1978). So there is today no practical problem (except that one would like to see independent investigators proceed to further extension and refining of the age curves).

In the case of *primary abilities* psychologists have investigated age curves on very few, and though the IQ takes care of *intelligence* comparisons in the growth period, it is *far* less reliably applicable to parents' scores in the middle- and late-adult range than those who have put weight on parent–child correlations have assumed. The reason, of course lies in that unequal exposure to learning fields from 18 through the professional years, which we have already sufficiently discussed, and which can only be technically met by first investigating "cultural maps" and applying to them the equipotent and isopodic principles for factor score comparisons. Even if we accept an approximate flat curve for g_c after age 20 we know that the g_f component follows a totally different, downturning pattern in those years. Moreover, we must recognize that the difference in IQ sigma between g_c (traditional) and g_f (culture-fair) tests is so large (50%) that comparisons of the two need to be carefully watched.

4. Psychometric Handling of Compositeness and Correlation in Calculating True Variances

Since every scale or battery mixes error of various kinds with the concept, or factorial trait, it is supposed to measure, behavior-genetic research at its best should correct the obtained concrete variances before proceeding to analysis.

In the last two decades I (Cattell, 1957, 1964c, 1973a, 1978a) have developed various psychometric concepts to clear up what had become a Tower of Babel, and to do justice to sophisticated factorial concepts. We shall use these, but the task we have set ourselves will still not be an easy one.

First let us define the variance fractions that go into any of the personality and ability scales and batteries, as follows:

σ_{wi}^2 The invariant, nonfluctuating part of the variance of the wanted factor, that is, what we want the scale to measure

σ_{wf}^2 The fluctuating, reversibly changing part of that factor variance, occurring with modulation or internal changes (some may like to speak of this as intraindividual variance)

σ_{ui}^2 The invariant part in the variance of the unwanted factor in the scale; this will consist of specific factor or factors and unwanted broad factors insufficiently excluded by suppressant design

σ_{uf}^2 The variance in the unwanted factor due to fluctuation

σ_e^2 The variance due to experimental error of measurement.

In the newer psychometric concepts *validity* has three dimensions and *reliability* two. We are concerned here with only one validity—*concept* (formerly construct) *validity*—which is the correlation of the test with the pure factor source trait as follows:

$$r_v^2 = \frac{\sigma_{wi}^2 + \sigma_{wf}^2}{\sigma_{wi}^2 + \sigma_{wf}^2 + \sigma_{ui}^2 + \sigma_{uf}^2 + \sigma_e^2}. \tag{8.3}$$

The fluctuant part has to be included in the numerator if, as usual, the validity is determined by a *single occasion* for factoring, when the fluctuant bit would be present in both the test and the extracted factor.

The two species within the genus *reliability* that we concentrate on here are *dependability* and *stability*. Incidentally, one should beware of calling *homogeneity* a form of reliability; for broadly the varieties of *test consistency* fall taxonomically thus:

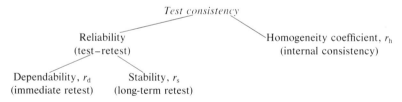

The dependability, r_d, is presumed to involve a close enough retest to avoid difference due to trait fluctuation, thus

$$r_d = \frac{\sigma_{wi}^2 + \sigma_{wf}^2 + \sigma_{ui}^2 + \sigma_{uf}^2}{\sigma_{wi}^2 + \sigma_{wf}^2 + \sigma_{ui}^2 + \sigma_{uf}^2 + \sigma_e^2}. \tag{8.4}$$

The denominator in all these cases is the full observed variance of the test and could be written σ_0^2 (o for observed), or 1.0 if in standard score. The stability coefficient respects change from trait fluctuation and is

$$r_s = \frac{\sigma_{wi}^2 + \sigma_{ui}^2}{\sigma_{wi}^2 + \sigma_{wf}^2 + \sigma_{ui}^2 + \sigma_{uf}^2 + \sigma_e^2}. \tag{8.5}$$

From the empirically obtained r_d and r_s, a derived (not directly observable) coefficient can be obtained *which is not a property of the test but of the trait mixture (w + u) involved,* namely, the *trait constancy* coefficient, r_c:

$$r_c = \frac{\sigma_{wi}^2 + \sigma_{ui}^2}{\sigma_{wi}^2 + \sigma_{ui}^2 + \sigma_{wf}^2 + \sigma_{uf}^2} = \frac{r_s}{r_d}. \tag{8.6}$$

What we now want depends on our behavior-genetic objectives. Is the fluctuant variance, σ_{wf}^2, now to be considered part of the trait or not? If the data we work with are the mean of say, a dozen occasions of measurement, σ_{wf}^2 is largely excluded (averaged out) already from the mean value across the 12, which is used to get σ_0^2. But if, as is likely to be true of 99 out of 100 researches, a single occasion measure is used, then fluctuation variance resides in the measure, and we may seek to get rid of it, if we want to, by statistical means. Incidentally, we do not know whether σ_{wf}^2—the fluctuation of the wanted trait—operates both in genetic and in the threptic part of the trait. So we best assume that it is in both and divided in certain proportions between them. With this assumption we shall get a more stable result from research to research, since σ_{wf}^2 and σ_{uf}^2 are forms of "error" peculiar to one study, if we correct for (eliminate) them. This we can do statistically, and we shall then be able to state that the H's and variances we get apply to the *stable,* σ_{wi}^2, part.

There are thus in theory several possible *degrees* of purification of the empirically obtained concrete variances. It must be confessed that most studies yet published have not attempted them at all, and that our own results at present have not got beyond corrections for error of measurement. The degrees of purification are:

1. Correction for error of measurement by use of the dependability coefficient, r_d.
2. Correction, further, for unwanted factor variance by a function of invalidity, $1 - r_v^2$.
3. Correction further for trait fluctuation by r_c, etc.

Pursuing a complete correction, that is, obtaining σ_{wi}^2 from an observed σ_0^2, knowing r_v, r_d, and r_s, we can begin by writing from (8.3), (8.4), and (8.5) three equations containing the unknown variances (note we attach the actual σ_0^2 in each, instead of calling it "unit variance," to keep values as raw score, concrete variances):

$$\sigma_{wi}^2 + \sigma_{wf}^2 + \sigma_{ui}^2 + \sigma_{uf}^2 = r_d\sigma_0^2, \tag{8.7}$$
$$\sigma_{wi}^2 + \sigma_{ui}^2 = r_s\sigma_0^2, \tag{8.8}$$
$$\sigma_{wi}^2 + \sigma_{wf}^2 = r_v^2\sigma_0^2. \tag{8.9}$$

From these we cannot solve for four unknowns without an assumption, namely, that the fluctuation magnitude of the wanted and unwanted factors bears the same ratio to their absolute variances thus

$$\frac{\sigma^2_{wf}}{\sigma^2_{wi}} = \frac{\sigma^2_{uf}}{\sigma^2_{ui}}. \tag{8.10}$$

This is probably not a strong assumption.
It can be shown algebraically that the solution for σ^2_{wi} is then

$$\sigma^2_{wi} = \frac{r_s r^2_v}{2r_s + 2r^2_v - r_d} \cdot \sigma^2_o. \tag{8.11}$$

Parenthetically, if a psychologist prefers to ignore trait inconsistency and include the fluctuation variance in the trait measure he is, of course, at liberty to do so. But his definition of a trait then becomes different, and, in our opinion (based on the modulation model $S_{kx} = s_{kx}L_x$, given in the following) a confused one. At any rate it would help the clarity of comparisons of results if one more subscript were used in the resultant heritabilities, namely, H_c with trait constancy (mean measure over repeated occasions, or with statistical correction) and H_f for a trait with fluctuation inconstancy. In the trait measure used for heritability estimation (contrary to the earlier argument) it is desirable that he or she subscript the resultant H in some way to distinguish it from that given earlier.

Additional arguments for using H_c and excluding σ^2_f are that behavior geneticists have in general preferred to work with "repeatable" measures (to use Falconer's term for the geneticist's equivalent of our psychometric definition). Thus with H_c our values would be more comparable. (The fluctuation of physical measures like height and weight is in any case so small compared to psychological measures that physical anthropological genetics is virtually operating with H_c even when based on single occasion measures). One should note, however, that usually the physical geneticists "repeatable measure" considers the reliability of a *composite* measure and is not concerned with some pure factor fluctuation within the composite.

Parenthetically let us note that the present psychometric concept integrates well with developments in the *state-trait* field (Cattell, 1971a; Nesselroade, 1973) in which a *state proneness* (or *liability*) trait, L_x, for trait x, is taken as a fixed trait, modulated by a modulation index, s_{kx}, deriving from situations k_a to k_b to k_c, etc., thus

$$S_{xik_a} = s_{xk_a}L_{xi}, \tag{8.12}$$

where the state x of individual i in situation k_a, namely, S_{xik_a}, is equal to the *state proneness* trait, L_x, multiplied by a modulation index, s_{xk_a}, for that situation. This has been shown to hold also for what have commonly been regarded as traits, that is, we can substitute T_{xik_a} for S_{xik_a}, T being a trait measured on a given occasion. So the more obvious case of states has led us to a model for distinguishing a latent trait, L, from a momentary trait level. Thus in the case of, say, anxiety, modulation theory does not suggest that we ask the virtually absurd question, "How far is a state of anxiety hereditary?" but "How far is L (anxiety liability, in the case of anxiety), derived from the state measure at a mean s_k, an inheritable trait?"

This model may prove particularly useful when researchers come to enquire about the heritability of drive strengths. For, as shown with more detailed formulas elsewhere (Cattell, 1980a; Cattell & Child, 1975), *drive strength* has to be deduced from observed *ergic tension* on a given occasion by a formula similar to (8.12), whereby fluctuation variance is removed.

Now, so far, in formula (8.11), we have reached the ability to extract the *true* ("wanted") trait score, and as *stable* variance, from a *single* (not a *difference*) measurement. For the next argument let us shorten Eq. (8.11) to a single transformation value which we will call δ (delta). Thus (8.11) becomes

$$\sigma^2_{wi} = \delta\sigma^2_0 \qquad (8.13)$$

(wi means wanted factor, as an invariant).

However, since we deal with variances of differences, our actual end goal is not correction of a *single* measure, but of the variance of a *difference* measure. The calculation is further complicated by the fact that the two measures are correlated. For in all concrete within-family variance measures of the type σ^2_{ST} there is a positive correlation of sibs or twins $r_{s_1s_2}$, so that, as regards the *literal, observed* measures, with each sib population having the ordinary population variance σ^2_0, the variance of the *difference* is

$$\sigma^2_{(s_2-s_1)o} = 2\sigma^2_0 - 2r_{s_1s_2}\sigma^2_0 \qquad (8.14a)$$

or

$$\sigma^2_{(s_2-s_1)o}/2\sigma^2_0 = 1 - r_{s_1s_2}. \qquad (8.14b)$$

We need to go beyond this, first by transforming from observed score, o, to the *pure* (wanted factor) score variance (8.13) and second by replacing the observed value, $r_{s_1s_2}$, by the *correlation of* the pure factor parts r_{s_1wi,s_2wi}, which might *not* be that of the observed measures. Thus what we want is

$$\sigma^2_{(s_2-s_1)wi} = 2\delta\sigma^2_0 - 2r_{(s_1wis_2wi)}\delta\sigma^2_0. \qquad (8.15)$$

If we could assume that the observed correlation of the sibs is due only to the true, wanted factor part then with the uncorrelating part removed the new r would be higher (by σ^2_0/σ^2_{wi}). But $r_{s_1s_2}$ is due to common variance in the unwanted as well as the wanted factor. Our safest assumption is that the ratio of common to total is the same for the total with unwanted as with wanted variance, since we know nothing definite about the resemblance in the unwanted. So as in the analogous case of assortive mating (where the two parts are, however, g and t not w and n) we conclude that the best estimate of the correlation between sibs over the true part is their observed correlation over the whole, corrected only by the dependability coefficient, r_d, for the error variance removed from the observed variance thus

$$\sigma^2_{(s_2-s_1)wi} = 2\sigma^2_0(1 - \delta r_{s_1s_2}/r_d).$$ (8.16a)

Let us finally go back to observables by substituting for δ its origins in (8.11) thus

$$\sigma^2_{(s_1-s_2)wi} = 2\sigma^2_0\left[1 - \left(\frac{r_s \cdot r^2_v}{2r_s + 2r^2_v - r_d}\right) r_{s_1s_2}/r_d\right].$$ (8.16b)

This derives the difference variance on the true and invariant scores of the sibs from the variance of single sibs, σ^2_0. But we have probably already calculated the within-family variance from the observed scores (or are given that value for someone else's published data) and wish to transform directly from that to the within-family variance on the true and invariant factor. In that case we use the relation to σ^2_0 of the difference variance on raw scores given in (8.14b), yielding a value ϕ to be used in transforming from the given to the desired pure and invariant within-family variance:

$$\phi = \frac{1 - [r_s r^2_v/(2r_s + 2r^2_v - r_d)] \cdot (r_{s_1s_2}/r_d)}{1 - r_{s_1s_2}}.$$ (8.17a)

Thus

$$\sigma^2_{(s_1-s_2)wi} = \phi \cdot \sigma^2_{(s_1-s_2)o}.$$ (8.17b)

Further algebraic examinations show the between-family and total population concrete variances would be multiplied in the same way. Like the simpler correction for reliability alone, it would not affect ultimate heritabilities if all equations were linear. The extent of effect on H values of a reliability correction can be seen in Chapter 10.

If one wishes to correct for reliability alone (dependability coefficient, short of unity only by error of measurement) then the errorless variance, which we will call σ^2_m, is for population scores

$$\sigma^2_m = r_d\sigma^2_0.$$ (8.17c)

The variance of differences in the pairs is now

$$\sigma^2_{m(s_1-s_2)} = 2r_d\sigma^2_0 - 2r_{\bar{s}_1\bar{s}_2}r_d\sigma^2_0,$$ (8.17d)

where $r_{\bar{s}_1\bar{s}_2}$ is the correlation of the sibs in the errorless part of their measurements which is

$$r_{\bar{s}_1\bar{s}_2} = \frac{r_{s_1s_2}}{r_d},$$ (8.17e)

whereby

$$\sigma^2_{m(s_1-s_2)} = 2\sigma^2(r_d - r_{s_1s_2})$$ (8.17f)

and the relation of corrected to uncorrected variance is

$$\theta = \frac{\sigma^2_{m(s_1-s_2)}}{\sigma^2_{o(s_1-s_2)}} = \frac{2\sigma^2_0(r_d - r_{s_1s_2})}{2\sigma^2_0(1 - r_{s_1s_2})} \tag{8.17g}$$

$$= \frac{r_d - r_{s_1s_2}}{1 - r_{s_1s_2}}.$$

In the present chapter, on inheritance of intelligence, a relatively conservative approach to correction of variances has been made, namely, for validity by the simpler method; but the personality factor measures in Chapters 9 and 10 have been corrected by other approaches too. The concept (construct) validities of the culture-fair tests, that is, the correlation of the unweighted sum of subtests with the pure factor, was found to be, by factoring, .71, while the concept validity of the much shorter crystallized intelligence in the HSPQ was down to .66. The raw concrete variances that are the basic data in this chapter can therefore be obtained by multiplying the corrected "observed" columns in Table 8.9 by the reciprocals of these values.

As pointed out elsewhere (p. 108) this multiplication of raw variances by a constant leaves the majority of derived values (see the MAVA solutions, Table 4.4) quite unaffected, and a minority only slightly different from those reached by operating on the raw variances. However, as the formulas indicate the "simple" use of the validity or reliability coefficent correction —ignoring the correlation of those between whom the difference variance is taken—is relatively crude and in later work on our data the more ideal formulae have been used. The ultimate differences in conclusion on heritabilities are nevertheless slight, especially in relation to sampling and other sources of variation necessarily affecting the final values to be discussed.

When the psychometric corrections judged most appropriate for error of measurement in the tests, as measures of the concept, have been made we still have to address ourselves to the second form of error in conclusions from experiment—sampling error. The sampling errors of the variances for the various constellations (no matter, of course, which algebraic computer avenue we use to get them) are themselves statistically straightforward. For example, the within-family variance—Eq. (8.1)—is an ordinary variance with an ordinary standard error of $\sigma^2/2N$. The problem concerns how we combine these concrete variance standard errors, in such equation solutions as in Table 4.7 which combine several concrete variances, to get the sampling error of each resultant abstract variance. In the twin method—at least when our answer comes from only two concrete variances—the significance of the result can be reached through the simple variance ratio, F, as Eaves (1972), Eaves and Gale (1974), Elston and Gottesman (1968), Mittler (1971), Vandenberg (1965a), and others have suggested and used. In the MAVA model with the raw variances of *many* constellations fixing each abstract variance (Table 4.7) the evaluation will depend on whether we follow the OSES approach of solving simultaneous equations or the least-squares fit and the maximum-likelihood methods.

The estimation of standard errors of abstract variances, from the equations in Table 8.3, unfortunately depends on getting standard errors of ratios, as well as of sums, occurring in the original concrete variances. One has to decide also whether the variances on the right are correlated. Usually within- and between-mean-squares are obviously uncorrelated for any one particular analysis. In this case, since they are quite different groups of subjects (Ss), it might be argued that the chances of systematic correlation are still less. However, the question is actually a very complex one, to be answered only by study of the structure of the data in each case. For example, in some instances of sibling pairs, they are *not* random subgroups from the larger group. The consensus of expert statistical opinion consulted nevertheless is that, in terms of possible correlation over a series of samples, we are justified in treating these as uncorrelated variances. Accordingly the variance of the abstract variance estimate is a simple sum of the independent variance functions for the experimentally obtained variance estimates on the right of our equations.

As stated earlier, it is likely in most actual experiments, unless one is prepared to throw away hard won data, that the sample numbers in the various experimentally obtained family constellation subgroups (sibs, twins, etc.) will differ. If the successive groups in Table 4.4 (p. 106) have n_1, n_2, etc., cases, then the estimate of the variance of the variance, as stated earlier (p. 108), is

$$\sigma_{\hat{\sigma}^2}^2 = \frac{2k_1^2\sigma_1^4}{n_1 - 1} + \frac{2k_2^2\sigma_2^4}{n_2 - 1} + \cdots + \frac{2k_n^2\sigma_n^4}{n_n - 1}. \qquad (8.18)$$

However, if n is small—less than 30 cases (unlikely in these experiments)—the degrees of freedom, which give an unbiased estimate, become $n_i + 1$, etc., according to Daniels (1939) and Welch (1956).

Before leaving Eq. (8.18) let us pause to rewrite it in condensed form, thus

$$\sigma_{\hat{\sigma}^2}^2 = 2 \sum_{l=n}^{N} \frac{k_i^2\sigma_{ai}^4}{(n_i - 1)} \qquad (8.19)$$

where i is any one of the k constellations (empirically used groups) N in number and n_i is the number of cases in the ith constellation.

An introduction to the required formulas here has already been given with the MAVA method in Eqs. (4.16) and (4.17), but here we shall somewhat extend the definitions. Thus the upper and lower confidence limits, at a desired level of confidence, P_j set by the experimenter are shown for the estimated abstract variance, $\hat{\sigma}_a^2$ by the following expression:

$$\text{Pr}\left\{\left[\sigma_a^2 - Z_j\frac{(2\Sigma k_i\sigma_i^2)^{1/2}}{n_i - 1}\right] < \hat{\sigma}_a^2 < \left[\sigma_a^2 + Z_j\frac{(2\Sigma k_i\sigma_i^2)^{1/2}}{n_i - 1}\right]\right\} = P_j, \qquad (8.20)$$

where Z_j is the normal standard deviate for a given probability, P_j, σ_a^2 is the "true" value for the population variance, and we assume that, through large

n or the use of an unbiased estimate, the distribution of the variance is normal.

Finally, to determine the confidence limits of heredity–environment variance ratio's, for example

$$N_b = \frac{\sigma_{bg}^2}{\sigma_{bt}^2}; \qquad H_w = \frac{\sigma_{wg}^2}{\sigma_{wg}^2 + \sigma_{wt}^2}$$

or, in general, σ_a^2/σ_b^2 we simplify the probability statements for the numerator and denominator as follows:

$$\Pr [A < \hat{\sigma}_a^2 < B] = P_1, \qquad (8.21)$$

$$\Pr [C < \hat{\sigma}_b^2 < D] = P_2. \qquad (8.22)$$

The combined probability for the ratio is not strictly the product $P_1 P_2$ because P_1 and P_2 are not entirely independent, but when using the extreme confidence interval ranges (.95 to .99) in which we are alone interested, the error from assumption of independence is very small and we may use

$$\Pr \left[\frac{A}{D} < \frac{\hat{\sigma}_a^2}{\hat{\sigma}_b^2} < \frac{B}{C} \right] = P_1 P_2. \qquad (8.23)$$

As with any confidence limits test, it is not possible to decide, without actual empirical variance data carrying experimental error, what the desirable number of cases should be in a well-planned experiment.

Following discussions with Kempthorne and Norton, we basically accepted in our own calculations later that the standard error of an abstract variance derived from observed, concrete variances, would be as given in Eqs. (8.18) through (8.22). However, just as the concrete variances ideally need correcting according to dependability, validity and stability coefficients as in (8.17) when estimating the abstract variances so they need, ideally, to be corrected when estimating the variance of the abstract variances.

Second, if the standard error (or, at any rate, the confidence limits) of the ratio of variances is to be determined, to be able to summarize the results as a nature–nurture (or, later, H) ratio, one must take account of the presence of common terms in numerator and denominator, as follows:

$$\sigma_R^2 = [\hat{\sigma}_g^2/\hat{\sigma}_t^2]^2 \, [\hat{\sigma}_{\sigma_g^2}^2/(\hat{\sigma}_g^2)^2 - 2 \, \text{Cov} \, \hat{\sigma}_g^2 \sigma_t^2/\hat{\sigma}_g^2 \hat{\sigma}_t^2 + \hat{\sigma}_{\sigma_t^2}^2/(\hat{\sigma}_t^2)^2] \qquad (8.24)$$

in which R is the ratio and σ_t^2 and σ_g^2 are the σ_{wt}^2 or σ_{bt}^2, and σ_{wg}^2 or σ_{bg}^2, or their sums, as usually derived from the concrete variances. This approximation neglects higher powers of the ratio and accepts the obtained $(\sigma_{\sigma_t^2}^2/\sigma_{\sigma_g^2}^2)^2$ as a substitute for the unbiased estimate which should be used. However, the magnitude of error likely to be introduced is small compared with that involved in the old formula which neglected the existence of common variance in the numerator and denominator. The practical result of this modification is to give us confidence in working with somewhat smaller samples than the forbiddingly large numbers which other arguments have so far seemed to suggest. Nevertheless, since empirical estimates in the 10-equation solution of the within-family sib variance and the between-sib-family variance must be experimentally independent if this formula is to be used (Cattell, 1960), it is still highly desirable to measure, to cover all constellations, some 2500 pairs of children. Fortunately, as Eq. (8.18) indicates, a shortage in one category can to some extent be made up by greater numbers in another. This has the important practical corollary that one might reduce a study to say, 100 pairs for rarer groups such as twins or sibs reared apart and at the same time use several hundreds of sibs normally reared together. Although the MAVA method thus has technical promise for yielding better values and previously insoluble values its full use will demand funds and fortitude.

The case of estimation in the least-squares method will be taken up in connection with results given later. Undoubtedly, the most satisfactory evaluation of standard errors of results is provided by the maximum-likelihood method (Kempthorne, 1957) as illustrated in Chapters 9 and 10.

Standard texts contain several excellent discussions of genetic significance testing (Eaves, 1969; Eaves et al., 1978; Kempthorne, 1957, pp. 164–250; McClearn & DeFries, 1973, p. 60; Loehlin, 1979; Van Abeelen, 1974) deals with the problems in the useful task of combining results from different samples and studies. The general conclusion (except for some defense of moderate samples in twin analysis, by Eaves [1969, 1972]) is that almost all studies to date have operated with insufficient samples, though Loehlin argues this is not the sole source of the remarkably high discrepancies of various studies. I would argue that if the condition holds that *the constellations are from the same racial–culture locality,* then 200 cases from each of half a dozen constellations, and thus about 2000 for the general population should be counted as a minimum for satisfactory work. In stating this I am aware of taking the position, "Do what I say and not what I do," since several groups in what follows in this chapter fall somewhat short of this. In the main MAVA findings, however, the samples do not fall short nearly to the extent of the data bases of most psychological conclusions published in the past 30 years.

5. The Heritability of Primary Abilities

The provision of evidence on the inheritance of primary abilities hinges very much on the issue of "What are primary abilities?" There appear to be only three researches of a basic kind aimed at a sufficiency of variables, and carried to discovery of the existing simple structure, to determine the primary correlations, namely, those of Thurstone (1938), of Hakstian and Cattell (1978), and Horn (1972b, 1980). They agree well, but the two latter, being

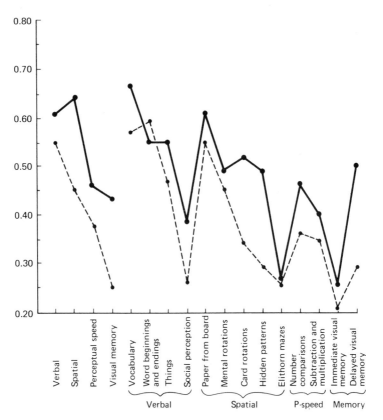

Figure 8.2 Regression of mid-child on mid-parent for primary and special abilities; ———— Americans of European ancestry (830 families); ----- Americans of Japanese ancestry (305 families).

This is modified from De Fries *et al.* (1979) by omission of tests likely to be largely measures of general intelligence. The grouping of the test into primary abilities is partly based on face validity, and the results are given here mainly to let the reader get at a glance the variability of resemblance on special abilities. It must not be forgotten that these are *phenotypic* resemblances, and could be high without high heredity if the ability is one on which parents and children interact a good deal environmentally. This would obtain more for verbal ability than most others. The especial interest of this work by De Fries *et al.* is that it compares two racial groups in largely the same environment. The almost uniformly lower parental resemblance in children of Japanese ancestry could be due to intergenerational shift in culture such as Meredith (1965) has shown for Japanese on the 16 Personality Factor Questionnaire, though other theoretically provocative possibilities might occur to the reader.

years later, carried exploration into new behavioral domains, especially with Horn and Stankow's extension into auditory primaries. Thus the latter deal with a wider spectrum of primaries, as shown in Table 8.1, with their second order structure developed on a wider basis.

All genetic evidence so far available is only by the twin method or in the form of correlations of parents and children, as in Figure 8.2, the latter per-

TABLE 8.2

Evidence on Heritability of Primary Abilities by the Twin-Method Ratios of Within-Pair Variances for Fraternal over Identical Twins

	Primary ability	Blewett, 1953	Thurstone et al., 1955	Vandenberg, 1962	Vandenberg, 1966a	Mean of 10 studies (Vandenberg, 1968)	
	Verbal	3.13[a]	2.81[a]	2.65[a]	1.74[a]	2.53	.61
	Space	2.04[a]	4.19[a]	1.77[a]	3.51[a]	2.25	.56
	Number	1.07	1.52	2.58	2.25[a]	1.91	.48
	Reasoning	2.78[a]	1.35	1.40	1.10	1.65	.39
	Word fluency	2.78[a]	2.47[a]	2.57[a]	2.24[a]	2.47	.60
	Memory	Not used	1.62	1.26	Not used	2.18	.54
Sample	FTT pairs	26	53	37	36		
Size	ITT pairs	26	45	45	76		

[a] Heritability differences between ITT and FTT pairs are statistically significant. (Tested by Vandenberg, 1967, 1978).

mitting only the very limited conclusion that there are upper limits to the possible heritability.

Table 8.2 is of the first (twin) origin, and reports Dixon and Johnson's survey (1980) of four different sources—to show extent of variability—plus Vandenberg's (1968) combined result of 10 studies. All these studies are experimentally well directed at Thurstone's original set of primaries. However, they have not all reported actual H_w heritabilities, which we have derived in Table 8.2 from Vandenberg's composite, allowing unit variance to σ^2_{FTT} in $H_w = (\sigma^2_{FTT} - \sigma^2_{ITT})/\sigma^2_{FTT}$. It will be seen here (and later indirect results are consistent) that verbal ability and word fluency have highest heritability, with spatial fairly close and number, reasoning, and memory well below. One possible inference is that verbal and spatial performances become most fully trained and exercised, so that most people approach their upper limit of performance and there is little environmental range to produce threptic differences.

Since the Hakstian Battery (Hakstian & Cattell, 1976) Table 8.4, of 20 primaries (22 with Horn and Stankow's addition) has not yet been used in genetic research we are reduced for evidence on the remaining primaries to oblique inferences, in two respects (a) inferences as to what primaries are involved in the assortment of actual tests used by DeFries et al. (1979), Stafford (1965) and others (in Table 8.3). They give grouping into primaries for some, but several more primaries would be involved in some measures and (b) inferences from phenotypic correlations only, of relatives of varying relation. Results with these limits are shown in Table 8.3.

To help (a), the grouping of types of actual tests among the 20 now-established primaries is given in Table 8.4, since it is necessary to be clear on

TABLE 8.3
Phenotypic Resemblance of Near and More Remote Relatives on General Ability, Primary Abilities, and Specific Abilities

I. *Within the inner family*
  (a) *On intelligence* (crystallized; WAIS and WISC)

Sample 55	Sample 55	Sample 55
Mother–son	Father–son	Midparent–son
.36	.43	.44

Reported by Williams (1975). The *factor equivalence* of the *adult* WAIS and *child* WISC, as Williams points out, is by no means proved. (Compare results on a wider variety of intelligence tests in Tables 5.10 and 5.11). It would be desirable to get r's for culture-fair tests, where the four subtests and their fluid intelligence (g_f) loadings are essentially the same for scales 1, 2, and 3 (i.e., the whole age range).

  (b) *On primary abilities*

	Sample 50	Sample 63	Sample 64	Sample 51
	Mother–son	Father–daughter	Mother–daughter	Father–son
Numerical	.62	.21	.25	.08
Spatial	.41	.36	.22	.03

(Assortiveness of mating: numerical .07, spatial .05)
Courtesy of Vandenberg (1965) and Stafford (1965) and Academic Press.

II. *More distant collateral relatives*

	Biological uncles–aunts[a] and nephews–nieces	Cousins[b]
(a) *Intelligence* (First principal component of measures below)	.18**	.29**
(b) *Primary ability factors*		
Verbal	.15*	.24**
Spatial	.14*	.11
Perceptual speed	.20**	.15*
Visual memory	.14*	.07
(c) *Tests*		
Vocabulary	.27**	.31**
Visual memory (immediate)	.11	.08
Things	.11	.26**
Mental rotations	.15*	.12
Subtraction and multiplication	.20**	.21**
Lines and dots (Elithorn mazes)	.05	.17**
Word beginnings and endings	.14*	.23**
Card rotations	.18**	.17**
Visual memory (delayed)	.12	.14*
Pedigrees	.20**	.26**
Hidden patterns	.23**	.19
Paper form board	.26**	.23**
Number comparisons	.21**	.18**
Social perception	.01	.23**
Progressive matrices	.20**	.22**

  * $p < .05$.
  ** $p < .01$.

[a] Regressions, pairwise deletions for missing data.
[b] Intraclass correlations; pairwise deletions for missing data Part II from Table 6.3, Resemblances Collateral Relatives in Cognitive Abilities. In R. C. Johnson, J. C. De Fries, F. M. Ahern, & M. P. Mi. Seventh Annual Meeting, Behavior Genetics Assoc., Louisville, Kentucky, 1977.

TABLE 8.4
The Definition of Primary Abilities by Actual Tests

Test	Test content	Primary factor
1	Vocabulary	
2	Proverbs	Verbal (V)
3	Verbal analogies	
4	Arithmetic calculation	
5	Arithmetic calculation	Numerical (N)
6	Arithmetic calculation	
7	Figures (see Thurstone & Thurstone, 1941)	
8	Figures	Spatial (S)
9	Surface development (Thurstone, 1938)	
10	Letter–number similarity	
11	Figure–figure similarity	Perceptual speed and accuracy (P)
12	Figure–figure similarity	
13	Incomplete pictures	
14	Mutilated words	Speed of closure (Cs)
15	Mutilated words	
16	Letter series	
17	Letter sets	Reasoning (I) (mainly inductive)
18	Letter sets	
19	Word–number memorization	
20	Figure–number memorization	Associative memory (Ma)
21	Figure–number memorization	
22	Physical principles	
23	Electrical–automotive facts	Mechanical (Mk)
24	Woodwork and shop	
25	Hidden figures	
26	Hidden figures	Flexibility of closure (Cf)
27	Hidden figures	
28	Digit span	
29	Digit span	Span memory (Ms)
30	Digit span	
31	Spelling	
32	Spelling	Spelling (Sp)
33	Spelling	
34	Design preference	
35	Design preference	Esthetic judgment (E)
36	Design preference	
37	Object-attribute memorization	
38	Object-attribute memorization	Meaningful memory (Mm)
39	Object-attribute memorization	
40	Multiple object grouping (heterogeneous)	
41	Multiple grouping (foods)	Originality I (Ol)
42	Multiple grouping (animals)	
43	Attribute listing—stimulus I	
44	Attribute listing—stimulus II	Ideational fluency (Fi)
45	Attribute listing—stimulus III	
46	Anagrams	
47	Anagrams	Word fluency (W)
48	Anagrams	

(*continued*)

TABLE 8.4 (*continued*)
The Definition of Primary Abilities by Actual Tests

Test	Test content	Primary factor
49	Object synthesis (see Guilford, 1971)	
50	Object synthesis	Originality II (*O2*)
51	Object synthesis	
52	Aiming	
53	Aiming	Aiming (*A*)
54	Aiming	
55	Drawing—straight lines	
56	Drawing—left-hand curves	Representational drawing (*Rd*)
57	Drawing—right-hand curves	

what we mean by primary abilities. The large 57 × 20 factor loading table for these is given in Hakstian and Cattell (1974). Grouped as DeFries *et al.* (1979) have grouped them, the Table 8.3 results agree with Table 8.2 at any rate to the extent of showing highest values for verbal (*V*) and spatial (*S*) and lower for memory (*M*), though Stafford's results disagree in giving equally high results for numerical ability. For the general result as regards phenotypic correlations is clearly that relatives, even of remoter degrees, are commonly significantly correlated on these primaries, numerical and memory again running below vocabulary and spatial (though no significance of these differences are given). Incidentally, Loehlin and Nichols (1976) found no firm evidence of different familial resemblance and heritability for different primaries, but Carter (on relatively crude measures, 1932) found a verbal–numerical difference, consistent with the preceding, as also did Park *et al.* (see also DeFries *et al.*, 1979).

The real problem in coming to a firm conclusion simply from parent–child correlation or regressions, as reached in Table 8.3 (or Figure 8.2) has already been clarified in the discussion in Chapter 5 on the "fragmentary" parent–child and adoptive-family methods. One's initial hope—that by contrasting the phenotypic correlation reached (8.25a) with genetic knowledge (at least for additive genetic action) of the theoretic purely genetic correlation to be expected—proves to be a mirage. We know that r_g—the purely genetic correlation in (8.25b) for mid parent with child or mid child—must be .707 (Table 5.10)

$$r_p = \frac{\sigma_{gc}^2 + \sigma_{tc}^2}{\sigma_g^2 + \sigma_t^2}, \tag{8.25a}$$

$$r_g = \frac{\sigma_{gc}^2}{\sigma_g^2}, \tag{8.25b}$$

$$r_t = \frac{\sigma_{ti}^2}{\sigma_t^2}. \tag{8.25c}$$

(Here g and t have the usual meaning, p is phenotypic, c is common variance for members of the family, e.g., $\sigma^2_{gc} = \sigma^2_{wg}$ and σ^2_g and σ^2_t are general *population*—parent or child—variances.) Incidentally one should distinguish between the purely genetic correlation, σ^2_{gi}/σ^2_g here and as used throughout Chapter 5, and, what some writers have also called or miscalled the genetic correlation:

$$r'_g = \frac{\sigma^2_{gc}}{\sigma^2_g + \sigma^2_t}, \qquad (8.26)$$

which for clarity we designate r'_g and call the *phenogenetic correlation*. As we consider Eqs. (8.25) and (8.26) we should note that only r_p is an observable (as in Table 8.3) and of the rest only (8.25b) can be known (from genetic theory). The others are abstract values we should like to know but do not. What we would like to know, of course is that $\sigma^2_{bg}/(\sigma^2_{bg} + \sigma^2_{bt}) = H_b$, but this cannot be derived algebraically from what we have: (8.25a) and (8.25b). As in the earlier examination of fragmentary designs, we find we can handle Table 8.3 only with auxiliary information, for example, from identical twin data. All we can say is that the values there represent limits or solutions granted special assumptions. Thus if there were no threptic variance in the trait at all we should know, because r_p would equal r_g. If we knew there was threptic variance but no *common* threptic variance, r_p could be interpreted as r'_g. If we could assume $r_g = r_t$—which might have moderate probability —then the obtained r_p equals the purely genetic correlation, r_g. In other words if r_p as obtained equals genetic r_g, then threptic variance is either absent altogether or $r_t = r_g$. Beyond that all we can say is that if r_g is larger than r_p then r_t must be smaller (and vice versa), but this permits no statement about the heritability.

Regardless of the method of analysis used—full MAVA or the partial MAVA designs of twin, adopted family, parent–child, etc.—it will become necessary in research in this coming generation to enter with measures consistent with some comprehensive theory of ability structure such as the *triadic model* above. The latter suggests batteries aimed at three levels, and at attempts to separate the estimates of each stratum from undue contamination by the adjoining stratum. The top stratum *general capacities, g_f, g_c, g_r,* and g_s, must inevitably contain specifics in *the score*, reducible in their individual influence only by using and adding *several* subtests. (For example, fluid intelligence should not be measured by Raven's matrices alone.) The *provincial powers, p's*, need sharper research definition. The *agencies, a's*, such as numerical skill, mechanical aptitude, need to be measured as "stubs" (if the SUD model is sustained) which can be done by partialing out from them, after scoring, the g and p's.

Doubtless the triadic theory will move on in the next few years, for example, under growing knowledge of the location in the right and left hemispheres. But on its present basis some hypotheses can be formed as to anticipated heritabilities of g's, p's and a's. The g's in general seem to express some property of the cortex as a whole. Thus g_f may be total myelinated cell count; g_r may be the total growth of axon connections; g_s may be the level of some chemical pacemaker; and so on. The sensory perception and association areas guiding sensory discrimination may well become better sustained as distinct cortical zone factors and sharpened in pattern, but thus relations to g's and a's may well become complex in path factor relations. On the triadic theory the behavior-genetic outcome might well be substantial inheritance for the g's, high for the p's in early life, falling as the size of the association areas is built up by experience, and quite low heritability for the a's such as numerical ability, reasoning ability, various dexterities and so on.

If the last hypothesis, about agencies, seems to fit poorly, as witnessed by substantial heritability for verbal ability and some other primaries (most primaries are considered agencies) it is because investigators have not yet faced (*a*) the psychometric problem of getting well-separated factor estimates (especially from the higher order intrusions) and (*b*) the implications of the *investment theory*. If we take the SUD (stratified unrelated determiner) model (Cattell, 1978a) of analysis of second- and higher-order factors, only a fraction (the independent "stub") of a primary like V, N, or S is the *real* determiner, possessing the unique quality of the primary. The rest is the variance from the investment, present in all of them, of fluid general intelligence, (See Figure 8.1), which produces the correlation among them. These stubs could be pure threptic gain from experience. Similarly, on a larger range of performances we have argued that crystallized intelligence, g_c, is g_f plus experience. If psychometrically we could separate this stub (though a broad one!) from the g_f in g_c we might expect to find it has zero heritability. The psychometric problem with the immediately reported V, N, R, etc., heritabilities is that the set of measures for the stub has to be the same as that for the oblique primary as it stands. There is at present no subset that will magically set aside the g and get hold of what could be a threptic part only.

The possibilities are (*a*) to develop, by continued exploration, tests that discriminate in themselves a more genetic or more purely threptic part; (*b*) to locate the builders of the threptic adjunct in the environment, as we hypothesize variance in length of uniform schooling to account for the threptic part of g_c; and (*c*) to find measures in the area of the eidolon factor which show high inheritance. If the last is achieved it could be partialed out. This could be done today by partialing g_f scores out of individuals' V, N, R, S, etc., primaries (if the investment theory is correct) and applying the twin or MAVA analyses to stub scores so obtained.

Our theory is that the heritabilities, in Tables 8.2, 8.3, etc., for primaries would thus be substantially reduced, though it is also possible that *some* local primary genetic element would exist, along with the effect of growth of primaries as agencies. For example, both evolution and brain anatomy suggest that the primaries V, S, and the hand dexterity factor are not merely developed as agencies but lean on brain genetic areas too. The heritabilities of V and S are as high as for g_f, and if the stubs were nothing but experience these H's would be lower. But in numerical and various other agencies H is decidedly lower, and the stub might show a heritability of zero once g_f is partialed out.

6. The Heritability of Fluid and Crystallized Intelligence Examined by the OSES Method

It is proposed now to begin the application of the methods and concepts studied to this point to personality and ability source traits. It is appropriate to begin with abilities, the structure of which has been studied longer, and on which far more data has been gathered. In each area the main exposition will concern the MAVA method, though the final summary will call in twin method results, with appropriate adjustments for the different meaning of the within-family H's.

The MAVA experiment on which the conclusions of this section are based covered 466 brothers reared together, 94 pairs of identical twins reared together, 124 pairs of fraternal twins reared together and, from the general population, 1973 boys in the case of fluid intelligence and 2973 for crystallized intelligence (in OSES; 1543 for least squares). All were thus uniformly made comparisons, since identicals had to be of the same sex, and all were in the 12–18-year range[3] (in OSES, but 2579 in least squares).

The theory of fluid and crystallized intelligence must be considered sufficiently defined earlier and in the literature (Buss & Poley, 1976; Butcher, 1968; Hakstian & Cattell, 1974; Horn, 1972b, 1976, 1980; Johnson & Dixon, 1980; Vandenberg, Meredith, & Keese, 1979). The tests used were, for fluid intelligence, *Forms A and B of the Culture Fair Intelligence Test,* Scale 2 (Cattell & Cattell 1949, 1968) and for crystallized intelligence the *B scale in the A and B forms of the HSPQ (High School Personality Questionnaire)* (Cattell & Cattell, 1969). The obtained concrete variances, for IQ's in the case of the CF test and for age-corrected raw scores in the case of the HSPQ are given in Table 5 of the article by Cattell, Klein, Graham, and Schuerger (1980).

With four constellations, and within and between variances available

---

[3] Throughout the three chapters now beginning, and special articles elsewhere, the reader may find occasional relatively small differences in the subject sample totals reported. These arise from some subjects not having completed certain tests, from some analyses having been begun before subjects were added for later analyses, etc. A stringent combing of the original gathering of over 3000 tested subjects was made by Frank Ahern, Jerry Brennan, and myself for inadequate test data and having family relationships in doubt, resulting in about a 25% reduction from the original card list.

for three of them, the seven equations shown in Table 4.2 (p. 103) for the most limited MAVA design alone become available in this case. Further, in an analysis by simultaneous equations it seemed safest to drop one of the three in the set (3), (6), and (7) though some statisticians have argued this is unnecessary since they are *experimentally* independent.

It turned out, as happens frequently in the overlapping sets of simultaneous equations (OSES) method, that for algebraic and arithmetical reasons only one six-set solution could be obtained, but solutions were available from four different sets of five equations (see Cattell, Klein, Graham, & Schuerger, 1981, Table 3). Since solutions for seven unknown abstract variances—σ_{wg}^2, $\sigma_{wt.t}^2$, $\sigma_{wt.s}^2$, σ_{bg}^2, σ_{bt}^2, r_{wgwt}, and r_{bgbt}—were our target for an adequate basic solution, some auxiliary aids were indicated to be necessary. (The third of these unknowns—$\sigma_{wt.s}^2$—was vital if we aim to avoid the prime weakness in the twin method of assuming the environment variance within *sibs*, $\sigma_{wt.s}^2$, is the same as within pairs of *twins,* $\sigma_{wt.i}^2$.

The first of these aids was to drop attempting to get σ_{bg}^2 as an independent solution and to derive it instead from the empirically obtained σ_{wg}^2 by the assumption that σ_{bg}^2 is related to σ_{wg}^2 by the geneticist's equation for assortive mating effects given in Eq. (5.14) on p. 139. The question then arose as to what degree of assortiveness at the genetic level ($r_{fm.g}$) we should assume. From our data (Cattell & Nesselroade, 1967) and others' (Jensen, 1973; Vandenberg, 1972) .25 seemed very modest, but we were persuaded by Rao and Morton and other geneticists who seem to doubt *any* significant *genetic* assortiveness, and by the trend arguments of Johnson, Ahern, and Cole (1980) to go as low—at least in an alternative trial solution—as .07. These give ratios of σ_{bg}^2 to σ_{wg}^2 of 1.67 and 1.15, respectively.

The algebraic solutions from Table 4.2, using these values, are illustrated in Table 4.4. There are, of course, different solutions for the other five sets of equations and for the six (equation) set. Since solutions are in some cases quadratic there is then a choice of two solutions that has been guided by the rules that (*a*) no negative variance is accepted, (*b*) the value must not be quite outside the range indicated by general scientific considerations, and (3) the value should essentially match that reached when more equations are available to pin down that value. In fact our experience was that we never had difficulty in rejecting the "absurd" quadratic alternative.

It has seemed desirable to report here both the abstract variances reached and the heritabilities derived from them, since the former are always important in themselves. Since, however, there are *two* assortive mating values, and *two* empirical estimates of the general population variance (one by taking one member of each pair, and one by taking the test general population standardization giving a larger but less local N) not to mention gross and net heritabilities, there are quite a number of alternative end values to be scanned.

The full range of these alternatives is set out in the original article, but

TABLE 8.5
Abstract Variances Derived by MAVA
for Fluid (g_f) and Crystallized (g_c)
Intelligence by OSES

	Fluid	Crystallized
$\sigma^2_{wt.t}$	83.09	3.74
$\sigma^2_{wt.s}$	147.69	6.54
σ^2_{wg}	214.76	10.68
$r_{wg.wt.s}$	−.59	−.52
σ^2_{bt}	116.32	11.39
σ^2_{bg}	246.97	17.84
r_{bgbt} (r_{wgwts})	−.59[a]	−.81

[a] No solution was possible. For reasons in text r_{wgwts} was taken in this case as best approximation

	Fluid	Crystallized
$\sigma^2_{wt.s}/\sigma^2_{wt.t}$	1.78	1.75

here we shall set out the two sets finally preferred, on the following grounds: First we accept a higher assortive mating for crystallized (.25) than fluid (.07), on the ground that the latter quality is less evident to guide assortiveness in regard to a prospective mate than is the education-level-tied g_c. The independent evidence available in the later analyses where $r_{fm.g}$ is found, instead of assumed, points to the same conclusion. Second, we take a single pool of sources for the general population in the case of g_c, but, because of different regional input in the g_f norms we have worked in that case also with that estimate of the general population variance found by taking one person from each of the pair collected. The resulting eight "raw" concrete variances as IQ's for the culture-fair and age corrected raw scores for the summed B scale in the HSPQ are reported in Table 5 of Cattell, Klein, Graham, and Schuerger (1981).

The values for the abstract variances thus derived are given in Table 8.5. (Incidentally it has not seemed appropriate in view of the complexity of the derivative formulas [Cattell, Klein, Graham, & Schuerger, 1981, Table 4] to attempt, by Eq. (8.14), any reliable estimation of the standard errors of these abstract variances. They must eventually be evaluated by comparison with other independent, experimental results.) Meanwhile it is at least evident that the values are well *within* the ranges commonly encountered in earlier research.

The heritabilities, calculated as net values[4] (Chapter 7, p. 273) are set

---

[4] It will be at once evident from calculating the two kinds of H_w's here that with the unusually high negative covariances the *gross* heritability would actually exceed unity. This is true

TABLE 8.6
Net Heritabilities of Fluid and Crystallized General
Intelligences by OSES Method (with Genothreptic
Correlations as Given)

Ordinary families	Fluid (g_f)	Crystallized (g_c)
H_w	.59	.62
H_b	.68	.38
H_p	.64	.44
r_{wgwt}	−.59	−.52
r_{bgbt}	−.59[a]	−.81
Calculated on twin $\sigma^2_{wt.t}$ from Table 8.5		
$H_{w.t}$	.72	.74

[a] In this case equal to r_{wgwt} by assumption.

out in Table 8.6. The main features of this table will be commented on after the second approach, by the superior, least-squares method, has had its results set out. But as a matter of immediate interest we may notice the following:

1. That for fluid intelligence all three heritabilities are within the usual range of reported values though somewhat on the low side. For crystallized intelligence, however, the between-family, H_b, and total population, H_p, values, are decidedly lower. An appreciable part of this difference from twin method calculations arises, as pointed out earlier, from the acceptance in the latter method of $\sigma^2_{wt.i}$ (or $\sigma^2_{wt.t}$) = $\sigma^2_{wt.s}$. If we shift to the classical twin method basis of calculation on our present data, according to the ratio of threptic variances found at bottom of Table 8.5 then, as the last row in Table 8.6 shows, the H values rise, for the within-family relation, to the values, .72 and .74. This result is higher than for MAVA as is the .64 for earlier twin method studies surveyed by Vandenberg (Table 8.12) and the .72 from Erlenmeyer-Kimmling and Jarvik (Tables 3.1 and 8.3.)

2. Since the magnitude of the difference of within-family threptic variance between identical twins and (a) fraternal twins, and (b) sibs, has been much in debate, it is of considerable interest to find what MAVA gives by its ability to solve independently for the values. As Table 8.5 shows the ratios are 1.78 and 1.75, and since the fraternal and identical twins are thrown together here, the ratio could well be higher for the latter, that is for $\sigma^2_{wt.i}/\sigma^2_{wt.s}$. The corresponding ratios in the Cattell, Stice, and Kristy (1957)

regardless of whether we "reconstruct" the total variance in the denominator from $\sigma^2_{wg} + \sigma^2_{wt} + 2r_{wgwt}\sigma_{wg}\sigma_{wt}$ values as found in the abstract solutions or from the available, slightly to moderately different, raw σ^2_{ST} values. This adds an "empirical" argument to the preceding arguments against the H_G as a scientifically useful concept rather than as a sometimes practically useful calculating value.

and Cattell, Blewett, and Beloff (1955) MAVA studies were 3.39 for g_f and 1.26 for g_c, averaging 2.33. As far as results on intelligence are concerned, therefore, it is a reasonably safe conclusion, from the OSES method, that the threptic difference variance environmental produced in ordinary sib families is about twice that of identical twins. In some personality traits examined in Chapter 9 and 10 the difference is even greater, but at least the point is made that it is a gross source of error to consider them as equal.

3. The results in Table 8.6 show a higher *general population* heritability, H_p, for fluid than crystallized intelligence which is due entirely to a higher heritability in the between-family situation. As of this date there is available in behavior-genetic research only one other MAVA analysis for g_f and one for g_c, and both might have the method (or data) disadvantage of being about 25 years old. But fortunately the tests used (culture fair for g_f and the B scale in the HSPQ in the 11–15-year range, for g_c) are almost identical with those used today. The age range too is almost identical (11–15 years with a mode at 12–13), and the sample numbers are substantial: 647 tested by g_f and 962 by g_c. (The latter while not as large as here: 1082 g_f and 3267 g_c are not so low as to produce a lot of sampling error.) The heritabilities with suitable allowance for making genothreptic correlations comparable with the present for the g_f measure (Cattell, Stice, & Kristy, 1957) were $H_w = .41$, $H_b = .93$, and $H_p = .53$, and for the g_c study (Cattell, Blewett, & Beloff, 1955) $H_w = .59$, $H_b = .39$ and $H_p = .52$. One has to allow for the fact that the state of the art then was not what it is now, in that although the OSES method was used it had no definite solution for the genothreptic correlations and the values are taken from the articles as a mean of (usually) four different tried genothreptic correlations adjusted to present values. The theoretically important convergence is on a distinctly higher H_b for fluid than crystallized intelligence.

Our inclination, in view of these uncertainties, is not to weight these pioneer MAVA studies any more than the numerous past twin studies in the final conclusions in this chapter on intelligence heritability. However, we see agreement of the older OSES analysis of MAVA in (*a*) that the population heritability for crystallized intelligence is either less than or barely equal to that for fluid intelligence, (*b*) that decidedly lower value for H_{g_c} obtains in the *between-family* heritability, (*c*) that the heritabilities (taking H_p as most comprehensive) are on the lower side of values commonly reported in the past, and (*d*) that if calculated as in most past studies, using the twin method, assuming $\sigma_{SA}^2 - \sigma_{ITT}^2 = \sigma_{wg}^2$ (i.e., that sib and twin environments are the same), the H_w value comes up into the .65 to .75 range which is about central for all past results. It is therefore desirable to point out that if we are speaking of typical sib families which, after all, constitute the bulk of the population, these latest H_w values, by MAVA rather twin design, point to .60 as about the best estimate, at least for an ethnically diverse U.S. population, though, as discussed later, perhaps not for older European populations.

7. The Heritability of Fluid and Crystallized Intelligence Examined by the Least-Squares Method

The drawbacks of the OSES method of analysis, on which the three earlier studies in the last section are based, have been pointed out. The main disturbing consideration is that different subsets of equations give somewhat different results, though with as many soluble equations available as unknowns this would not arise. To proceed by averaging them with equal weights is not entirely satisfactory since each rests on a different set of samples of concrete variances, differing perhaps in population representativeness (e.g., in the adoptive, unrelated reared together cases) as well as in the total size of that collection of constellations. Also we encounter with real data the fact that some sets are algebraically soluble, but arithmetically insoluble, because of chance extreme piling up of sampling errors. And there are assumptions in rejecting one of two quadratic solutions as "absurd," which we have already mentioned. Finally in this list of "doubts" we must mention that except with a six or larger equation set (permitting independent solutions for σ_{wg}^2 and σ_{bg}^2) an assumption has had to be made about the magnitude of the genetic assortive mating correlation.

Because of these shortcomings, we shall give greater weight in final evaluation of the inheritance of intelligence to the different method now to be described and applied. At the same time let us not overlook some genuine advantages of the OSES method. It has one main advantage over such "push-button" program uses as in the least-squares or maximum-likelihood methods, namely, that one can see at each step what is happening. One can see where a result is strongly affected by inclusion of a particular concrete variance, about the relative experimental soundness of which one has an experimenter's insight, and one can watch the effect closely of the various assumed "experimental" sizes of assortive mating.

Indeed, in the debates that will arise over the weight to be given to these two methods and the maximum-likelihood method let us not lose a broader perspective on scientific method. The fact is that in many scientific results which have stood the test of time, and broader comparisons, a part has generally been played by "artistic" research skills and experience, which could not be reduced to mechanical use of scientific procedures. The history of science can present instances where the tentative, but ultimately definitely confirmed results of investigators like Copernicus, Curie, Dalton, Edison, Harvey, Koch, and Pasteur could not be immediately sustained by mechanical procedures of experimenters without "flair." Incidentally an important instance of this in current psychology has been the incapacity of some researchers to replicate unique factor resolutions obtained by those with years of experience in simple structure rotations. This has led to skepticism in untrained onlookers about the status of primary and secondary trait structures.

It is not argued that any results here are of that nature, But it *is* claimed that the OSES method, despite difficulty over standard errors and the inherent inability to check back from abstract to concrete variances (when the mean of several equation sets is used) can yield significant contributions in the hands of skillful users. Its final checking lies in independent experiments. And in the extreme, infrequently attained, experimental condition where there are just as many constellations available as there are unknowns, for example, a set of nine that will solve directly for all the nine main unknowns one desires (Table 4.7)—σ^2_{wg}, $\sigma^2_{wt.i}$, $\sigma^2_{wt.f}$, $\sigma^2_{wt.s}$, $r_{wg.wt.f}$, $r_{bg.wt.s}$, σ^2_{bg}, σ^2_{bt}, and r_{bgbt}—the preceding drawbacks disappear and it would then be worthwhile to work out the meaningful standard errors by the formulas on pp. 108 and 293 for each.

Let us, however, pass on to the solution of the same data independently, by *the method of least squares.* In that method, as described in a behavior-genetics setting by Kempthorne (1957), and others, one proceeds by iteration of abstract variance values to reproduce a set of concrete variance values that have the best (minimum) least-squares fit to the empirical values. This is done by taking second derivatives of the variances to discover maxima. The computer program used here (which can, rarely, hang up at a local maximum) was devised by D. C. Rao and checked by J. M. Brennan, to whom we are much indebted. The only difference from the earlier data basis was that we ran the local and national GP samples together in the case of g_f measures for a single general population estimate. The sample size in all were the same, namely, 94 male identicals reared together, 470 brothers together, 124 fraternals together, and a general population sample of 1973 for fluid intelligence (culture fair) and 2579 crystallized intelligence, all in the 12–18-year range and age corrected.

Using all seven equations above, it still seemed desirable to compare solutions with trials of different assortive mating values, namely, $r_{fm.g} = .0$, .10, and .25. Another extension of design was of a more radical kind. It consisted in trying three models: the full MAVA and two more parsimonious models. The latter were (*a*) the environmentalist's dream in which it was sought to explain the results *wholly* by environment, cutting out all genetic terms from the equations in Table 4.2 and (*b*) a less parsimonious model in which only both genothreptic correlations, r_{wgwt} and r_{bgbt}, and any assortiveness of mating were dropped. Table 8.7 shows the test of these models. Clearly the general MAVA is the best, by the Q test of fit. Q is a derivative of the chi-square distribution, the evaluation of which, according to D. C. Rao who suggested it, is not yet exactly determined, but which might in this case make the first degree of parsimoniousness (second column) also acceptable. It will be observed that the least-squares method also gives a standard error (here expressed as a mean deviation [error]) which points to high accuracy on the variances, but not on the genothreptic correlation r_{wgwt}.

For crystallized intelligence the three models yield the values in Table 8.8. Here again any purely environmental explanation proves quite impossi-

TABLE 8.7
Abstract Variances Reached and Goodness of Fit of Three Models: Fluid Intelligence[a]

		Estimates ± measures of error under various hypotheses					
		General MAVA		Parsimonious		Purely threptic	
	Parameter	Estimate	Mean error	Estimate	Mean error	Estimate	Mean error
	$\sigma^2_{wt.t}$	58.31	±.16	50.42	±.06	90.67	±.04
	σ^2_{wg}	77.45	±.03	46.62	±.06	0	0
(Assortive	r_{wgwt}	−.33	±.62	0	0	0	0
mating,	$\sigma^2_{wt.s}$	133.03	±.11	96.83	±.03	137.08	±.02
r_{ga}, at	σ^2_{bg}	94.66	±	46.62	±.06	0	0
+.10)	σ^2_{bt}	0	0	47.90	±.06	100.89	±.02
	r_{bgbt}	0	0	0	0	0	0
	Q^*		.48		.90		7.13

[a] The mean error conveniently giving the 50/50 range × standard error.

ble; the condition of parsimony on correlations is somewhat closer to the full MAVA, and the full MAVA is the best. The degree of accuracy of reconstruction of the seven empirical concrete variances from the abstract variances above by back-calculation is shown in Table 8.9, and although the fit is better in fluid than in crystallized intelligence (which if sustained might be an interesting indication of some peculiarity in the g_c model) it is actually statistically good in both.

Let us now take the best fitting model which, being the full MAVA, offers no obstacles to comparison with the preceding section on the OSES basis. It yields the heritabilities presented in Table 8.9.

It is at once evident that the eight heritabilities here, with one excep-

TABLE 8.8
Abstract Variances Reached and Goodness of Fit of Three Models: Crystallized Intelligence[a]

		Estimates ± measures of error under various hypotheses					
		General MAVA		Parsimonious		Purely threptic	
	Parameter	Estimate	Mean error	Estimate	Mean error	Estimate	Mean error
	$\sigma^2_{wt.t}$	2.57	±.12	3.11	±.02	5.85	±.02
	σ^2_{wg}	.96	±.19	3.17	±.02	0	0
(Assortive	r_{wgwt}	1.00	±0	0	0	0	0
mating,	$\sigma^2_{wt.s}$	1.62	±.14	1.91	±.03	4.65	±.02
r_{ga}, at	σ^2_{bg}	1.61	±0	3.17	±.02	0	0
+.25)	σ^2_{bt}	5.14	±.04	1.22	±.06	4.83	±.02
	r_{bgbt}	−.41	±0	0	0	0	0
	Q^*		8.95		10.42		28.56

[a] The mean error conveniently giving the 50/50 range × standard error.

TABLE 8.9
Goodness-of-Fit of the MAVA Model to Fluid (g_f) and Crystallized Intelligence (g_c)

Variance	Sample size (n_i)	g_f Observed	g_f Experimental	g_c Observed	g_c Experimental
σ^2_{ITT}	94	58.99	58.31	2.47	2.57
σ^2_{FTT}	124	82.03	91.29	5.21	6.69
σ^2_{ST}	470	147.33	143.42	5.66	5.09
σ^2_{BITTF}	44	341.29	339.65	17.60	17.75
σ^2_{BFTTF}	124	288.12	280.62	17.12	15.48
σ^2_{BNF}	470	331.88	332.74	13.63	13.88
σ^2_{GP}	1973 and 2579	236.96	238.08	9.42	9.48
Q^* (Validity)		0.48		8.95	

tion, are lower by the least-squares method (Table 8.10) than by the OSES method (Table 8.6). We do not have sufficient experience with the method to suggest reasons for so noticeable a discrepancy, except to note that the considerable discrepancies in the genothreptic correlations would affect the H's in this way. These correlations look unusual in the least squares method. However, there is consistency in that the OSES and the least squares follow the same pattern (compare Tables 8.6 and 8.10) in that: (a) the total inheritability is decidedly lower for g_c than g_f; (b) *within* family, however, g_c and g_f are in both cases practically identical; and (c) on the other hand a very large difference of heritability of g_f and g_c—of the order of 1.5 or 2 to 1—exists in the *between*-family value, H_b.

TABLE 8.10
Net Heritabilities of Fluid and Crystallized General Intelligences by
Least-Squares Method (with Genothreptic Correlations as Given)

	Fluid (g_f)	Crystallized (g_c)
Ordinary families		
H_w	.37	.37
H_b	1.00	.24
	(alternative .75)[a]	
H_p	.56	.28
r_{wgwt}	−.33	1.0
r_{bgbt}	0	−.41
From identical twins compared with (Table 8.9)		
Fraternal twins	$H_{w.t}$ .57 ⎫ .59	.27 ⎫ .42
Sibs (same sex)	$H_{w.t'}$ .60 ⎭	.56 ⎭

[a] Since 100% inheritance suggests some extreme coincidence of sampling or measurement error we have noted a second possibility of estimation. The difference of fit of the first two models in Table 8.6 is slight. We have therefore averaged the estimates of the σ^2_{bg} and σ^2_{bt} values from them and calculated a second H_b on that basis.

TABLE 8.11
Averaged Heritabilities for MAVA
Method, Giving Least Squares Twice
the Weight of OSES Methods

	Net heritabilities	
	Fluid (g_f)	Crystallized (g_c)
H_w	.44	.45
H_b	.89	.29
H_p	.59	.33

As to genothreptic correlations, we find decidedly larger standard errors of estimate than for variances but if we do a straight averaging across four values (the refinement of Fisher's z is not called for in such initial comparison) we find

$$r_{wgwt} = -.11,$$
$$r_{bgbt} = -.41.$$

(The second is averaged over three only, to avoid leaning on the assumption in Table 8.5.) The discrepancy of sign between r_{wgwt} in fluid and crystallized is discussed later.

Meanwhile we shall briefly summarize for overview the three existing MAVA analyses for fluid and three for crystallized (in each case two recent and one old). In Table 8.11, in accordance with "state of the art" arguments, and comments on relative reliabilities earlier, we have given least squares a weight equal to the two OSES combined.

8. Final Hypotheses with Suggestions for Further and Crucial Research Designs

In view of the discrepancy in some *absolute* levels of the heritability coefficients by the two OSES studies on the one hand and the least squares on the other, our theoretical inferences and discussions will rest on the consistent findings in the area of *relative* values as just discussed. However, before proceeding to those, a glance at absolute values, past and present, is nevertheless called for. Table 3.1 (p. 73) has given us a first glance at results on ordinary intelligence tests (g_c) by the twin method, yielding a mean H_w of .57. Other investigators have named .70 as a best estimate of the central value, and Burt, on contested data, claimed .80 or a bit over.

Data on intelligence inheritance, covering five of the same constellations as were more recently used in the MAVA design, was published several years ago by Erlenmyer-Kimmling and Jarvik (1963), though the MAVA design was not then used. Their main conclusion was that the "correlations closely approach the theoretical value predicted on the basis of genetic relationship *alone*." As we have seen (p. 143) this can be true, however, with certain relations of threptic to genetic variance when the former is nevertheless substantial. This survey is a classic one, especially in its extent, and its results, but one need not slip into false conclusions from the equality of r's (correctly just stated). The results are presented in Figure 8.3. (Incidentally, with the relations given earlier here of variances to correlations its data could be used to give a partial MAVA solution.)

Using the twin method data on 14 studies from Figure 8.3 these authors settled on a mean H_w of .72. A more recent—and culturally still broader—survey, with careful evaluation of sources, is that of Vandenberg (1971), presented in Table 8.12. His mean H_w is .64. If we keep strictly to the twin analysis part of our own MAVA research we reach a value of .74. If on the other hand, we take the full MAVA value, for sibs, by the OSES, of .47, and alter it to the higher value when $\sigma^2_{wt.t}$ is allowed to equal $\sigma^2_{wt.s}$ we get about .56.

Keeping to crystallized intelligence, g_c, by traditional tests, and to the

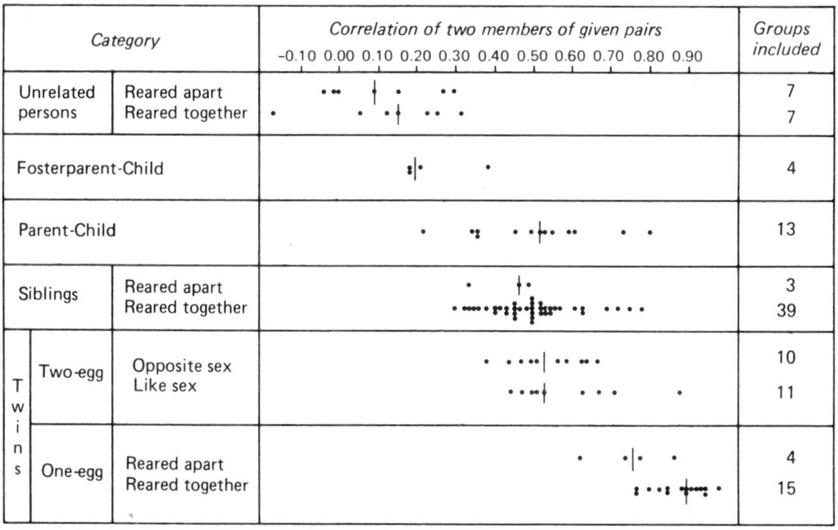

Category		Correlation of two members of given pairs -0.10 0.00 0.10 0.20 0.30 0.40 0.50 0.60 0.70 0.80 0.90	Groups included
Unrelated persons	Reared apart		7
	Reared together		7
Fosterparent-Child			4
Parent-Child			13
Siblings	Reared apart		3
	Reared together		39
Twins	Two-egg	Opposite sex	10
		Like sex	11
	One-egg	Reared apart	4
		Reared together	15

(Median indicated by vertical line)

Figure 8.3 Intelligence correlations of relatives in the Erlenmyer-Kimmling and Jarvik survey. This total of 113 ascertained correlations is from 56 publications from 1911–1962. (From one experimental group two or three different r's are sometimes derived. The measures would all be defined as crystallized intelligence, being on traditional intelligence tests and sometimes from a group of primary ability tests. Note each study value is a dot and the vertical line is a median not a mean. (From "Genetics and intelligence: A review," by Erlenmeyer-Kimmling and Jarvik, L. F., *Science,* 142, 1477–1479, 13 December 1963. Copyright 1963 by the American Association for the Advancement of Science.)

TABLE 8.12
Summary of Heritability Values for Intelligence as Reached from Fifty Years of Twin
Design Research

(a) *Vandenberg's summary* (1971), using correlation as calculating basis (see p. 69)

Country	Researcher	r_{ITT}	r_{FTT}	H
United States	(1932) Day	.92	.61	.80
England	(1933) Stocks & Karn	.84	.65	.54
United States	(1937) Newman, Freeman, & Holzinger[a]	.90	.62	.74
Sweden	(1952) Wictorin[a]	.89	.72	.61
Sweden	(1953) Husen	.90	.70	.67
England	(1954) Blewett	.76	.44	.57
England	(1958) Burt	.97	.55	.93
France	(1960) Zazzo	.90	.60	.75
United States	(1962) Vandenberg[b]	.74	.56	.41
United States	(1965) Nichols	.87	.63	.65
England	(1966) Huntley	.83	.66	.50
Finland	(1966) Partanen, Bruun, & Markkanen[c]	.69	.42	.51
United States	(1968) Schoenfeldt[d]	.80	.48	.62
				Mean H_w = .64

(b) Recent studies		r_{ITT}	r_{FTT}	H
United States	(1980)[e] Cattell	.88	.77	.53
	(N_{MZ} = 94; N_{DZ} = 124)			
United States	(1980) Osborn Male	.85	.65	.66
	(N = 175)			
	Female	.83	.54	.62
	(N = 252)			
United States	(1978) Plomin & DeFries			
	(N_{MZ} = 1300; N_{DZ} = 864)	.86	.62	.63
	Mean of recent studies			Mean H_w = .61

Source. Part (a) Vandenberg (1971, p. 197).
Source. Reprinted from *Intelligence: Genetic and environmental influences* R. Cancro (Ed.) by permission of Grune and Stratton Inc. 1971.
[a] Average of two tests.
[b] Average of six tests, recalculated from twin differences.
[c] Average of eight tests.
[d] Data for both sexes combined.
[e] Taking the crystallized intelligence, g_c, as most equivalent to WAIS, WISC, etc., studies. Value here
taken more accurately from variance (Table 8.9) not r's. Mean of old and new H's = .63 (covering more than
a score of independent researches). If weighted as in Table 8.11 this falls to .43.

twin method—which together cover by far the most data to this date—we
reach a range of values: .53 (Cattell), .57 (survey, Table 3.1), .62 (least-
squares method, Table 8.10), .63 (Plomin & DeFries, 1979), .64 (Vandenberg,
1971), .64 (Osborne, 1980), .72 (Erlenmeyer-Kimmling and Jarvik, 1963), and
.74 (OSES method, Table 8.6). Since judgments as to weightings would be
abstruse we present the straight average of these eight composite sources—
.64—as *the best final estimate today of H_w by the twin method and on g_c*. In

accepting this value we remind the reader that it rests on the unhappy assumption that $\sigma^2_{wt.t} = \sigma^2_{wt.s}$ and that for many purposes the between-family H_p and the total population H_p heritabilities are in any case more significant than H_w.

With this glance at the upshot of the numerous twin researchers let us turn to a survey of the newer results by MAVA and the OSES and least-squares methods. The first thing we find that we did not know before is that if one separately measures g_f and g_c they have very different properties. In particular we note that the between-family heritability of *fluid* intelligence is very high. MAVA gives us also the first indications of genothreptic correlations, and we note that in general the correlation of genetic and threptic deviations tends to be *negative*, especially between families. The straight heritability values, in final conclusion from MAVA, are given in Table 8.11 and it will be seen that as *within-family* values they run appreciably lower—about .15—than by the twin method, for reasons sufficiently discussed. For the typical sib family which largely constitutes our population, however, these are the values by which we would stand.

Since it can be argued that some traditional intelligence tests used in the numerous past studies contained an appreciable content of pure g_f measures it is probable that treating these as pure g_c measures as we have done in summarizing the twin method conclusions, has led to too high an estimate of g_c heritability. On the other hand the undue brevity of Scale B (Forms A and B together, however) of the HSPQ in our present research has probably, through the effect of a lower reliability, led to an *underestimate* of heritability here. With careful consideration of biasing influences we conclude that our values of $H_w = .37$ and $H_p = .28$ (Table 8.10) are quite low limits of estimates probably to be obtained in future MAVA studies of g_c.

The point has been made earlier above that g_c is an ill-defined entity, because there has been no research to establish agreement on the diverse types of common cultural core, particular in different classes and at different ages. In the 12–18-year range this problem is not at all serious, but if g_c has lower heritability here it should have still lower heritability, measured by standard school investments at later ages. In any case, we should expect more variability of H_{g_c} than H_{g_f}, in studies on different populations.

The very high H_b for fluid intelligence has social implications in suggesting that differences of social status and educational background have little influence on fluid intelligence,[5] which is not to say that there are not genetic

---

[5] If the within-family variability of fluid intelligence from environmental causes cannot be accounted for to the degree we have hypothesized by physical and physiological events, then an as yet undefined source of psychological influence remains to be discovered. That source is not the differences of education, family vocabulary, or whatever differences of nutritional resources still may exist in the American welfare state. For if these acted on g_f between individuals *within* families they would act still more patently *among* families, which possess different social status; and judging by the H's this does not happen. In the present study we have already recognized that the brevity of the g_c test *might* either increase or decrease the differences from

social status differences (a status correlation of about .15 to .20 has been indicated for culture-fair tests, but it is higher—.3 to .35—for traditional tests, as would be expected from cultural content). This has somehow to be reconciled with the fact that the *within*-family heritability of g_f is certainly no higher than for g_c, and indeed, if $H_w = .41$ (Table 8.11) is supported by equally low values in later work it might even be slightly lower. Since the $\sigma^2_{wg}/\sigma^2_{bg}$ ratio is relatively fixed, the differences we are discussing must hinge on and derive from environmental variances. That the cultural differences *between* families, and the relatively uniform atmosphere *within* a family should make H_b lower and H_w higher in crystallized intelligence is readily understandable. What is new, and demanding of explanation, is that the effect of environmental differences *within* families is productive of so extensive a threptic variance in *fluid* intelligence, relative to between-family environmental influence. Plomin and DeFries (1978) have also noted indications of this "anomaly."

If differences of sib rank, parental attachment, etc., could cause differences in demands to exercise g_f (assuming exercise helps g_f) the finding might be so explained. But surely such differences of *mental* environmental stimulation within the family should be significantly less than between families. Consequently we are forced by this new finding to the relatively novel position that *physical and physiological* influences on the cortex must differ appreciably within families. We have no space to explore these sources extensively but we can state a few hypotheses as follows.

1. That age of the mother at gestation has appreciable effects. (It is known that IQ declines on an average from earlier to later children, but this has been put down to mental environment (less parent; more peer contact). That σ^2_{wt} is found above almost twice as large for sibs as twins supports this.

2. That the effects of smoking, alcohol, drugs, transient ill health, and changes in nutrition of the mother on intelligence are significant now supports our new finding. The finding of Pencavel (1976) that birth order is related to the size of the offspring; of Scarr (1969) that birth weight and intelligence are positively related in twins; of Brackbill (1976) and Kraemer (1972) on medication of the mother, all point to the environment of gestation being important.

3. Brain injury (not at a clinically recognizable level) to the child, from brain trauma, anesthesia of the mother (Scanlon *et al.,* 1974), high temperature fevers, blows on the head, encephalitis, anesthesia, exposure to insecticides, etc.

4. What have come to be recognized in chromosomal genetics as

---

g_f, and we shall not venture to build superstructures of theory until (*a*) longer tests (higher r_d) and (*b*) test sophistication practice by the children involved, have been added. But at least results as now available point to a possible unknown source of "exercise" growth of g_f (related to "thoughtfulness" of personality?).

"copying errors." The unfolding of the gene message through the processes studied as *epigenesis,* from DNA through enzymes, etc., is known to fail of completion at times through as yet unknown small "accidental" influences. Whatever they are they fall in statistical analysis in the category of environmental influences since they occur after the genome is fixed. Incidentally in other contexts, notably the age curves of fluid intelligence (Cattell, 1971; Horn, 1979), it is beginning to be recognized that some degrees of brain injury, from varied sources, are decidedly more common in life than was first realized. It is to sources such as these that environmental effects on fluid intelligence may principally be due. Parenthetically, it will be understood that the *H* values we are here discussing are derived from children in the normal range of intelligence. The hereditary bases of 1–2% of special conditions of imbecility and idiocy, such as Down's syndrome, phenylketonuria, etc., are fully handled in other texts and do not concern the broad spectrum of polygenic inheritance involved here.

While we would argue that the influence upon the cortical substrate of intelligence of physical factors, especially those affecting gestation, has been disregarded and insufficiently researched, we do not contend that *threptic* variations in fluid intelligence are entirely from this cause. Presumably exercise can produce some effect even on a *generalized* relation-perceiving capacity, and speed, at least, in such an activity, may be assisted by having much experience in having to perceive relations. Horn (1980) has argued, and produced some evidence along these lines, that ordinary experiences, as contrasted with the cultural and scholastic experiences that produce differences between families in g_f, contribute to g_f performance. Whereas the cultural and status differences of families would be of a kind contributing to g_c variance, this individual exercise would be the larger cause of difference within families in g_f, and H_w for g_f would not be expected to differ from H_w for g_c. The latter holds because g_c is an investment product of g_f, and differences in development of g_f would be expected, with a lag, to be followed by corresponding differences in g_c. Since the experiences *within* families, as far as g_c content is concerned, are presumably slight, we might expect similarities in behavior of g_f and g_c *within* families to be marked, relative to other situations.

Not the least interesting (though unfortunately not the most reliable!) of our findings is in the genothreptic correlations. In one analysis the within-family correlation is positive for g_c, but the central tendency of results is a zero or slightly negative *r*. An important point to keep in mind is that made earlier to the effect that the variances and covariances of children *found* within the family have their origins in part in environmental influences actually *outside* the family. Since the between-family correlations are negative or (in one case, g_f by least squares, 0) it is reasonable to conclude that but for the world intrusion into the family the correlation there would be positive, probably more definitely so in g_c.

A positive genothreptic correlation for intelligence is what psychologists have expected, on the grounds that the display of a desirable quality like intelligence would be encouraged. Since the growth of g_c depends on other investments than fluid intelligence and school experience, namely, on such aid as super ego (G), self-sentiment (Q_3), and self-sufficiency (Q_2) (Cattell & Butcher, 1968, p. 186; Johnson, 1980) it is likely that in the reliably discerning circle of the family the sib who shows more g_c will be encouraged to show still more, and given, for example, more advanced books to read than the "dumber" brother. The environmental influences on g_f, we have hypothesized, would be more incidental and accidental, and thus show less positive correlation.

Results (Chapters 9 and 10) for personality traits show a frequent, indeed predominant tendency for between-family r's (r_{bgbt}) to be negative, but it will probably surprise many to see that this holds also for intelligence, most powerfully in g_c. One needs to know whether this instance of apparent "coercion to the biosocial mean" shows simply as a negative regression of cultural influence on natural capacity throughout the whole range, or only on either of the upper or lower halves. In the lower half it would appear through strong efforts to bring the backward up to the average, with no such correlation expected to exist as a reverse tendency, above the average, to bring the bright down to the mean.

However, despite the ideal of a meritocracy, and "To him that hath shall be given," it is easy to see that institutions and forces exist that *could* upset any tendency to a positive correlation of intelligence with greater opportunity across society as a whole.

Among the influences denying a proportionality of stimulus and expanded experience to above average intelligence are such widespread practices as the unavoidable classroom lockstep in pace of discussion and advance, school promotion by chronological age, rather than mental age, and the aim of television and other media to cater to the more numerous and remunerative median adult level. Naturally the genothreptic conclusion itself —that the family adjusts learning opportunities and expectations at least slightly more sensitively to perceived capacities, but that society as a whole pulls to the center—needs verifying before social psychology researches the mechanisms. But such independent findings as the reduced standard deviation of IQ on traditional compared to culture-fair tests support this conclusion reached independently from genothrepic r's.

It behooves us, in conclusion, to indicate questions for further research in the field and the improvements that now seem desirable. They cover the theory for entering experiment, the design of experiment, and the analysis.

1. In theoretical approach the use of measures pointedly indicated by the triadic theory is suggested. In intelligence we need, first more good g_f

measures, and, secondly, as such, an altogether sharper definition of g_c, *constant in factor reference tests across researches to be compared.* This consistency and validity might be achieved by covering all 20 or so primary abilities, weighting each according to its saturation (loading) in the general factor among them. (This is an operational way to define, and confine experiment, to, Jensen's Type 1 rather than Type 2 performances.)

2. A better evaluation of the effect of varieties of error, which we presently conclude (*a*) tends to bias slightly toward lower values from the true heritabilities, and (*b*) tends to reduce the degree of difference among traits in size of their apparent heritabilities. The first arises notably in the twin method if any misclassification of cases occurs, or elsewhere if, say, adoptees are not adopted *from birth*. The difference [see Eq. (8.27)] of σ^2_{FTT} and σ^2_{ITT}, which is σ^2_{wg}, is then falsely reduced. All forms of error will, on an average, operate to produce some bias as well as blur. The effect of undependability and invalidity of tests, along with trait fluctuation, will be to expand measured variances above true variances, by a coefficient $1/\phi$ as defined in Eq. (8.17a). If we bring forward the old twin method equation (3.20), adding the ϕ correction as (8.27) here, it will be seen that provided ϕ is the same for the different constellations, $H_{w.t}$, the heritability will remain the same, as with cancellation in

$$H_{w.t\,(corrected)} = \frac{\phi\sigma^2_{FTT} - \phi\sigma^2_{ITT}}{\phi\sigma^2_{FTT}}. \tag{8.27}$$

But even if the testing situation is maintained so constant in the two constellations, that nothing worse than sampling differences occur in r_v, r_s, and r_d, the fact that $r_{s_1s_2}$ enters ϕ, and is different for FTTs and ITTs will mean that the corrected H_{wt} will differ from the uncorrected, that is, the ϕ's in (8.27) will actually *not* cancel out, being unequal.

The need for improvement in experimental measurement therefore calls first for determining r_v, r_d, and r_s (or, in lieu of the last, making several repeated measures extended over, say, some months). And although they will permit an estimate of the true and steady variance, that estimate will be better if the choice of tests gives a large r_v and the length, etc., a large r_d.

3. The help of leading statisticians is called for, progressing in the direction indicated by Eaves (1972) and Eaves *et al.* (1978), for example, toward greater precision and satisfactoriness of estimates of *sampling* error. Meanwhile the degrees of instability in twin and MAVA results presently available suggests it is undesirable to proceed with less than a total across the constellations of 1500 to 2500 individuals (exclusive of the general population sample).

4. As regards design and analysis it would be a considerable gain to use the full MAVA design, with no assumptions such as have had to be made in the most limited MAVA here. Three within-family threptic variances—$\sigma^2_{wt.i}$, $\sigma^2_{wt.f}$, and $\sigma^2_{wt.s}$—need to be used, and some attempt should be made to solve

for the two possibly different values which Loehlin's criticism shows we have collapsed into one. In analysis it should be desirable to use least squares and maximum likelihood, over and above OSES, since only by trial of different assortive mating values therein can one converge on the best fit.

5. If it is not daydreaming in the realm of funding one might plan using comparative MAVA studies on such sufficiently reliable samples as would permit calculations deriving from H_w, H_b, and H_p values for different ages, races, and cultures. Therein lie the possibilities of tying threptic developments in intelligence to environmental variances; of discovering the ages when interaction of learning and maturation is most powerful; etc.

6. At one pole of behavior genetics—that occupied by the molecular geneticist—there is the ultimate aim of tracing the genetic variance parts to genes of various kinds and magnitudes of effect. There has been speculation on the number and nature of genes affecting intelligence, going back to Hurst (1935), Burt and Howard (1956), and many others. Perhaps the nearest to a testable hypothesis so far is Lehrke's (1978) argument, from the greater intelligence variance in men than women, that some gene contributors are on the sex chromosome. More precise determination of the differences of correlation of father–son, father–daughter, etc., on the genetic component of intelligence would constitute a useful pursuit of this theory. Meanwhile, every source of data points to intelligence being determined by many genes, large and small in effect, some with dominant and recessive alleles, and most probably affecting, pleiotropically, other features of the organism.

9. Summary

1. The stage of development of knowledge about ability structure is such that behavior geneticists need to give closer attention to using meaningful pure factor measures. They need to move in this direction, first, to avoid getting lost in a wilderness of specifics. Second, it would be desirable to enter with some initial theory, such as the triadic model, offering concepts of abilities as pure genetic or pure threptic, or overlap clusters, or eidolon structures. Third, the first attacks need to be made on important structures wherein knowledge regarding genetic contribution would directly help personality and ability theory.

2. In the ability area a nutshell statement is given of the triadic theory of abilities, covering general capacities, g's, provincial powers, p's, and agencies or primary abilities, a's. Neurological evidence would lead us to expect higher heritability in g's and p's than a's.

3. Although most methodological problems have been handled *prior* to the present chapters on psychological findings, those psychometric issues connected with *handling of actual measurements* have seemed best treated in connection with this first examination of actual data.

Error of measurement is comprehensively handled under three headings: (a) random *experimental error,* referring to the dependability coefficient, r_d; (b) *function fluctuation* in the trait itself, assessed through the stability coefficient, r_s; and (c) *unwanted factor variance* in the scale, evaluated by the validity coefficient, r_v. A new expression, ϕ, is developed, involving these three basic psychometric parameters of tests and $r_{s_1s_2}$, the correlation of the offspring pairs, to correct observed concrete variance, properly recognized as a variance of difference of correlated scores.

Equations for sampling error are also presented, tentatively. If accurate, they unfortunately suggest that a majority of the published research to date has operated on samples too small to give a precision of result suitable for the foundation of further hypothesis building. The sigma of heritabilities from several independent researches on a given trait, relative to the mean, is at present the best guide to the firmness of the foundation of the central heritability value.

4. The heritabilities of *primary abilities* at present rest on twin methods, with some help from adoptive studies. In interpreting the results the causal and factorial model called the Stratified Uncorrelated Determiners (SUD, Cattell, 1978a) is most useful, reminding us that the primaries are conceptually and quantitatively properly understood when reduced to *stub factors.* There is some suggestion that in general their heritabilities as non-stub "tests" are related to the amount of general intelligence (g_f or g_c) in them, so that if treated as stub factors they would show largely environmental determination. However, the stubs could also have genetic components and there are certain instances, for example, spelling, sense of absolute pitch, spatial ability, that are perhaps tied to the role of highly heritable provincial power stubs, p's, or agency stubs, a's.

5. The design of the MAVA method with its broad basis in several constellations has several definite advantages over the twin method; but it also introduces complexities and encounters algebraic limitations in solving its simultaneous equations. In four researches here reported, for intelligence, the algebraic difficulties have first been met by the OSES (*overlapping simultaneous equation subsets method*) procedure and second by a program of iterative approach to a least-squares fit. The first has to proceed with two assumptions: (a) on the degree of assortive mating (genetic) and (b) on alternative solutions for quadratic equations. The second also starts with trial assortive mating values, but settles on the best in terms of goodness of final fit.

6. By the OSES method, using a total of 2424 boys, 12–18 years old, from midwestern schools for the fluid intelligence measures, and 3423 for crystallized intelligence, it was found (a) that g_f has a larger ($H_p = .64$) heritability than g_c ($H_p = .44$) over the *whole population;* (b) that g_f has its highest heritability *between families* and g_c its lowest there; and (c) *genothreptic correlations are substantial and negative* for r_{bgbt}; but less significantly so for r_{wgwt}.

7. Analysis was next made by the least-squares method. Three models

—full MAVA, MAVA without covariances, and MAVA with threptic variance only—were tried on both g_f and g_c data. In each case the full MAVA was best, but MAVA without use of covariances was close behind. This method yielded standard errors for the abstract variances, which proved trivial, except for the genothreptic correlations.

8. The agreement over the two methods and the four studies amounted to consensus (a) in giving higher population heritability to g_f; (b) in giving roughly equal within-family heritability to g_f and g_c; (c) in indicating decidedly greater *between*-family than within-family heritability for g_f; (d) in giving significantly greater between-family heritability for g_f than g_c; and (e) in giving negative correlations of genetic and threptic deviations *between* families, probably more marked for g_c, and correlations within family which, if allowance were made for part arising from intrusion of effects from outside the family, suggest that parents and sibs exert a zero or even slightly positive (especially for g_c) environmental pressure to encourage the achievement of the more achieving.

Although agreeing on patterns of relationship the two modes of analysis yielded uniformly lower absolute heritabilities (except for H_g for fluid intelligence) for the least-squares method. The OSES levels are actually closer to previously published levels. In deference to statisticians favoring the least-squares method we accordingly gave it double weight in combining the two to a final estimate (Table 8.11). This final estimate suggests that perhaps due to greater ethnic environmental variability in these Cleveland and Chicago populations the intelligence heritabilities are somewhat lower than those previously reported in the literature, which mainly come from more uniform cultures. However, that H_w should be lower than in previous results, all by the twin method, should not surprise us, since MAVA is able to show that σ^2_{wt} for sibs is up to twice that for twins. Even corrected to the same basis, however, that is, calculating from $\sigma^2_{wt.t}$ instead of $\sigma^2_{wt.s}$ on present data, an H_{wt} of .53 rather than .64 is indicated.

9. The fact that g_f has a between-family heritability which is high absolutely and very high relative to g_c, well fits all theory to this point on the nature of g_f and g_c, but the equality of g_f and g_c in within-family heritability is surprising and suggests high uniformity of environment in the family for g_c content and an unexpected degree of some environmental difference effects upon g_f. It is hypothesized that environmental differences at the physical level—age of gestation in the mother, head injuries at birth, febrile illnesses, etc.—are mainly responsible for this threptic g_f variability. According to the investment theory of crystallized intelligence, in which g_c arises by interaction of g_f with the cultural environment, the between-family $H_{b.gc}$ would be lower than $H_{b.gf}$. The similarity of g_f and g_c heritability within the family points to a high similarity of whatever environmental "exercise" is there common to g_f and g_c growth performances. It is hypothesized that g_f is not altogether insusceptible to mental exercise effects, as well as to the admitted

physical influences. This suggests there must be in general everyday life reasoning experiences other than the more specifically cultural and scholastic patterns which constitute g_c.

10. The substantial negative correlation of genetic and threptic deviation within society could be an example of the law developed elsewhere of "coercion to the biosocial norm," (which, in its simplest form, is shown in designing everything for right-handed persons!) The push may be toward one end of the trait but more commonly toward the mean. In intelligence there clearly seems more of the latter, and the finding is consistent with the 50% larger IQ sigma of g_f relative to g_c. Lockstep education and mass media catering are considered as possible environmental–cultural pressures toward the mean.

11. Suggestions are made for research improvement in this area by attention to (a) theory, in choice of variables (b) reduction of experimental and sampling error, the effect of which at present is probably a bias toward (i) reduction of the true spread of heritabilities of various traits, and (ii) some reduction of the average magnitude of heritabilities.

12. We anticipate that with attention to the preceding the heritability found for a well-defined crystallized intelligence measure (not the ever-changing mixture in traditional intelligence tests) will prove lower than previously supposed (perhaps H_p about .4) in this age range (and lower still later) and the heritability of improved fluid intelligence measures (Cattell & Horn, 1978) will be higher (perhaps .7 for H_p and .9 for H_b). In stating this theoretical estimate we would add that the environmental effects on g_f are likely to arise from physical sources more than investigation has yet recognized.

9

The Heritability of Nine Primary and Five Secondary Source Traits in Q-Data

1. The Properties of Questionnaire Data

The ability field has seemed to most psychologists, perhaps rightly, a simpler field than that of personality. The poor agreement among theorists like Jung, Freud, McDougall, Sullivan, Kelly, Ericson, Fromm, Rogers, Maslow, Moreno, and others, characteristic of the premetric phase of the psychology of personality has had much to do with this. And when metric, statistical, and multivariate experiment of methods were applied the complex results in personality structure were less readily grasped than in abilities. Between Spearman's "Intelligence objectively measured" in 1905 and the achievement of some confidence in the nature of primaries and secondaries, after Thurstone, Guilford, Horn, and others, some 40 years had elapsed. Pursuit into the finer fastnesses of ability structure and development is indeed still going on in the researches of Horn, Baltes, Nesselroade, Hakstian, and others, and thus 80 years after Spearman we can probably claim a truly useful science of abilities.

By contrast the factor-analytic research on personality did not really begin until 1935, and when surveyed by the present writer (Cattell, 1946) was still beset by intolerable methodological inconsistencies. Thirty-five years of strategic, programmatic research since then, with more sophisticated developments of method, have led to well-replicated areas of knowledge (though poorly informed criticism still gets lost therein). Despite the slowness of follow-up in some areas, technically well-informed critics will probably agree, with minor reservations, with the essentials of the "nutshell" statements which we now give here for personality, paralleling those given in the previous chapter for the abilities. At the same time all researchers will recognize that there is much more yet to do here, especially in the dynamic trait area, and in developmental studies, than in the ability field.

The *observational* methods on which personality trait experiment and calculation has rested are of three kinds, called "media" or "panels": (*a*) behavior rating in the life situation, called *L-data*, (*b*) life situation behavior, and introspection, reported by the subject in consulting room or questionnaire, called *Q-data*, (*c*) laboratory or behavioral measurement data not involving self assessment (as Q does) called *T-data*. The use of factor analysis on such data has involved R-, dR-, and P-techniques. The dR technique is an R factoring of individual differences on *difference* scores, say January to February. It has shown that people change uniformly on the *same* unitary patterns as appeared in R-technique. For example, surgency, ego strength, g_f, and g_c show each a simultaneous change on their various expressions, and independently of one another. In P-technique one factors measures on a single person on, say, 40 behavioral expressions over 100 days. These strategic experiments across methods show that the shape of person-unique *individual* patterns, for example, of surgency, of intelligence expression, or exvia cluster pretty closely around the *common* (R) pattern (though *some* unique traits have no common factor equivalent). This is analogous to saying that people have the same bodily organs but with some differences of form. The uniqueness of the individual finally lies both in his or her P-technique pattern variations, and in his or her uniqueness of combination of *common* trait scores.

It has been shown (Cattell, Pierson, & Finkbeiner, 1976) that, as one might expect from their both being couched in everyday life behavior, the factors from observer ratings or measures of behavior in situ (*L-data*) and from Q-data (also *in situ*) are essentially the same in number and nature. They include at the primary level such traits as ego strength (C), super ego (G), surgency (F), radicalism–conservatism (Q_1), guilt-proneness (O), and the self-sentiment (Q_3), as, for example, in the 16 P.F. scales. At the secondary level—that is, as second-stratum factors acting as broad organizers or products of the primaries, we find exvia–invia, anxiety (as for example in Eysenck's scales) and cortertia, independence, and control. We shall be concerned with their heritabilities in this chapter. Incidentally, it is scarcely possible to carry out heritability research on these same factors through the rating medium (*L-data*) because of the large element of observer projection (as found in attribution process research and trait view theory (Cattell, 1981) and the poor comparability of numerical standards among different raters.

Although certain systematic relations have been found between the aligned Q- and L-source trait factors, on the one hand, and those in T-data, on the other, it is better to consider the latter on their own in the next chapter, as independent evidence on personality structure. Meanwhile we may note that the main primaries and secondaries in Q-data have been found (as uniquely simple structure-rotated factors) extending across: (a) populations tested on translations of the 16 P.F. in Germany, France, Italy, Britain, Australia, Japan, South America, India, and some other countries, showing that we are dealing with structures *basic to human nature;* and (b) in age cross-sectional analyses in the 16 P.F. (Adult), HSPQ (High School, PQ, 12–18 years), CPQ (Child PQ, 8–12 years) and the Early School PQ (ESPQ, 6–8 years). Dreger's recent work with nonreading, tape-recorded questions and responses shows several of the child factors present also at preschool level. Naturally the behavioral content alters with age, but experiments with age groups at the "joints" (overlaps) between these scales are extending the checks on identity.

The situation today is that some 23 primaries in Q-data covering "normal" behavior, and some 11 or 12 more in abnormal behavior (depressions, schizophrenic, psychopathic, manic) have been checked (Kameoka, 1981a), 28 of which total are in the *Clinical Analysis Questionnaire* (research on which is now clearly called for in the behavior-genetic field). Meanwhile, the most frequently replicated 16 normal source traits are in the 16 P.F., and 14 of them in the HSPQ, with parallel forms extending enough to give whatever reliability–validity the time limits of most researchers permit them to aspire to. Exact weights are documented from second-order factoring for permitting scores on the primaries, in a normal population, to be combined to give scores on 4 to 8 secondaries—exvia–invia, anxiety, cortertia, independence, control, etc. There are, of course, almost countless questionnaires for this or that labeled, supposed unitary "trait" (Buros, 1972) that could be of interest specifically to some "applied" investigator. As Vaughan (1973), Cattell (1969, 1973, 1977), De Young (1972), Nesselroade, Baltes, and Labouvie (1971), Horn (1972b), and others have pointed out, the factoring, if any, underlying these subjective choices of trait scales has not met the six main technical criteria, and we see no point in mixing here the heritability studies on the best replicated factors with any earlier results that may exist today (in any case, all appear with the restrictions of the twin method) on such ephemeral bases.

In what follows we shall use the index symbols (Cattell, 1978a; Cattell and Warburton, 1967, Chapter 3) of A, B, C, etc., for the primaries and the usual Roman numerals—QI for exvia, QII for anxiety, etc.—for the secondaries, as well as the technical terms[1] such as surgency, ego strength, prem-

---

[1] When the physical sciences discover a new fact, for example, a chemical element or property, or a new concept based on an emerging law, it is invariably tied down with a new term.

sia, autonomic tension, protension used in the main textbooks and the numerous research article references. A considerable literature of findings has accumulated around these established, and indexed source traits, measurable by scales of known concept validity. These findings include age curves, predictive values for many educational, clinical, and industrial criteria, magnitudes of change under life experiences, and, of course, validities and reliabilities in batteries relevant to the present work. All such evidence up until 1973 is gathered in *Personality and Mood by Questionnaire* (Cattell, 1973a). The evidence we shall see here on heritability adds a valuable new approach to the initial interpretation of the nature and properties of these factors made in writings before 1981.

In what follows the nine most important primaries are studied, three to each of the three sections here. The grouping together is compelling in the case of the three control factors in personality—C, G, and Q_3—but only a convenience for section length in the others.

2. The Personality Control Triumvirate: C, G, and Q_3, by OSES, Least-Squares, and Maximum-Likelihood Methods

Clinical psychology and general personality theory have long—since Freud's delineations—shown good reason to believe in the structure of an organizing ego and a morally controlling super ego. These unitary structures have been factor-analytically verified, precisioned, and rendered measurable, in the Q- and L-data media, as C and G in the standard index (Cattell,

---

The new elements are commonly named by their discoverers, as Madame Curie named Polonium, or Rayleigh named Neon. A new property may get a Greek derived word, as in lyophobic or isotropic, and a new concept a term like entropy or an acronym like laser.

Psychologists, however, perhaps through being fooled so often in the prescientific phase by elaborate new terms for stale ideas (William James's, "What everyone knows in language that no one understands") have been abnormally slow in employing the correct technical terms for the *new* source traits, such as surgency–desurgency (F), premsia–harria (I). Using popular terms—such as "sociable" for surgent (F) and again for exviant (QI) and yet again for affectia (A)—has been a serious obstacle to both precision of thought and technical effectiveness in applied psychology. It has been the ideal of myself and my colleagues and discussants never to use a new term unless there is a demonstrable new pattern entity, or a new relation that can be mathematically tied down, for example, ipsative scoring, a trait constancy coefficient, a modulator index. Only half a dozen of the 30 or more source traits revealed by factor analysis had been described by the unaided eye of the clinician as, for example, in extraversion (Jung), ego strength and super ego (Freud), tender-mindedness (William James), etc., just as only a minority of planetary satellites and outer planets were identified and named before Galileo's telescope. The new terms therefore describe new patterns visible to the computer, and the new names given to them—affectia (A), surgency (F), parmia (H), premsia (I), protension (L), autia (M), etc.—are either classical-language-based or acronym-derived terms best connoting the properties as presently perceivable.

1973a, 1980a). Additionally, factor analysis has revealed a third unitary factor which may also be qualitative described as in the control field, namely, Q_3, the self-sentiment, (''self esteem''), not previously perceived by the approaches of classic clinical theory.

The measurement of these source traits was carried out for the present investigations by the HSPQ (two forms A and B) and on the numbers and samples of children (from the schools of Cleveland, Chicago, and downstate Illinois) already recorded in the previous chapter for the intelligence measures (the B scale in the HSPQ). However, instead of using only the OSES and the least-squares methods, we decided to pursue analysis further by the *maximum-likelihood method*. The latter is increasingly used (Eaves, 1972, 1974; Eaves *et al.*, 1978); Fulker, 1979; Kempthorne, 1957) in behavior-genetic research and it seemed timely to make the research on the primary personality factors here readily comparable with what may follow in research by others. For details of the method the reader is referred to our introductory definition earlier (p. 120) and to the statistical sources (Jöreskog & Lawley, 1968 Kempthorne, 1957; Rao & Morton, 1974); cited. As stated its purpose is to find the population values from which the results in the current sample are most likely to have been taken.

Because of space the reader must be referred—in this chapter and Chapter 10—for the detailed infrastructure of testing and variance calculations, to the several articles by the team of Cattell, Rao, Schuerger, Vaughan, Schmidt, Klein, Kameoka, Graham, Ahern, and Brennan, at places noted. It has been pointed out in the preceding chapter that possible corrections for experimental error in the variances can take five forms: (*a*) no correction: an acceptance of the raw variances as they stand, (*b*) by a simple application of test reliability (dependability coefficient) directly to variances, (*c*) simple application of factor (concept) validity to variances, (*d*) a more sophisticated use of the dependability coefficient, as in earlier formulas (p. 284), and (*e*) a more sophisticated allowance for test validity and dependability coefficients paying respect to the *correlations* of subjects in sum and difference variances.

Practically all published studies (that of DeFries *et al.* [1978] correcting for unreliability attenuation [b] preceding) (being one of the few exceptions) have not got beyond (*a*). Here we have conservatively stopped at (*c*), not to get too far away into untried methods from comparisons with most previous studies. However, in the personality data of the next chapter, Vaughan and Cattell have explored the effect on final heritabilities of different corrections. The validity correction (*c*) has been used here, but comparisons of other corrections are reported in the three articles in which Vaughan is co-author. The differences are in general small, but should be followed up by future researchers.

Our experience suggests that larger differences than those from the above corrections arise from differences in models and methods of analysis. The former—differences between full MAVA and simpler models—are

TABLE 9.1

Heritabilities for the Control Triumvirate Ego Strength, Super Ego Strength, and Self-Sentiment

Main results	C (ego strength)				G (super ego strength)				Q_3 (self-sentiment)			
	OSES	LS	ML	Mean[a]	OSES	LS	ML	Mean	OSES	LS[b]	ML	Mean
H_w	.29	.34	.29	.31	.03	.10	.15	.09	.65	.55	.38	.53
H_b	.66	.73	1.00	.80	.06	.21	.22	.16	.46	.86	1.00	.77
H_p	.41	.47	.35	.41	.04	.15	.18	.12	.52	.68	.54	.58
r_{wgwt}	-.09	-.12	(0)	-.07	.26	1.00	1.00	.75	-.26	1.00	1.00	.58
r_{bgbt}	-.78	-.95	(0)	-.58	-1.00	-.41	-.50	-.64	-.95	-.41	0	-.45
Assortive mating	.25	0	0	.08	.25	.10	0	.12	.25	.25	0	.17
Auxiliary data												
H_{wt}	.40	.44	.37	.40	.05	.09	.18	.11	.64	.24	.21	.35
$\sigma^2_{wts}/\sigma^2_{wtt}$	1.59	1.51	1.46	1.52	1.54	.85	.83	1.07	.94	.25	.45	.57

[a] Parsimonious hypothesis.

[b] Best of .25 assortive mating assumption.

evaluated later. The results of different methods of analysis as such (OSES, least squares, and maximum likelihood) are shown in Table 9.1. Some idea of where fit is poor or good is obtainable from Table 9.2a. This shows the typical sizes of the mean errors in the least-squares method, and, in Table 9.2b of the size of error in the reconstitution of concrete from abstract variances by the maximum-likelihood method, as well as the associated chi-square calculations of total fit. A second estimation effect the psychologist will be interested in—the effect of assumption of different sizes of assortive mating r's—is given in Table 9.4.

There has been little or nothing published before to show investigators the magnitude of differences of estimate from the same data and samples by different analysis methods. The difference in Table 9.1 should surprise no one aware of statistical procedure differences in comparisons of a combination of a wide spectrum of data, or familiar with the wide differences (Eysenck, 1956; Mittler, 1971; Vandenberg, 1965a) actually reported in the past, even on such a familiar trait as extraversion.

Actually, in the present case of C, G, and Q_3, all three methods put H_w and H_p in the same rank order—highest for self-sentiment (Q_3), middling for ego strength (C), and very low for super ego strength, (Q_3). As far as H_b is concerned ego strength, though middling, shows an unexpectedly high between-family heritability relative to its within-family heritability.

The genothreptic correlations typically have much larger standard errors than the variances themselves, as the least-squares method brings out, and so they quite obviously vary more from method to method. There is, nevertheless, an interesting consensus that the between-family correlations (when not zero, through the parsimonious model) are large and negative. As we shall see later, it is somewhat unusual to have such large *positive* within-family r's (.75 and .58) with such substantial between-family *negative* r's ($-.64$ and $-.45$) as occur in the case of super ego and self-sentiment, and we shall discuss later a contribution which this makes to psychological theory.

As to the assortive mating r's, these are reached independently of any direct assumption only in the case of least squares and maximum likelihood, being *assumed* at the most likely value in OSES (but maintained internally consistent, thereafter, with the resulting other parameters). As Table 9.3 shows, appreciable variation of $r_{fm.g}$ (genetic correlation of f = father and m = mother) over ranges used here actually has only a trivial effect on the heritabilities (except for H_p on Q_3). What has an appreciable effect, as the auxiliary data part of Table 9.1 shows, is the use of the twin method, instead of MAVA, to estimate heritability (note H_w on Q_3). That is to say, if we assume the threptic variance of sibs is the same as that of twins [as in twin method equation (3.20)] and calculate $H_{w.t}$ (heritability on the twin method) as $\sigma^2_{wg}/(\sigma^2_{wt.t} + \sigma^2_{wg})$, then $H_{w.t}$ turns out decidedly larger in C and, on the other hand, smaller in Q_3, the self-sentiment. This result is a function of the

TABLE 9.2
Illustrations of Standard Errors of Estimation, on Factors C and G^a

(a) *Of abstract variances (for factor* C) *by least-squares method*

	Estimate	Standard error
σ^2_{wg}	8.30	.06
$\sigma^2_{wt.s}$	15.94	.03
r_{wgwt}	$-.12$	.33
σ^2_{bg}	8.30	.00
σ^2_{bt}	3.00	.13
r_{bgbt}	$-.95$	.00
$\sigma^2_{wt.t}$	10.54	.02

(b) *Of concrete variances as observed and as experimentally reconstructed from abstract variances (for factor* G)

	Observed	Derived from experimental analysis
σ^2_{ITT}	8.98	9.13
σ^2_{FTT}	18.16	17.55
σ^2_{ST}	15.58	15.40
σ^2_{BITTF}	26.67	29.08
σ^2_{BFTTF}	25.77	24.79
σ^2_{BNF}	23.21	22.65
σ^2_{GP}	18.90	19.02

^{a} These two source traits are taken for illustration at random from the 10 investigated. $\chi^2 = .76$; range from .76 to 1.50.

ratio of sib to twin threptic within-family variance set out in the last line. The substantial and consistent difference of C and Q_3 could have considerable psychological significance.

Viewing the significance of these behavior-genetic results for personality theory we get a major jolt from the high heritability of the self-sentiment, which, like sentiments to other objects, has always been considered a largely acquired structure—at least until Cattell, Blewett, and Beloff (1955)[2] found a strikingly high between-family heritability for it, namely, .69, not far from the present, .77. One can only suppose that the ergic satisfactions built into this sentiment—notably narcissism, self-assertiveness, and need for secu-

[2] We are not giving more than passing references here to the Cattell *et al.* (1955) study, but its results can be seen elsewhere systematically collated against the later studies. (Cattell, Rao, & Schuerger, 1981; Cattell, Rao, Schuerger, & Kameoka, 1981; Cattell, Rao, Schuerger, & Klein, 1981) The reason for putting those results in a lower class of significance from the present is not the smaller sample size—which was actually appreciably bigger than most used even to this date—but that the factorial definition of the primary factors in Q-data in 1955 was still insufficiently checked. The factors there that have needed little progressive rectification procedures (Cattell, 1973a) since are B, C, E, H, Q_4, and possibly Q_3. Accordingly it is on the 1955 results on these that we shall comment in passing.

rity (Cattell & Child, 1975, Table 3.3, p. 84)—are themselves so highly inherited, and that the attaching of them to the self-concept is so universal and "natural," that learning plays little part in creating the variance. As the lower within-family heritability and the positive within-family genothreptic correlation further suggest, the self-sentiment, with its self-respect and concern for personal reputation, is a trait uniformly encouraged by parents. Those naturally more endowed with the ergic needs concerned thus grow more. At the same time the H_w lower than H_b suggests that much of the differential environmental learning, in this trait popularly called "self esteem," is given within the family. Finally, a psychologist may ponder the quite unusual situation in Q_3: namely, that in this case twin threptic variance is actually decidedly larger than that of sibs. Is this what we are calling the "Brazil nut effect" wherein the closeness of twins forces greater narcissistic and assertive learning in one to be accompanied by an adjusting lesser development in the other (to an extent absent in sibs of different ages, and in larger sibships)?

The clinician has long considered ego strength (C), as something largely acquired, damaged by traumas, and aided by successfully controlled expression. The recent multivariate experimental work of Birkett (1979) on the ego, however, has led to the concept (Cattell & Cattell, 1982) that *one* of the contributions to this integrative learning comes from an innate capacity to control impulse, long enough for alternative expressions to be considered and learnt. Such innate differences could reside in differences of neural, anatomical structure of the hypothalamus, and available projection paths to and in the frontal lobes.

Clinical psychology would best proceed in its theorizing henceforth, therefore, by taking into consideration the undoubted finding that ego strength has an hereditary component, and social–psychological theory should recognize that the between-family differences in control capacity could be even substantially genetic. That is to say, alternatively viewed, the differences in social class and family position have a much smaller environmental effect on this trait than do the events *within* the family. This is probably support for the psychoanalytic position that the intrafamily situations of early childhood are the big determiners; but until there is experimentation to find H_w and H_b values at different, later, age levels this last remains uncertain.

It will be evident in perusing this chapter and the next that the heritabilities for super ego strength are, uniformly across all relations, H_w, H_b, and H_p, the lowest of all personality factors. For some reason, at least in H_w, this did not hold in the earlier study (Cattell, Blewett, & Beloff, 1955) and provoked some discussion on possible evolutionary genetic developments of altruism. An emphatic message here is that of a difference between a large positive genothreptic correlation within the family and a large negative one between families. Slight though the genetic differences are, this suggests that within the family the atmosphere is in general such that those who show tendencies to conscientiousness are encouraged. In the world at large, on the other hand, the indications are that the more conscientious individual is more punished than rewarded, possibly because long-term evaluation is lost in a geographically mobile society relative to the appreciation in an intimate family group.

As everyone realizes, since at least Stanley Hall's *Adolescence,* complex developments in altruism occur at adolescence, and the data of Cattell, Eber, and Tatsuoka (1970) shows G increasing into early adolescence. However, the factor studies support, by the general constancy and continuity of the pattern at childhood age cross sections, that Freud was correct in saying the values are fixed in fundamental nature, if not in detailed content, at an early age. Again, we see a real need, in personality theory, for H determinations carefully made for G at, say, 6, 12, 18, and 30 years of age.

3. Heritabilities of Dominance (E), Premsia (I), Surgency (F), and Self-Sufficiency (Q_2), by the Three Methods

Before proceeding let us consider the promised illustration of the effects of different assumptions of assortive mating upon obtained heritabilities, in the case of the four diverse traits considered here. Since the value taken does not affect H_w, but only H_b, H_p, and the larger resulting genothreptic correlation, these latter are exhibited in Table 9.3, and specifically in relation to the OSES method.

The main parameter affected by $r_{fm.g}$ is that of σ_{bg}^2, which, in its derivation from σ_{wg}^2 can vary from 1.15 to 1.67 times the latter, as $r_{fm.g}$ is raised.

TABLE 9.3
Magnitudes of Effects of Two Different Assumptions of Genetic Assortive Mating Magnitudes (by OSES Method)

	Personality source traits							
	E (dominance–submissiveness)		F (surgency–desurgency)		I (premsia–harria)		Q_3 (self-sufficiency)	
Assortive								
$r_{fm.g}$	.25	.07	.25	.07	.25	.07	.25	.07
H_w unaffected								
H_b	.57	.53	.52	.53	.60	.66	.54	.50
H_p	.37	.35	.50	.51	.48	.51	.53	.44
r_{bgbt}	−.94	−.90	−.95	−.92	−.81	−.70	−.93	−.89

However, the remaining adjustments throughout the equations result in the changes in final heritabilities being minor, as shown in Table 9.3. In the other two methods the value of $r_{\text{fm.g}}$ is found, not assumed, inasmuch as diverse values are tried for it and that yielding the best least-squares or maximum-likelihood overall fit is accepted. As Tables 9.1, 9.4, and 9.5 show our assumption of .25 is found by the latter methods to be somewhat too high in G, Q_3, E, F, and H, and especially in C, O, and Q_4, but not on I. No relation is evident between these modifications of $r_{\text{fm.g}}$ and the magnitudes of phenotypic correlations found by Cattell and Nesselroade (1967)—p. 131—but as C, O, and Q_4 are the principal primary source traits in anxiety the indication needs exploration in terms of a hypothesis that genetically anxiety-prone people do *not* seek partners similarly endowed but follow rather the "completeness" (compensatory) principle.

The traits whose heritabilities we focus on in Tables 9.3 and 9.4 are together only because they are neither control traits as in Table 9.1 nor traits of clinical importance as in Table 9.5. They have distinct individuality and are best discussed separately. As to dominance (E), its significant sex difference (Cattell, *et al.* 1970; Saville, 1972), and its susceptibility to physiological manipulation by endocrines in animals, have started an assumption that it would have a large genetic determination. If we accept $H_p = .18$ as the overall statement (Table 9.4), we need to reverse our ideas *considerably,* for it is the second lowest of 10 traits considered here. This agrees better, with the clinical evidence of rise of E under therapy, and its fall under physical or mental illness, as well as with Mowrer's animal experiments showing decided change in dominance behavior due to placement in more or less dominating social environments.

In two other traits here—I and Q_2—there is a tendency for H_b to be higher than H_w or H_p, but in none is this so marked as in E. As pointed out throughout, such a difference is bound to be determined more by an "anomaly" of environmental variance than genetic variance. It means that the environmental influences that determine an individual's dominance are far more powerful and varied within the family than between families, that is, than socioeconomic position and atmosphere. Psychologists will probably at once see this—and rightly so—as belated support for Adler's connection of the role of family position (such as birth order) with the "inferiority complex." Since the family influences mainly occur earlier in life than the SES awarenesses, and economic advantages, it is also an argument for the nongenetic influences on personality (for several traits have this pattern, as in Table 9.6) occurring early in life. However the large negative ($-.63$) r_{bgbt} estimate definitely suggests that in adult life and in the world generally, the environmental pressures are to reward the humble and "cut down to size" those with unusually dominant behavior.

In surgency–desurgency (F), the theory accepted for some time that desurgency represents a history of punishing, depriving, and inhibiting influ-

TABLE 9.4
Heritabilities of Personality Factors E, F, I, and Q_2 by Three Different Analysis Methods

	E				F				I				Q_2			
	OSES	LS	ML	Mean	OSES	LS	ML^a	Mean	OSES	LS	ML	Mean	OSES	LS	ML	Mean
H_w	.25	.04	.04	.11	.48	.78	.62	.63	.41	.33	.23	.32	.36	.20	.16	.24
H_b	.51	1.00	1.00	.84	.52	.79	1.00	.77	.60	1.00	1.00	.87	.54	.86	1.00	.80
H_p	.37	.08	.08	.18	.50	.76	.69	.65	.48	.49	.44	.47	.53	.32	.27	.37
r_{wgwt}	-.33	.24	.21	.04	-.30	-.46	(0)	-.25	-.44	-.10	-.05	-.19	-.35	-.12	-.06	-.18
r_{bgbt}	-.94	-.95	00	-.63	-.95	-.45	(0)	-.47	-.81	-.95	0	-.59	-.93	-.41	-.50	-.61
Assortive	.25	.10	.10	.15	.25	00	.10	.12	.25	.25	.25	.25	.25	0	0	.08
H_{wt}	.32	.04	.04	.13	.59	1.00	.56	.72	.51	.32	.25	.36	.40	.21	.16	.26
$\sigma^2_{wt.s}/\sigma^2_{wt.t}$	1.39	1.06	1.08	1.18	1.57	1.53^b	1.53^b	1.54	1.53	.99	1.13	1.22	1.20	1.04	1.03	1.09

[a] Better fit parsimonious.
[b] Because of one zero value these were pooled and averaged.

ences now calls for revision more radical even than those of largely genetic theories of dominance (*E*). Since its overall heritability, H_p (.65), is the highest among 10 source traits here, and indeed on a par with that found for fluid intelligence, it must be regarded more as "susceptibility to inhibitory presses" (or "inhibitability" if one can pronounce it!) rather than life *experience* of an undue amount of inhibition. The fact that *F* level changes with alcohol intake has perhaps been regarded as evidence of environmental influence when actually it should be regarded as indication of a temperamental, physiological basis. ("Some are born three drinks ahead of others.") Among various attempts to find temperament and body build associations in the Sheldon tradition, one of the few that the present writer has encountered as significant is body breadth, and especially breadth of face, with surgency (see also Kretschmer [1929] re body breadth and manic tendency). Some confusion, here, has come from regarding depression as a simple opposite of mania, and regarding desurgency (which is sober, prudent, inhibited behavior) as depression. (The CAQ, Kameoka [1981a] scales for the seven depression factors are, incidentally, all distinct from, i.e., uncorrelated with, desurgency).

Our present evidence therefore points to a heritability of surgency–desurgency (*F*), among the highest in personality, and of essentially the same level as for intelligence. A search for physiological or neural–anatomical associates of surgency is now indicated in personality research.

Premsia (*I*)—the word is an acronym for over-*pr*otected *em*otional *se*nsitivity—is of interest as having been spotted by William James and some predecessors (Jordan, Spranger) as "tendermindedness." Eysenck (1970) has shown that it projects powerfully into sociopolitical opinions (anticapital punishment, anticompulsory vaccination, etc.). In both Britain and the United States it is decidedly higher for women than men and is higher in neurotics, manic-depressives, etc., than in normals (Cattell, 1973a). There are indications of association with the A blood group (Cattell, Young, & Hundleby, 1964) and other suggestions that an hereditary component could be appreciable. As Table 9.4 shows, the total heritability, .47, is above average for personality traits (Table 9.6) and the between-family heritability, .87, is actually the highest of 10 traits here. Results by the twin method (Table 3.4, p. 77) also place it as the most heritable of personality traits examined. However, compared with the rival for high inheritance, surgency, it shows much lower heritability in the within-family situation. Evidently the existing theory of *premsia*—as due to an upbringing characterized by overprotected emotional sensitivity—has to be modified. It remains true that premsia is associated with chronic childhood delicacy of health, etc. (Cattell, 1973a), with the greater protection of girl children, etc., but we must now admit an appreciable genetic basis. There may be false correlations. Perhaps the premsic child makes more of slight ailments, as the questionnaire items suggest, and the greater protection of girl children may be a passing tradition. As usual the conclusion from the low H_w/H_p ratio could be (a) that whatever is environmental in origin is connected more with the intense intrafamilial traumatic and other molding influences that clinicians see, rather than the family position and ethnic traditions stressed by anthropologists and sociologists, and/or (b) I is affected by forces early in life, rather than those forces outside the family which, even in the pre-adult period, the individual meets later than the intenser family events. Like surgency, it is affected by forces both within and between families in the direction of bringing the individual toward the norm. Thus both the harsh toughess of harria ($I-$) and the impractical degree of sensitivity in premsia ($I+$) at the other extreme are evidently behaviors that society seeks to modify, as the $r_{bgbt} = -.59$ (Table 9.4) suggests.

In total heritability, self-sufficiency (Q_2), falls between such relatively high H's as we find in F, I, and Q_3 on the one hand, and such low ones as dominance, E, and super ego, G, on the other. The twin method (Table 3.4, p. 77) also agrees in putting it as of middling heritability (.38). Its H_w, H_b, and H_p pattern is rather close to that of ego strength (C) and suggests that these two could conceivably be instances of the genothreptic model which we called the eidolon, or "box and lid." In defining the eidolon model the effect of environment has been depicted as a set of additions to the genetic components of the elements of that pattern, that are proportional to the genetic components concerned. In this way a factor is generated in individual differences that is the same for the genetic and threptic components (box and lid form) and is hard to separate factor analytically.

The "response" of or to environment need not, however, be a positive addition but could also be a proportional subtraction, and, indeed, in the probable large negative genothreptic correlations found here there is indication that this is so. In the initial equation (3.10, p. 65) we naturally used a plus sign to indicate added action, but it could equally have been written:

$$a = v_g g - v_t t \qquad (9.1)$$

which would result in the variance expression, (dropping the v weights to keep to essentials used before):

$$\sigma_a^2 = \sigma_g^2 + \sigma_t^2 - 2r_{gt}\sigma_g\sigma_t$$

That is to say, the subtraction of actual scores on elements would still lead to a summation of variances, but with a genothreptic correlation that would reveal the general opposition of environment to the genetic character. This, however, would not prevent the factor in the threptic measures as such being just the same, inverted in sign, as the genetic, and following the eidolon model of genothreptic interaction.

With the caveat that the standard error of estimate of the genothreptic correlations in Tables 9.1, 9.4, and 9.5 are poorer than for variances, we may contingently note that the biggest negative r's (within and between combined) are for C, F, Q_2, Q_4 and, less so, E and I. This condition, combined with the biggest H_b/H_w heritability ratios over all being for $E(8.0)$, $I(8.3)$, $O(4.0)$, $H(3.6)$, $C(2.6)$, and $Q_2(2.5)$ may suggest that E, I, C, and Q_2 fit the negative eidolon model of trait development. In the case of Q_2, we know of several significant environmental associations. Positive correlations are found with being the oldest child and being an only child, both of which might reduce self-sufficiency due to parental nearness. Being seclusive among peers might also be a reaction to social pressure against natural self-sufficiency.

4. OSES, Least-Squares, and Maximum-Likelihood Analysis of Guilt Proneness (O), Parmia (H), and Ergic Tension (Q_4)

Whether guilt proneness, O, is what the psychoanalysts talk about as guilt (it is uncorrelated with G, super ego strength) is uncertain. Its factor pattern is one of a sense of inadequacy, worry, unworthiness, depression, and oversensitivity to disapproval by others, and it often leads to phobic, overscrupulous, and compulsive behavior. It normally falls slightly through life (Cattell *et al.* 1970). As Table 9.5 shows, it appears to have only a low moderate heritability overall, but a high one between families. The latter, indicating substantial family heritability, could fit a theory that it corresponds to what psychiatrists describe as endogenous depression (responsive more to drugs than psychotherapy), but the H_w of only .21 nevertheless indicates that differences of treatment, sib position, etc., within the family are powerful in bringing the potential trait out. The twin studies (Table 3.4, p. 77) agree in giving it a low within-family heritability (.34). Within families there is some suggestion that the inadequate are made to feel more so (by their sibs?), but between families there is the usual normalizing tendency, which is perhaps more marked in "unhealthy" traits, to bring children from genetically high and low families more toward the mean.

The primary source trait H is one of considerable importance in psychology as being prominent in both the second-order factor of exvia (QI, loading .50) and of anxiety (QII, loading $-.38$). Since the finding that this boldness-versus-shyness factor is associated with sympathic response to the cold pressor test (not to mention numerous daily life criteria, e.g., air pilot choice, indicating calmness-versus-agitation under pressure) the theory was developed that $H(-)$ is high sympathetic system reactivity, hence *threctia*, defined as high susceptibility to threat. The opposite pole was hypothesized to be a predominance of *parasympathetic* control, hence parmia, for $H+$.

On this theoretical basis we should expect from the present research a high heritability, and indeed its H_b value of .83 is the third highest in the set. On the other hand its H_w and H_p are moderate, at about the mean of Figure 9.1. Again we are forced to the view that it must have a high modifiabil-

TABLE 9.5
Heritabilities of H, Parmia, O, Guilt Proneness, and Q_4, Ergic Tension, by Three Methods

	H				O				Q_4			
	OSES	LS	ML[a]	Mean	OSES	LS	ML	Mean	OSES	LS	ML[a]	Mean
H_w	.30	.22	.28	.27	.26	.24	.12	.21	.36	.30	.28	.31
H_b	.53	.95	1.00	.83	.63	.87	1.00	.83	.43	.43	—	.43
H_p	.41	.34	.36	.37	.43	.38	.21	.34	.40	.35	.28	.34
r_{wgwt}	−.23	.18	(0)	−.02	−.17	.07	.44	.11	−.37	−.09	0	−.15
r_{bgbt}	−.91	−.95	(0)	−.62	−.92	−.95	0	−.62	−.93	−.95	0	−.63
Assortive r	.25	0	.10	.12	.25	0	0	.08	.25	0	0	.08
$H_{w.t}$	.37	.25	.33	.32	.30	.27	.14	.24	.50	.29	.27	.35
$\sigma^2_{wt.s}/\sigma^2_{wt.t}$	1.38	1.25	1.19	1.27	1.23	1.18	1.13	1.18	1.80	.94	.94	1.23

Note. Ratio of $H_{w.t}$ (twin basis) to the H_w (MAVA) heritability values over 10 traits = 1.15. Change from a .25 to a .07 assortive r changes: OSES H_w not at all; H_b to .56, .57; and .44; and r_{bgbt} (most change) to −.87, −.89, and −.90.

[a] For H and Q_4 the best fit was for the parsimonious mode assuming no genothreptic interaction. For O the difference was so trivial that we retained the full model.

ity within the family situation. The twin method (Table 3.4, p. 77) places it slightly above average heritability, at H_w = .40, while the twin basis of calculation of H_w applied here (Table 9.5) places it at .32. The Cattell *et al.* (1955) MAVA study, in which the *H* scale is essentially the same as here, gave H_w = .40, H_b = .78, and H_p = .59, agreeing well with the Canter twin study, but stepping up appreciably the H_p value reached here. In the light of such other findings we may speculate that the present Chicago–Cleveland sample might be getting a greater environmental range in regard to anxiety-provoking environmental threat. Contingently we would conclude parmia-versus-threctia has a heritability slightly above the average of all personality primaries, but not as high as previously hypothesized.

Finally, in Q_4, we also have a moderate clash with the Cattell *et al.* (1955) result, which again happened to be based, even at that early stage of research, on a Q_4 scale of relatively good validity. The result in Table 9.5, pretty consistent in the OSES and least-square: solutions, points to a vir-tually average heritability level in H_w and H_p, but a below-average inter-family genetic effect. The clash with the pioneer MAVA finding con-sists, therefore, in the latter showing quite a large H_b, namely, .90. At pres-ent we have no explanation for this discrepancy but are inclined to discount our earlier work. Meanwhile new data are eagerly awaited.

The theory of Q_4—*ergic tension* as it has been designated in the litera-ture to this point—is that it represents *the sum total of all undischarged (frustrated) ergic (drive) tensions.* The numerous life criteria relations now known for it (Cattell *et al.*, 1970) support this conclusion. Its very high load-ing in the second-order anxiety factor (*Q*II), fits well the psychoanalytic con-cept of transformation of undischarged libido into anxiety. However, it is noticeable that many items in the Q_4 scale have to do with *autonomic* signs of being overwrought, jumping at sudden sounds, suffering physical and muscular tensions, etc., which might suggest not *only* a high level of ergic frustration but also a characteristic proneness *to discharge through high au-tonomic reactivity.* Since it seems reasonable to assume that most people are equally endowed genetically in total strength of ergic needs their differences in ergic tension as frustration would be expected to be due to differences in situational stresses and thwartings and thus show no genetic component. However, getting into environmental difficulties could be mediated by other personality traits, for example, low intelligence, or low ego strength, that have substantial heredity, so that even if ergic tension is wholly (immedi-ately) environmentally produced we should expect it to show *some* heredity from indirect associations. Accordingly, we see no compelling reason from the present behavior-genetic results to modify the main theory of Q_4—that it is the level of stimulated but undischarged ergic tension—though a genetic component through a tie to an autonomic mode of expression (a labile sym-pathetic system) now seems indicated.

5. The Heritabilities of the Secondaries:
Exvia–Invia (QI), Anxiety (QII), Cortertia (QIII),
Independence (QIV), and Control (QVIII)

The earlier reference to one factor creating (in the same person) situations affecting another source trait development reminds us that there is need to give attention, in behavior genetics, to a new aspect of analysis. This is tactically best approached at present though a hypotheses from observing the correlation among primaries, which means that we should now study second-order personality factors. There is interaction of trait with trait in one person and there is also interaction of different traits among persons (mainly in the family) which we have looked at so far by path coefficients confined to one trait. We recognized, however, that in truth dominance in the parent might affect super ego development in the child, and high intelligence in one sib might augment guilt proneness (O) in another.

We can dismiss briefly in the present section the problems of influence of one family member upon another, since models and methods are straightforward. As we have seen, the influence of heredity of parent upon child is not only directly through heredity upon child genes, but also as an environmental influence of the parental phenotype (which contains part of that heredity) on the threptic part of the child's trait. A whole spectrum of such phenotypic traits in the parent would normally influence the threptic part of each *particular* trait in the child. The same would be true between sibs or twins. Investigation of such connections presents no truly new problem. It can be carried out by the same two methods as have been described earlier (see comparative MAVA, pp. 224, 251) and which are applicable to *any* measurable environmental feature — personal or physical. It would, needless to emphasize, be a very valuable contribution to developmental psychology to find what parts of the threptic variance of a trait are contributed by human and nonhuman environmental features, and, particularly, what the specification equation weights are for *all* phenotypic source traits of a parent or sib in estimating a child's threptic portion of a given trait. Theories of family interaction with more substance and accuracy than those now current could be built from such evidence.

Meanwhile we turn here to those correlations, stably recurrent across populations, among the primary traits of any one individual, which have yielded second-stratum personality factors. Some eight such secondary, broader patterns have now been replicated a dozen times in this country (Cattell, 1973a) and also in England (Saville, 1972), New Zealand (Adcock & Adcock, 1977), Germany (A. K. S. Cattell, Krug, & Schumacher, 1980), Japan (Tsujioka & Cattell, 1965a,b), and other cultures that could be listed in detail in an intensive study. Secondaries increase to as many as 14 or 15 where factoring is extended to span both normal and abnormal primaries

TABLE 9.6
Broad Secondary Source Traits: Replicated Loadings on Primaries in these Experiments

(a) *Evidence of unique simple structure.*

	QI (exvia)	QII (anxiety)	QIII (cortertia)	QIV (independence)	QVIII (inhibitory control)
C	07	$\boxed{-66}$	10	-01	05
E	07	00	01	$\boxed{56}$	$\boxed{-18}$
F	$\boxed{51}$	-03	01	$\boxed{30}$	$\boxed{-27}$
G	06	03	01	-03	$\boxed{67}$
H	$\boxed{50}$	$\boxed{-38}$	$\boxed{-14}$	$\boxed{34}$	00
I	07	00	$\boxed{-73}$	-03	02
O	-02	$\boxed{78}$	-06	-05	03
Q_2	$\boxed{-65}$	-01	-04	03	01
Q_3	-05	$\boxed{-43}$	01	02	$\boxed{47}$
Q_4	00	$\boxed{80}$	-05	10	-02

(b) *Estimation weights for obtaining secondaries.*[a]

QI	QII	QIII	QIV	QVIII
	-.20	.06		
.16		.08	.40	
.35		.15		-.11
		-.06	-.11	.61
.30	-.07	.08	.11	
		-.32		
	.26	-.08	-.16	
-.19			.22	
	-.09	-.10		.28
	.38	.07		

Note. It will be seen that the high simple structure count in this average of 10 researches (Cattell, 1973, p. 116) leaves only the boxed numbers out of the ±10 hyperplane, and it is these that have used here in defining the second-stratum, "broad secondary" factors.

[a] As explained in text these are adjusted from the usual weights, in Cattell, Eber, & Tatsuoka (1970, p. 128), derived from the usual loadings with usual regard to factor correlations by being made to add to unity in covering the set of primaries for which heritabilities are known.

(Cattell, 1973a; Kameoka, in press, 1981a). Patterns of secondaries alter only slightly across cultures, and within one nation (the United States) the weights therein of the primaries can now be stated with considerable accuracy. We shall consider here only the five listed in Table 9.6, which are the most stable of all, and best known in clinical and other connections.

If we consider the most universal practicable model accounting for higher-order factors—the SUD model—there are five possible interpretations of them as far as genothreptic relations are concerned. The SUD (Cattell, 1978b, p. 307) considers that a second order appears among correlated (oblique) primaries because of one *broad* factor *contributing to the variance of all* of them, plus *stub* factors each peculiar in its variance to one of the primaries.

The five genothreptic possibilities are (see Figure 10.1, p. 382): (*a*) That both the broad factor and the stubs are threptic; (*b*) that the broad (secondary) factor only is genetic and the stubs are threptic acquisition patterns; (*c*) that both are genetic, (*d*) that all are in some degree both genetic and threptic, as in the box-and-lid model (p. 233), for example, and (*e*) that the secondary is threptic and the stubs are mixed.

Since the score of a secondary factor can only be estimated by adding the various primary (or the many variable) scores, *H* for the first type should approach zero, and for the fourth 1.0; while (b) and (d) and (e) would be middling but leaving us unable to know which is operating. There have been suggestions that the Thurstone primaries, *V*, *S*, *N*, etc., are a case of (*b*) in which fluid intelligence is the largely genetic second order, and the stubs left when it is partialled out of *V*, *S*, *N*, etc., represent acquisitions by imposed patterns of learning (Cattell, 1971(b). There is nothing to cause us to reject this in regard to *some* primaries, for example, *N* and *R*; but *V* and *S* may well have part of even their stub variance genetic, for their *H* values are high.

Let us consider the further problems in relation to what we now obtain as results. Of the two ways in which we can get *H* values for second orders, probably the better is to *score* each individual on a secondary by weighting his or her primary scores, as in Table 9.6, and then find the secondary concrete variances and proceed to *H*'s. But the second way, which we follow here, is simpler, namely, to weight the *H* values of the primaries as in Table 9.7 and average *them* to get the *H* for the given secondary.

To do this the loadings in Table 9.6 have first to be transformed to factor estimation weights by the usual formula: $V_{fe} = R_{fI}^{-1} V_{fpII} R_{fII}$, where R_{fI} is the correlation matrix among primaries, R_{fII} among secondaries, and V_{fpII} the factor pattern loadings of secondaries on primaries. These weights have then been divided in each column by the arithmetic (not algebraic, since variances are involved) sum of that column, so that, if all primary *H*'s were unity the secondary *H* would be unity, if all .5 then the secondary would be .5, and so on. There could be other assumptions, but these seem the most defensible of simple procedures for getting the heritability of secondaries from those of constituent primaries.

Taking the total population *H*, Table 9.7, as the main criterion we see that cortertia and exvia–invia have the highest inheritance, anxiety proneness follows next, and independence, and control are lowest. There is, however, an interesting effect in independence, namely, that H_b is farther above H_p than in other traits. This is sustained in the objective, *T*-data of Chapter 10. However, the general question of agreement of *Q*- and *T*-data is deferred till we have inspected *T*-data results in Chapter 10. To the findings of decidely higher H_b in independence, exvia, and anxiety we shall consistently apply the same explanation as with primaries: that in these three traits we have substantial heritability, but that it is modified by strong environmental influence within the family constellation, and very little by influences in society as a whole. The net result, however, is that the total population heritability is only middling among personality traits for exvia and cortertia which means also that it is rather lower than for our accepted average of all estimates for fluid intelligence heritability (58, Table 8.11), but about the same as for crystallized intelligence (.41).

If we compare these results with those available from other sources and methods we find no other MAVA results available, but twin method results exist from Canter, Eysenck, and Shields in Table 3.5, of .39 for exvia and .41 for anxiety. Since these depend on the twin method assumption (that within-family twin threptic variance equals sib or fraternal twin $\sigma^2_{wt.s}$) they would correspond to our transformed H_{wt}'s, which average 1.15 times H_w (Tables 9.1, 9.4, 9.5). Thus our within-family values, made comparable, would be .41 and .32, which agree for exvia, but give a somewhat lower value for anxiety.

On present evidence, therefore, the two largest secondaries, exvia and anxiety, have substantial genetic determination, but, in the total population heritability slightly less genetic than environmental susceptibility. Particular interest at this juncture in personality theory is attached to cortertia, for it is a firm pattern among the "new," secondary personality source traits, that is, those previously unknown in clinical parlance, and has substantial predictive power on real-life criteria (Cattell & Schuerger, 1978). It happens that Nesselroade (1967) has shown it to have unusual (developmental) stability over time, which is consistent with its now being found to rival exvia for the highest heritability (and exceeding it in T-data measures, see Chapter 10). Regarding criteria in life, Cattell and Scheier (1961) found it stood out among a dozen or more factors as one of the two most powerful in distinguishing neurotics from normals. (Normals are cortertic, neurotics are high at the opposite pole—pathemia.) In regard to the moderate tendencies that have been found for heritability of "neuroticism" as a single "syndrome," our hypothesis would be that pursuit of the source traits contributing would show cortertia (negative, as pathemia) to be the main genetic ingredient, with U.I.16 assertiveness-versus-weakness (negatively) as the second largest genetic predisposition for neurosis. Unfortunately, little attention has been given to other life criteria associates of cortertia–pathemia though it is known to be associated with success in "demanding" (stress) occupations. Pawlik and Cattell (1965) found it quite significantly related (− .48) to alpha index prevalence and (.27) to alpha frequency, which again agrees with its higher heritability. Incidentally the fact that cortertic individuals have much alpha interruption and higher frequency is consistent with the high alertness (surgency), toughness (harria), and assertiveness in questionnaire loadings (see Table 9.7) and generalized speed of response in behavioral measures (Cattell & Schuerger, 1978, p. 278).

Thus among the well-replicated and validly measured secondaries, cor-

TABLE 9.7

Heritabilities of Secondary Personality Factors by Q-Data

	Primary loading	Corrected[a] primary weight	H_w Singly	H_w Joint	H_b Singly	H_b Joint	H_p Singly	H_p Joint
QI (exvia–invia)								
E	.18	.16	.11		.84		.18	
F	.38	.36	.63		.77		.65	
				.37		.83		.44
H	.33	.30	.27		.83		.37	
Q_2	−.21	.19	.24		.80		.37	
QII (anxiety)								
C	−.22	.20	.31		.80		.41	
H	−.07	.06	.27		.83		.37	
O	.29	.26	.21	.30	.83	.67	.34	.38
Q_3	.10	.09	.53		.77		.58	
Q_4	.43	.39	.31		.43		.34	
$QIII$ (cortertia)								
C	.10	.07	.31		.80		.41	
E	.13	.09	.11		.84		.18	
F	.25	.16	.63		.77		.65	
H	.12	.08	.27		.83		.37	
				.36		.81		.46
I	−.52	.34	.32		.87		.47	
O	−.13	.09	.21		.83		.34	
Q_3	−.16	.11	.53		.77		.58	
Q_4	11	.07	.31		.43		.34	
QIV (independence)								
E	.42	.37	.11		.84		.18	
G	−12	.11	.09		.16		.12	
H	11	.10	.27		.83		.43	
				.18		.74		.28
O	−17	.15	.21		.83		.34	
Q_2	23	.20	.24		.80		.37	
Q_4	−09	.08	.31		.43		.34	
$QVIII$ (good upbringing [control])								
F	−.14	.12	.63		.77		.65	
G	.66	.56	.09	.30	.16	.43	.12	.33
Q_3	.38	.32	.53		.77		.58	

[a] Note there will be no signs on the score contribution weights in this column, as these deal with variance contributions. The weights of primaries are so corrected that if each and all had a heritability of X the mean of their weights-times-heritabilities would also be X (see Table 9.6b).

tertia (if we include T-data evidence, Chapter 10) is probably the most heritable, followed by exvia and anxiety, while independence, and, especially inhibitory control show a predominantly environmental threptic component from environmental influences.

6. General Considerations in Evaluating Results and Their Implications

Our final evaluation will not be made on the secondaries until their expression in equivalent T-data measurements has been inspected in the next chapter. At that place the question of which of the models of secondaries is best considered operative (p. 378) in relation to primaries will also be debated.

Meanwhile, it is necessary to take stock of the questionnaire findings on primaries with consideration of Q-data properties mentioned in the earlier, and the nature of our population. One must consider:

1. That Q-media measures may be subject to *motivational distortion* not operative in T-data. The only real answer to this is application of *trait view theory* corrections (Cattell, 1973a) appropriate to the given role situation of testing. Since this was a school testing situation, in which subjects were told results would be anonymous, the probability of low distortion was such that the elaborate task of *trait view correction* was not undertaken. The result of this influence would be a slight increase in the cross-factor blurring discussed in (2).

2. We have already noted that questionnaire scales are polluted (as to *factor trueness,* Cattell, 1973a) not only by (*a*) *specific factors,* in each item and scale but also (*b*) by other *common* factors, since suppressor action is never complete. Let us ask what the effects in (1) and (2) are likely to do to H values.

An argument can be made that specifics are likely to be more environmentally determined than broad factors. For example, H factor, parmia-versus-threctia, can be considered at the threctic end to be a generalized timidity, fearfulness, and susceptibility to threats and phobias. The particular directions these "phobias" take, however, are largely a matter of particular environmental events ("conditionings"), as any reflexological therapist recognizes. (It is, however, only as recently as the last decade that reflexologists have recognized, in borrowing the term "agorophobia" for "multiple phobias," the existence of a personality trait of general susceptibility to phobias, which is probably largely our threctia factor!) Although in scale design we try to choose life situations widely to make these "accidents" even out, they exist as a total, and in sum the more *specific* behaviors are a story of environment. If this is generally so in all scales with accumulating specifics, and the correction for scale validity (Chapter 8) is not applied, the degree of invalidity of any factor scale through accumulated specifics will show up in H (according to its amount) through (*a*) some shift from the true H values toward lower heritabilities and (*b*) some movement of all trait H's toward their mean, since each will have a little of all other factors in its measurement. A perhaps extreme result of the latter is Loehlin's (1965) exposure of the fact that no significant differences of heritability could be shown over many personality scales. But the majority of these scales had not been constructed for factor uniqueness in the first place, that is, they were subjective and arbitrary (e.g., the CPI, the Tellegen scales). However, it must be recognized that even in scales uniquely rotated to factor source traits, such as in the HSPQ (used here), the 16 P.F., CPQ, and CAQ, blurring of H values will remain until *progressive rectification* (Cattell, 1973a, p. 292) has been carried further.

The effect of sources of error on heritabilities derives from its effect on variances, as examined in Eq. (8.11). There it will be seen that the form of correction is similar for error of measurement $(1 - r_d)$, invalidity $(1 - r_v^2)$, and function fluctuation $(1 - r_f)$, all ending in a ϕ ratio by which the raw variances need to be multiplied to get a "pure" estimate. But the absolute contribution to this ratio is probably greatest (in most tests forms) from r_v and r_f. The work of Brennan (1979), Cattell and Scheier (1961), Nesselroade (1967, 1976) and Kameoka (1981b) using *dR-factoring* clearly shows that quite strong function fluctuation variance (see analysis in Cattell (1973a) occurs in many factors hitherto treated as fixed traits. It occurs, for example, in such source traits as exvia–invia (QI), anxiety, (QII), independence-versus-subduedness (QIV), the seven depression factors (Kameoka, 1981a), and even intelligence (Horn, 1972a). If Jung were alive today he would have to reconcile his theories with the existence of extraversion (exvia) as a *state*.

However, if one looks at the simpler form of calculation of H, as in H_w for the twin design [Eq. (3.20), p. 69], it will be seen that the multiplying of all variances by ϕ, as would be done to transform each to an estimate of the true variance, would leave the heritability, H_w, unchanged. But if the estimate happened to be in the following (imaginary illustration) form:

$$H_w = \frac{(\sigma_{c1}^2 - \sigma_{c2}^2)^{1/2}}{\sigma_{c3}^2 + \sigma_{c4}^2}, \tag{9.1}$$

where c_1 through c_4 are concrete variances, then multiplying each variance by ϕ would no longer leave H_w the same corrected as uncorrected. And, as pointed out in Chapter 8, since several of the MAVA solutions in Table 4.4 are nonlinear, the ϕ correction *would* affect the variance and the heritability values. Chapter 10, however, shows the effects to be small.

When we consider the joint effect of error in experimental measurement, test invalidity, function fluctuation, we must accept that the *standard error* of the σ^2 and H estimates is likely to be enlarged by these influences. There is thus no escape from the conclusion that since statistically estimated "correction" for intruded error is not an ideal answer, it becomes vital to work with *longer testing*, to raise r_d, with more *factor pure tests* to raise r_v, and with *repeated measurements* to reduce function fluctuation.

Ten years ago, when the present researches were started, it was possible only to make some approach to these goals by choosing well-factored tests and employing two forms instead of one. On this basis the present results are not treated to an estimated correction. Perspective can be given to the findings by presenting them all together, along with the intelligence results of the preceding chapter in Figure 9.1. For condensation and easy view the heritabilities are given as population heritabilities, H_p's, though in subsequent discussion we shall give some prominence to H_b's.

							34	37			58		
	12	18				O	35	H	.41	47	g_f	65	
	G	E				Q_4	C	Q_2	g_c	I	Q_3	F	
0	10		20		30		40		50		60		70

Figure 9.1. The distribution of population heritabilities over 12 traits (intelligence H's from Table 8.10).

As hypotheses for further exact testing by MAVA, and for checking in other domains of personality research, we would finally state (with an eye to the sources of error discussed):

1. The heritabilities for g_f, Q_3, and F might, if corrected, reach up into the 80s and 90s. That is to say their factor patterns would be largely due to influences in the form of genetic maturation forces with only minor (unless traumatic) effects from environment. If we consider the very high H_b of premsia–harria (*I*), it could perhaps also be placed among the substantially inheritable.

2. One may conclude for the present that C, g_c, H, O, and Q_2 would be in the middle range, but more separated than now seen. Probably, therefore, we should hypothesize that these fit the eidolon model.

3. That G, E, and possibly Q_4 (with its low H_b value) will turn out to be patterns largely due to environmental influence, with relatively trivial effect from the heritability of the drives which are expressed in the genetic material that goes into the acquired sentiments that G represents. They are therefore hypothesized to be due, respectively, to the pattern of moral values taught from childhood (*G*), the fortune of position and circumstances encouraging dominant or submissive behavior (*E*), and the extent to which basic ergic expressions are circumstantially frustrated at a given time (Q_4). Although radicalism–conservatism (Q_1) has not been measured here (but only in twin study) one would surmise that it also belongs in this category.

At this stage of personality inheritance research, it is evident from the unevenness of values from different sources, in Tables 9.1, 9.4, and 9.5, that we must consider results as exploratory. They create hypotheses as initial beacons to follow with more precision, perhaps for one source trait at a time. Despite unevenness in certain absolute values there is at least an orderliness in the rank order of heritability given for the 12 traits (this count including intelligence) by the different methods. And there is order also in the nature of the discounts and relative significances delivered by the different methods. As we have said, the statistician per se will probably prefer the maximum likelihood. But in fact it proves here more frequently to give extreme values (1.00 and 0), that seem less probable to the psychologist than, say, some OSES values, in which correlations never reach either 0 or 1 and heritabilities always fall short of perfect heritability or complete environmental creation. The OSES method yields compomise values, on the other hand, that probably understate the real extent of difference among traits. So in due perspective, we should recognize that the degree of convergences may give us confidence at least in the final *rank order,* in Figure 9.1, of the heritabilities of primary source traits.

Finally, let us look at the features of heritability and genothreptic correlation that seem to characterize personality traits in general, in the fairly representative group of independent traits that we took for experiment. The overall averages are presented in Table 9.8.

In the first place, we see a central tendency for about three-fifths of the population variance of the observed traits to come from environmental sources and two-fifths from genetic variance.

Second, we note that the between-family inheritability is decidedly greater than the within. Before accepting this let us ask whether any artifact, for example, of mode of calculation, could have caused it. The genothreptic correlations are, according to the least-squares results (the illustration in Table 9.2), beset (especially when small) by larger standard errors than the variances, but not when they reach such a value as $-.58$ in Table 9.8.

Now the value for the between-family variance is concretely given, and we know it equals $\sigma_{bg}^2 + \sigma_{bt}^2 - 2r_{bgbt}\sigma_{bg}\sigma_{bt}$. But in calculating H_b, by the agreed method, we use only the first two terms in the denominator. The existence of a negative genothreptic correlation would therefore enhance, not reduce, the value used for the denominator of H_b. That is to say, if we *had* overestimated $(-)r_{bgbt}$, the H_b value would be smaller, not larger than it should be; and our H_b's are perhaps underestimates as far as this source effect is concerned.

Bias in the other direction could arise from the method of obtaining σ_{bg}^2.

TABLE 9.8

Some General Properties in Personality Trait Inheritance (by Q-Data)

Mean H_w overall traits	= .30
Mean H_b overall traits	= .71
Mean H_p overall traits	= .38
Mean r_{wgwt} overall traits	= .06
Mean r_{bgbt} overall traits	= $-.58$
Mean $r_{fm.g}$ assortive mating (genetic)	= .13
Mean ratio of between-sib threptic variance to between-twin threptic variance	= 1.19
Mean $H_{w.t}$ (i.e., within-family inheritance calculated by twin method)	= .32

In the OSES contributory data σ^2_{bg} is obtained from σ^2_{wg} by a ratio (ranging from equality to 1.67) that is fixed by genetic laws, but only in regard to an assumed value for assortive mating. However, we took either very small $r_{fm.g}$ values or those central to the best fit in the least-squares method. Unless, therefore, some bias can be shown toward underestimating σ^2_{bt} in the denominator of H_b—and we can find none—the systematically larger H_b than H_w must be accepted.

Since, as we have pointed out, some between family differences of environment, for example, of social status, enter into the within-family σ^2_{wt} estimate, insofar as sibs of one and the same family wander into different social associations outside home, any correction for this contamination would increase, not reduce, the $H_b - H_w$ difference. The final implication is, therefore, that the $H_b - H_w$ difference must arise from the inherent nature of family and social life and in fact from within-family threptic variance being larger than the between, as in the illustration of ego strength in Table 9.2. That is to say, *where broad, universal personality traits* are concerned (not superficial attitudes, skills, political opinions, etc.) what happens from events and positions, etc., in the family is far more important than position in society. Despite the claims of many sociologists regarding status, economics, etc., in shaping personality, it seems that as regards development of basic traits the clinicians have been much nearer the mark. At least up through the 12–18-year age range (mid value 15 yrs.) of our subjects, what happens *within* the family is (on an average over 10 traits) rather *more than twice as important* as their *differences in social position*. A repetition of MAVA with 45-year-old subjects might show something different, and it would be most interesting to have such comparative results.

Although we have agreed not to digress into the entanglements of social implications, I cannot well pass up the support for my 1937 book on the role of inheritance in regard to national intelligence and birth rate. In the intelligence domain the present results show an H_b for crystallized intelligence (traditional, school learning based intelligence tests like WAIS and WISC) of only .37, but (Table 8.10) for fluid intelligence (culture-fair tests) an H_b value of .94. In short, in estimating the fluid intelligence level of the next generation, the differential birth rate *between families* is as important as it was then asserted to be on the basis of culture-fair intelligence tests. The differential marriage and death rates which at that time (1936) apparently offset the birth rate effect completely (see check in Cattell, 1953) may or may not be doing so in a welfare society today.

Finally, since the larger genothreptic correlations are significant, we should ask about the implications for personality and social psychology of the fact that the generally negative response of environmental influences to initial genetic deviation is decidedly larger outside than inside the family (Table 9.8). Although children may doubt it, parents are, apparently, generally more "fair" than society. The average within-family genothreptic correlation (r_{wgwt}, Table 9.8) is near zero (.06). However, in certain traits — so far it seems to be the desirable ones, such as superego (G), and self-sentiment (Q)$_3$ — r_{wgwt} is quite large and positive (.75 and .58). These latter r's point to a strong fostering by the family atmosphere of whatever signs the child shows of more conscientious and self-respecting behavior. The slight negatives uniformly across the rest might well be due to the intrusion, just mentioned, of between-family into within-family relations measures.

Probably our best conclusion at present regarding what happens within the family on most traits is that rearing influences, producing threptic variance, are relatively uniformly applied as far as *differences in sib genetic level* are concerned. However, this conclusion must not be thought inconsistent with our conclusion that considerable differences of environmental impact exist within the family. They exist but (with the exceptions noted) are not related to genetic differences. The 6:5 ratio (Table 9.8) of sib to twin threptic variance, found over this representative set of traits, suggests that part of the difference is due to sibs being born into different peer culture epochs and differences of the age and experience of parents when each was born. (Let us remember the scores were corrected for age as such) But that is not all, and the most probable origin is surely where clinicians like Freud, Jung, and Adler have long said it is, namely, in almost random, unpredictable events in the domain of accidental traumas and sib relations, small in themselves but large in the perception and emotional reaction of the infant and the young child. Comparative MAVA, permitting the formulae relating *threptic* to *environmental* measures to be discovered, may in time bring clinical hunches to the level of quantitative statements of environmental effects on threptic development.

Across various methods and traits the most central value for genetic assortive mating for husband and wife is .13 (Table 9.8). As pointed out earlier, it seems to be higher for more obvious traits like stature and intelligence. We therefore reject the conclusion of some behavior geneticists that the observed homogamy applies purely to the acquired part of the *phenotype*, though we would conclude that $r_{fm.g}$ in at least the less obvious personality traits is decidedly smaller than for some physical features. Not only physical features but behaviors more obvious in the interests of congeniality, such as education, verbal intelligence, political attitudes, acquired recreational interests, may reach a higher assortiveness coefficient. But intuitions of mutual responsiveness based on broad genetic temperament endowments demonstrably play a real if minor part. And as Cattell and Nesselroade (1967) have shown, at least at the gross phenotype level, these personality congenialties are higher in stably than unstably married couples and hence more likely to affect a sample of the next generation. In any case, in view of possible social trends in homogamy, any conclusion about population heritability should have tagged to it (as in Table 9.8) an assortive mating r, since this fixes the ratio of σ_{bg}^2 to σ_{wg}^2, which is appropriate to the epoch and culture.

Finally, we come to the *law of coercion to the biosocial mean*, formulated in several places by myself on the basis of such evidence as the substantial negative r_{bgbt} value for most traits, the 33% reduction of IQ sigma in traditional compared to culture-fair intelligence tests, the relation of popularity to trait scores in small group dynamics situations (Cattell & Stice, 1960), and much else. One must notice that the negative r_{bgbt} is largest from the OSES method and that occasionally the maximum likelihood has given a better fit for the parsimonious hypothesis that no genothreptic correlations at all exist. Nevertheless by all methods over all traits negatives predominate, the only instances where $r_{wgwt} + r_{bgbt}$ sum to a positive value being intelligence, (by least squares) self-sentiment, and super ego, all of which are effective and desirable traits.

Since (without comparative MAVA findings, p. 222) one cannot make correlation plots of *individual* scatter based on endowments of the abstract components, it is not possible to decide whether a genothreptic correlation (*a*) extends over the whole range or shows only (*b*) bringing down the genetically above average toward the average or (*c*) raising the genetically low toward the average. Nor can we check the curvilinear correlations which we suspect exist sometimes when the linear *r* is zero (Chapter 7). The probability is that all three effects occur, because there is no obvious socially desirable pole to most traits, for example, premsia, parmia, and surgency; and because the correlation is so large that it could scarcely be reached by a regression holding only over half a range.

Social psychologists seem not yet to have studied sufficiently—or even to have recognized—what on behavior-genetic evidence is a quite powerful tendency of society to teach and reward in personality learning in such a way as to bring individuals toward a biosocial central norm. One says *bio*social because the mean is in part biologically determined. It is interesting here to note both the parallel and the divergence between society and the larger natural environment. The larger natural environment which determines stability or trend is obviously constantly exercising natural selection mainly by trimming off deviant mutations. For example, the remarkable tendency of races all over the world to have fluctuated for perhaps ten million years around a certain central stature for mankind suggests there are reasons having to do with the tensile strength of bone, the maintaining of blood pressure, etc., which check the deviations toward very tall stature which social competition might otherwise favor. In such temperamental traits as risk taking (parmia-versus-threctia) and social inhibitability (desurgency), both of which are among the more inherited, one can see how the optimum value would tend to the mean value, from which neither society nor the outer world would leave deviations undiscouraged. In other traits, such as intelligence and emotional stability (sanity), there could be biological encouragement, from the natural world's demands, toward movement in *one* direction, e.g., to larger brains. Society, less wise than nature, seems disposed automatically to favor convergence toward the middle—on almost all traits here studied. (Just as fear of nature is older than any systematized religious awe, so the tendency we here reveal could be older, as pity and envy, than any conscious rationalist and rationalizing philosophy of egalitarianism.)

To keep properly to our intended behavior-genetic sphere we must desist from going further into hypotheses for research concerning the mechanisms by which the law of coercion is made to work. But they must constitute quite a powerful array of forces, operating largely between rather than within families, and springing from motives as diverse as individual dislike of strangeness, on the one hand, and practical needs of institutions, on the other for example, the school, the mass media, the army, with their needs to handle humanity efficiently en masse.

7. Summary

1. It is most strategic to base behavior-genetic research on factorially well-replicated personality structures, both because other researchers will have no difficulty in checking these and because these source traits now have theoretical meaning and criterion associations. Agreement on these structures has come later than on abilities but, since 1940, sufficient progress has been made in tying down the forms of primary and secondary structures, systematically across the possible L-, Q-, and T-data observation media. Flexible and sophisticated use of factor analysis by R-, dR-, and P-techniques has now replicated some 35 normal and abnormal primaries, and some 15 secondaries. The major source traits cross-check in L- and Q-data but the latter is alone practicable for measurement. Consequently the choice here has been made of a dozen or more of the most frequently replicated; most validly measurable; most richly clinically, educationally, and socially criterion-associated; and most standardized of these.

2. In what clinicians and personality theorists think of as the control area, three factors—ego strength (C), super ego strength (G), and the self-sentiment (Q_3) —are statistically and clinically most prominent and are first studied here. Using the same samples of around 3000 boys, 12–18 years old, as in the previous chapter on abilities, the MAVA analysis was pursued through three distinct analytical approaches: OSES; the least-squares fit; and the maximum-likelihood method. Although individual H values from different methods show some degree of scatter all three methods agree in ranking, giving highest heritability (total population) to Q_3, (self-sentiment, esteem) moderate to C (ego strength), and very low to G (super ego). In ego strength, heritability is decidedly lower within the family than between families, indicating that events within the family, presumably in early years, exercise the bulk of environmental influence. The unexpected result on Q_3 il-

lustrates the importance of behavior-genetic findings in shaping concepts in personality theory.

3. As to reliability of final outcomes the standard errors of estimated abstract variances and the chi squares of the restored concrete variances point to good dependability, that is, values within a narrow range of standard error, for variance solutions by least squares and maximum likelihood. The least-squares and maximum-likelihood analyses in general, but not always, agree mutually somewhat better than with the OSES. The reliability of estimate of the genothreptic correlations, notably of the r_{wgwt}, is notably less, in the lower values, than that for variances. Where an assumption has to be made about the magnitude of genetic assortive mating of father and mother it is shown that variations in assumed value over the usual ranges considered possible, make comparatively small alterations in the H values. The best fit, over all traits, is for $r_{fm.g}$ values that range from 0 to .25, with .10 as most common and .13 as a mean.

4. The results for dominance (E), surgency (F), premsia (I), and self-sufficiency (Q_2) place surgency as most highly heritable and premsia next, the former indeed being in the same class as (fluid) intelligence. Self-sufficiency (Q_2) follows, and (E) dominance is lowest, except between families. Dominance, premsia, and self-sufficiency alike have the pattern of high heritability between families and low within, again suggesting that it is through events within the family that environment exerts its main threptic effects.

5. Of three primary factors—guilt proneness (O), parmia–threctia (H), and ergic tension (Q_4)—which have been of clinical interest (since O and Q_4 are the highest associates of anxiety (Table 9.6) and H is also high), H and O show high between-family inheritance. The heritability of Q_4 is low, agreeing with previous psychological theory that it represents a largely circumstantially determined degree of total ergic frustration. O and H show the same extreme discrepancies of within- and between-family inheritance as were noted for E and Q_2, suggesting that H (threctia, shyness) and O, guilt-proneness, are much affected by family position and early experience, but otherwise have appreciable heritability.

6. Of the 5 (among 9 to 15 recognized as tolerably known today) most widely replicated second-stratum factors (across age and cultures), cortertia (QIV) and exvia–invia (QI) have the highest heritability, but again less within than between families. Former studies on twins, if discounted for the untenable assumption of equal twin–sib environment inherent in that method, agree reasonably well with the within-family heritabilities for exvia and anxiety obtained here.

7. The secondary $QVIII$, previously called good upbringing or control, has quite low heritability, while independence, $QIII$ has moderately high between-family H, but quite low within-family and total population heritability. The former agrees well with previous theory, but the latter does not (at least in that the expression of independence called "perceptual field inde-

pendence" was considered by Witkin, *et al.*, (1962) as largely constitutional. Other research approaches have led to concepts of QIII as a substantially innate, sex linked assertiveness, but molded in the direction of a rigid, compulsive mode of expression. Perhaps this particular molding (a dominant parent?) results in its being placed only at about the same middling level of heritability as for exvia–invia.

8. When we pass from specific trait findings to general features of the present research outcomes we find:

(a) A comparison with past twin-method studies and one earlier MAVA shows moderate variations (greater than within the present three analyses) in absolute H values, but, wherever the same trait measurements are involved an agreement in rank order over most of 10 traits.

(b) Among the present methods the least squares and the maximum likelihood—particularly the latter—tend to give more extreme values (1.00 and 0) both for variances and correlations, than does the OSES. The maximum likelihood also more frequently gives within family genothreptic r's of 0.

(c) The between-family heritability on an average runs more than twice as high as the within, which we explain as due to the environmental influences on basic personality traits occurring largely within the family, since the result implies that σ^2_{wt} runs higher generally than σ^2_{bt}.

(d) As mentioned, the central tendency for well-fitting assortive mating (genetic) r is .13, but it runs higher for more socially obvious traits like intelligence (g_c), the self-sentiment (Q_3), dominance (E), surgency (F), premsia (I), and exvia (QI) than for more "subtle" traits. At the *pheno-typic* level higher assortive r's are found also for autia, (M), radicalism–conservatism (Q_1), and independence (QIV).

(e) The within-family threptic variance for sibs averages 1.2 times that for twins (identical and fraternal together). Quite conceivably the ratio would be higher if a separate solution for identicals (which should be lower than for fraternals) were available. When we need to compare present values for within-family heritability for sibs with those previously found by the twin method (which assumes $\sigma^2_{wti} = \sigma^2_{wt.f} = \sigma^2_{wt.s}$) we find that a MAVA sib $H_{w.s}$ of .29 becomes an $H_{w.t}$ of .33, and a MAVA $H_{w.s}$ of .56 becomes a twin $H_{w.t}$ of .60. Even thus made comparable, our heritabilities tend to run a little below those of the twin method (made on smaller samples) for some traits.

(f) The average population heritability for the array of varied personality traits runs at .38, distinctly below the average for fluid intelligence, but still assigning two-fifths of observed personality variance to genetic causes.

9. If we apply the general discussion on error in the previous chapter (suggesting that present uncorrected results may reduce the H differences among traits and possibly lower H somewhat below the true value, we would hypothesize that (a) fluid intelligence, self-sentiment (self-esteem), Q_3, sur-

gency, F, and premsia, I, represent behavior patterns essentially expressing unitary genetic patterns, only superficially modified by environment; (b) ego strength, C, crystallized intelligence, g_c, parmia, H, guilt proneness, O, and self-sufficiency, Q_2, are in the middle range of roughly equal genetic and environmental influence, and may be examples of unitary trait development on what we have called the eidolon model; and (c) super ego strength, G, and, to slightly lesser degree, dominance, E, and ergic tension, Q_4, are factors largely determined by environment, though the last may have some genetic autonomic reactivity component.

10. It is seen that with the advent of knowledge on heritabilities, previously completely missing from general theoretical discussions on the nature of source traits, the formulation of hypotheses about their origins and natures becomes radically affected. In the first place the existence of an appreciable genetic influence in ego strength, C factor, requires reconstruction of psychoanalytic ideas. Desurgency, $F(-)$, previously ascribed to a punishing, inhibiting environment is now shown to be instead largely a temperamental susceptibility to inhibition. In human society, as shown by Mowrer in some animal societies, hereditary dominance endowments are demonstrated to be powerfully changed by within-family variability, from early experience of position, etc. Although premsia-versus-harria I factor is still regarded as affected in level by sensitive protectiveness in the environment, far more of it is due to hereditary sensitivity than was previously supposed.

Super ego development level, as hypothesized, is largely learned, and ergic tension, as hypothesized, is borne out to be a level of frustration, which could be largely situationally determined, though enough genetic determination remains to fit the hypothesis that Q_4 is a frustration expressed through a partly genetic channel of autonomic high reactivity. The new knowledge in general supports or refines hypotheses, but the finding of very high inheritance of Q_3, self-sentiment strength, is a shock that almost overturns the theory of its development. The finding can be reconciled with the existing concept of the self-sentiment as aquired attitudes about the self only by the added concepts (a) that the environmental pressure to form a self-concept is so uniform and strong that little individual difference arises therefrom and (b) that the ergs involved, which have been shown to be especially loaded: fear, sex, self-assertion (self esteem), and narcissism (Cattell & Child, 1975, p. 84), are themselves highly inheritable in their drive strengths, thus accounting for the high genetic variance in Q_3.

Another valuable lead to personality theory, and also to social psychology, is given by the genothreptic correlations. There is a tendency for these to be more positive for desirable traits, suggesting (if we consider the correlation to hold mainly in the upper half) that genetic deviations toward, for example, higher super ego and self-sentiment are encouraged, whereas those toward premsia, surgency, self-sufficiency, and ergic tension are discouraged.

Between families, that is, in society at large, however, the genothreptic correlations are decidedly larger and always negative. Along with some evidence from outside behavior genetics (the U and I components in motivation [Cattell & Child, 1975], the lower standard deviation for g_c than g_f) this is taken as support for the author's *law of coercion to the biosocial mean* (or the sovereignty of the norm) the mechanisms and roots of which social psychologists might well investigate.

10

Heritability and Conceptual Advances for Source Traits in *T*-Data

1. Properties of Objective Test (*T*-Data) Source Trait Measurement

The aim of research on personality structure in *T*-data has been to devise laboratory-type tests—if possible rendered group administrable for convenience—in which the unitary structures known in *Q*- and clinical *L*-data can be caught in *objectively* measurable form. (A questionnaire with multiple choice is only *scorable* "objectively," that is, it is what is properly called *conspective* because two scorers will see it alike. It is not *objective,* in the correct sense, because the subject is estimating himself, and we know several forms of distortion can then occur.

The factor-analytic structuring of personality in objective measures has occurred only in the last 35 years, though as *T*-data it may be considered preceded by ability testing, which is also objective. As far as theoretical perspective and scientific method were concerned—but not the momentum of custom—it cleared the slate of whatever fragmentary validity claims had been made for *ad hoc* devices, such as the Rorschach, the Downey, the Bender, (reviewed by Cattell, 1946). Then, with some knowledge of the main personality dimensions from *L*- and *Q*-analyses by then completed, it started with the established concept of a total *personality behavior sphere,* until then lacking in objective test design. Thus with an extremely wide variety of human behaviors, caught with varying success according to ingenuity in devising objective situations with measurable responses, it approached by 1965 a harvest of some 20 or more defined and measurable dimensions.

The patterns thus found were collated by Hundleby, Pawlik, and Cattell (1965) showing from 4 to 20 studies replicating the pattern of each factor. The factor source traits were indexed U.I. (Universal Index) 16 through 36, the number U.I. 1 through U.I 15 being kept for *T*-data factors in abilities, as already collated by French (1951). Although exact description of the subtests most valid for each factor were given in some 30 articles and the book by Cattell and Warburton (1967) it was not until 1978 (Cattell & Schuerger, 1978) that the O–A (Objective–Analytic) Battery was issued in a cut-and-dried, tape-administered, standardized kit, permitting half an hour's reliable testing of each factor, much as for the usual testing time of the general intelligence factor.

Meanwhile, however, these factor patterns and measures had been:

1. Checked for simple structure, unique, factorial resolution and identity at each of several *ages,* by Baggaley, Coan, Damarin, Gruen, Haworth, Hundleby, Nesselroade, and others.
2. Shown to have the same source-trait identity of pattern in diverse *cultures* (United States, Japan, Austria, Britain, Germany) by Eysenck, Häcker, Ishikawa, Pawlik, Schmidt, and others, and across clinical subgroups by Patrick, Price, Rickels, Scheier, Schmidt, and others.
3. Demonstrated to be significantly predictive of clinical, educational, occupational, and other life *criteria* by Barton, Birkett, Bjersted, Brogden, Cartwright, Dielman, H. J. and S.B.G. Eysenck, Häcker, Hundleby, Ismail, Killian, Klein, Knapp, Patrick, Price, Reuterman, Rickels, Scheier, Schmidt, Schuerger, Tatro, Tomson, Ustrzycki, Wardell, Yeudall, and others. The references to researchers in this paragraph are in Cattell and Schuerger (1978).
4. Investigated as to age trends and other characteristics needed for standardized, comparable measures in children and adults.

Considering the systematic knowledge thus accumulated around these persistent and definite behavior response patterns, behavior geneticists may feel that psychologists have been tardy in supplying the undoubted empirical patterns with clear theoretical meanings for them. It is likely, of course, that the 20 known factors do not cover all potentially recognizable dimensions of the personality sphere. But that is not the problem. Indeed, average psychologists are "stumped" more by the fact that these source traits, after confirming and sharpening the traits they believe they know such as anxiety, extraversion, regression, psychoticism, cortertia, independence, and asthenia, present them with a dozen more they have never heard of. Like alchemists who might have been brought forward three centuries to look at Mendeléef's periodic table, they suffer from a conceptual *embarras de richesse*.

Determining *H* values for the generally understood source traits, as in the preceding Chapter of *Q*-data traits, is important because it is practically useful, but determining them for the still enigmatic patterns here is more important still, because it helps us toward a first theoretical breakthrough on their meaning. Though U.I. 20, 26, . . ., 31 may be as puzzling as say, the aberrations in the orbit of Neptune which led to finding Uranus, or in the atomic weight of nitrogen which led to the discovery of the noble gases, they are real enough. That is to say they appear repeatedly as unitary traits and they predict everyday life behavior. "There are more things in heaven and earth than in some psychologists' philosophies," and the O−A Battery traits now rightly call for quite serious hypothesis formation and the help therein of behavior-genetic investigation.

For years the demonstrable lack of simple alignment between the *L−Q*-data factors *A*, *B*, *C*, *D*, etc., and the U.I. factors 16, 17, 18, etc., was extremely puzzling to psychologists adopting the likely theory of *instrument-transcending personality* structures. But in the last 10 years, the explanation for at least some cases has appeared: the *T*-data factors U.I. 17 (inhibitory control), U.I. 19, (independence), U.I. 24 (anxiety), and U.I. 32 (exvia−invia) align with *second-order Q*-data factors, rather than with the primaries. In fact they align, respectively, with *Q*VIII (control), *Q*IV (independence), *Q*II (anxiety), and *Q*I (extraversion), while there are indications of a fifth line up, in U.I. 22 (cortertia) in *T*-data and *Q*III (cortertia) in *Q*-data. This alignment occurs not only in factor analysis but receives support (*circumstantial validity*, Cattell, 1973a) from the *Q* and the *T* factors correlating in the same way with clinical and other criteria.

As we shall see, some light may be thrown through this connection on the genesis and internal structure of second-order factors. It must be made clear, however, that the *primary* evidence on alignment is that the marker variables for the *Q* and *T* factors fall together when *Q* and *T* variables are factored together. It does *not* mean that we have yet succeeded (except on U.I. 24−*Q*II, anxiety) in demonstrating any *substantial* correlation between the two factor score measures. If the *Q* factor score has a validity of .6 and the *T* factor battery the same (reasonable in early stages) they would be expected to correlate only .36, which is what has happened in some cases. The low value does not at all deny the conclusion that the *Q* and *T* markers fall together in a common factoring, on one and only one factor. But it does suggest either that instrument factors (the parts peculiar to *Q* or *T* tests), are large or that a new phenomenon, in the form of the SUD model, must be recognized, as discussed in the following.

2. The Heritability of Assertiveness (U. I. 16), Inhibitory Control (U.I. 17), and Independence (U.I. 19) by OSES, Least Squares, and Maximum Likelihood

The *T*-data factors, from U. I. 16 on, have been indexed in various textbooks in order of diminishing variance (on a personality sphere of variables) and of lesser frequency of replication. Thus U.I. 16, 17, and 19 are among the most important personality factors on this basis and also in prediction of criteria such as susceptibility to neurosis (marked by 16−, and 19−), achievement (16+, 17+, 19+), and so on. U.I. 18 is a pathological manic factor (Tatro, 1968) which was not included in this normal survey.

The samples of adolescent males (12–18 years old) in the studies in this chapter are from the same areas (Chicago, Cleveland, downstate Illinois) as the last chapter but are different sets and include a large enough sample of unrelated boys reared together in adopting families, which was absent from the IQ and *Q*-data. The sample sizes are shown in Table 10.1 and it will be seen that they sum to 2923, 1160 from five family constellations and 1763 from a general population standardization (Cattell & Schuerger, 1978, p. 148).

The objective subtest devices used are shown in Table 10.2 and are essentially the same (also as to administration, scoring and weighted addition to a factor score) as in the published *O–A Kit* (Cattell & Schuerger, 1978).

The basis of data here goes beyond that for the *Q*-data, inasmuch as we have five constellations and eight resulting equations—or nine if we reckon the general population variance to be experimentally independent of the sum of σ_{ST}^2 and σ_{BNF}^2. We felt entitled to do this in the maximum-likelihood method, but, with certain limitations already existing in the OSES method, we kept there to eight equations rather than the full nine in Table 4.6 (p. 110). There are now seven unknowns—five variances and two genothreptic correlations—and the OSES solutions (equations in Table 4.7, p. 112), for overlapping sets of six, seven, and eight equations are given in detail elsewhere (Cattell, Schuerger, & Klein, 1981b, in press). As stated earlier, the OSES unfortunately permits no estimation of error of abstract variance estimates, and in this case their absolute values diverge appreciably from the other two methods though fortunately the resulting heritabilities do not (Table 10.4). The least-squares and maximum-likelihood methods show the same general order of size, except for the genothreptic correlations, as shown in Table 10.3.

The reader will note that in this *objective* test (*T*-data) personality factor analysis we have introduced and explored the effect of correction of raw score variances for error of measurement. This is done by using only the effect of the reliability coefficient (*not* validity and stability) which means multiplying the concrete variances by $\theta = (r_d - r_{s_1 s_2})/(1 - r_{s_1 s_2})$ given in Eq. (8.17g) on page 292.

TABLE 10.1

Sample Sizes and Maximum Likelihood Fit of Reverse Calculated Concrete Variances to Observed (Goodness-of-Fit of the General Model)[a]

Variance	Sample size (N)	U.I. 16		U.I. 17		U.I. 19	
		Observed	Experimental	Observed	Experimental	Observed	Experimental
s^2_{ITT}	67	.37	.37	.39	.38	.17	.17
s^2_{FTT}	81	.37	.45	.54	.55	.25	.28
s^2_{ST}	371	.64	.56	.88	.68	.39	.37
s^2_{BITTF}	67	1.55	1.55	.70	.63	.87	.87
s^2_{BFTTF}	81	1.69	1.10	.49	.55	1.03	.79
s^2_{BNF}	371	1.28	1.22	.67	.68	.86	.88
s^2_{UT}	61	.78	.78	.88	.93	.77	.78
s^2_{BSF}	61	.89	.89	.98	.91	.92	.92
s^2_{GP}	1763	.85	.89	.64	.68	.62	.63
χ^2		14.00		17.06		3.86	
pa		<.01		<.01		>.10	

[a] In considering these p values as seemingly denying the null hypothesis it should be remembered that the very large GP sample size exempts us from this conclusion. s rather than σ is used here because we are specifically dealing with concrete sample values. Note these entering variances are validity-corrected.

TABLE 10.2
The Standard O–A Batteries for U.I. 16, U.I. 17, and U.I. 19[a]

M.I. (*master index*) number	Direction of score
U.I. 16 (ego standards)	
M.I. 244	Quicker social judgment
M.I. 61	Higher coding speed
M.I. 282	More seen in unstructured drawings
M.I. 2409	More logical assumptions detected
U.I. 17 (control)	
M.I. 21	Fewer questionable reading preferences
M.I. 7	Slower speed in gestalt completion
M.I. 43	Larger G.S.R. response to threatening stimuli
M.I. 117	More responses indicative of "highbrow" educated taste
U.I. 19 (independence-versus-subduedness)	
M.I. 1207	Higher ratio of accuracy to speed
M.I. 206	More correct Gottschaldt figures
M.I. 167c	Better immediate memory on reading passage

[a] These are cut short of the full batteries, which are $\frac{1}{2}$ an hour each in length (see Cattell & Schurerger, 1978).

	Dependability–reliability	Concept validity
U.I. 16	.92	.92
U.I. 17	—	—
U.I. 19	.92	.79

In Table 10.3 paragraphs (a) and (b) give a comparison of maximum likelihood results from validity-corrected and dependability-corrected variances, and it will be seen that difference due to mode of correction (comparing the NG model in both the upper and lower parts, [a] and [b] of Table 10.3) are trivial. The differences in (a) between the least squares and the maximum likelihood method, despite the former using the full model (FM) and the latter the nongenothreptic (NG) are also small—except, of course that the NG is committed to zero values for the correlations. Conversely, however, the differences of FM and NG become quite substantial when maximum likelihood is applied to the *dependability-corrected* values. This greater consistency with validity-corrected scores may be kept in mind as an argument for its preferred use.

For a final weighing of conclusions from different assumptions and methods we shall wait until results from all personality factors have been set out. But we can note here at the outset that in comparisons of the four alternative models we have tried—full MAVA (FM), MAVA with no genothreptic correlation (NG), purely genetic (PG), and purely threptic (PT)—the greater number of terms on the first will permit a literally better fit. When each of these has its significance examined according to degrees of freedom,

TABLE 10.3
Abstract Variances with Genothreptic Correlations with Mean Error[a]

(a) *Based on validity corrected concrete variances*

	U.I. 16				U.I. 17				U.I. 19			
	LS	Mean error	ML	Mean error	LS	Mean error	ML	Mean error	LS	Mean error	ML	Mean error
$\sigma^2_{wt.t}$	.41 ±	.02	.35 ±	.02	.53 ±	.01	.38 ±	.09	.21 ±	.04	.17 ±	.02
σ^2_{wg}	.21 ±	.04	.14 ±	.06	.00	.00	.14 ±	.22	.03 ±	.03	.11 ±	.04
r_{wgwt}	−.01 ±	.12			1.00	.00			.69 ±	.05		
σ^2_{bg}	.30 ±	.05	.27 ±	.04	.02 ±	.01	.00	.00	.16 ±	.06	.22 ±	.03
$\sigma^2_{wt.s}$	.29 ±	.02	.42 ±	.02	.68	.00	.54 ±	.06	.22 ±	.06	.27 ±	.02
σ^2_{bt}	.00	.00	.06 ±	.12	.02 ±	.02	.00 ±	.00	.01 ±	.03	.04 ±	.07
r_{bgbt}	1.00	.00			−1.00	.00			1.00	0		

			L.R. χ^2 (NG model)	1.83	5.07	5.33
			df	2	2	2
			p	>.25	>.05	>.05

(b) *Based on correction of test scores for level of test reliability* (Maximum likelihood only)

Abstract variance and genothreptic correlation	U.I. 16 (Rel. 75)		U.I. 17 (Rel. 71)		U.I. 19 (Rel. 75)	
	FM	NG	FM	NG	FM	NG
$\sigma^2_{wt.t}$	.32	(.31)	.42	(.42)	.20	(.20)
σ^2_{wg}	.61	(.12)	.01	(.16)	.05	(.12)
r_{wgwt}	−.60	(.00)	.99	(.00)	.45	(.00)
$\sigma^2_{wt.s}$	.66	(.37)	.56	(.60)	.31	(.34)
σ^2_{bt}	2.63	(.06)	.00	(.00)	.11	(.05)
σ^2_{bg}	1.31	(.24)	.01	(.00)	.24	(.27)
r_{bgbt}	−.98	(0)	−.01	(.00)	−.09	(.00)

The difference of full MAVA (FM) and non-genothreptic (NG) is nowhere significant but approaches significance in U.I. 19.

L.R. χ^2	1.84	2.47	4.62
df	2	2	2
p	>.10	>.10	>.05

however, this advantage naturally vanishes. In the case of the least squares the full MAVA (top left in Table 10.3a) gave a better fit than any more parsimonious model, but in maximum likelihood the model with no genothreptic correlations gave the better fit, and it is this that is utilized to contribute to the heritabilities averaged in Table 10.4. The restriction of the model to a purely genetic or a purely threptic explanation in no case came anywhere near significance.

Results in final heritability values are given in Table 10.4. Although absolute differences (ignoring r's) are sometimes appreciable, as in U.I. 16, it is

TABLE 10.4
Mean Heritabilities of Assertiveness (U.I. 16), Inhibitive Control (U.I. 17), and Independence (U.I. 19) across Three Methods[a]

	U.I. 16				U.I. 17				U.I. 19			
	OSES[b]	LS	ML	Mean	OSES[b]	LS	ML	Mean	OSES[b]	LS	ML	Mean
H_w	.30	.42	.26	.32	.33	.00	.21	.18	.32	.12	.26	.23
H_b	.29	1.00	.81	.70	.27	.03	.00	.10	.55	.14	.84	.51
H_p	.29	.64	.46	.46	.29	.03	.21	.18	.44	.45	.50	.46
r_{wgwt}	−.92	−.01	(.00)[c]	−.31	−.33	1.00	(.00)	.22	−.17	.69	(.00)	.17
r_{bgbt}	−.93	1.00	(.00)	.02	−1.00	−1.00	(.00)	−.67	−.86	1.00	(.00)	.05
On twin method only	.77	.34			.56	.00			.47	.13		

[a] Results on validity-corrected basic data.
[b] OSES based on assortive r of .25.
[c] Values in parens are assumed 0 by the model.

noteworthy that relative values across the methods show order. Even were this not so there would be no alternative, in attempting an estimate at this stage of research, to that of averaging the values reached. It should be noted here and in later tables that the maximum likelihood was first run by Rao's program on the concrete variances corrected by the validity coefficients (Cattell & Schuerger,1978). It was then repeated by Vaughan's program on concrete variances corrected for test unreliability by the preceding formula. (Some contributory articles give variances on these factors *before* such further transformation.) In almost all cases (U.I. 19 was an exception) the effect of the validity and reliability corrections was small on the abstract variances and gave no change or a trifling .01 to .04 in the heritability finally calculated. The heritabilities given in this chapter are, however, in all cases those from corrected variances.

On the discrepancies of results from different methods, and what weight one might think to give to alternatives, we shall comment later. Except for the least squares-versus-maximum likelihood on U.I. 19, the OSES is on an average, the most out of line here (though not necessarily in later tables). Taking the average, H_w, H_b, and H_p, it is of interest to compare these results with those of Cattell, Stice, and Kristy (1957), the only other study available with MAVA design and objective, T-data source traits. It cannot be given equal weight (and is not included in our table) because it was a pioneer study, using early test batteries, though on a tolerably acceptable sample of 878 cases, and of the same age and sex as here. Just as here, the U.I. 16 results by OSES used there are erratic, but U.I. 17 and U.I. 19 are not so different. U.I. 17 there shows definitely low H values, as here, (H_w, .31; H_b, .24; and H_p, .31) while the correlations are .10 instead of .22 within and $-.51$ instead of $-.67$ between families. The H_w and H_p values are there, as here, higher for U.I. 19 than U.I. 17. The genothreptic values there for U.I. 19 are negative, as here, but low ($-.20$, $-.28$).

The conclusion on averages is that U.I. 16 and U.I. 19 are about equally heritable, and precisely so on the "overall" population value (H_p = .46 and .46), while both show, consistently, more than twice as high a between as a within inheritance. As before we must interpret this as meaning that the within-family *environmental* variance is much more important than the between, and Table 10.3 supports this strongly.

The psychologically important important indication here is that U.I. 17 (inhibitive control) sometimes called "good upbringing" shows a far lower inheritance than the other two, and its low H's are much more even across the situations. This is strongly supportive of the interpretation given to it, both as a T-data trait and in its matching Q-data (QVIII) pattern, namely, that it represents a pattern of restrained and moral behavior acquired to different levels by different persons through differences in home and school teaching. The indications for the nature of U.I. 19 (independence) will be discussed later in this chapter.

3. The Heritability of Anxiety (U.I. 24), Regression (U.I. 23), and Narcissistic Self-Sentiment ("Self Will") (U.I. 26)

The basis of constellations analyzed here is precisely the same as in the other sections of this chapter. The measurement instruments are again from the O–A Kit, and are set out with validities in Table 10.5. Again we use all three of the possible analysis methods. The details of

TABLE 10.5
Subtests in the Batteries for Source Traits U.I. 23, U.I. 24, and U.I. 28

M.I.[a]	Title, with direction of score	Time (min:sec)
U.I. 23 (capacity to mobilize-versus-regression)		
M.I. 242	Higher ratio social–nonsocial annoyances	2:00
M.I. 69b		
& 120	Higher ratio of accuracy to speed (letters)	1:00
M.I. 609	Higher perceptual coordination	2:30
M.I. 401	Less preference for competitive associations	2:30
M.I. 36	Higher ability to state logical assumptions	3:00
M.I. 105	Fewer threatening objects seen	1:30
M.I. 167	Better immediate memory	1:30
M.I. 2a(1)	Lower perceptual motor rigidity	2:00
U.I. 24 (anxiety)		
M.I. 2404	Preference for outright not inhibited humor	2:00
M.I. 219	More common frailties admitted	2:00
M.I. 205	More emotionality of comment	2:00
M.I. 218	More willingness to play practical jokes	2:00
M.I. 1370	Less willing compliance in unpleasant task	2:00
M.I. 211b	Higher susceptibility to ego threat annoyance	2:00
M.I. 321	More questionable book preferences	2:00
M.I. 473	Fewer friends named	2:00
U.I. 26 (narcissistic ego)		
M.I. 1414	Faster speed of cancellation	2:00
M.I. 13a	Greater variance of accuracy in cancellation	2:00
M.I. 35	Less susceptibility to authority suggestion	3:00
M.I. 273	Higher verbal fluency on own dreams	2:00
M.I. 2a	Lower motor rigidity	2:00
M.I. 24a	Higher ratio of final to initial performance	2:00

[a] The M.I. (master index number of performance) numbers are those in the Cattell and Warburton encyclopedia (1967), and in the Cattell and Schuerger (1978) description of the O–A Battery. The *T* number identifying the test from which the defined M.I. performance was scored can be read from the M.I.-*T* Index in Cattell and Warburton (1967). The choice of tests in the present research for each source trait is not exactly identical with that in the final O–A batteries, as in the last work, but, as will be seen, is very close to it. Data are not available for U.I. 26.

	Dependability–reliability	Concept validity
U.I. 23	.91	.76
U.I. 24	.97	.92

the OSES and maximum likelihood are reported, respectively, in Cattell, Schmidt, Klein, and Schuerger (1981, in press) and Cattell, Rao, and Schuerger (1981, in press) so we shall proceed here to the final heritabilities.

In the maximum likelihood the restoration of the concrete variances from the abstract variance gave, χ^2 significances for the full MAVA model, for U.I. 23, $p < .10$, for U.I. 24, $p < .02$, and for U.I. 26, $p < .01$. However, the extreme parsimonious hypothesis—that nothing but threptic variance exists—curiously gave a better significance by chi square in the case of U.I. 26. In detail χ^2 for the general MAVA was 25.94 with 2 degrees of freedom, and for the pure threptic model, 34.66 with 6 degrees of freedom. There was essentially no difference (second decimal place only) between validity and dependability-corrected results. Accordingly, as far as the maximum-likelihood method is concerned we have felt compelled to take the extreme position that there is no genetic component whatever in U.I. 26. Least squares gave a very poor fit ($Q = 57.02$) on any other model and on this too, so its results therefrom are completely ignored. In traits U.I. 23 and 24 a fine difference of significance, in favor of the parsimony of the model to the extent of supposing no genothreptic interaction, was found. The results are shown in Table 10.6, wherein the agreement of methods is better than in Table 10.4.

The heritability of anxiety proneness (U.I. 24) runs consistently (H_w, H_b, H_p) higher than the other two traits, and reaches an unusually high value among all personality traits. It is interesting to note that it runs evenly within and between families, unlike the other clinically important traits we have seen. Since this implies larger differences in families in the anxiety-provoking situations to which *all* members are uniformly exposed, two hypotheses occur to the psychologist: (*a*) That anxiety is so emotionally contagious that each family has its own characteristic "anxious atmosphere" and (*b*) that, since we know anxiety increases with lower economic level (Lynn, 1971; Spielberger, 1972; Cattell, 1942, 1945), the child's awareness of economic stringency affects all sibs about equally.

The regression factor (U.I. 23) is known to have some close functional relation to anxiety. Like anxiety it is significantly higher (Cattell & Scheier, 1961) in neurotics (Eysenck calls it *the* neuroticism factor) and it shifts toward normal values with therapy (Cattell, Rickels, Weise, Gray, & Yee, 1966). The only other MAVA evidence (Cattell *et al.*, 1957) gave a somewhat lower overall population inheritance, namely, .25 (compared to .32). It agreed in negative genothreptic correlations, higher between family, but less so than here namely, $- .50$ (within) and $- .56$ (between). Its somewhat lower value in each situation compared to anxiety would fit the theory (Cattell & Schuerger, 1978) that is represents a long-term fatigue, as an outcome of exposure to anxiety and conflict. For its greater environmental dependence would arise from the circumstances which decide how long the individual is exposed to anxiety-provoking situations.

The finding that U.I. 26 has a very low heritability agrees well with the

TABLE 10.6
Mean Heritabilities of Regression (U.I. 23), Anxiety (U.I. 24), and Self Will (or Narcissistic Self) (U.I. 26) across Three Methods

	U.I. 23				U.I. 24				U.I. 26			
	OSES[a]	LS	ML	Mean	OSES[a]	LS	ML	Mean	OSES[a]	LS	ML	Mean
H_w	.35	.58	.34	.42	.33	.55	.47	.45	.27	—	.27	.27
H_b	.30	.26	.27	.28	.29	.46	.68	.47	.25	—	.00	.13
H_p	.32	.36	.32	.33	.30	.50	.52	.44	.25	—	.27	.26
r_{wgwt}	-.54	-.35	.00	-.30	-.31	-.02	.00	-.11	-.61	—	.00	-.32
r_{bgbt}	-.79	-.93	.00	-.57	-.71	-.55	.00	-.31	-.90	—	.00	-.45
From twin method only												
H_{wt}	.69	.63	.69	.60	.60				.60			

[a] The OSES values are all on assortive mating $r = .25$; the alternatives, differing little, on the basis of $r = .07$, are given elsewhere (Cattell, Klein, & Schuerger, in press). Present results on validity-corrected variances. Note ML uses parsimonious, non-genothreptic model.

earlier (Cattell *et al.*, 1957) finding (H_p = .25 there and .26 here). The existing hypothesis that U.I. 26 represents a pattern of the "spoiled" or at least
"overvalued" child is entirely consistent with this largely environmental determination. The U.I. 26 performances of "high fluency on self" and "belated warming up to a task" are combined, however, with several indicators
of self-confidence, competence, and desire to excel, suggesting more than
"spoiling". The life criteria relations combine positive high school achievement and getting into trouble, in delinquency. With the earlier H determination, the field is now cleared for research by developmental psychologists on
the environmental features—presumably mainly in the home—which we
now know largely produce the U.I. 26 pattern.

4. The Heritability of Exuberance (U.I. 21) and Asthenia (U.I. 28)

These two traits, and their many criterion relations, have excited
enough curiosity in psychologists for their meaning to have debated in special articles (Meredith, 1966, 1967; Cattell, 1964b)

The O–A battery subtests used in measuring them are given in Table
10.7.

The values obtained by the three methods are shown in Table 10.8. The
least-squares method seems slightly lower in agreement in this case. The details of the findings by the OSES and maximum-likelihood analyses are given
in Cattell *et al.* (1981, in press) and Cattell *et al.* (1981, in press), respectively, and the empirical concrete variances except for the dependability
correction are, of course, the same basis in all three.

Both of these factors are of moderate low heritability. As usual, the
maximum likelihood is more emphatic, and in this case it initially offered the
statistical best fit for U.I. 28 with a model so far supported in only one other
trait, U.I. 26. It offered an explanation in which *no* genetic terms were required to account for the variances. However, with the dependability error
corrected values by Vaughan the outcome is definitely less extreme than in
Rao's validity corrected analysis. Furthermore, there is good convergence
of the Vaughan calculation with the earlier pioneer MAVA study of Cattell
et al. (1957) (which gave .30, .16, and .22 for the H's, and − .30 and − .96 for
the correlations). The unusual result for asthenia, U.I. 28 is that heritability
is most distinctly low *between* families, with implications to follow.

For U.I. 21 (exuberance) the Q value indicated a relatively poor fit for
the least-square solution, but, combined with the maximum-likelihood outcome, it points to a low heritability level, with even distribution over H_w,
H_b, and H_p. This is also found in the OSES solution, at a level showing that
narrowing of H range of all traits we have recognized as an OSES characteristic before. Nevertheless, some doubt must persist about low H in U.I. 21,

372 Heritability and Conceptual Advances for Source Traits in *T*-Data

TABLE 10.7
Source Traits U.I. 21 and U.I. 28 and the Chief Subtests Used in Obtaining Scores on Them

M.I.	Direction and nature of performance	Time (min:sec)
U.I. 21 Exuberance		
M.I. 3356	Faster speed following directions	
M.I. 271	Higher ideational fluency	
M.I. 853	More concrete completion of drawings	
M.I. 699	More garbled words guessed in tautophone	
M.I. 7	Faster closure in perceiving incomplete drawings	
M.I. 8	High frequency of alternating perspective	
U.I. 28 Asthenia		
M.I. 152	Higher tendency to agree	1:40
M.I. 125	More institutional values	1:40
M.I. 100	More cynical pessimism	2:00
M.I. 97	Longer estimates of real waiting time	2:10
M.I. 364	Preference for stories with external control	2:00
M.I. 116	Lower guilt and severity of judgment	1:40
M.I. 1160(2)	More grudging scepticism regarding success	3:00

	Dependability–reliability	Concept validity
U.I. 21	.98	.80
U.I. 28	.86	.64

since the pioneer study of 1957 gave decidedly higher heritability, though agreeing with the final maximum-likelihood verdict of trivial genothreptic correlations. Incidentally, it will be noted that here, as in the other tables, the calculation by the twin method, which assumes that intersib variance, $\sigma^2_{wt.s}$ has the twin value $\sigma^2_{wt.t}$, raises the heritabilities appreciably, as it did with the *Q*-data trait measures.

The theories formerly built around U.I. 21 have been consistent with earlier indications of higher heritability, and these consequently now need some revision. The fact that it shows (Cattell, 1978b) an age curve much resembling the growth curve of intelligence (up to 19 years, when observations

TABLE 10.8
Mean Heritabilities of Exuberance U.I. 21 and Asthenia U.I. 28 across Three Methods

	U.I. 21				U.I. 28			
	OSES	LS	ML[a]	Mean	OSES	LS	ML	Mean
H_w	.38	.08	.29	.25	.42	.60	.37	.46
H_b	.36	.44	.45	.42	.37	.28	.23	.29
H_p	.37	.16	.32	.28	.39	.41	.34	.38
r_{wgwt}	−.59	73	0	.05	−.46	−.44	0	−.30
r_{bgbt}	−.97	1.00	0	.01	−.90	−.97	0	−.62
By twin method only	.72	12	.43	.42	.64	64	0	.43

[a] ML results on dependability-corrected variances.

ended) was also taken as supporting substantial heredity. Wardell and Royce (1975) note high enthusiasm, energy, dominance, and excitability in high U.I. 21 persons, supporting earlier studies. It is very significantly related in its negative direction (subduedness) to neuroticism, psychotism, and all depressions. Getzels and Jackson (1962) found tests from its battery related to creativity—and unpopularity! Knapp (1961, 1962) confirmed the latter. Considering it as an "irrepressible temperament," I speculated that the high fluency ("divergent thinking") would be found related to a cortical pacemaker. If so it looks as if we must recognize that the cortical physiology or its behavioral expression must change with environmental influences. The appreciable between-family heredity (H_b = .42) usually goes with significant heritability in the ordinary sense, and probably we should look here for some powerful environmental influences *within the family*.

As a unitary trait U.I. 21 is well defined and has powerful associations, but shows, as we have seen, some conflicting indications on the role of heredity and environent in its determination. With U.I. 28 (asthenia) on the other hand, all evidence to date points uniformly to environment being the major determiner. Asthenia's oppposite pole has been called self-assurance, and it climbs (but in a different shape from U.I. 21) to around 17–19 years of age. There seems no need to modify the general nature of the theory put forward (Cattell, 1964b), namely, that U.I. 28+ represents a perhaps too early imposition of super ego and other standards, so that the individual, in positively acquiring and manifesting these standards, nevertheless, at the same time, develops a certain resistive counteraction, expressed in asthenia, surliness, autism, and a half-hearted negativistic and uncooperative general outlook. Conceivably it relates to traits of dominance and high standards in parents. Despite a certain paralysis of enterprise and ambivalence to authority that ensues, the facts are that higher U.I. 28 individuals are less associated with delinquency, alcoholism, and drug addiction. They outwardly accept authority and do better than others (Knapp, 1962) in military school, but apparently not in school generally.

Evidently the U.I. 28 pattern is an important one, where research can profitably look for roots in early teaching forces in the environment, and study their time of onset relative to the development of the child.

5. The Heritability of Reality Contact (Realism-versus-A Psychotic Break) (U.I. 25), Exvia–Invia (U.I. 32), and Sanguineness-versus-Discouragement (U.I. 33)

The tests in the U.I. 25 battery (Table 10.9) are the markers for the factor which Eysenck and Eysenck (1968) called "psychoticism" and showed significantly to distinguish psychotics. However, Tatro (1968), Killian (1960), and Cattell, Schmidt, and Bjersted (1972) have found other factors,

TABLE 10.9
O–A Battery Subtests for Personality Factors U.I. 25, U.I. 32, and U.I. 33

M.I.	Direction and nature of performance
U.I. 25 (realism-versus-tensidia)	
M.I. 1006	Less pessimistic insecurity in life attitudes
M.I. 2411	Better immediate memory
M.I. 2408	Great accuracy in digit span reproduction
M.I. 144	More agreement with homely wisdom perspectives
U.I. 32 (exvia-versus-invia)	
M.I. 733	Greater willingness to decide on vague data
M.I. 1169	Less influenced on assigned punishments by extenuating circumstances
M.I. 15	More excessive use of "trump cards" in the CMS (Cursive miniature situation) test
M.I. 2a	Lower motor-perceptual rigidity
U.I. 33 (discouragement-versus-sanguine temperament)	
M.I. 108	Less confidence in unfamiliar situations
M.I. 473	Fewer people who appreciate one as a friend written down
M.I. 212	Less belief in attainability of goals

	Dependability–reliability	Concept validity
U.I. 25	.85	.74
U.I. 32	.92	.71
U.I. 33	.95	.85

notably U.I. 23 and U.I. 21, *also* to distinguish psychotics, equally, while U.I. 25 was found to distinguish also neurotics, cases of frontal lobe damage, and persons performing poorly in practical situations. One conclusion (Cattell & Schuerger, 1978; Hundleby *et al.*, 1965) has been that it is (in its negative direction) *one component* in psychoticism, namely, a withdrawal or breakdown of reality contact. From test content and other associations one notes also at this pole an inner tension and rigidity, whence the term *tensidia* has been given to the negative pole and, in brief designations, *realism* to the positive pole.

This preliminary clarification of the concept is necessary for U.I. 25, because its conceptual history is a bit tangled, but U.I. 32 (exvia–invia) the pure factor core in the publicly battered and overextended notion of "extraversion–introversion," is sufficiently familiar (also as *Q*I), and now sufficiently defined to need no comment. The remaining factor here, U.I. 33, as its index number indicates, is a late comer among *T*-data factors. The reader should note it has often been scored in reverse, as discouragement-versus-sanguineness (Cattell & Schuerger, 1978) but in accordance with a natural preference in naming psychological traits by a "positive" pole, for example, g_f is intelligence not stupidity, we have finally scored it (See Table 10.9) in the sanguine direction. The reasons for not calling its negative pole "depression" are, first, that as many as six other factors have been found in the depression area, (Kameoka, in press; Cattell *et al.*, 1981) and, secondly, a low score significantly distinguishes schizophrenics (Tatro, 1968; Cattell *et al.*, 1972). The schizophrenic has (since Kretschmer, 1929), for example, been recognized, and found in tests (Cattell, 1969; Cattell & Schuerger, 1978) to show little anxiety or depression, but is undoubtedly in a state of high general frustration. Hence "discouragement" better defines the negative pole of U.I. 33.

The trait genetic analyses, of the data on the constellation samples used throughout this chapter, are shown in Table 10.10.

For U.I. 25 the results are of the order of internal agreement to which we have been accustomed. The H_p value is higher for the least-squares method, except in U.I. 32 where OSES is distinctly larger. In U.I. 25 and U.I. 33 we see (*a*) a somewhat higher between- than within-family H; (*b*) a rather high degree of inheritance, but neither as low as we have met in U.I. 17, 26, and 28 nor quite as high as for U.I. 16, U.I. 24, and U.I. 1 (intelligence); and (*c*) the highest negative genothreptic correlations we have yet encountered.

For U.I. 25, realism-versus-tensidia, there is some indication (H_b = .57 against H_w = .33) that environment operates more in the family situation. That is to say the innate predisposition to better or poorer reality contact is more affected—in what direction we do not know—by events in the family than by family status and experience outside the family.

The main addition to our knowledge of U.I. 25, however, is that in between family and total population settings its heredity component is distinctly high. Indeed, as Figure 10.2 summarizes, it just comes within the top four traits in this regard. With the corrections that have been discussed, the genetic contribution could well exceed half the phenotypic variance, and approach the values found for such traits as intelligence and surgency. A figure of this kind would be compatible with the generally high genetic role found in the psychoses (see p. 87) as contrasted with the neuroses, if we agree that its role in the former is greater. Though we have questioned Eysenck's conceptualizing U.I. 25 as *the* psychoticism factor, it is evident that whereas it plays a lower role than U.I. 16, 19, 22, and 24 in distinguishing neurotics, it plays a higher role (along with U.I. 23 and perhaps U.I. 21, incidentally) in distinguishing schizophrenics and other psychotics (see the tables on pp. 250, 251, etc. in Cattell & Schuerger, 1978).

The present results therefore encourage us to continue with the theory of U.I. 25 as realism-versus-tensidia, considering it as an appreciably genetic endowment in a capacity which bolsters retention of contact with reality (rather than retreat into tense and rigid inner escape) under prolonged stress and frustration. The strong negative reaction of society, both within (− .51) and between (− .81) the family, to deviation in the trait would support our hypothesis *if* the greater part of the regression should plot in the lower range. That is to say, both family and society intuitively see behavior tending to break with reality as so serious that they exert all educative efforts to reduce it.

TABLE 10.10

Mean Heritability of Reality Control, Exvia–Invia, and Sanguineness

	U.I. 25				U.I. 32				U.I. 33			
	OSES	LS	ML[a]	Mean	OSES	LS	ML[a]	Mean	OSES	LS	ML[a]	Mean
H_w	.24	.43	.32	.33	.43	.04	.14	.20	.32	.09	.04	.15
H_b	.35	.66	.69	.57	.43	.00	.00	.14	.30	.96	.50	.59
H_p	.32	.61	.43	.45	.43	.02	.14	.20	.31	.35	.13	.26
r_{wgwt}	-.96	-.44	-.12	-.51	-.40	-.10	.00	-.10	-.81	.12	.00	-.23
r_{bgbt}	-.96	-1.00	-.47	-.81	-.93	-1.00	.00	-.64	-.98	-1.00	.00	-.66
From twin method only												
$H_{w.t}$	.83	.65	.61	.70	.55	.04	.31	.30	.70	.10	.11	.30

[a] Note the values are somewhat different here from Cattell, Rao, Vaughan, and Schmidt (1981) because I have preferred the fuller model based on values in parens in Table 7 there, and the mean of results from corrected concrete variances (Rao's and Vaughan's analyses) which in these factors, alone, show noticable differences of magnitude.

Unless we reject the general statistical view that least squares and maximum likelihood are superior analytical approaches to the admittedly less statistically controlled OSES, the results on exvia-versus-invia are quite upsetting. Extraversion, since Jung, has been regarded as a temperament factor, and Eysenck, working with questionnaire measure of it that are highly consistent with our defined second-order, QI, factor measure, has reported twin-method results (Table 3.5, p. 81) of $H_w = .62$. The results of Canter, Shields, and others, however, bring it down to an average of .39, and our own result in the Q-data second order from the 16 P.F. drops to $H_w = .37$ (Table 9.7, p. 346). If changed from an $H_{w.s}$ to $H_{w.t}$ basis by the degree of inflation we have typically found for the reduced size of $\sigma^2_{wt.t}$ relative to $\sigma^2_{wt.s}$, (i.e., if twin and MAVA results are made comparable) this would become about .42, and the mean $H_{w.t}$ for our own and the earlier twin result could be considered about .40. This seems the most reliable value.

This Q-medium value, however, remains *well* above our T-data medium H_{wt} of .20 in Table 10.10. From this surprising and challenging finding there arises a new hypothesis about exvia better discussed in the space of the next section. But contingently we must take the position that extraversion as measured in T-data is of lower heritability than previously supposed, and, on between-family value, H_b, stands ninth from the top (and third from the bottom) among 11 traits. (In population value, H_p, it stands with U.I. 17 at the bottom [see Figure 10.2].)

U.I. 33 (sanguineness-versus-discouragement), being relatively new in factor replication, has been the center of little more theory than that given at the opening of this section. Schizophrenics are lower on it, good school achievers are slightly higher, and child delinquents distinctly higher. Does sanguineness arise from a history of success, and discouragement arise from the opposite, or are U.I. 33 $(-)$ individuals by natural disposition more susceptible to discouragement and suspicion of life? The unusually large discrepancy between quite high H_b and low H_w (except by the OSES) suggests that heredity would decide much if it were not for quite powerful environmental effects *within* the family. Certainly it is a trait whereon research should look for effects of family position, especially the privilege of the first born compared to the lowly importance in the sib group of the last born. Or again, it might be a case for examining cross-trait effects, that is, to see if greater dominance or intelligence in one sib means lower U.I. 33 in the other.

6. Structural and Genetic Models to Explain the Heritability of Secondary Personality Traits

Four personality source traits—U.I. 17, QVIII (inhibitory control); U.I. 19, QIV (independence); U.I. 24, QII (anxiety proneness); and U.I. 32,

TABLE 10.11
Contrasting Heritabilities from Factors in *Q*- and *T*-Media of
Measurement[a]

		H_w	H_b	H_p
Inhibition control	As U.I. 17	.18	.10	.18
	As *Q*VIII	.30	.43	.33
Independence	As U.I. 19	.23	.51	.46
	As *Q*IV	.18	.74	.28
Anxiety proneness	As U.I. 24	.45	.47	.44
	As *Q*II	.30	.67	.38
Exvia-versus-Invia	As U.I. 32	.20	.14	.20
	As *Q*I	.37	.83	.44

[a] The heritabilities of the *Q* values are repeated from Table 9.7.

*Q*I (exvia-versus-invia)—have been examined as to heritability, here, through *both* *T*- and *Q*-media of observation and measurement. We plan in this section to compare the results through the different instrumentalities, having so far commented in passing only on the strange discrepancy in extraversion. The question will raise other issues important to considering personality structure when designing behavior-genetic research, and to personality structure itself. For convenience the results in several tables are put side by side in Table 10.11.

The fact that the values in Table 10.11 have come a long way from the concrete variances, in terms of abstraction processes, might justify our expecting some discrepancies in *H*'s arising from varying test validities, random sampling, and rounding calculation effects, but the *Q* and *T* differences are not of that random kind. In some cases they are negligible; in others very substantial and systematic. *Q*VIII is in all situations decidedly higher in *H* values than the U.I. 17 measure of the same control factor, and the same is true of exvia–invia, where, additionally, this *Q*I factor puts heritability far more strongly in the between-family situation. Except for some tendency in that same direction, there is no notable difference in the *Q*- and *T*-data placing anxiety proneness at a middling high heritability (.38, .44) as found also in twin studies. Except for H_b, independence seems to reverse this relation, reaching higher heritabilities in the *T*-data, so even these four instances suggest that no simple empirical generalization of *Q* being higher than *T* or vice versa is possible.

It behooves us at this point to look at hypotheses about the structure of second-stratum factors. According to the Stratified Uncorrelated Determiners, SUD, model (Cattell, 1978b) their structure is essentially that of one broad factor covering the areas in which several specific, more local, "stub" factors operate, which are initially correlationally independent both of each other and of the second-stratum factor (hence "uncorrelated"). The result of the joint action of the stub factors and the broad second stratum is a series of primary factors, which are correlated (oblique) because they share the broad, secondary factor variance. Although the hypothesis is such that all these determiners are uncorrelated it is necessary in order to *find* the stub factors that we first *admit* oblique factors and *exactly determine their obliquities* by the usual procedures of simple structure or confactor rotation.

Now the causal action—the path coefficient analysis if one likes—leading to the oblique structure we find can arise either by the second-stratum factor contributing additively to the action of the stubs or, conversely, by the stubs generating the second-stratum entity through some mutual action of among them in relation to environment (Cattell, 1965a). The former model has been proposed by myself (Cattell, 1971b) to account for certain primary abilities. In that area the second-order factor is fluid intelligence. The stubs represent amounts of environmental experience—investment of intelligence —in different areas the amounts of which are uncorrelated with intelligence. Thus the oblique verbal ability factor, V, is the investment of the relation-perceiving capacity of g_f in exercises in verbal material, the numerical factor, N, is similarly due partly to individuals differentially investing their g_f in numerical exercise, and so on.

The converse structure, in the model that the stubs generate the broad second-order factor, has been proposed for exvia (QI), and anxiety (QII). I have put forward this *spiral emergence* theory rather reluctantly, because the whole sense of simple structure factor analysis as such is that higher-order factors contribute to the variance of lower-order factors (and these in turn to variables, as concretely measured). However, psychological observations point to path coefficient causal action in the opposite direction in these two cases (QI and QII), though no physical example of such action can yet be cited. The argument in the case of anxiety (QII), with its loadings on $C(-)$, $H(-)$, O, $Q_3(-)$, and Q_4 is that ergic frustration (Q_4) weakens ego strength, through removal of its normal reinforcements by success, and that inability to execute decisions as planned weakens the self regard of Q_3 and generates guilt O. Anxiety per se develops primarily according to the formula (Cattell, 1980a, p. 412; and in Spielberger, 1972) for threat to the ego, and, finally, threctia, $H(-)$, as temperamental timidity, magnifies the response to this internal threat just as it would to any external threat. Since Q_4 is the prime mover it would be expected to have the highest loading on QII, and, at least for males, it does (Cattell, Eber, & Tatsuoka, 1970, p. 121) with $C(-)$ as a close second, as it should be in the chain. However, an unexplained anomaly is that O is also in the trio of highest loaded primaries (Cattell, 1973a, p. 118).

Whereas the preceding would excellently fit the psychoanalytic theory of the orgin of anxiety, the spiral emergency model for exvia–invia stands on less discussed ulterior theories or observations. Of the four primaries— A, F, H, and $Q_2(-)$—we have evidence for highest heredity in surgency (F)

and self-sufficiency (Q_2), (Figure 9.1), and these also have highest loadings on the secondary, QI. Let us consider this from the inviant (introvert) pole. An individual who combines the low fluency and social ineptness of desurgency with some turn for self-sufficiency (Q_2) is likely to increase by "spiral interaction" in both. His or her social slowness will cause him or her to take refuge more in self-sufficiency, and self-sufficiency will deprive him or her more of exercise in social situations. The loading on $H(-)$ might then be explained as expressing the magnifying effect of temperamental susceptibility to threat (shyness) upon punishment in social situations due to $F(-)$ and Q_2. The principal feature of the fourth component in $QI - A+$ to $A-$ continuum —is a declining faith in human nature, with critical, skeptical, and aloof behavior, which thus arises (as rationalization?) from the inviant's sense of failure in social relations.

It will be noted that the spiral emergence theory evokes doubts, as mentioned earlier, from the standpoint of usual factor-analytic thinking. There we are accustomed to a collection of primary factors acting jointly on a concrete variable, for example, intelligence, super ego, and memory capacity, producing high achievement in, say, a statistics paper. But there are also situations where factors act similarly on a whole array of variables, producing a correlation cluster, at least, as broad as a second-order factor among them.

A fact not to be forgotten in the spiral emergence theory of a second-order factor is that it requires suitable environmental situations and experiences for it to take place. Q_4 does not act *directly* to reduce C, nor does $F(-)$ act directly to produce $H(-)$. The first requires experiences of defeat in ego control, in life situations, because of the undischargeable ergic tension. The second requires experiences of unreadiness and gaucherie at parties and such to push the individual toward greater shyness. Without contact with environments in which these rewards and punishments can occur the correlations among the primaries, that is, the appearance of the secondary factor, would not arise. And individuals will differ in the extent to which they encounter such environments. Robinson Crusoe, alone, might have remained desurgent without becoming shy.

Since the reader has patiently borne this digression, wondering what it has to do with behavior genetics, let us now note that the first *contributory* theory could be compatible with the secondary being largely hereditary, whereas the *spiral emergence* theory definitely demands appreciable threptic variance in the secondary. Thus if we switch to a contributory theory for anxiety, QII (as we already did for g_f among primary abilities) we should suppose say, some innate autonomic susceptibility to strong anxiety response. This would contribute to Q_4 by heightening sensitivity to emotional frustration. It would weaken C by paralyzing its action (Cattell *et al.* [1966] have actually shown some support for this in ego strength increasing in proportion to the time the individual is sheltered from anxiety by pharmaceutical means). And low autonomic response would also take some of the power out of guilt (O).

In the behavior genetic findings we have a new avenue, along with other data, for throwing light on the structure of higher strata factors. To approach this goal let us first consider the theoretically possible ways in which the genetic and threptic components could be distributed among primary and secondary stubs and their consequence in observed oblique primaries. Figure 10.1 shows the nine possibilities. There are three levels in each, as will be evident from the figure, without further description. However, the direction of causal action will be different according to whether contributory or emergent properties are added as machinery among the strata. In the contributory case the bottom row will add to the middle to produce the top. In the emergent case interaction of the middle row with environment (not shown) will produce the bottom row, which will then add to the middle to produce the top. In the first case the bottom (secondary) exists from the beginning and is combined with the stubs to produce the top—the oblique primaries. In the second the secondary factor (bottom) is first produced by the primary stubs (and may no longer be entirely uncorrelated with them) and then joins them to produce the oblique primaries. In Figure 10.1b to bring out the causal difference, the secondary is placed in the middle row, deriving from the bottom.

In Q-data the oblique primaries are the only variables we can directly measure and the H's we calculate are for them and for the score on the secondary which is a weighted sum of them and therefore not a pure factor measures. One says "in Q-data," because it is possible that in T-data the second order can be more directly measured. The fact that first order T's align with second order Q's means at least that we do not have to contend with contamination of score by primary Q-data stubs, but one must psychometrically admit that the O-A battery tests have appreciable specifics in them, though almost certainly narrower and more varied ones that in the primary Q stubs.

The available data evidence we may use in solving for the alternative models are (1) H's for the primary Q scales, (2) H's for the secondaries from loading-derived sums of the primaries, (3) the loadings of secondaries on primaries, and (4) the H's of the secondaries derived from the second, distinct medium—T-data. In no case can we get direct measures on orthogonal (stub) components, in the middle and bottom row, with the possible exception that a T-data primary (e.g., U.I. 24, anxiety, equivalent to secondary QII, may *approach* an uncontaminated measure of the bottom row factor because its specifics are so narrow and so many.

The arguments from measures to conclusions are complex and in this space we cannot hope to state all assumptions. At this point in research we can simplify by dropping discussion on models (a3), (a4), (b6), and (b7) in Figure 10.1, because no instances are known of primary scales that are *purely* genetic or threptic. Let us then consider:

(a) *The contributory model applied to (a1), (a2), and (a5).* The first has been instanced by Thurstone's primaries being possibly derived from a genetic g_f combined with purely environmental associated stub experiences. In

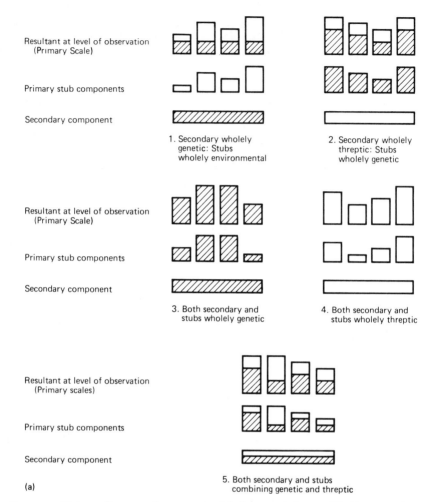

Figure 10.1 The contributory and emergent models for secondaries amplified in terms of models of broad secondary and narrow stub interaction products: SUD factor method (a) with the secondary a real independent contributor; (b) with the secondary as an emergent product of the primary state factors.

this case (a1) the loadings of the secondary on the primaries should be essentially the same as their heritabilities.

The second (a2) should show the converse, loadings being proportional to (1-H) for each primary.

The third (a5) is indeterminate, because both kinds of stubs are genothreptically mixed.

(b) *The emergent model, applied to b8 and 9.* It is impossible for purely environmental stubs to produce other than purely environmental second-

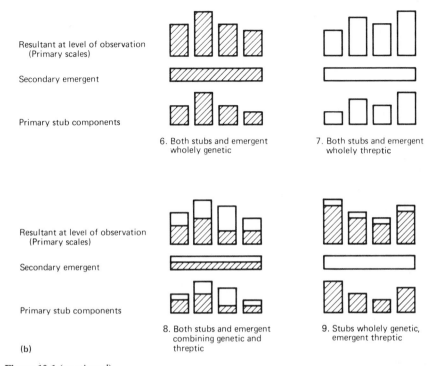

6. Both stubs and emergent
wholely genetic

7. Both stubs and emergent
wholely threptic

8. Both stubs and emergent
combining genetic and
threptic

9. Stubs wholely genetic,
emergent threptic

(b)

Figure 10.1 (*continued*)

aries and scales which is why we do not consider (b7). On the other hand purely genetic stubs could produce a purely environmental secondary.

This case, (b9), is at least a possible model for introversion as we shall see.

Secondly, in (b8)—and also (b6) we must consider the perhaps far out possibility that genetic stubs (partly genetic in (b8)) can produce in the course of maturation a genetic secondary by a confluence of physiological influences.

Cases (a3), (a4), (b6), and (b7) are *logically* possible, but (a1), (a2), (a5), (b8), and (b9) are what we are much more likely to meet.

Considering the four sources of evidence above the (a1) and (a2) models should be supported (1) by relations of loadings to H's, as indicated, on Q primary scales, and (2) (If T-data factors may be considered relatively pure secondaries) by the T-data being of higher (a1) and lower (a2) heritability than the H reached from the ordinary summing of primary scales to get the secondary, as in Table 9.7.

Since (a2) and (b9) both involved a threptic common secondary factor with genetic stubs the two tests above would work similarly and leave them indistinguishable and probably extra evidence would help from the emergent model causing the threptic component to become greater with time and experience. This observation of H change with age should help also to support (a1).

The remaining models, (a5) and (b8), would demand longer discussion but at least we can conclude that loadings and heritabilities would not be expected to align in special ways, as in (a1), (a2) and (a9).

In regard to the four sources of evidence above a refinement in (3) is to set out both the *loadings* of the secondaries on the primaries and the *weights* of the primaries in estimating the secondaries, as in Table 10.12. For it is reasonable to assume that if the contributory model is at work the H's of the primary scales should, when the second order is purely genetic, resemble the loadings (the factor contribution) rather than the weights. In the case of the emergent theory we need two assumptions to reach an inference: (a) that the larger a primary is the more it will contribute to the emergent threptic secondary, by its interactions with other primaries; and (b) that early in life the size of factors is decided more by their genetic component than any threptic influence. Thus if, say, the extraversion secondary is an emergent then factors like F with large H_p (.65) should have a large weight in the factor production (estimation weight). Parenthetically this relation would need more side calculations because the degree of heredity in a primary is not decided (in the contributory case) by that secondary only, but also by some other secondary which happens to be largely hereditary and contributes to (loads) this same primary. For the present we shall assume we look at a main contributor.

If we apply these conceptions to the data on the secondaries (QI; U.I. 32) exvia, (QII; U.I. 24) anxiety, (QIV; U.I. 19) independence and ($QVIII$; U.I. 17) covered by the Q-data of Chapter 9 and the present T-data it is possible to approach certain hypotheses, admittedly speculative, on the basis of present rough data, which is illustrated in detail for two secondaries in Table 10.12.

The scales are too few for a correlation coefficient distinguishing the relative fit for weights and loadings, but in inhibitory control, U.I. 17, both are in *inverse* order of size to the heritabilities. This should therefore be a good example of (a2) in Figure 10.1, in which the secondary is *threptic* and *contributory*. This is supported by the comparison of Q- and T-data heritabilities, for the former is contaminated, as an estimate of the second order, by hereditary stubs (specifics) in the primaries. We therefore infer, consistent with diverse development and criterion evidence on U.I. 17, that this threptic second order represents the purely environmental good upbringing of a family with positive behavior standards, which simultaneously contributes to the super ego, G, the self sentiment, Q_3 and the restraint shown in desurgency $F(-)$.

Our surprise over extraversion in Table 10.11 was that the T-data points to a decidedly lower heritability than has customarily been accepted from

TABLE 10.12
A Test of Contributive and Emergent Models on Exvia and Inhibitory Control

| | Exvia–Invia (QI, U.I. 32) | | | | Inhibitory control (QVIII, U.I. 17) | | |
Primary	Loading[a]	Weight[b]	H_p	Primary	Loading[a]	Weight[b]	H_p
(a) *Relation to primaries*							
E	(−).03	.16	.18	F	(−1).51	.12	.65
F	.56	.36	.65	G	.64	.56	.12
H	.51	.30	.37	Q_3	.60	.32	.58
Q_2	(−).76	.19	.37				
(b) *Heritability of secondaries*							
Q data heritability, H_p			.44				.33
T data heritability, H_p			.20				.18

[a] Loading values from HSPQ (not 16 P.F.) as given p. 120, Cattell, 1973a.
[b] Weights from Table 9.7, second column. Note these are adjusted downward as there described: the rank order is the important character. The negative sign on loadings is included where present, but it is irrelevant to the present variance contribution which would be the square of the loading.

Q-data. We can now begin to explain this by the observation that contrary to U.I. 17, we now have a positive relation of weights and heritabilities. If the *Q*- and *T*-heritabilities were reversed, that is, the latter showed higher H_p, we should infer model (a1) Table 10.1, that is, simply that there is a basic, heritable broad extraversion factor, which contributes itself as an investment in *F, A, H, Q₂* and *E* (slightly). But by the fact of decidedly lower *T*-data (U.I. 32) heritability we are forced to conclude that exvia–invia as a secondary is largely threptic, and to fit this to the other data we must shift from a contributory to an emergent theory. The formation of this emergent would begin very early, when the child's endowments in *F, H*, etc. were largely fixed by heredity, and their interactive contribution to the emergent would therefore be dependent on their heritability, yielding the positive relation we see in Table 10.12. The emergent model would call for this agreement to be closer with the weights than the loadings, since the direction of contribution is from primaries to secondaries—and so it is, though with too few variables for appraisal of significance.

The data for anxiety and independence secondaries are not so clear, but from the *T*-data heritability being somewhat larger we would be inclined to infer that we have in Independence, U.I. 19 at least a secondary that preexists as a temperamental tendency investing itself in dominance, self-sufficiency and other primaries.

The emergent theory of anxiety is just possible from this data, but was put forward (Cattell, 1972d) mainly on other grounds, which psychoanalysts, for example, seem to favor. Most likely it is a combination, as in Figure 10.1, (a5) and (b8) of two way action. It is in the case of our exvia hypothesis that we expect the greater opposition of opinion, since the arguments for its being a temperamental, physiologically determined tendency have long been accepted. Our explanation of H_p for QI having been relatively high is that researches estimating it have dragged in the hereditary components in the stubs of F, H, Q_2, etc.

A piece of evidence that is useful in the above inferences about the nature of secondary factors is whether the Q- or the T-data derived H is higher. Although the *corrective estimates* for validity and reliability have not made much difference to H results, a real difference in validity of factor estimates from Q and T might do so.

Here we first ask whether the loadings of Q primaries on the one hand and T subtests on the other differ enough to be responsible for the H differences. The answer is that joint factorings are few and give if anything higher Q loadings, whereas separate factorings (compare Cattell, Schuerger, Klein, & Finkbeiner [1976] with Cattell [1973a]) give probably a slight edge to T-data loadings.

A second consideration concerning the same fact—that we deal with estimated scores not pure factors—turns our attention to the nature of the "specifics" in the nonvalid portion. How would these be likely to distort heritability? Cattell and Warburton's (1967) analyses of the basic parameters of psychological tests—especially of instrument factors—would remind us that questionnaires range over particular life situations, whereas T-data subtests deal with relatively abstracted performances. (For example, U.I. 32, exvia, Table 10.7, typically uses such a process subtest as motor perceptual rigidity, essentially contentless.) There is a sense in which Q and T tests are latitude and longitude cuts across a factorial area of behavior, with the T "coarser grained." The former located separate primaries which the latter "averages across" or even—if they are situational—misses altogether. In effect this argument is that the stubs of the Q primaries could be heavily genetic or threptic in themselves, and that the particular set of four to eight stubs that go into a secondary Q-data estimate could bias the heritability of the secondary more than would the more numerous and miscellaneous (but equally large) specifics in the T-data estimate. In short the best conclusion at the moment is that the T-data estimate of H for a secondary is likely to be less biased than the Q-data.

Using this conclusion in relation to Table 10.11 we would conclude that the general population heritability of independence and anxiety is among the higher levels in source traits, whereas that for inhibitory control and exvia–invia is quite low. We fit this to the further conclusion that inhibitory control is a family–parental–cultural atmosphere which uniformly contributes threptically to G, Q_3 and $F(-)$, as stated above, and that the "generality" one seems to see in extravert behavior is an emergent from the joint interaction of A, F, H and $Q_{2(-)}$ with environment and each other.

7. Some Perspective on Regularities and Irregularities of Findings

If we now take stock of the distribution of heritabilities of personality traits measured in the T-medium, by O-A batteries, as we did with the questionnaire primaries we finish with a picture as in Figure 10.2.

As with the Q-data we have cut the presentation to H_p, which is the "final" figure, and to H_b which is of interest as showing the higher values when special treatment within the family is cut out.

Throughout the exposition of H's from four different approaches; three methods of analysis of the MAVA model (sometimes with backing from the fourth, early, exploratory study) and results of the twin method, we have been disturbed, on statistical grounds, at the degree of "wobble" of particular values, compared to overall *ranking* steadiness. We have felt it important —no matter how critical a pure statistician may become—fully to set out these discrepancies. They bother the general psychologist, one suspects, less than the genetic statistician, because the former realizes, the inevitable complexities of derivation of psychological scores and is happy if he gets, dependably, a mere rank order of heritabilities where previously all was chaos.

That is to say, one must take the average value, at present, as the best scientific-empirical conclusion yet available, and consider comprehensively the numerous realities that could account for these discrepancies. We recognize that in the first place, despite larger samples (in the 2000–3000 region originally) here than have been used in nine-tenths of previous studies in this personality–ability region, we are operating at what researchers are at length painfully admitting is the lower fringe of sample size for acceptable standard errors. Acceptable for single variance estimates, these previous sample sizes are insufficient for such complex derivatives as the special H's and, unfortunately for the much needed genothreptic correlations.

As to experimental error we have just recognized, in perceiving the cause of discrepancy of two instrumentalities (Q and T) for studying secondary source traits, that even with batteries and scales at usually acceptable concept (construct) validities of .75 to .90, the genetic or threptic contribution to the factor estimate from differing specifics and "stubs" can throw the H values to systematically different levels. However, as often in science, the "perturbations" of observations that begin as a nuisance end as a signpost to new knowledge. In this case a discrepancy may help tell us whether a second-stratum factor is a broad contributor or a broad consequence or emergent from primary interactions, and we have hypothesized that it is the former in anxiety and the latter in exvia–invia.

Some of the error variation even within results from one medium of observation would be reduced a little by any corrections that could be applied to the concrete variances for differing validities, reliabilities, and, a good deal by new measures of differing *liabilities to function fluctuation*. We have discussed how the usual psychometric application of such test parameters to single testings needs to be modified for variances from the pair difference scores used in this kind of behavior genetic research (p. 287). However, there has been only a limited application of the new corrections in the data here, and it remains for research on our data proceeding by Vaughan (in press) and others to present such improvements. Actually, since the changes produced in *H* values by reliability corrections have proved quite small, our present conclusions on relative heritability are essentially unaffected.

Meanwhile, we must repeat that despite what appear at first as rather disturbing variations of absolute numerical values appreciable *order* nevertheless prevails across different methods. Despite somewhat different absolute values, they nevertheless tend to put all the traits in the same *rank order* of heritability. Second, the least-squares and maximum-likelihood methods consistently tend to give more extreme—all or nothing—values than OSES, and also rather more frequently give zero genothreptic correlations. Statisticians with whom we have discussed results largely agree in putting the analytic methods we used in increasing order of preference from OSES, through least squares to maximum likelihood. Yet general psychological knowledge on these traits has several times agreed better, for some mysterious reason, with the OSES result. And the acceptance of a genetic assortive mating *r* of .25 in OSES has seemed to give more acceptable results than the .10 or 0 assortive mating in the maximum-likelihood best fit.

On the other hand, the delivery by the two latter methods of more easily and reliably obtained standard errors of the estimates is a substantial gain. There are several instances, however, of the results from the two latter standing further apart than would be possible (except beyond one in a thousand) from the separate standard errors given. This brings us to a major theoretical question, "Is the MAVA model as presently constucted within the biometric genetic (continuous variable) modes of analysis the correct model?" Let it be noted that we have not balked at asking this basic question in the present research and have gone so far as to try model fitting (*a*) with no genothreptic terms, and (*b*) even with no genetic terms. In one case the latter actually offered a better overall fit, while the former was accepted in two or three cases on the argument that though the significance of fit was a tie between full MAVA and the correlation-free MAVA, the latter was more parsimonious. We are not convinced that this last, purely statistical argument should be allowed to dominate the final inference. There are instances in science—several could be cited in factor analytic research—where the model giving most parsimonious fit in one experiment is different from that giving most parsimonious fit over several. From *general* psychological principles we would consider it impossible—as LS and ML modified models sometimes for a phenotypic trait require—for it to be *absolutely* all threptic (or genetic), or for genetic and threptic effects to be *absolutely* uncorrelated. The deviations from absolute may be small, and when we have combined the other two methods in an average with maximum likelihood a small value has resulted which fits general evidence on genothreptic relations better than an absolute zero.

However, we have recognized that the MAVA model is certainly in need of extension and modification. For example, the "six unknown" solution, with σ^2_{wg}, σ^2_{wt}, σ^2_{bg}, σ^2_{bt}, r_{wgwt}, and r_{bgbt} would commit the obvious and now proved error of considering within-family threptic variance the same for twins and sibs. The "seven unknown" still makes the *almost* certain error of considering $\sigma^2_{wt.t}$ to be the same for identical and fraternal twins. Even the full MAVA (Table 4.8) with 19 equations and 9 constellations is forced to make the approximation of considering that adopted children have the same σ^2_{wt} as others, though it recognizes in its 15 unknowns the genetic resemblance of adoptees to the adopting parents, and the special σ^2_{bg} due to sociobiological selection of adoptees as such.

Most of the poorer fits we here meet could be overcome by finding and using more constellations, with more specialized unknowns, but there remain other possible weaknesses even in an adequate set of constellations to which we must not be blind and which the occasional statistical failures of fit of the model indicate to be possible causes of trouble. One is the acceptance of simple additive genetic effects (without dominance or epistacy), for example, in calculating the effect of assortive mating. A more fundamental and serious one is absence of allowance for interaction (not correlation) of genetic and environmental influences. There can surely be little doubt that, say, 6 months' study of mathematics applied to a 10 year old of IQ 140 will give a bigger increment in math score than when applied to one of IQ 80, and the same tendency to substantial interaction must be true of many personality endowments. The rather awesome complications of the MAVA model through interaction which we contemplated on page 189 must be faced by the next generation of researchers if the probable *main cause* of poor fit, and of irregularities of results is to be overcome.

Granted that some beating of breasts is inevitable over the results being less regular than we had hoped, it can nevertheless be said that as an exploratory study its eight years of work have not been wasted. Despite the intrusion of varied incalculable sources of error, that would be expected to bring all *H* values toward an indistinguishable heap or central value, the heritabilities reached actually exhibit, here and in *Q*-data, a considerable spread in the values obtained, (despite being method averages) and yield a fairly steady position for each trait in that wide range. As Figure 10.2 shows, U.I. 16 (assertion), U.I. 19 (independence), U.I. 24 (anxiety proneness), and U.I. 25 (realism-versus-tensidia) give evidence of a substantial heritability, not much short of values found for fluid intelligence, and above those found for

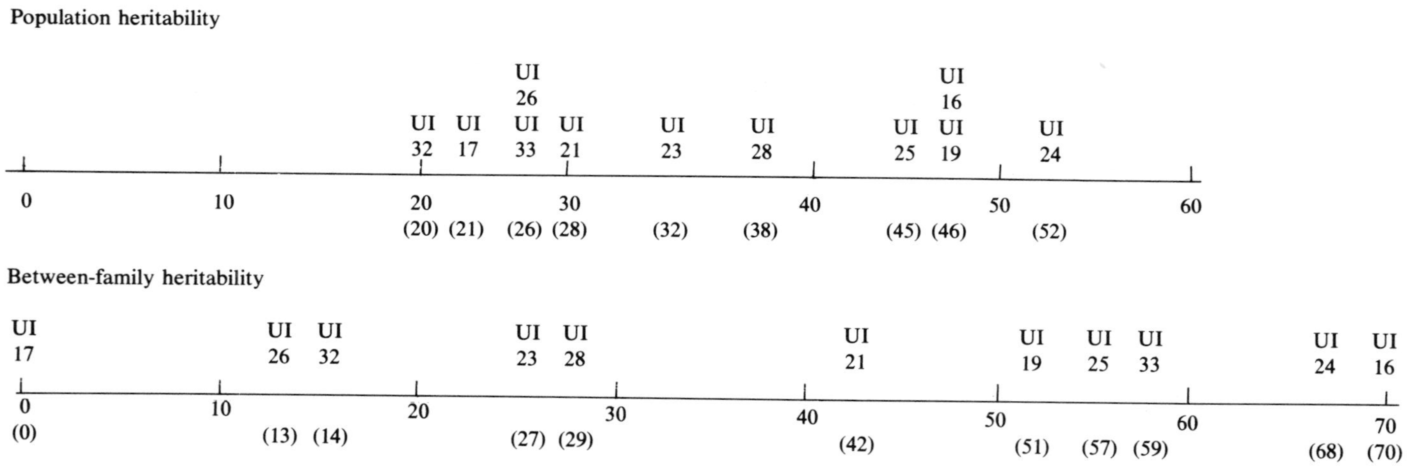

Population heritability

| 0 | 10 | 20 | 30 | 40 | 50 | 60 |

UI 32 UI 17 UI 26 / UI 33 UI 21 UI 23 UI 28 UI 25 UI 19 / UI 16 UI 24
(20) (21) (26) (28) (32) (38) (45) (46) (52)

Between-family heritability

| 0 | 10 | 20 | 30 | 40 | 50 | 60 | 70 |

UI 17 UI 26 UI 32 UI 23 UI 28 UI 21 UI 19 UI 25 UI 33 UI 24 UI 16
(0) (13) (14) (27) (29) (42) (51) (57) (59) (68) (70)

Figure 10.2. Distribution of population (H_p) and between-family heritabilities (H_b) for source traits measured in the T-medium (O–A Battery).

crystallized intelligence (when due transformation is made to compare MAVA with twin-method values found in the earlier intelligence studies). U.I. 17 (inhibitory control), U.I. 26 (narcissistic ego), U.I. 32 (exvia-versus-invia), and to a lesser degree, U.I. 23 and U.I. 28 (asthenia),on the other hand, show quite low heritabilities. In U.I. 17, 26, 28 and 23 these results fully fit the theories put forward on other psychological ground about these source traits (Cattell & Schuerger, 1978).

On the other hand we are forced to a theory about exvia–invia that is entirely different from the long standing traditional view, for example in Jung and Eysenck,that it is basically an *innate* temperament dimension.Previous support for this, it would seem, has been due to large heritabilities in the *stubs* of A, F, H and Q_2, as measured in Q-data, contaminating the estimate of the Q-data secondary, QI. The theory of exvia-invia as such which I venture to propound on present, admittedly limited data, is that its genesis follows the emergent model. That is to say, initial genetic possession above average, of A, H, F or $Q_2(-)$, produces, through the effects from environmental interaction of one trait on another (spiral interaction) an increase in the threptic component of all. This common threptic addition is the second order exvia–invia factor.

The roots of second order factors are thus envisaged as (a) a common environmental influence from outside; (b) a common genetic influence from within; and (c) a common threptic effect of spiral interaction (Figure 10.1). But the creation of what the psychometrist encounters as primary factors needs to be clearly distinguished from this secondary birth, and concerning these the source we have recognized are also essentially three: (*a*) a genetic influence acting on its own; (*b*) a molding, institutional learning pattern on its own; and (*c*) the eidolon model in which the environmental learning for certain reasons shapes itself on the emerging genetic maturation pattern (box-and-lid). The latter would give middle H value ranges, the others values close to each end of the scale. In all we have supposed that error will (*a*) shift values toward the middle, and (*b*) reduce the true heritability. The results in Figure 10.2 and H_b thus corrected, would fit (a), (b), and (c) with only U.I. 19, 21, 25, and possibly 23 staying in the middle type. We have to recognize, however, that even admitting the "drift" from error—(*a*) and (*b*)—the environment events that happen *within the family* are powerful enough to throw environmental variance even into such patterns as U.I. 16, 19, and 33. It would follow that if these are patterns primarily gaining unitariness from genetic sources, they should (if they are not also eidolon examples) nevertheless lose some unitariness as factors, i.e. show incomplete loadings on expressions, due to random impact of environmental sources on particular expressions.

As to regularities beyond those of overall heritabilities, we note in these T-data results as in Q-data, that most traits are pretty consistent in showing much the same order of heritabilities between as within families.[1] However, a few—U.I. 16, 19, and 33—show almost startlingly higher values between families. Another regularity with both *T*- and *Q*-data results is that the genothreptic correlations are (*a*) larger between than within families, (*b*) yielding between-family values almost invariably negative.

8. Objectives for the Future in Behavior Genetics: In Research and in Psychological Practice

The injunction to "Do what I suggest, not what I did" is more excusable in science than in most conduct, even though researchers invoke it too often. If we turn to improvements in procedures and directions, it is obvious from passing comments that we need (*a*) larger samples, and across at least as many constellations as here and (*b*) more valid, longer batteries and scales. The latter suggests that, with the present exploratory indications of where a combination of debatable *H*'s and trait importance exists researchers would now do well to concentrate on one or two source traits at a time, using longer batteries. An important improvement on this psychometric area would be to test each individual on three or four occasions scattered weeks apart to eliminate (or at least reduce and estimate) the variance due to that function fluctuation which we now know (Cattell & Scheier, 1961; Brennan, 1977; Horn, 1972a; Nesselroade, 1976) exists appreciably in most traits.

---

[1] The causes of difference between H_w and H_b *must* lie in the environmental area, since the genetic part of these ratios is fixed (either identical or modified to a predictable value by assortive mating, $r_{fm.g}$). We recognize, however, that the *causes* (the environmental sources as depicted in path coefficients) of threptic variation within and between families are in part common. That is to say the variance found within the family, σ^2_{wt}, is not entirely due to sib and parent interactions in the family but partly also to sibs' different experiences in the outer world, for example, one associating with peers of a different social class from those the other sib encounters. The precise separation of contributions into those from family happenings and those from society conditions, given σ^2_{wt} and σ^2_{bt}, is for further debate and research, but, granted a one-way intrusion (some σ^2_{bt} into the true, original σ^2_{wt}), we can at least infer that when H_b is larger than H_w, a better estimate of the latter as a "family happening" value would be lower than that presently reached by σ^2_{wt} and vice versa. For example, the much higher H_b than H_w for independence (U.I. 19) suggests that *within* the family there are strong environmental situations adding and subtracting from sibs' levels on independence. The same interfusion of σ^2_{we} and σ^2_{be} (using *e* now to represent the *environmental influences* from within and without, so that $\sigma^2_{wt} \neq \sigma^2_{we}$ and $\sigma^2_{bt} \neq \sigma^2_{be}$) will affect the genothreptic correlations obtained. Thus a zero r_{wgwt} for a trait along with a large negative r_{bgbt} suggests a hypothesis that the true r_{wgwt} is actually somewhat positive. There are examples of this, notably where we hypothesize (Table 8.6), that the moderate negative correlation within the family for crystallized intelligence ($-.41$, Table 8.6) is actually an effect of what is happening outside the family and that the discerning parents within the family actually provide more intellectual challenge for their brighter offspring. In this case the alternate, least-squares solution (Table 8.10) actually settles on $r = +1.0$ as the best estimate of within family encouragement of intelligence.

With this improvement in procedure the more ambitious researcher will also want to make an improvement in model, regarding which two principal directions of development are indicated:

1. Entering interaction terms at first as simple products, as on page 189, unless ulterior data give a more specific hunch. Though this will require some inspiration, our preliminary inspection suggests it should not be impossible to find enough constellations for a MAVA model equation solution for the new terms.
2. Proceeding from discovered variances, interactions, and correlations to causal networks, as currently indicated by path coefficients (Cattell, 1978a, p. 486; Cloninger, 1981; Rao *et al.* 1974, 1976; Wright, 1934). This almost certainly needs, for convincing solution, auxiliary evidence from developmental studies (Nesselroade & Reese, 1973), since any set of correlations and variances can be explained by a variety of path coefficient networks.

It has perhaps been sufficiently emphasized that both for the geneticist and for the learning theorist (to whose special needs we will turn in a moment) what we urgently require for several theoretical advances is results from *comparative* MAVA. That is to say we need variances and H's systematically determined in one given society at, say, successive 5 or 10-year age cross-section of levels, and also systematically compared across cultures in relation to different racial gene pools and cultural ranges in the ("teaching") environment. This is a "tall order" in itself, but hopes of fulfilling it diminish to zero unless investigators first agree on the few "significant variables" to be uniformly retained across studies. In the personality–ability area they can be found, in my opinion, in a careful perusal of the evidence on the best replicated and most validly and objectively measurable source trait factors.

Finally, as in any area of science that has attained a reasonable foundation of inductive laws it is time for experiments to be guided by more theory. A small instance of this has just been given in terms of checking theories of second-order personality structure development. But in more general terms personality theory is capable now of giving us theories about a variety of source traits to check in behavior-genetic terms. We have achieved some of this in this book where theories have been previously published about particular source traits, by Horn, Eysenck, Nesselroade, Meredith, and myself on fluid (g_f or U.I. 1) and crystallized (g_c or U.I. 2) intelligence, and on U.I. 17, U.I. 21, U.I. 25, U.I. 28, and U.I. 32, for example. As we have seen, genetic investigation strongly supports the theories on g_f, g_c, U.I. 17, and U.I. 28, and is compatible in U.I. 25, with clinical evidence on the appreciable inheritance of psychosis. Regarding U.I. 21 and U.I. 32 (extraversion), however, it throws a monkey wrench into the machinery of existing theories.

Theories of temperament traits (taking the common connotation of "temperament" as more innate) are likely to raise their heads more clearly as behavior genetic evidence piles up on certain source traits, that links up for example, with findings in brain physiology and anatomy, or studies of alcoholism and addiction. The area where we are most lacking in behavior genetic findings to rest theoretical expectations is in the dynamic trait modality. Factoring has there yielded, in objective, laboratory motivation, and interest measures, two distinct kinds of source traits—ergs (drives) and sentiments (Cattell & Child, 1975). The former are the presumably innate propensities to reaction we share with the primates. The latter are acquired patterns, mixed as to ergic basis, deriving from reinforcements provided by particular institutions, and linked in a unity of cognitive activation.

At first thought one would expect MAVA research (or twin-method research) with measures of these factors (as available in the MAT, Cattell, Horn, Sweney, & Radcliffe, 1964) to show virtually complete inheritance of the former and complete environmental determination of the latter. However, ergic patterns show in ergic tension measures, and these show a degree of function fluctuation altogether higher than for, say, abilities. Further, as to sentiment structures, unless they are measured by cognitive measures only (I, not U, components in MAT) they would be contaminated by the genetic quality of the ergic satisfactions involved in them, so that their entirely environmental origin might stand in doubt. Some clear theoretical planning is therefore essential before setting up expectations and experiments in the genetics of traits of dynamic modality.

An emphasis in this book that may be new to psychologists has been the claim that behavior-genetic research is as important for learning theory as for genetics. It has been pointed out that the aim of finding the correlation (in life and over life epochs, as distinct from an hour spent in a narrow laboratory experiment) of environmental forces with trait learning is aborted by developmental learning studies as commonly pursued. They fail to avoid a confounding of maturation with learning in the measured trait change. What we need is the correlation of a particular environmental range *with the purely threptic component of the trait*. As shown earlier (p. 224) the possibility then exists, by multiple correlation, of finding how much of a threptic endowment in a given trait is due to this and that element in life environment, and how much is still to be accounted for by unknown environmental elements.

Leaving the learning theorist to pursue the implications of this avenue of behavior-genetic research let us turn and ask what the geneticist gets out of it. It must be freely admitted that the present MAVA and twin divisions of biometric genetics do not get him or her very far toward the pinnacle to which he or she aspires. That remote goal is the determination of the number and nature of individual genes, their chromosomal locations, their dominant and epistatic actions, and perhaps the protein productions through which they act. There being enough for psychologists to handle in the area of the present book already, we have left Mendelian molecular biology and population genetics algebra to other sources. But some steps that the geneticist can initially take from data of the present kind may be briefly indicated.

Geneticists are actually not so much interested in H values as in the genetic variances themselves. (Just as learning theorists reciprocally can get most out of the sheer threptic variance magnitudes in relation to concrete environmental variances magnitudes.) If they could get these variances for offspring from MAVA studies on a sufficient diversity of parental biological relationships (first cousin, second cousin, brother–sister, uncle–niece, etc. marriages) they would be able, by the complex and subtle methods of population genetics, to find out something of the gene action as such. In this connection we should note the conceptual problem that arises when we come to the question (provoked by our second-order factor investigation) as to whether the same genetic variance and the same gene set enter into the genetic part of the variance of several distinct primaries (diagram [a1] in Figure 10.1).

There are complex problems in arguing from present data to second-order structure. First, a given population variance in a given gene set *per se* may *genetically* contribute differently to variance in different traits affected by the same genes. For example, if gene set X contributes to stature and also to hand breadth it may contribute with different potency in terms of separate gene effects to the two. Second, to go beyond the action in this example, the majority of traits are measured in different *units*, so that even if the gene effects were identical, the absolute numerical magnitude of the genetic variance would not be the same. In short, just as we have argued that a particular *threptic* variance is related by different rates (regressions) to various outer *environmental* variances, so we argue that a particular *genotypic* variance is related at different rates to the *genetic variances* of various traits in which it is genetically involved. These complex considerations may to some extent vitiate, or at least modulate, the generalizations we sought in Table 9.7 that (a) a correspondence of the heritabilities of primaries with their second-order weights connotes that their heritability comes from a highly heritable second order, and (b) since T-data factors may be argued to have left out Q data primary stubs any higher heritability found from the Q primary estimates than from T-data connotes that the former is due to genetic parts of the primaries. This situation in exvia led to the conclusion that we probably deal with a spiral emergent model in which the pure second order is largely dependent on environmental interactions.

Finally, we must find a brief space to move on from the interests of the researcher to those of the practicing psychologist. The latter can get help in the form of (a) general psychological fact and theory about psychogenetics, and (b) diagnostic information about the genetic makeup of a given client.

The latter is an ideal for the future, not presently very practicable. Yet a psychologist blind to the genetic part of an endowment can, like a surgeon before X rays, do considerable damage. One sees this in the nineteenth century pressures on mental defectives and some present day ignoring of a schizophrenic diathesis. The theory of obtaining the genetic part of a total trait score for a given individual has been explained, in terms of two possible methods, on page 225. It involves getting the threptic part first and subtracting it from the total. To calculate the former it is necessary for science to have established regression coefficients of threptic upon environmental features, and to know the environmental features in the client's life. Occasionally it may be possible to take the shorter route of using a specially devised measure of a trait capable of getting at the more innate part, as a culture-fair intelligence test gets at the more innate part included with g_c in a traditional intelligence test, like the WISC. But either way of estimating the innate component in a given individual belongs to the future,[2] and meanwhile let us consider what help can come from a practicing psychologist's training in general genothreptic principles and knowledge.

First and foremost psychologists need to have at their fingertips the heritability values for a common spectrum of traits. Incidentally, they would do well to keep in mind our estimate that values from existing research, with the defects indicated in them, are probably insufficiently high and insufficiently differentiating among traits. Mainly, the object of such knowledge is to know when a trait may be expected to change appreciably, under therapy, education, etc., and when it may not. A man who finds embedded in the garden of the house he has just bought a 40-ton granite boulder is probably correct (*a*) to decide not to try to get it out, and (*b*) to think of some useful thing he can do with it as it is, such as using it as base for an elevated fountain or summer house. Similarly the trained psychologist will not ask for trouble trying to change a largely genetic trait, but will shrewdly use it as a fulcrum for levering some other trait that *is* susceptible. Educational psychologists know they would be well advised not to send a young man of IQ 80 into a university math course. Clinicians know that in a client of IQ 140 they can use the high intelligence as a fulcrum for leverage of insights into emotional problems and new cognitive corrections. Vocational psychologists, in view of the high innateness of surgency, would be unwise to direct a very desur-

___

[2] In clinical psychology to know whether a schizophrenic client has a considerable or only a minor predisposition would guide one to different prognoses and treatment. If Seguin—"The Apostle of the Idiot"—had possessed presently existing culture-fair tests to distinguish a merely backward referral from a true defective, he might have avoided driving the latter to distraction by intensive "training of the intellect" by form boards. Incidentally, the notion of separating out an individual genetic part score by looking at relatives and making an estimate for the client by regression from their score is not effective because *their* scores also are mixed genetic and threptic. For example, the phenotypic regression of intelligence on a brother is about .6, so if you could not test John and knew his brother had an IQ of 120 you would estimate his as 112. But it would still be a crystallized intelligence, g_c, traditional intelligence test measure.

gent person to a job known to require surgency—in the mistaken expectation that training will change all that. At the same time one must make due allowance for the fact that H's represent the ratio for that trait under the given ranges of environment and racial mixture existing in our given society. Explicit intensive teaching should be able effectively to change levels where H is high because the present range of environmental impact is probably smaller than it need be. Probably no one comes near the ceiling to which intensive training could move him or her with his or her given genetic potential. But one must also realistically recognize that with only 24 hours in a day intensive training on one trait will tend to mean neglect of another.

If a trait such as G or U.I. 17 (super ego strength), or U.I. 26 (narcissistic ego) or U.I. 28 (asthenia), is involved, as too high or too low for a particular adjustment, the behavior geneticist should be able to reassure the psychologist that there is no known reason why well planned reeducation and reconditioning should not change such a trait to a major degree. In view of the substantial, but far from complete, degree of genetic determination of anxiety, therapists should not consider themselves failures if their patients never get down to a truly low level, but neither, perhaps should the parent expect a lethargic, unreactive offspring to catch fire or get strongly motivated through parental attempts to stimulate anxiety.

Knowledge of differing expectations of possible change through manipulation of various traits is not all that the behavior geneticist can give the psychologist. The genothreptic correlations indicate where the client typically will or will not experience social pressures toward the biosocial norm. Finally, as the differences of H_w and H_b values are more carefully determined the practicing psychologist will receive some dependable guidance as to where the forces that typically mold certain traits lie. It is interesting that, on the whole, the "deeper" traits considered most important by clinicians have here shown greater shifts from situations *within* the family.

In these matters we should conclude by pointing out that in psychological practice as in medical practice the most progressive steps are made by interaction of the researcher and the practitioner—and the researcher has now thrown the ball of broad exploratory studies to see where the practitioner can carry it.

9. Summary

1. Behavior-genetic investigation through the instrumentality of objective (nonquestionnaire) batteries is more recent than Q-data, but rests on an equally comprehensive, replicated, age and cross-culturally explored array of factor analytic findings. The relation of Q- to T-data factors is established in four or five instances a second-order matching first-order T. Thus

QI = U.I. 32, QII = U.I. 24, QIV = U.I. 24, QVIII = U.I. 17, and perhaps QIII = U.I. 22.

2. The data analyzed for the *T*-medium in this chapter are on a somewhat larger sample of constellations than for the *Q*-data and permit solution for one more unknown. The same OSES, least-squares, and maximum-likelihood methods are used as before. The fit to the model for U.I. 16, 17, and 19 is good, though the last two methods indicate that no genothreptic correlations are necessary. The outcome supports the theory of U.I. 17 as a home and school taught pattern of inhibitory control, but points to substantial inheritance (highest among all traits in H_b) for U.I. 16, ego assertiveness, and U.I. 19, independence.

3. The universal index (U.I.) system has indexed factors in declining order of variance contribution to the personality sphere of behavior variables (Cattell & Warburton, 1967). The next trio in that order studied here covers U.I. 23, 24, and 26. The first two have medium value heritability (U.I. 24 slightly higher) but U.I. 26 is very low. All three have even levels across H_w, H_b, and H_p, and between-family genothreptic correlations are uniformly negative, and higher, but still moderate, than within families. Narcissistic ego (U.I. 26) is evidently a largely environmentally acquired pattern, possibly what is popularly designated the "spoiled child." The rather high inheritance of anxiety proneness (U.I. 24) fits the degree of "inheritance" of actual neurosis, in which anxiety is a major component. The finding that U.I. 23, capacity to mobilize-versus-regression, is in the same region of heritability as anxiety fits their close functional interrelation, and supports the clinical theory that a fall in U.I. 23 is a long-term fatigue following excessive anxiety stimulation.

4. On the other hand the three analysis methods show distinctly poor agreement on U.I. 28 and only moderate on U.I. 21. An average of methods for U.I. 28 gives low heritability, and if we accept the superior statistical power of maximum likelihood, a virtually zero H_w, H_b, and H_p. This fits well the previously published theory of U.I. 28, as an asthenia brought on by too early an imposition of social standards. (Freud's anal-erotic pattern may be an anticipation of the U.I. 28 factor pattern). In the case of U.I. 21, on the other hand, the rather low H value, if sustained, threatens to demolish the previously published theory, which viewed exuberance, with its high fluency and rapid decisions, as the result of a chemical pacemaker. The decision on whether it is mainly a temperamental "irresponsibility" or a personal history of being little suppressed awaits independent research evidence on H.

5. The H and genothreptic values for U.I. 25, 32, and 33 are tolerably consistent, especially by least squares and maximum likelihood on the last two. U.I. 25, reality contact-versus-tensidia, which has been shown one of the most powerful discriminators of psychotics and normals (Cattell, 1969b; Cattell *et al.*, 1972; Eysenck & Eysenck, 1958; Killian, 1960) has rather high

heritability between families, consistent with findings on clinical syndromes of schizophrenia and manic-depression. U.I. 33, sanguineness-versus-discouragement, a little discussed or explored personality factor associated with schizophrenia, has a discrepancy of high between-family and a low within-family heritability, suggesting a basic heritability but one particularly susceptible to early within-family environmental situations.

6. Comparison and contrast of heritabilities for traits U.I. 17, 19, 24, and 32 as found through (a) the questionnaire medium, as second-order factors, and (b) the objective test instrumentality, is potentially capable of contributing to a decision between two theories of second-order factor structure. The alternatives are that correlation among primaries is produced by (i) a broad second-order influence (genetic or environmental) contributing to all of them, or (ii) a mutual interaction of primaries in life situations producing the "spiral emergence" of an additional behavior pattern fed by all primaries. No fewer than nine models are possible when hypotheses of heritable and nonheritable factors are introduced into these two theories. Since an actual estimate of a second order can only be obtained from first-order measures, inferences remain at a probability level. The probability seems to be, since the T-data values for H are lower than the Q values (in which the stub heritabilities enter) that the essential second orders in U.I. 17, and 32 are of quite low heritability, whereas in U.I. 19, independence, and U.I. 24, anxiety, they are high. The relation of heritabilities of primaries to their loadings in the second orders further suggests that U.I. 32, exvia, is a spiral emergent factor, while the other three are instances (like intelligence and primary abilities) of a preexisting second-order influence contributing to primary "stubs." The exvia finding reverses current reaching about that trait.

7. If examined with regard to the limits of standard errors one must recognize that different methods here, twin methods elsewhere, and other data by the OSES method yield significantly different *absolute* heritabilities, for reasons not yet fully apparent.[3] Although one must regard all present results,

---

[3] A truly comprehensive evaluation of a source and magnitude of errors at the present stage of the art of behavior-genetic research is much needed today, but would belong to a more advanced text than this. Many of the ingredients may be found in the writings of Eaves (1973), Eaves *et al.* (1978), Falconer (1960), Fulker, Wilcock, and Broadhurst (1972), Jencks *et al.* (1972), Kempthorne (1957), Mather (1949), Loehlin (1979a), Rao, Morton, Elston, and Yee (1977), and others. Thus, in the area of intelligence data Loehlin (1978) writes:

> Three different published heredity-environment analyses of Jenck's summary correlations for I.Q. have yielded strikingly different results. It is shown empirically that differences in selection of data, computational procedures, and logical inconsistencies in specifying equations are not responsible for the differences in results. Rather, the differences trace to the underlying assumptions made by the various authors. The analyses suggest that the assumptions concerning genetic dominance, assortative mating, and special twin environments were especially critical, while those regarding selective placement and different modes of environmental transmission were not.

In our present studies the problems of dependable assortative mating estimates, and environment difference ratios of sibs to twins seem tolerably solved, but we have done nothing about

therefore, as exploratory, yet there are convincing consistencies in (a) the rank order of heritability of traits, (b) what the different methods indicate about the predominance in some traits of high H_b relative to H_w values (or vice versa), (c) the typical existence of small negative within-family and larger negative between-family genothreptic correlations, (d) the finding that the full MAVA gives a better fit, in least-squares and maximum-likelihood tests, to the data than alternative models, except in a small minority of traits, (e) the finding that a moderate positive assortive mating correlation (genetic) consistently gives the best fit for most traits.

8. The basic question here has been whether the general MAVA model fits a wide range of new data with acceptable significance. It does so generally, but in a few instances a good fit is not found. The improvement needed in forthcoming research is first to obtain larger samples and more constellations, permitting less gross lumping together of potentially different unknowns. Second, to give longer (more valid and reliable) source trait tests, directed to factors agreed by research to be best replicated and important. Such measures should be based on several retestings, to obviate function fluctuation variance. Third, a solution is needed to the fact that present genothreptic correlations do not yield the full information we need about the form of the relation, notably as to whether the regression is the same above as below the bio-social mean and whether curvilinear relations exist. Fourth, and most fundamentally, the MAVA equations and solutions need to be worked out allowing *interaction* terms on the model.

9. The obtained distribution of heritability values is concluded, from our discussion of the effect of errors, to be probably (a) less dispersed than the true range, and (b) shifted toward lower levels than the true heritabilities. Although no statistically defensible specific correction can be given to support this view of the distribution we take the position, with some support on wider grounds, that our earlier model is correct, namely, that (a) some traits are virtually wholly environmentally produced patterns, (b) some express almost wholly a genetic, maturational pattern, and (c) some follow what we called the eidolon model, of acquired characters shaping on an existing genetic pattern. As far as H_b results are concerned there are empirical indications of such a trimodel distribution.

10. Future research would do well to have more *theoretical guidance*, from general personality theory, to get the best harvest from behavior-genetic experiments. In particular, in extending research to dynamic trait (MAT measurable) heritabilities some clear concepts need to be worked out for the kinds of effects to be expected in regard to measures of ergs and sentiments.

---

evaluating selective placement. However our main unsolved challenge has been that of differences from different analytic methods. We have also pointed out that sample sizes are on the lower edge of acceptability, and that the absence of interacting terms in the model could have accounted for the few instances of unacceptable fit to the data.

11. Psychologists are beginning to recognize that in behavior genetics research they have an avenue as important for personality learning theory as for genetics. Every behavior-genetic research yields a measure of a threptic (environmentally determined) variance as well as a genetic variance. As regards learning in the life situation it is the former variance, rather than the total phenotypic variance, that the learning theorist needs to relate to environmental experience. For otherwise he is confounding and confusing his learning measures with maturational effects; and there is no way but behavior-genetic research to yield estimates of the pure threptic change.[4]

Reciprocally, the genetic variances which the present method supplies are, for the geneticist's interests, only a beginning, though an indispensable beginning. Geneticists naturally wish to proceed thence to knowledge of number of genes, to Mendelian laws, chromosomal positions, etc. And unlike their designs in animal research, those possible in the human domain enforce a long and complex transit to genetic bases. Briefly, their additional task is one of developing equations in population genetics, which must at present rest on genetic variances, as discovered here, but derived additionally from more rare populations of biologically related parents. Incidentally, the further work of both the personality learning theorist and of the chromosomal geneticist is less concerned with the H values here, then an accurate evaluation of the variances as such.

12. Just as researchers in personality now see the strong relevance of behavior genetics to structural and developmental theory, so practitioners are perceiving its aid in reducing the clumsiness of earlier, uninformed diagnosis, treatment, and educational and vocational guidance methods. Its present use concerns general application of the knowledge of relative modifiability of traits, the situations in which teaching is most potent, the likelihood of systematic influences on traits by community forces, and the use of favorable hereditary endowments in certain traits for leverage on others. In the future there are methods, requiring cooperation of psychologists for get-

---

[4] Actually a whole new domain of possibilities of environmental analyses based on behavior-genetic findings has been opened up in this book, but it has too many parts to summarize here. Besides the purification of learning laws, that is, of threptic–environmental relations, however, we may mention especially (a) the hypotheses on formation of unitary trait structures by "spiral interaction" creating second-order factors; this supposes genetic endowments either lead to more of certain environmental experiences affecting other traits, besides the one concerned or that there is some more direct trait interaction in a product relation to produce new trait variance, (b) the multidimensional treatment of path coefficients, which supposes that the properties in parents, or other environmental features, interact with ambient environmental situations, in the form of several factors in a specification equation, to account for the contribution to trait T_a in the child from traits T_a, T_b, T_c, etc., in the parent. One consequence of this would be that the heritability of a trait—say, C (ego strength)—would be different (because of environmental difference) in families high on, say, g_c (crystallized intelligence) than in a population at a lower level on the latter. These relations can be experimentally attacked, but, as in comparative MAVA generally, only with sufficient research endowment to get proper samples.

ting the necessary parameters, by which an estimate can be made of the magnitudes of the genetic and threptic components in the existing trait score of a given individual, but this is probably still far away.

Meanwhile psychologists whose knowledge encompasses an appreciation of estimated within- and between-family heritabilities for the chief primary and secondary personality traits and abilities are equipped better to handle many clinical, educational, and vocational problems. They will not falsely depend on expectations of great change in some traits or underestimate what major changes can be brought about by good methods in others, and their prognoses will carry farther into the future.

References

Adcock, N., & Adcock, C. J. The validity of the 16PF personality structure: A large New Zealand sample item analysis. *Journal of Behavioral Science,* 1977, *2,* 227–237.

Ahern, F. M. *Family resemblances in intelligence.* Unpublished doctoral dissertation, University of Hawaii, 1979.

Asher, H. B. *Causal modelling.* Beverly Hills, California: Sage, 1980.

Baker, J. R. *Race.* New York: Oxford University Press, 1974.

Baltes, P. B. Longitudinal and cross sectional sequences in the study of age and generation effects. *Human Development,* 1968, *11,* 145–171.

Baltes, P. B., & Nesselroade, J. R. The development analysis of individual differences on multiple measures. In J. R. Nesselroade & H. W. Reese (Eds.), *Life-span developmental psychology: Methodological issues.* New York: Academic Press, 1973.

Baltes, P. B., Reese, H. W., & Nesselroade, J. R. *Life span developmental psychology: Introduction to research methods.* Monterey, Calif.: Brooks-Cole, 1977.

Baltes, P. B., & Schaie, K. W. The myth of the twilight years. *Psychology Today,* 1974, *8,* 35–40.

Baltes, P. B., & Schaie, K. W. On the plasticity of intelligence in adulthood and old age: Where Horn and Donaldson fail. *American Psychologist,* 1976, *31,* 720–725.

Barton, K., & Cattell, R. B. Personality factors affected by job promotion and turnover. *Journal of Counselling Psychology,* 1972, *19,* 430–435.

Barton, K., & Cattell, R. B. Personality factors of husbands and wives, as predictors of own and partner's marital dimensions. *Canadian Journal of Behavioral Science,* 1973, *5,* 83–92.

Barton, K., & Cattell, R. B. Changes in personality over a five-year period: Relationship of change to life events. *JSAS Catalog of Selected Documents in Psychology,* 1975, *5,* 283–283.

Bentler, P. M. Multistructure statistical model applied to factor analysis. *Multivariate Behavioral Research,* 1976, *11,* 1–18.

Bereiter, C. Genetics and educability: Educational implications of the Jensen debate. In J. Hellmuth (Ed.), *Disadvantaged child* (Vol. 3. *Compensatory education: A national debate).* New York: Brunner-Mazel, 1970.

Bernhard, V. W. Psychische Korrelate der Augen- und Haarfarbe und ihre Bedeutung fur die Sozialanthropologie. *Homo,* 1965, *16,* 1–31.

Birkett, H. *The ego and its distinction from the superego and self sentiment structures.* Unpublished doctoral dissertation, University of Hawaii, 1979.

Birkett, H., & Cattell, R. B. Diagnosis of the dynamic roots of a clinical symptom by P-technique: A case of episodic alcoholism. *Multivariate Experimental Clinical Research,* 1978, *3,* 173–194.

Birkett, H., & Cattell, R. B. *The ego: Its structure, action and measurement.* In press.

Blacker, C. P. *The chances of morbid inheritance.* London: H. K. Lewis and Co., Ltd., 1934.

Blakeslee, A. F., & Fox, A. L. Our different taste worlds. *Journal of Heredity,* 1932, *23,* 96–110.

Bleuler, M. The delimitation of influences of environment and heredity on mental disposition. *Character and Personality,* 1933, *1,* 286–300.

Blewett, D. B. An experimental study of the inheritance of intelligence. *Journal of Mental Science,* 1954, *100,* 922–933.

Böök, J. A. A genetic and neurophychiatric investigation of a North-Swedish population. *Acta Genetica Statistical Medicine,* 1953, *4,* 345.

Brackbill, Y., Kane, J., Maniello, R. L., & Abramson, D. Obstetrical premedication and infant outcome. *American Journal of Obstetrical Gynecology.* 1974, *118,* 377–384.

Bramwell, B. S. Observations on racial characteristics in England. *Eugenic Reform,* 1924, *15,* 556–571.

Brennan, J. *A test of confactor rotation on orthogonal and oblique cases.* Unpublished doctoral dissertation, University of Hawaii, 1977.

Broadhurst, P. S. Application of biometrical genetics to behavior in rats. *Nature,* 1959, *184,* 1517–1518.

Broadhurst, P. L. The biometrical analysis of behavioural inheritance. *Scientific progress, Oxford,* 1967, *55,* 123–139.

Broadhurst, P. L. New lights on behavioural inheritance. *Bulletin of the British Psychological Society,* 1971, *24,* 1–8.

Broadhurst, P. L. Methods of psychogenetic analysis. In *International Symposium: Finnish Foundation for Alcohol Studies,* 1972, *20,* 97–103.

Broadhurst, P. L., & Jinks, J. L. What genetical architecture can tell us about the natural selection of behavioural traits. *Genetics of Behaviour,* 1974, 2, 43–63.

Broadhurst, P. L., & Jinks, J. L. Psychological genetics from the study of animal behavior. In R. B. Cattell & R. M. Dreger (Eds.), *Handbook of modern personality theory.* New York: Wiley, 1977.

Bulmer, M. G. *The biology of twinning in man.* Oxford: Clarendon, 1970.

Bulmer, M. G. *Genetics of quantitative variability.* New York: Oxford University Press, 1975.

Burke, W. E., Tuttle, W. W., Thompson, C. W., Janney, C. D., & Weber, R. J. The relation of grip strength and grip endurance to age. *Journal of Applied Psychology,* 1953, *5,* 628–630.

Burks, B. S. The relative influence of nature and nurture upon mental development: A comparative study of foster parent–foster child resemblance and true parent–true child resemblance. *Nature and nurture: Their influence upon intelligence, Twenty-Seventh Yearbook of the National Society for the Study of Education,* 1928, Part 1, 219–316.

Burks, B. S., & Roe, A. Studies of identical twins reared apart. *Psychological Monographs,* 1949, *63,* No. 5.

Buros, O. K. *Seventh mental measurement yearbook.* Highland Park, N.J.: Gryphon Press, 1972.

Burt, C. intelligence and heredity: Some common misconceptions. *Irish Journal of Education*, 1969, *3*, 75–94.

Burt, C., & Howard, M. The multifactorial theory of inheritance and its application to intelligence. *British Journal of Statistical Psychology*, 1956, *9*, 95–131.

Burt, C., & Williams, E. L. The influence of motivation on the results of intelligence tests. *British Journal of Statistical Psychology*, 1962, *15*, 127–136.

Buss, A. R. A recursive–nonrecursive factor model and developmental causal networks. *Human Development*, 1974, *17*, 139–151.

Buss, A. H., & Plomin, R. A. *Temperament theory of personality development*. New York: Wiley, 1975.

Buss, A. R., & Poley, W. *Individual differences: Traits and factors*. New York: Gardner Press, 1976.

Butcher, H. J. *Human intelligence: Its nature and assessment*. London: Methuen, 1968.

Campbell, J. F. *Natural classification in science*. Unpublished master's thesis, Wichita State University, 1979.

Canter, S. *Personality traits in twins*. Paper presented at the annual meeting of British Psychological Society, 1969. (See results in Mittler, 1971, and Claridge, Canter, & Hume, 1973.)

Carey, G., Goldsmith, H. H., Tellegen, A., & Gottesman, I. I. Genetics and personality inventories: The limits of replication with twin data. *Behavior Genetics*, 1978, *8*, 299–313.

Carlier, M. La methode des adoptions dans l'analyse genetique des psychoses schizophreniques. *Psychiatrie de l'enfant*, 1980, 23, 4.

Carter, H. D. Family resemblances in verbal and numerical abilities. *Genetic Psychology Monographs*, 1932, *12*, 1–104.

Carter-Saltzman, L., & Scarr, S. MZ or DZ? Only your blood grouping laboratory knows for sure. *Behavior Genetics*, 1977, *7*, 273–280.

Cartwright, D. S., Tomson, B., & Schwartz, H. *Gang delinquency*. Monterey, Calif.: Brooks-Cole, 1975.

Cattell, A. K. S., Krug, S. E., & Schumacher, G. Sekundare Persönlichkeitsfaktoren im Deutschen HSPQ und ihr Gebrauchswert fur die Daignose, fur interkuturelle Vergleich, fur eine empirische Uberprufung tiefen-psychologisches Modellvorstellungen sowie fur die Konstruktvalidat des HSPQ. *Praxis der Kinderpsychologie und Kinderpsychiatrie*, 1980, *2*, 47–52.

Cattell, H. E. P. *Personality resemblances of parents and children on measured primary and secondary personality factors*. Unpublished master's thesis, Michigan State University, 1980.

Cattell, R. B. *The fight for our national intelligence*. London: King, 1937.

Cattell, R. B. The concept of social status. *Journal of Social Psychology*, 1942, *15*, 293–308.

Cattell, R. B. The cultural functions of social stratification I: Regarding the genetic bases of society. *Journal of Social Psychology*, 1945, *21*, 3–23.

Cattell, R. B. *Description and measurement of personality*. New York: World Book Co., 1946.

Cattell, R. B. Ethics and the social sciences. *American Psychologist*, 1948, *3*, 193–198.

Cattell, R. B. The dimensions of culture patterns by factorization of national characters. *Journal of Abnormal and Social Psychology*, 1949, *44*, 279–289.

Cattell, R. B. *Personality: A systematic theoretical and factual study*. New York: McGraw-Hill, 1950.

Cattell, R. B. Research designs in psychological genetics, with special reference to the multiple variance method. *American Journal of Human Genetics*, 1953, *5*, 76–93.

Cattell, R. B. Formulae and table for obtaining validities and reliabilities of extended factor scales. *Educational and Psychological Measurement*, 1957, *17*, 491–498.

Cattell, R. B. The multiple abstract variance analysis equations and solutions: For nature–nurture research on continuous variables. *Psychological Review*, 1960, *67*, 353–377.

Cattell, R. B. Formulating the environmental situation and its perception in behavior theory. In S. B. Sells (Ed.), *Stimulus determinants of behavior*. New York: Ronald, 1963. (a)

Cattell, R. B. The interaction of hereditary and environmental influences. *British Journal of Statistical Psychology*, 1963, *16*, 191–210. (b)

Cattell, R. B. Beyond validity and reliability: Some further concepts and coefficients for evaluating tests. *Journal of Experimental Education*, 1964, *33*, 133–143. (a)

Cattell, R. B. Objective personality tests. A reply to Dr. Eysenck. *Occupational Psychology*, 1964, *38*, 69–86. (b)

Cattell, R. B. The parental early repressiveness hypothesis for the authoritarian personality factor, U.I.28. *Journal of Genetic Psychology*, 1964, *106*, 333–349. (c)

Cattell, R. B. Higher order factor structures and reticular-vs-hierarchical for their interpretation. In Banks, C. & P. L. Broadhurst (Eds.), *Studies in Psychology*, London. University of London Press, 1965. (a)

Cattell, R. B. Methodological and conceptual advances in evaluating hereditary and environmental influences and their interaction. In S. G. Vandenberg (Ed.), *Methods and goals in human behavior genetics*. New York: Academic Press, 1965. (b)

Cattell, R. B. *Handbook of multivariate experimental psychology*. Chicago: Rand McNally, 1966.

Cattell, R. B. Comparing factor trait and state scores across ages and cultures. *Journal of Gerontology*, 1969, *24*, 348–360. (a)

Cattell, R. B. The diagnosis of schizophrenia by questionnaires and objective personality tests. In D. Siva Sankar (Ed.), *Schizophrenia: Current concepts and research*. Hicksville, N.Y.: PJD Publications, 1969. (b)

Cattell, R. B. The isopodic and equipotent principles for comparing factor scores across different populations. *British Journal of Mathematical and Statistical Psychology*, 1970, *23*, 23–41. (a)

Cattell, R. B. Separating endogenous, exogenous, ecogenic and epogenic component curves in developmental data. *Developmental Psychology*, 1970, *3*, 151–162. (b)

Cattell, R. B. *Abilities: Their structure, growth and action*. Boston: Houghton Mifflin, 1971. (a)

Cattell, R. B. Estimating modulator indices and state liabilities. *Multivariate Behavioral Research*, 1971, *6*, 7–33. (b)

Cattell, R. B. The nature and genesis of mood states: A theoretical model with experimental measurements concerning anxiety, arousal, and other mood states. In C. D. Spielberger (Ed.), *Anxiety: Current trends in theory and research* (Vol. 1). New York: Academic Press, 1972. (a)

Cattell, R. B. *A new morality from science*. New York: Pergamon Press, 1972. (b)

Cattell, R. B. The 16 P.F. and basic personality structure: A reply to Eysenck. *Journal of Behavioral Science*, 1972, *1*, 169–187. (c)

Cattell, R. B. *Personality and mood by questionnaire*. San Francisco: Jossey Bass, 1973. (a)

Cattell, R. B. Unravelling maturational and learning developments by the comparative MAVA and structured learning approaches. In J. R. Nesselroade & J. Reese (Eds.), *Life Span Developmental Psychology*. New York: Academic Press, 1973. (b)

Cattell, R. B. How good is the modern questionnaire? General principles for evaluation. *Journal of Personality Assessment*, 1974, *38*, 115–129.

Cattell, R. B. Personality and culture: General concepts and methodological problems. In R. B. Cattell & R. M. Dreger (Eds.), *Handbook of modern personality theory*. Washington, D.C.: Hemisphere, and New York: Halsted, 1976.

Cattell, R. B. Lernfahigkeit, Persönlichkeitstruktur und die Theorie des Strukturieten Lernens. In G. Nissen (Ed.), *Intelligentz, Lernen und Lernstorungen*. Berlin. Springer Verlag, 1977. (a)

Cattell, R. B. Structured learning theory applied to personality change. In R. B. Cattell & R. M. Dreger (Eds.), *Handbook of modern personality theory*. New York: Wiley, 1977. (b)

Cattell, R. B. Adolescent age trends in primary personality factors measured in T-data: A contribution to use of standardized measures in practice. *Journal of Adolescence*, 1978, *1*, 1–16. (a)

Cattell, R. B. *The scientific use of factor analysis in behavioral and life sciences.* New York: Plenum, 1978. (b)

Cattell, R. B. Are culture fair intelligence tests possible and necessary? *Journal of Research and Development in Education,* 1979, *12,* 3–13. (a)

Cattell, R. B. *Personality and learning theory, Vol. 1: Structure of personality in its environment.* New York: Springer, 1979. (b)

Cattell, R. B. *Personality and learning theory, Vol. 2: A systems theory of maturation and structural learning.* New York: Springer, 1980. (a)

Cattell, R. B. The heritability of fluid, g_f, and crystallized, g_c, intelligence, estimated by a least squares use of the MAVA method. *British Journal of Educational Psychology,* 1980, *50,* 253–265. (b)

Cattell, R. B. The determiners of the genothreptic correlation of heredity—environment interaction within and between families. In press. (a)

Cattell, R. B. The relations among the Phenotypic r_{pa}, Genetic r_{ga}, and Threptic r_{ta} assortive mating correlations of husbands and wives. In press. (b)

Cattell, R. B. From attribution theory to spectrad theory via the general perceptual model. *Multivariate Behavioral Research.* In press. (c)

Cattell, R. B., & Bartlett, H. W. An R-dR technique operational distinction of the states of anxiety, stress, fear, etc. *Australian Journal of Psychology,* 1971, *23,* 105–123.

Cattell, R. B., & Birkett, H. The known personality factors aligned between first order T-data and second order Q factors: Inhibitory control, independence, cortertia and exvia. *Personality and Individual Differences,* 1980, *1,* 229–238.

Cattell, R. B., Blewett, D. B., & Beloff, J. R. The inheritance of personality: A multiple-variance analysis determination of approximate nature–nurture ratios for primary personality factors in Q-data. *American Journal of Human Genetics,* 1955, *7,* 123–146.

Cattell, R. B., Bolz, C. R., & Korth, B. Behavioral types in pure bred dogs objectively determined by Taxonome. *Behavioral Genetics,* 1973, *3,* 205–216.

Cattell, R. B., Brackenridge, C. J., Case, J., Ropert, D. N., & Sheehy, A. J. The relation of blood types to primary and secondary personality traits. *Mankind Quarterly,* 1980, *21,* 8–20.

Cattell, R. B., & Brennan, J. Population intelligence and national syntality dimensions. *Mankind Quarterly,* 1981, *22,* 1–17.

Cattell, R. B., & Brennan, J. The culture pattern groupings of modern nations by objective statistical procedures. In press.

Cattell, R. B., & Brennan, J. M. Fluid and crystallized intelligence in the aged: A re-interpretation. *Developmental Psychology.* In press.

Cattell, R. B., Bruel, H., & Hartman, H. P. An attempt at more refined definition of the cultural dimensions of syntality in modern nations. *American Sociological Review,* 1952, *17,* 408–421.

Cattell, R. B., & Butcher, J. *The prediction of achievement and creativity.* New York: Bobbs Merrill, 1968.

Cattell, R. B., & Cattell, A. K. S. *The Culture Fair Test of Intelligence (and Manual)* (3d ed.). Champaign, Ill.: IPAT,

Cattell, R. B., & Cattell, M. D. *Handbook for the Jr.–Sr. High School personality questionnaire.* Champaign, Ill.: IPAT, 1969.

Cattell, R. B. & Child, D. *Motivation and dynamic structure.* London: Holt, Rinehart and Winston, 1975.

Cattell, R. B., Coulter, M. A., & Tsujioka, B. The taxonomic recognition of types and functional emergents. In R. B. Cattell (Ed.), *Handbook of Multivariate Experimental Psychology.* Chicago: Rand McNally, 1966.

Cattell, R. B., & Cross, K. P. Comparison of the ergic and self-sentiment structures found in dynamic traits by R- and P-techniques. *Journal of Personality,* 1952, *21,* 250–271.

Cattell, R. B., & Dickman, K. A dynamic model of physical influences demonstrating the necessity of oblique simple structure. *Psychological Bulletin,* 1962, *59,* 389–400.

Cattell, R. B., Eber, H. W., & Tatsuoka, M. *Handbook for the sixteen personality factor questionnaire.* Champaign, Ill.: IPAT, 1970.

Cattell, R. B., Graham, R. K., & Woliver, R. E. A reassessment of the factorial cultural dimensions of modern nations. *Journal of Social Psychology,* 1979, *108,* 241–258.

Cattell, R. B., & Horn, J. L. A check on the theory of fluid and crystallized intelligence with description of new subtest designs. *Journal of Educational Measurement,* 1978, *15,* 139–164.

Cattell, R. B., Horn, J. L., Sweeny, A. B., & Radcliffe, J. A. *The motivation analysis test.* Champaign, Ill.: IPAT, 1964.

Cattell, R. B., Kameoka, V., Klein, T. W., & Schuerger, J. M. The heritability of the clinically relevant personality factors: Guilt proneness, O; parmia, H; and ergic tension, Q_4, by questionnaire and the MAVA and OSES methods. *Indian International Journal of Psychology.* In press.

Cattell, R. B., Kawash, G. F., & De Young, G. E. Validation of objective measures of ergic tension: Response of the sex erg to visual stimulation. *Journal of Experimental Research in Personality,* 1972, *6,* 76–83.

Cattell, R. B., Klein, T. W., & Schuerger, J. M. Heritabilities by the MAVA method of objective test measures of personality traits U.I. 23, capacity to mobilize, U.I. 24, anxiety, U.I. 26, narcissistic ego, U.I. 28, asthenia. In press.

Cattell, R. B., & Malteno, V. Contributions concerning mental inheritance II: Temperament. *Journal of Genetic Psychology,* 1940, *57,* 31–47.

Cattell, R. B., & Nesselroade, J. R. Likeness and completeness theories examined by the 16 P.F. measures on stably and unstably married couples. *Journal of Personality and Social Psychology,* 1967, *7,* 351–361.

Cattell, R. B., & Nesselroade, J. R. Note on analyzing personality relations in married couples. *Psychological Reports,* 1968, *22,* 381–382.

Cattell, R. B., Pierson, G., & Finkbeiner, C. Alignment of personality source trait factors from questionnaires and observer ratings: The theory of instrument-free patterns. *Multivariate Experimental Clinical Research,* 1976, *2,* 63–88.

Cattell, R. B., Price, P. L., & Patrick, S. V. Diagnosis of clinical depression on four source trait dimensions—U.I. 19, U.I. 20, U.I. 25, and U.I. 30—from the O-A Kit. *Journal of Clinical Psychology,* 1981, *37,* 4–11.

Cattell, R. B., Rao, D. C., Kameoka, V., & Brennan, J. M. The heritability of the clinically relevant personality factors: Guilt proneness, O, parmia, H, and ergic tension, Q_4, by questionnaire, the MAVA method, and maximum likelihood analysis. In press.

Cattell, R. B., Rao, D. C., Schmidt, L. K., & Vaughan, D. S. Heritability of personality source traits U.I. 21, exuberance, U.I. 25 realism, U.I. 32 extraversion, and U.I. 33 sanguine temperament by MAVA and maximum likelihood. *L'Année Psychologique.* In press.

Cattell, R. B., Rao, D. C., & Schuerger, J. M. Heritability in the personality control system: Ego strength (C), super ego strength (G) and the self-sentiment (Q_3); by the MAVA model, Q-data, and maximum likelihood analyses. *Journal of Clinical Psychology.* In press.

Cattell, R. B., Rao, D. C., Schuerger, J. M., & Klein, T. W. The heritability of dominance, E, surgency, F, premsia, I, and self-sufficiency, Q_2, by questionnaire measurement, MAVA and the maximum likelihood method. In press.

Cattell, R. B., Rao, D. C., Schuerger, J. M., & Vaughan, D. S. Source traits U.I. 16, ego assertion, U.I. 17, inhibition control, and U.I. 19, independence, analyzed for heritability by MAVA, on O-A Battery measures, by the maximum likelihood method. *Journal of Human Heredity.* In press.

Cattell, R. B., Rao, D. C., Vaughan, D. S., & Ahern, F. Heritability by the MAVA method and objective test measures of personality traits U.I. 23, capacity to mobilize, U.I. 24 anxiety, U.I. 26, narcissistic ego, and U.I. 28, asthenia, by maximum likelihood methods. *Behavior Genetics.* In press.

Cattell, R. B., Rickels, K., Weise, C., Gray, B., & Yee, R. The effects of psychotherapy upon measured anxiety and regression. *American Journal of Psychotherapy,* 1966, *20,* 261–269.

Cattell, R. B., & Scheier, J. H. *The meaning and measurement of anxiety and neurosis.* New York: Ronald, 1961.

Cattell, R. B., Schmidt, L. R., & Bjersted, A. Clinical diagnosis by the objective-analytic personality batteries. *Journal of Clinical Psychology Monograph Supplements,* 1972, *34,* 239–312.

Cattell, R. B., Schmidt, L. R., Klein, T. W., Schuerger, J. M., & Vaughan D. S. Heritabilities by the MAVA method of objective test measures of personality traits U.I. 23, mobilization vs. regression, U.I. 24, anxiety, U.I. 26, narcistic ego and U.I. 28 asthenia. In press.

Cattell, R. B., Schmidt, L. R., Klein, T. W. & Schuerger, J. M. Anlage—und Umweltkomponenten von Persönlichkeitsfaktoren (U.I. 21, 25, 32 und 33)—ermittelt mit der MAVA methode. *J. für Differentielle und Diagnostische Psychologie,* 1980, *1,* 275–288.

Cattell, R. B., Schmidt, L. R., & Pawlik, K. Cross cultural comparison (U.S.A., Japan and Austria) of the personality structure of 10–14 year olds in objective tests. *Social Behavior and Personality,* 1973, *1,* 182–211.

Cattell, R. B., & Schuerger, J. M. *Personality theory in action: Handbook for the Objective-Analytic (O-A) Test Kit.* Champaign, Ill.: IPAT, 1978.

Cattell, R. B., Schuerger, J. M., Klein, T. W., & Ahern, F. M. The degree of inheritance of the control triumvirate: Ego strength, C; superego strength, G; and the self-sentiment, Q_3 by MAVA. *Journal of Clinical and Consulting Psychology.* In press. (a)

Cattell, R. B., Schuerger, J. M., & Klein, T. W. Heritabilities of source traits U.I. 16, ego standards; U.I. 17, inhibition control; and U.I. 19, independence, measured by the O-A Batteries and analyzed by MAVA. *Journal of Personality and Social Psychology.* In press. (c)

Cattell, R. B., Schuerger, J. M., & Klein, T. W. The heritability of dominance, E, surgency, F, premsia, I, and self-sufficiency, Q_2, by questionnaire measurement and the MAVA method. *Multivariate Experimental Clinical Psychology.* In press. (d)

Cattell, R. B., Schuerger, J. M., Klein T. W., & Finkbeiner, C. A definitive large sample factoring of personality structures in objective measures, as a basis for the high school objective analytic battery. *Journal of Research in Personality,* 1976, *10,* 22–41.

Cattell, R. B., Schuerger, J. M., Klein, T. W, & Kameoka, V. The heritability of fluid and crystallized general intelligence. *Australian Journal of Psychology,* 1980.

Cattell, R. B., & Stice, G. F. *The dimensions of groups and their relations to the behavior of members.* Champaign, Ill.: IPAT, 1960.

Cattell, R. B., Stice, G. F., & Kristy, N. T. A first approximation to nature–nurture ratios for eleven primary personality factors in objective tests. *Journal of Abnormal and Social Psychology,* 1957, *54,* 143–159.

Cattell, R. B., & Sullivan, W. The scientific nature of factors: A demonstration by cups of coffee. *Behavioral Science,* 1962, *7,* 184–193.

Cattell, R. B., & Vogelmann, S. Second-order personality factors in combined questionnaire and rating data. *Multivariate Experimental Clinical Research,* 1976, *3,* 40–64.

Cattell, R. B., & Warburton, F. W. Objective personality and motivation tests: A theoretical introduction to practical compendium. Champaign, Ill.: University of Illinois Press, 1967.

Cattell, R. B., & Willson, J. L. Contributions concerning mental inheritance I: Of intelligence. *British Journal of Educational Psychology,* 1938, *8,* 129–149.

Cattell, R. B., Woliver, R. E., & Graham. R. K. The relations of syntality dimensions of modern national cultures to the personality dimensions of their populations. *International Journal of Intercultural Relations.* 1980, *4,* 15–41.

Cattell, R. B., Young, H. B., & Hundleby, J. D. Blood groups and personality traits. *American Journal of Human Genetics,* 1964, *16,* 397–402.

Cavalli-Sforza, L. L., & Bodmer, W. F. *The genetics of human populations.* San Francisco: W. H. Freeman, 1971.

Chalmers, E. M. Personal communication, 1980.

Claridge, G., Canter, S., & Hume, W. J. *Personality differences and biological variations: A study of twins.* New York: Pergamon Press, 1973.

Clarke, C. A. Blood groups and disease. In A. G. Steinberg (Ed.), *Progress in medical genetics, Vol. 1.* New York: Grune and Stratton, 1961.

Cloninger, C. R. Interpretation of causation and association by path analysis (assortative mating/multifactorial inheritance). In press.

Cloninger, C. R., Rice, J., & Reich, T. Multifactorial inheritance with cultural transmission and assortive mating, II: A general model of combined polygenic and cultural inheritance. *American Journal of Human Genetics,* 1978, *30,* 10–44.

Cockerham, C. C. An extension of the concept of partitioning hereditary variance for analysis of covariance among relatives when epistasis is present. *Genetics,* 1954, *39,* 859–882.

Cockerham, C. C. Effects of linkage on the covariance between relatives. *Genetics,* 1956, *41,* 138–141.

Cohen, J. The factorial structure of the WAIS between early childhood and old age. *Journal of Consulting Psychology,* 1957, *21,* 283–290.

Conrad, H. S., & Jones, H. E. A second study of familial resemblance in intelligence: Environmental and genetic implications of parent–child and sibling correlations in the total sample. *39th Yearbook of the National Society for the Study of Education,* pt. 2, pp. 97–141. Bloomington: Public School Publishing Co., 1940.

Cooke, F., Finney, G. H., Rockwell, R. F. Assortative mating in lesser snow geese (*Anser caerulesceus*). *Behavior Genetics,* 1976, *6,* 127–143.

Coon, C. S. What is race? *The Atlantic Monthly,* 1958.

Coon, C. S. *The origin of races.* London: Cape, 1963.

Cronbach, L. J. How can instruction be adapted to individual differences? In R. M. Gagné (Ed.), *Learning and individual differences.* Columbus, Ohio: Merrill, 1967.

Cronbach, L. J., & Furby, L. How should we measure change—or should we? *Psychological Bulletin,* 1970, *74,* 68–80.

Cronbach, L. J., & Snow, R. E. *Aptitudes and instruction methods* New York: Irvington, 1977.

Crook, M. W. Intra-family relationship in personality test performance. *Psychological Record,* 1937, *1,* 479–502.

Crow, J. F., & Felsenstein, J. The effect of assortative mating on the genetic composition of a population. *Eugenics Quarterly,* 1968, *15,* 85–97.

Crow, J. F., & Kimura, K. *An introduction to population genetic theory.* New York: Harper, 1970.

Curran, J. P., & Cattell, R. B. *Handbook for the 8 state battery.* Champaign, Ill.: IPAT, 1976.

Daniels, H. E. The estimation of components of variance. *Journal of the Royal Statistical Society,* 1939, *6,* 186–197.

Darlington, C. D. *The evolution of man and society.* New York: Simon & Schuster, 1969.

DeFries, J. C. *The heritable nature of group differences as a function of the within-group heritability.* Paper presented at the annual convention of the American Psychological Association, Symposium on Human Behavior Genetics, Honolulu, Hawaii, Sept. 4, 1972.

DeFries, J. C., Johnson, R. C., Kuse, A. R., McClearn, G. E., Polovina, J., Vandenberg, S. G., & Wilson, J. R. Family resemblance for specific cognitive abilities. *Behavior Genetics,* 1979, *9,* 23–43.

DeFries, J. C., Weir, M. W., & Hegmann, J. P. Differential effects of prenatal maternal stress on offspring behavior in mice as a function of genotype and stress. *Journal of Comparative and Physiological Psychology,* 1967, *63,* 332–334.

De Young, G. E. Standard of decision regarding personality factors in questionnaires. *Canadian Journal of Behavioral Science,* 1972, *4,* 253–255.

De Young, G. E., & Fleischer, B. Motivational and personality trait relationships in mate selection. *Behavior Genetics,* 1976, *6,* 1–6.

Dixon, M. E., & Johnson, R. *The roots of individuality.* Belmont, Calif.: Brooks-Cole, 1980.

Dobzhansky, T. *Mankind evolving: The evolution of the human species.* New Haven: Yale University Press, 1962.

Dreger, R. M., Development structural changes in the child's personality. In R. B. Cattell & R. M. Dreger (Eds.), *Handbook of modern personality theory.* New York: Wiley, 1977.

Dugdale, R. L. *The Jukes.* New York: Putnam, 1877.

Duncan, O. D. Path analysis: Sociological examples. *American Journal of Sociology,* 1966, *72,* 1–16.

Dunn, L. C. *Cross currents in the history of human genetics.* Presidential address presented at the meeting of the American Society of Human Genetics, Atlantic City, New Jersey, May 3, 1961.

Eaves, L. J. The genetic analysis of continuous variation: A comparison of experimental designs applicable to human data. *British Journal of Mathematical and Statistical Psychology,* 1969, *22,* 131–147.

Eaves, L. J. Computer simulation of sample size and experimental design in human psychogenetics. *Psychological Bulletin,* 1972, *77,* 144–152.

Eaves, L. J. Assortative mating and intelligence: An analysis of pedigree data. *Heredity,* 1973, *30,* 199–210.

Eaves, L. J. The effect of cultural transmission on continuous variation. *Heredity,* 1976, *37,* 41–57.

Eaves, L. J., & Eysenck, H. Genotype and age interaction for neuroticism. *Behavior Genetics,* 1976, *6,* 359–362.

Eaves, L. J., & Gale, J. S. A method for analysing the genetic basis of covariation. *Behavior Genetics,* 1974, *4,* 253–267.

Eaves, L. J., Last, K. A., Young, P. A., & Marten, N.Y. Model fitting approaches to the analysis of human behavior. *Heredity,* 1978, *41,* 249–320.

Elasser, G. *Die Nachkommen geisterkranker Eltenpaare.* Stuttgart: Thieme, 1952.

Ellis, H. *A study of British genius.* Boston: Hougton Mifflin, 1926.

Ellis, H. *Man and woman: A study of secondary sex characteristics.* Boston: Houghton Mifflin, 1929.

Elston, R. C., & Gottesman, I. I. The analysis of quantitative inheritance simultaneously from twin and family data. *American Journal of Human Genetics,* 1968, *20,* 512–521.

Erlenmeyer-Kimmling, L. Studies on the offspring of two schizophrenic parents. In D. Rosenthal & S. S. Kety (Eds.), *The transmission of schizophrenia.* New York: Pergamon Press, 1968.

Erlenmeyer-Kimmling, L., & Jarvik, L. F. Genetics and intelligence: A review. *Science,* 1963, *142,* 1477–1479.

Eysenck, H. J. The inheritance of intraversion-extraversion. *Acta Psychology,* 1956, *12,* 95–110.

Eysenck, H. J. *Manual of the Mandsley Personality Inventory.* London: University of London Press, 1959.

Eysenck, H. J. *The structure of human personality* (2nd ed.). London: Methuen, 1965.

Eysenck, H. J. *The biological basis of personality.* Springfield, Ill.: Thomas, 1967.

Eysenck, H. J. *Crime and Personality* (2nd ed.). Paladin, 1970.

Eysenck, H. J. *The I.Q. argument.* New York: The Library Press, 1971.

Eysenck, H. J. *The inequality of man.* London: Temple Smith, 1973.

Eysenck, H. J. National differences of personality as related to the blood group polymorphism. *Psychological Reports,* 1977, *41,* 1251–1258.

Eysenck, H. J., & Prell, D. B. The inheritance of neuroticism. *Journal of Mental Science,* 1951, *97,* 441–465.

Eysenck, S. B. G., & Eysenck, H. J. The measurement of psychoticism: A study of factor stability and reliability. *British Journal of Social and Clinical Psychology,* 1968, *7,* 286–294.

Falconer, D. S. *Introduction to quantitative genetics.* New York: Ronald Press, 1960.

Falconer, D. S. The inheritance of liability to certain diseases, estimated from the incidence among relatives. *Annals of Human Genetics,* 1965, *29,* 51–71.

Falconer, D. S. The inheritance of liability to diseases with variable age of onset, with particular reference to diabetes mellitus. *Annals of Human Genetics,* 1967, *31,* 1–20.

Ferguson, G. A. On transfer and the abilities of man. *Canadian Journal of Psychology,* 1956, *10,* 121–131.

Feuer, G., & Broadhurst, P. L. Thyroid function in rats selectively bred for emotional elimination. *Journal of Endocrinology*, 1962, *24*, 385–396.

Finch, F. H. A study of the relation of age interval to degree of resemblance of siblings in intelligence. *Pedagogical Summary and Journal of Genetic Psychology*, 1933, *43*, 389–403.

Fischer, M. *Schizophrenia in twins*. Paper presented at the Fourth World Congress of Psychiatry, Madrid, 1966.

Fisher, R. A. The correlation between relatives on the supposition of Mendelian inheritance. *Transaction of the Royal Society (Edinburgh)*, 1918, *52*, 399–433.

Fischer, R. A. *The genetical theory of natural selection*. Oxford: Clarendon, 1930.

Fischer, R. A. The coefficient of racial likeness. *Journal of the Royal Anthropological Society*, 1936, *66*, 57. (a)

Fischer, R. A. Heterogeneity of linkage data for Friedreich's ataxia and the spontaneous autigens. *Annual of Eugenics*, 1936, *7*, 1–21. (b)

Floderus-Myred, B., Pedersen, N., & Rasmuson, I. Assessment of heritability for personality, based on a short form of the Eysenck Personality Inventory. *Behavior Genetics*, 1980, *10*, 153–161.

Fonseca, A. F. *Da analise heredoclinica das pertubacoes afectivas atraves de 60 pares de gemeos*. Oporto: Facultade de Medicina, 1959.

Freedman, D. G., & Freedman, N. C. Behavioural differences between Chinese-American and European-American newborns. *Nature*, 1969, *224*, 1227.

Freeman, F. N., Holzinger, K. J., & Mitchell, B. C. The influence of environment on the intelligence, school achievement, and conduct of foster children. *27th Yearbook of the National Society for the Study of Education*, pt. 1, pp. 103–217. Bloomington: Public School Publishing Co., 1928.

French, J. W. The description of aptitude and achievement tests in terms of rotated factors. Chicago: University of Chicago Press, 1951.

Freud, S. *The ego and the id*. Trans. by J. Riviere. Collected Papers. London: Hogarth, 1924.

Fulker, D. W. Mating speed in male *Drosophila melanogaster*: A psychogenetic analysis. *Science*, 1966, *153*, 203–205.

Fulker, D. W. A biometrical genetic approach to intelligence and schizophrenia. *Social Biology*, 1973, *20*, 266–275.

Fulker, D. W. Review of "The science and politics of I.Q." *American Journal of Psychology*, 1975, *88*, 505–519.

Fulker, D. W. *A unified approach to behavior genetic analysis*. Paper presented at the 9th Annual Meeting of the Behavior Genetics Association, 1979.

Fulker, D. W., Wilcock, J., & Broadhurst, P. L. Studies in genotype–environment interaction, I: Methodology and preliminary multivariate analysis of a diallel cross of eight strains of rat. *Behavior Genetics*, 1972, *2*, 261–287.

Fuller, J. L. Suggestions from animal studies for human behavior genetics. In S. D. Vandenberg (Ed.), *Methods and goals in human behavior genetics*. New York: Academic Press, 1965.

Fuller, J. L., & Thompson, W. R. *Foundations of behavior genetics*. St. Louis: Mosby & Co., 1979.

Galton, F. *Hereditary genius: An inquiry into its laws and consequences*. London: MacMillan, 1969.

Galton, F. *Natural inheritance*. New York: MacMillan, 1894.

Gary, A. L., & Glover, J. *Eye color, sex and children's behavior*. Chicago: Nelson Hall, 1976.

Gellis, S. Brain size, grey matter and race—factor or fiction? *Mankind Quarterly*, 1977, *17*, 243–282.

Gershon, E. S., Bunney, W. E., Leckman, J. F., Erdeweg, M. W., & Debauche, B. A. The inheritance of affective disorders: A review of data and hypotheses. *Behavior Genetics*, 1976, *6*, 227–262.

Getzels, J. W., & Jackson, P. W. *Creativity and intelligence: Explorations with gifted students*. New York: Wiley, 1962.

Glass, G. V. Educational Piltdown man. *Phi Delta Kappan,* Nov. 1968, 148–151.

Goddard, H. H. *The Kallikak family.* New York: Macmillan, 1912.

Goldberg, L. R. The comparative validity of questionnaire data (16 P.F. scales) and objective test data (O-A battery) in predicting five peer-rating criteria: A brief report. *ORI Research Bulletin,* 1972, *12*(1).

Goldberger, A. S. Pitfalls in the resolution of intelligence inheritance. In N. E. Morton & C. S. Chang (Eds.), *Genetic epidemiology.* New York: Academic Press, 1978. Pp. 30–45.

Gorsuch, R. L. *Factor analysis.* Philadelphia: Saunders, 1974.

Gottesman, I. I. Heritability of personality: A demonstration. *Psychological Monographs,* 1963, *77,* 1–47.

Gottesman, I. I. Severity-concordance and diagnostic refinement in the Maudsley–Bethlem schizophrenic twin study. *Journal of Psychiatric Research,* 1964, *6,* 37–48.

Gottesman, I. I. Personality and natural selection. In S. G. Vandenberg (Ed.), *Methods and goals in human behavior genetics.* New York: Academic Press, 1965.

Gottesman, I. I. Schizophrenia and genetics. In R. Wynne, Cromwell & G. E. Matthysse, *Schizophrenia and genetics.* New York: Academic Press, 1978.

Gottesman, I. I., & Shields, J. Schizophrenia in twins: 16 years' consecutive admissions to a psychiatric clinic. *Diseases of the Nervous System,* 1966, *27,* 11–19.

Gottesman, I. I., & Shields, J. A polygenic theory of schizophrenia. *Proceedings of the National Academy of Sciences,* 1967, *58,* 199–205.

Goudenough, D. R., Gandini, E., Olkin, I., Pizzamiglio, L., Thayer, D., & Witkin, H. A. A study of X chromosome linkage with field dependence and spatial visualization. *Behavior Genetics,* 1977, *7,* 373–387.

Guilford, J. P. *The nature of human intelligence.* New York: McGraw-Hill, 1967.

Gulbrandsen, C. L., Morton, N. E., Rao, D. C., Yee, S., *et al.* Behavioral, social, and physiological determinants of lipo protein concentrations. *Social Biology,* 1979, *26,* 168–190.

Guttman, R. Parent–offspring correlations in the judgment of visual number. *Human Heredity,* 1970. *20,* 57–65.

Guttman, R., & Shoham, I. Intrafamilial invariance and parent–offspring resemblance in spatial abilities. *Behavior Genetics,* 1979, *9,* 367–378.

Haggard, E. A. *Intraclass correlation and the analysis of variance.* New York: Dryden, 1958.

Hakstian, A. R., & Cattell, R. B. The checking of primary ability structure on a broader basis of performances. *British Journal of Educational Psychology,* 1974, *44,* 140–154.

Hakstian, A. R., & Cattell, R. B. An examination of adolescent sex differences in some ability and personality traits. *Canadian Journal of Behavioral Science,* 1975, *7,* 295–312.

Hakstian, A. R., & Cattell, R. B. *The Comprehensive Ability Battery.* Champaign, Ill.: IPAT, 1976.

Hakstian, A. R., & Cattell, R. B. Higher stratum ability structures on a basis of 20 primary abilities. *Journal of Educational Psychology,* 1978, *70,* 657–660.

Haldane, J. B. S. The combination of linkage values and the calculation of distance between the loci of linked factors. *Journal of Genetics,* 1919, *8,* 299–309.

Haldane, J. B. S. *The causes of evolution.* New York: Harper, 1932.

Haldane, J. B. S. Some theoretical results of continued brother–sister mating. *Genetics,* 1937, *34,* 265–274.

Haldane, J. B. S. The estimation of the frequencies of recessive conditions in man. *Annual of Eugenics,* 1938, *8,* 255–262.

Haldane, J. B. S. A defense of beanbag genetics. *Perspectives in Biological Medicine.* 1964, *7,* 353.

Haldane, J. B. S. The implications of genetics for human society. In S. J. Geortz (Ed.), *Genetics today: Proceedings of the XIth International Congress of Genetics.* New York: Pergamon Press, 1965.

Hall, C. S. The inheritance of emotionality. *Sigma Xi Quarterly,* 1938, *29,* 17–27.

Halperin, S. L. Human heredity and mental deficiency. *American Journal of Mental Deficiency,* 1946, *51,* 153–163.

Halperin, S. L., Rao, D. C., & Morton, N. E. A twin study of intelligence in Russia. *Behavior Genetics*, 1975, 5, 83–86.

Happy, R., & Collins, J. K. Melanin in the ascending reticular activating system and its possible relationships to autism. *Medical Journal of Australia*, 1972, 2, 1484–1486.

Hardin, G. *Promethean ethics.* Seattle: University of Washington Press, 1980.

Harman, H. H. *Modern factor analysis* (Rev. ed.). Chicago. University of Chicago Press, 1976.

Harrington, L. P. The relation of physiological and social indices of activity level. In Q. McNemar & M. A. Merrill, *Studies in Personality: In Honor of Lewis M. Terman*. Stanford: Stanford University Press, 1942.

Harrison, G. A. *Genetical variation in human populations.* New York: Pergamon Press, 1961.

Harvald, B., & Hauge, M. Hereditary factors elucidated by twin studies. In J. V. Neel, M. W. Shaw, & W. S. Schull (Eds.), *Genetics and the epidemiology of chronic diseases*. Washington, D.C.: U.S. Dept. H.E.W., 1965.

Hasseman, J. K., & Elston, R. C. *The estimation of genetic variance from twin data.* Unpublished manuscript, 1971.

Hegmann, J. P., & De Fries, J. C. Maximum variance linear combination from phenotypic, genetic, and environmental covariance matrices. *Multivariate Behavioral Research*, 1970, 5, 9–18.

Hendrickson, D. E., & Hendrickson, A. E. The biological basis of individual differences in intelligence. *Personality and Individual Differences*, 1980, 1, 3–33.

Herrnstein, R. J. *I.Q. in the meritocracy.* Boston: Little, Brown, 1973.

Heston, L. C. Psychiatric disorders in foster home reared children of schizophrenics. *British Journal of Psychiatry*, 1966, 112, 819–825.

Heston, L. C. The genetics of schizophrenia and schizoid disease. *Science*, 1970, 167, 249–256.

Higgins, J. V., Reed, E. W., & Reed, S. C. Intelligence and family size: A paradox resolved. *Eugenics Quarterly*, 1962, 9, 84–90.

Hirsch, J. (Ed.). *Behavior genetic analysis.* New York: McGraw-Hill, 1967.

Hogben, L. The correlation of relatives on the supposition of sex linked transmission. *Journal of Genetics*, 1932, 27, 379–406.

Holzinger, K. J. The relative effect of nature and nurture on twin differences. *Journal of Educational Psychology*, 1929, 20, 241–248.

Honzik, M. P. Developmental studies of parent–child resemblance in intelligence. 1957, 2. 215–228.

Hooton, E. A. *Up from the ape.* New York: Macmillan, 1959.

Horn, J. L. State, trait and change dimensions of intelligence. *British Journal of Educational Psychology*, 1972, 42, 159–185. (a)

Horn, J. L. The structure of intellect: Primary abilities. In R. M. Dreger (Ed.), *Multivariate personality research*. Baton Rouge, La.: Claitor, 1972. (b)

Horn, J. L. Psychometric studies of aging and intelligence. In S. Gershan & A. Raskin (Eds.), *Aging, Vol. 2: Genesis and treatment of psychologic disorders in the elderly*. New York: Raven, 1975.

Horn, J. L. Human abilities, a review of research and theory in the early 1970's. *Annual Review of Psychology*, 1976, 27, 437–485.

Horn, J. L. Personality and ability theory. In R. B. Cattell & P. M. Dreger (Eds.), *Handbook of modern personality theory*. New York: Wiley, 1977.

Horn, J. L. The nature and development of intellectual abilities. In R. T. Osborne, C. E. Noble, & N. Weyl (Eds.), *Human variation: The biopsychology of age, race and sex*. New York: Academic Press, 1978.

Horn, J. L. Concepts of intelligence in relation to learning and adult development. *Intelligence*. In press.

Horn, J. L., & Cattell, R. B. Age differences in primary mental ability factors. *Journal of Gerontology*, 1966, 21, 210–220. (a)

Horn, J. L., & Cattell, R. B. Refinement and test of the theory of fluid and crystallized intelligence. *Journal of Educational Psychology*, 1966, *57*, 253–270. (b)

Horn, J. L., & Donaldson, G. On the myth of intellectual decline in adulthood. *American Psychologists*, 1976, *31*, 701–719.

Horn, J. L., & Knapp, J. R. Thirty wrongs don't make a right: A reply to Guilford. *Psychological Bulletin*, 1974, *81*, 502–504.

Horn, J. L., & Little, K. B. Isolating change and invariance in patterns of behavior. *Multivariate Behaviorial Research*, 1966, *1*, 219–228.

Horn, J. M., Loehlin, J. C., & Willerman, L. Intellectual resemblance among adoptive and biological relatives: The Texas Adoption Project. *Behavior Genetics*, 1979, *9*, 177–208.

Horn, J. L., & McArdle, J. J. Perspectives on mathematical-statistical model building. In L. W. Poon (Ed.), *Aging in the 1980's: Psychological issues*. Washington, D.C.: American Psychological Association, 1980.

Horn, J. L., & Stankow, L. *Second order abilities among auditory and visual primaries*. Unpublished manuscript, 1979.

Humphreys, L. G. Statistical definitions of validity for minority groups. *Journal of Applied Psychology*, 1973, *58*, 1–4.

Humphreys, L. G., & Taber, T. Ability factors as a function of advantaged and disadvantaged groups. *Journal of Educational Measurement*, 1973, *10*, 107–115.

Hundleby, J. D., Pawlik, K., & Cattell, R. B. *Personality factors in objective test devices*. San Diego: Knapp, 1965.

Huntington, E. *The character of races*. New York: Scribner, 1927.

Hurst, C. C. *Heredity and the ascent of man*. London: Cambridge University Press, 1935.

Husén, T. *Psychological twin research*. Stockholm: Almquist & Wiksell, 1959.

Huxley, L. R. Heredity-east and west; Sysenko and world science. New York. Schuman, 1949.

Huxley, L. R. Evolution: The modern synthesis. New York. Hafner, 1979.

Huxley, T. H. *Science and education* in *Collected Essays*, Vol. 5. London: MacMillan, 1894.

Huyen, T. Talent, opportunity, and career: A twenty-six year follow up. *School Review*, 1968, *76*, 190–209.

Inouye, E. Similarity and dissimilarity of schizophrenia in twins. *Proceedings of the Third World Conference on Psychiatry*, 1961, *1*, 524–530.

Institute for Personality and Ability Testing. *The Culture Fair Intelligence Tests: Scales 1, 2 and 3* (Rev. ed.). Champaign, Ill.: IPAT, 1973.

Ishikawa, A. The structure of personality in objective tests. In R. B. Cattell & R. M. Dreger., *Handbook of Modern Personality Theory*. New York: Century, 1977.

Jaspers, J. M. F., & de Leeuw, J. A. Genetic–environment covariation in human behavior genetics. In L. J. Th. van der Kamp, W. F. Langerak, & D. N. M. de Gruijter (Eds.), *Psychometrics for educational debates*. New York: Wiley, 1980.

Jencks, C., Smith, M., Aclaud A., Bane, M. J., Cohen G., Gintes, H., Heyns, B., & Michelson, M. *Inequality*. New York: Harper & Row, 1972.

Jensen, A. R. Implications for education. *American Educational Research Journal*, 1968, *4*, 1–42.

Jensen, A. R. How much can we boost I.Q. and scholastic achievement? *Harvard Educational Review*, 1969, *39*, 1–123.

Jensen, A. R. I.Q.'s of identical twins reared apart. *Behavior Genetics*, 1970, *1*, 133–148. (a)

Jensen, A. R. Race and genetics of intelligence: A reply to Lewontin. *Bulletin of the Atomic Scientists*, 1970, *20*, 17–23. (b)

Jensen, A. R. The race x sex x ability interaction. In R. Canero (Ed.), *Intelligence: Genetic and environmental influences*. New York: Grave & Stratton, 1971.

Jensen, A. R. *Genetics and education*. New York: Harper & Row, 1972.

Jensen, A. R. *Educability and group differences*. New York: Harper & Row, 1973.

Jensen, A. R. Kinship correlations reported by Sir Cyril Burt. *Behavior Genetics*, 1974, *4*, 1–28.

Jensen, A. R. A theoretical note on sex linkage and race differences in spatial visualization ability. *Behavior Genetics*, 1975, *5*, 151–164.

Jensen, A. R. Twins' I.Q.'s: A reply to Schwartz and Schwartz. *Behavior Genetics*, 1976, *6*, 369–371.

Jensen, A. R. The current status of the IQ controversy. Unpublished manuscript, Institute of Human Learning, University of California, Berkeley, 1977.

Jensen, A. R., & Marisi, D. Q. A note on the heritability of memory span. *Behavior Genetics*, 1979, *9*, 379–387.

Jensen, A. R. Genetic and behavioral effects of non-random mating. In R. T. Osborne, C. E. Noble & N. Wegl (Eds.) *Human variation.* New York: Academic Press, 1978. Pp. 51–105.

Jinks, J. L., & Broadhurst, P. L. How to analyse the inheritance of behavior in animals. The biometrical approach. In J. H. F. Van Abeelen (Ed.), *The genetics of behaviour.* Amsterdam. North Holland, 1974.

Jinks, J. L., & Eaves, L. J. I.Q. and inequality. *Nature*, 1974, *248*, 287–289.

Jinks, J. L., & Fulker, D. W. Comparison of the biometrical genetical, MAVA and classical approaches to the analysis of human behavior. *Psychological Bulletin*, 1970, *73*, 311–349.

Johnson, R. C. Family background, cognitive ability and personality as predictors of educational and occupational attainment. Unpublished manuscript, University of Hawaii.

Johnson, R. C., Ahern, F. M., & Cole, R. E. Secular change in degree of assortive mating for ability? *Behavior Genetics*, 1980, *10*, 1–8.

Jones, H. E. A first study of parent–child resemblance in intelligence. *Nature and nurture: Their influence upon intelligence, Twenty-Seventh Yearbook of the National Society for the Study of Education*, 1928, *27*(Pt. 1), 61–72.

Jones, H. E. Perceived differences among twins. *Eugenics Quarterly*, 1955, *21*, 98–102.

Jones, H. E., & Conrad, H. S. The growth and decline of intelligence. *Genetic Psychology Monographs*, 1933, *13*, 223–298.

Jordan, J. J. Eye color and behavior: A revised annotated bibliography. Department of Psychology and Counselling Center, Georgia State University, Atlanta, 1980.

Jordheim, G. D., & Olsen, I. A. The use of a nonverbal test of intelligence in the trust territory of the Pacific. *American Anthropologist*, 1963, *65*, 5.

Jöreskog, K. G., & Lawley, D. N. New methods in maximum likelihood factor analysis. *British Journal of Mathematical and Statistical Psychology*, 1968, *27*, 85–97.

Jöreskog, K. G., & Sorbom, D. *Advances in factor analysis and structural equations.* Cambridge, Mass.: ABT Associates, 1979.

Kahn, E. Studien uber Vererbung und Enstehung geistiger Storungen. IV. Schizoid und Schizophrenie im Erbang. *Monographien aus dem Gesamptgebiete der Neurologie und Psychiatrie*, 1923, No. 36.

Kallmann, F. J. *The genetics of schizophrenia.* New York: J. I. Augustin, 1938.

Kallmann, F. J. The genetic theory of schizophrenia. *American Journal Psychiatry*, 1946, 103, 309–322.

Kallmann, F. J. The genetics of psychoses. An analysis of 1,232 twin index families. In *Congress International de Psychiatrie, Paris.* VI. *Psychiatrie Sociale Rapports.* Paris: Herman & Cie, 1950.

Kallmann, F. J. *Heredity in health and mental disorder.* London: Chapman & Hall, 1953.

Kameoka, V. *A check on the independence and second order structure of Cattell's 7 depression factors.* Unpublished doctoral dissertation, University of Hawaii, 1979.

Kameoka, V. A check on the factor trait structure of the Clinical Analysis Questionnaire. In press. (a)

Kameoka, V. The relation of state factors to trait factors in the Clinical Analysis Questionnaire. In press. (b)

Kameoka, V., & Sine, L. An examination of adequacy of simple structure in published factor analytic researches. In press.

Kamin, L. J. *The science and politics of I.Q.* Potomac, Md.: Erlbaum, 1974.

References

417

Karlson, J. L. Genealogic studies of schizophrenia. In D. Rosenthal & S. S. Kefy (Eds.), *The transmission of schizophrenia*. New York: Pergamon Press, 1968.

Karlson, J. L. A two-locus hypothesis for inheritance of schizophrenia. In A. R. Kaplan (Ed.), *Genetic factors in schizophrenia*. Springfield, Ill.: Thomas, 1972.

Karson, S. Personal communication, 1980.

Kelley, E. L. Consistency of the adult personality. *American Psychologist*, 1955, *10*, 659–681.

Kelley, T. L. *The influence of nurture upon native differences*. New York: Macmillan, 1926.

Kempthorne, O. The theoretical values of correlations between relatives in random mating populations. *Genetics*, 1955, *40*, 153–167.

Kempthorne, O. *An introduction to genetic statistics*. New York: Wiley, 1957.

Kempthorne, O., & Tandon, O. R. The estimation of heritability by regression of offspring on parent. *Biometrics*, 1953, *9*, 90–100.

Kety, S. S., Rosenthal, D., Wender, P. H., & Schulsinger, F. Mental illness in the biological and adoptive families of adopted schizophrenics. *American Journal of Psychiatry*, 1971, *128*, 302–306.

Kety, S. S., Rosenthal, D., Wender, P. H., Schulsinger, F., & Jacobsen, B. Mental illness in the biological and adoptive families of adopted individuals who have become schizophrenic. *Behavior Genetics*, 1976, *6*, 219–225.

Killian, L. R. *The utility of objective test personality factors in diagnosing schizophrenia and the character disorders*. Unpublished master's thesis, University of Illinois, Urbana, 1960.

King, R. C. *Genetics*. New York: Oxford University Press, 1965.

King, R. C. *A dictionary of genetics*. New York: Oxford University Press, 1974.

Klein, T. W., & Cattell, R. B. *A preliminary analysis of the inheritance of personality traits in male sibs*. Paper presented at the Third Annual Meeting of the Behavior Genetics Association, Chapel Hill, North Carolina, 1973.

Klein, T. W., & Cattell, R. B. *Heritability of personality and ability measures obtained from the High School Personality Questionnaire and the Culture Fair Intelligence Scale by the Twin Method*. Paper presented at the Seventh Annual Meeting of the Behavior Genetics Association, 1977.

Klein, T. W., & Claridge, G. S. Determining two zygosity through stepwise discriminant function analysis of a physical resemblance questionnaire. In press.

Klein, T. W., De Fries, J. C., & Finkbeiner, C. T. Heritability and genetic correlations: Standard errors of estimates and sample size. In press.

Knapp, R. R. The effects of time limits on the intelligence test performance of Mexican and American subjects. *Journal of Educational Psychology*, 1960, *51*, 14–20.

Knapp, R. R. Criterion predictions in the Navy from the objective-analytic personality test battery. Paper read at Annual Meeting of American Psychological Association, New York, September, 1961.

Knapp, R. R. The validity of the objective-analytic personality test battery in Navy settings. *Educational and Psychological Measurement*, 1962, *22*, 379–387.

Kraemer, H. C., Karnes, A. F., and Thoman, E. Methodological considerations in evaluating the influence of drugs used during labor and delivery on the behavior of the new born. *Developmental Psychology*, 1972, *6*, 128–142.

Kretschmer, E. *Korperbau und Charakter*. Berlin: Springer, 1929.

Kringlen, E. Schizophrenia in twins, an epidemiological-clinical study. *Psychiatry*, 1966, *29*, 172–184.

Kuse, A. R., & DeFries, J. C. Multivariate assortative mating: A cross correlational analysis. Paper presented at the Tenth Annual Meeting Behavioral Genetics Association, June, 1980.

Kuttner, R. E. *Race and modern science*. New York: Social Science Press, 1967.

Kuttner, R. E. Use of accentuated environmental inequalities in research on racial differences. *Mankind Quarterly*, 1968, *8*, 147–160.

Lane, P. J., & Mendelsohn, R. J. The relationship between eye color and field-dependence-independence. Paper presented at the Southeastern Psychological Association Meeting, Hollywood, Florida, 1977.

418

Lange, J. *Crime as destiny*. London: Allen & Unwin, 1931.

Lashley, K. S. *Brain mechanisms and intelligence*. New York: Dover, 1963.

Leahy, A. M. Nature–nurture and intelligence. *Genetic Psychology Monographs*, 1935, *17*, 236–308.

Lehman, H. G. The creative years in science and literature. *Scientific Monthly*, 1936, *43*, 151–162.

Lehrke, R. G. Sex linkage: A biological basis for greater male variability in intelligence. In R. T. Osborne, C. E. Noble, & N. Weyl, (Eds.), *Human variation. The biopsychology of age, race, and sex*. New York: Academic Press, 1978.

Lerner, I. M. *Heredity, evolution and society*. San Francisco: Freeman, 1968.

Lester, O. *A physiological basis for personality traits*. Springfield: Charles C. Thomas, 1974.

Lewis, A. Inheritance of mental disorders. *Eugenics Review*, 1933, *25*, 1–12.

Lewontin, R. C. (Ed.) *Population biology & evolution*. Syracuse, N.Y.: Syracuse University Press, 1968.

Lewontin, R. C. Further remarks on race and the genetics of intelligence. *Bulletin of the Atomic Scientists*, 1970, *26*, 23–25.

Li, C. C. *Human genetics*. New York: McGraw-Hill, 1961.

Light, R. J., & Smith, F. V. Statistical issues in social allocation models of intelligence: A review and a response. *Review of Educational Research*, 1971, *41*, 351–367.

Lindzey, G., Lykken, D. T., & Winston, H. D. Infantile trauma, genetic factors, and adult temperament. *Journal of Abnormal and Social Psychology*, 1960, *61*, 7–14.

Loehlin, J. C. A heredity–environment analysis of personality inventory data. In S. G. Vandenberg (Ed.), *Methods and goals in human behavior genetics*. New York: Academic Press, 1965. (a)

Loehlin, J. C. Some methodological problems in Cattell's multiple abstract varaince analysis. *Psychological Review*, 1965, *72*, 156–161. (b)

Loehlin, J. C. Psychogenetics from the study of human behavior. In R. B. Cattell & R. M. Dreger (Eds.), *Handbook of modern personality theory*. New York: Wiley & Sons, 1977.

Loehlin, J. C. Are CPI scales differently heritable: How good is the evidence? *Behavior Genetics*, 1978, *8*, 381. (a)

Loehlin, J. C. Heredity–environment analyses of Jenck's I.Q. correlations. *Behavior Genetics*, 1978, *8*, 415–436. (b)

Loehlin, J. C. Combining data from different groups in human behavior genetics. In J. R. Royce (Ed.), *Theoretical advances in behavior genetics*. Leiden, The Netherlands: Sijthoff and Noordhoff, 1979.

Loehlin, J. C., Horn, J. M., & Willerman, L. Personality resemblance in adoptive families. *Behavior Genetics*, in press.

Loehlin, J. C., Lindzey, G., & Spuhler, I. N. *Race differences in intelligence*. San Francisco: Freeman, 1975.

Loehlin, J. C., & Nichols, R. C. *Heredity, environment and personality: A study of 850 sets of twins*. Austin: University of Texas Press, 1976.

Loehlin, J. C., Sharan, S., & Jacoby, R. In pursuit of the "spatial gene": A family study. *Behavior Genetics*, 1978, *8*, 27–41.

Loehlin, J. C., & Vandenberg, S. G. Genetic and environmental components in the covariation of cognitive abilities. In S. G. Vandenberg (Ed.), *Progress in human genetics*. Baltimore: Johns Hopkins Press, 1968.

Loehlin, J. C., Willerman, L., & Horn, J. M. Personality resemblances between unwed mothers and their adopted-away offspring. *Journal of Personality and Social Psychology*. In press.

Lombroso, C. *L'uomo deliquente*. Pavia University Press, 1895.

Loranger, A. W. *Genetic independence of manic-depression and schizophrenia: A replication*. Paper presented at the Ninth Annual Meeting of the Behavior Genetics Association, 1979.

Lorenz, C., & Leyhausen, G. *Motivation in animals and men*. New York: Van Nostrand, 1970.

Lush, J. L. Intra-sire correlations or regressions of offspring on dam as a method of estimating heritability of characteristics. *Thirty-third Annual Proceedings of the American Society of Animal Production*, 1940, 293–301.

Lush, J. L. Genetic unknowns and animal breeding a century after Mendel. *Transactions of the Kansas Academy of Sciences*, 1968, *71*, 309–314.

Luxemburger, H. Vorlaufiger Bericht uber psychiatrische Serienuntersuchungen an Zwillingen. *Zeitschrift fuer die Gesamte Neurologie und Psychiatrie*, 1928, *116*, 297–326.

Lykken, D. T. The diagnosis of zygosity in twins. *Behavior Genetics*, 1978, *8*, 437–473.

Lynn, R. *Personality and national character*. Oxford: Pergamon Press, 1971.

Lynn, R. The intelligence of the Japanese. *Bulletin of the British Psychological Society*, 1977, *30*, 69–72.

Lynn, R. Social ecology of intelligence in the British Isles. *British Journal of Social and Clinical Psychology*, 1979, *18*, 14–35.

Lynn, R., & Dziobon, J. On the intelligence of the Japanese and other Mongoloid peoples. *Personality and Individual Differences*, 1980, *1*, 95–96.

Mai, F. M., & Beal, R. W. A study of the personality of voluntary blood donors. *Medical Journal of Australia*, 1967, *2*, 156–159.

Mai, F. M., & Pike, A. Correlation of Rhesus (Rh) and personality factors. *British Journal of Social and Clinical Psychology*, 1970, *9*, 83–84.

Malécot, G. *Les mathematiques de l'Heredité*. Paris: Masson, 1948.

Malécot, G. *Probabilité et heredité*. Paris: Presse Universitaire de France, 1966.

Markle, A. Color and form perception on the Rorschach as a function of eye color. *Perceptual and Motor Skills*, 1975, *41*, 831–834.

Markle, A. Eye color and responsiveness to arousing stimuli. *Perceptual and Motor Skills*, 1976, *43*, 127–133.

Marx, K. *Capital*. London: Lawrence & Wishard, 1967. (Originally published 1890)

Masters, A. B. The distribution of blood groups in psychiatric illness. *British Journal of Psychiatry*, 1967, *113*, 1309–1315.

Matheny A. P. Personal communication 1972.

Matheny, A. P., Jr., Wilson, R. S., & Dolan, H. B. Relations between twins' similarity of appearance and behavioral similarity: Testing an assumption. *Behavior Genetics*, 1976, *6*, 343–351.

Mather, K. *Biometrical genetics*. London: Methuen, 1949.

Mather, K. *Human diversity*. New York: The Free Press, 1964.

Mather, K., & Jinks, J. L. *Biometrical genetics* (2nd ed.). London: Chapman & Hall, 1971.

May, A. E., & Stirrup, W. Neuroticism and extraversion in ABO blood group. *British Journal of Psychiatry*, 1967, *113*, 281–282.

Mayr, E. The emergence of evolutionary novelties. In S. Tax (Ed.), *The evolution of life*. Chicago: University of Chicago Press, 1960.

McArdle, J. J., McConnell, J. P., & Goldsmith, H. H. *Multivariate analysis of stability and genetic influence: Some results from a longitudinal study of behavioral style*. Paper presented at the 10th Annual Meeting of the Behavioral Genetics Association, June, 1980.

McAskie, M., & Clarke, A. M. Parent–offspring resemblances in intelligence: Theories and evidence. *British Journal of Psychology*, 1976, *67*, 243–273.

McClaren, A. *Mendelian inheritance in man* (2nd ed.). Baltimore: Johns Hopkins Press, 1968.

McClearn, G. E., & DeFries, J. C. *Introduction to Behavior Genetics*. San Francisco: Freeman, 1973.

McClearn, G. E., Wilson, J. R., & Meredith, W. The use of isogenic and heterogenic mouse stocks in behavioral research. In G. Lindzey & D. D. Thiessen (Eds.), *Contributions to behavior-genetic analysis: The mouse as a prototype*. New York: Appleton-Century-Crofts, 1970.

McDougall, W. *National welfare and national decay*. London: Methuen, 1926.

McGaw, B., & Jöreskog, K. G. Factorial invariance of ability measures in groups differing in intelligence and socioeconomic status. *British Journal of Mathematical and Statistical Psychology*, 1971, *24*, 154–168.

McGee, M. G. Intrafamilial correlations and the heritability estimates for spatial ability in a Minnesota sample. *Behavior Genetics*, 1978, *8*, 77–80.

McGue, M. The use of twin data for the detection of genotype and environment interaction. Paper presented at the 10th Annual Meeting of the Behavior Genetics Association, June, 1980.

McGurk, F. C. J. The cultural hypothesis and psychological tests. In R. E. Kuttner (Ed.), *Race and modern science*. New York: Social Science Press, 1967.

McKeown, T., & Record, R. G. Early environmental influences on the development of intelligence. *British Medical Bulletin*, 1971, *27*, 48–51.

McKusick, V. A. Genetics in medicine and medicine in genetics. *American Journal of Medicine*, 1963, *34*, 594–599.

McKusick, V. A. *Human genetics*. New York: Prentice Hall, 1964.

McKusick, V. A. *Mendelian inheritance in man* (4th ed.). Baltimore: Johns Hopkins Press, 1974.

McNemar, Q. A critical examination of the University of Iowa studies of environmental influence upon I.Q. *Psychological Bulletin*, 1940, *37*, 63–92.

McNemar, Q. *Psychological statistics*. New York: Wiley, 1962.

Meade, J. E., & Parkes, A. S. *Genetic and environmental factors in human ability*. Edinburgh: Oliver & Boyd, 1966.

Meredith, G. M. Observations on the acculturation of sansei Japanese Americans in Hawaii. *Psychologia*, 1965, *8*, 41–49.

Meredith, G. M. Contending hypothesis of ontogenesis for the exuberance-restraint personality factor U.I. 21. *Journal of Genetic Psychology*, 1966, *108*, 89–104.

Meredith, G. M. Observations on the origins and current status of the ego assertive personality factor U.I. 16. *Journal of Genetic Psychology*, 1967, *110*, 269–286.

Meredith, W. A model for analyzing heritability in the presence of correlated genetic and environmental effects. *Behavior Genetics*, 1973, *3*, 271–277.

Miles, C. The influence of speed and age on intelligence scores of adults. *Journal of General Psychology*, 1934, *10*, 208–210.

Miles, W. R. Changes in movement time and reaction time with age. In E. V. Cowdry (Ed.), *Problems of aging*. Baltimore: Williams and Wilkins, 1942.

Mill, J. S. *Autobiography*. London: Longmans, 1874.

Mittler, P. *The study of twins*. London: Penguin, 1971.

Morton, N. E. The detection of major genes under additive continuous variation. *American Journal of Human Genetics*, 1967, *19*, 23–34.

Morton, N. E. Analysis of family resemblance, I: Introduction. *American Journal of Human Genetics*, 1974, *26*, 318–330.

Morton, N. E., & Chang, C. S. *Genetic epidemiology*. New York: Academic Press, 1978.

Morton, N. E., Chang, C. S., & Mi, M. P. *Genetics of inter-racial crosses in Hawaii*. Basel: Karger, 1967.

Mosher, L. R., Pollin, W., & Stabenhau, J. R. Families with identical twins discordant for schizophrenia: Some relationships between identification, thinking styles, psychopathology and dominance. *British Journal of Psychiatry*, 1971, *178*, 29–42.

Mourant, A. E. *The distribution of the human blood groups*. Oxford: Blackwell, 1954.

Mowrer, O. H. Animal studies in the genesis of personality. *Transactions of the New York Academy of Science*, 1938, *56*, 273–288.

Muller, H. J. Mental traits and heredity as studied in a case of identical twins reared apart. *Journal of Heredity*, 1925, *16*, 433–448.

Muller, H. J. Artificial transmutation of the gene. *Science*, 1927, *66*, 84–87.

Munsinger, H. The adopted child's I.Q.: A critical review. *Psychological Bulletin*, 1975, *82*, 623–659. (a)

Munsinger, H. Children's resemblance to their biological and adopting parents in two ethnic groups. *Behavior Genetics,* 1975, *5,* 239–254. (b)

Nance, W. E., Krieger, H., Azevedo, E., & Mi, M. P. Human blood pressure and the ABO human blood group system: An apparent association. *Human Biology,* 1965, *37,* 238–244.

Nesselroade, J. R. A comparison of cross product and differential R-factoring regarding cross study stability of change patterns. Unpublished Ph.D. Dissertation. Urbana, Ill.: University of Illinois, 1967.

Nesselroade, J. R. Application of multivariate strategies to problems of measuring and structuring long-term change. In J. B. Nesselroade & H. W. Reese (Eds.), *Life span developmental psychology.* New York: Academic Press, 1973.

Nesselroade, J. R. Application of multivariate analysis procedures to the study of aging and behavioral changes. In J. E. Birren & K. W. Schaie (Eds.), *The handbook of the psychology of aging.* New York. Van Nostrand Reinhold, 1976.

Nesselroade, J. R., Baltes, P. B., & Labouvie, E. W. Evaluating factor invariance in oblique space: Baseline data generated from random numbers. *Multivariate Behavioral Research,* 1971, *6,* 233.

Nesselroade, J. R., & Reese, H. W. *Life span developmental psychology: Methodological issues.* New York: Academic Press, 1973.

Newman, H. H., Freeman, F. N., & Holzinger, K. J. *Twins: A study of heredity and environment.* Chicago: University of Chicago Press, 1937.

Nichols, K. E. Collation of second order personality studies. Unpublished Master's Thesis, University of Illinois, 1973.

Nichols, R. C. The national merit twin study. In S. G. Vandenberg (Ed.), *Methods and goals in human behavior genetics.* New York: Academic Press, 1965.

Nichols, R. C. The resemblance of twins in personality and interests. In M. Manosowetz, G. Lindzey, & D. D. Thiessen, *Behavioral genetics. Method and research.* New York: Appleton, Century, Crofts, 1969.

Nichols, R. C. Twin studies of ability, personality and interests. *Homo,* 1978, *29,* 158–173.

Nichols, R. C. Heredity and environment: Major findings from twin studies of ability, personality and interests. New York, P.O. Box 3495: International Association for Advancement of Eugenics, 1979.

Omenn, G. S. Inborn errors of metabolism: clues to understanding human behavioral disorders. *Behavior Genetics,* 1976, *6,* 263–284.

Onslow, H. Fair and dark: Is there a predominant type? *Eugenic Reform,* 1920, *12,* 212–230.

Osborne, R. T. *Twins: Black and white.* Athens, Ga.: Foundation for Human Understanding, 1980.

Osborne, R. T., & Gregor, A. J. *Racial differences in inheritance ratios for tests of spatial ability.* Paper presented to the Instituto Internacional de Sociologia, XXII Congreso, Madrid, October 1967.

Osborne, R. T., Noble, C. E., & Weyl, N. *Human Variation: The biopsychology of age, race, and sex.* New York: Academic Press, 1978.

Osborne, R. T., & Suddick, D. E. *Blood type gene frequency and mental ability.* Georgia Academy of Science, 1971.

Outhit, M. C. A study of the resemblance of parents and children in general intelligence. *Archives of Psychology,* 1933, *149,* 1–60.

Parker, J., Theilie, A., & Spielberger, C. D. Frequency of blood groups in a homogeneous group of manic-depressive patients. *Journal of Mental Science,* 1961, *107,* 936–942.

Pauls, D. L., Crowe, R. R., & Noyes, R., Jr. *The genetics of anxiety neurosis: A pedigree study.* Paper presented at the 9th Annual Meeting of the Behavior Genetics Association, 1979.

Pawlik, K., & Cattell, R. B. Third order factors in objective personality tests. *British Journal of Psychology,* 1964, *5,* 1–18.

Pawlik, K., & Cattell, R. B. The relationship between certain personality factors and measures of cortical arousal. *Neuropsychologia*, 1965, *3*, 129–151.

Pearson, K. On the relationship of intelligence to size and shape of head and to other physical and mental characters. *Biometrica*, 1906/07, *5*, 105–136.

Pencarel, J. H. A note on the I.Q. of monozygotic twins raised apart and the order of their birth. *Behavior Genetics*, 1976, *6*, 455–460.

Penrose, L. S. *Outline of human genetics*. New York: Wiley, 1959.

Penrose, L. S. *The biology of mental defect* (3rd ed.). New York: Grune and Stratton, 1963.

Plomin, R., & DeFries, J. C. *Genetics and intelligence: Recent data.* Unpublished manuscript, Institute for Behavioral Genetics, University of Colorado, 1978.

Plomin, R., & DeFries, J. C. Multivariate behavioral genetic analysis of twin data on scholastic abilities. *Behavior Genetics*, 1979, *9*, 505–517.

Plomin, R., Willerman, L., & Loehlin, J. C. Resemblance in appearance and the equal environments assumption in twin studies of personality traits. *Behavior Genetics*, 1976, *6*, 43–52.

Poley, W., & Royce, J. R. Behavior genetic analysis of mouse emotionality: II. Stability of factors across genotypes. *Animal Learning & Behavior*, 1973, *1*, 116–120.

Porteus, S. D. Ethnic groups and the maze test. In R. E. Kuttner (Ed.), *Race and modern science*. New York: Social Science Press, 1967.

Prehm, H. J., Hamerlynck, L. Q., & Crosson, J. E. (Eds.), *Behavioral research in mental retardation*. Eugene, Oregon: University of Oregon Press, 1968.

Price, R. A., & Vandenberg, S. G. Spouse similarity in American and Swedish couples. *Behavior Genetics*, 1980, *10*, 59–71.

Rao, D. C., MacLean, C. J., Morton, N. E., Yee, S. Analysis of family resemblance, V. height and weight in Northern Brazil. *American Journal of Human Genetics*, 1975, *27*, 509–520.

Rao, D. C., & Morton, N. E. Path analysis of family resemblance in the presence of gene–environment interaction. *American Journal of Human Genetics*, 1974, *26*, 767–772.

Rao, D. C., Morton, N. E., & Cloninger, C. R. Path analysis under generalized assortive mating, I: Theory. *American Journal of Human Genetics*, in press.

Rao, D. C., Morton, N. E., Elston, R. C., & Yee, S. Causal analysis of academic performance. *Behavior Genetics*, 1977, *1*, 147–159.

Rao, D. C., Morton, N. E., & Yee, S. Analysis of family resemblance, II. A linear model for familial correlation. *American Journal of Human Genetics*, 1974, *26*, 331–359.

Rao, D. C., Morton, N. E., & Yee, S. Resolution of cultural and biological inheritance by path analysis. *American Journal of Human Genetics*, 1976, *28*, 228–242.

Reed, S. *Counseling in medical genetics* (3rd ed.). New York: Liss, 1981.

Reed, T. E., & Neel, J. V. Huntington's chorea in Michigan. *American Journal of Human Genetics*, 1959, *11*, 107–136.

Reich, T., Rice, J., & Cloninger, C. R. The detection of a major locus in the presence of multifactorial variation. In Gershon, E. S. (Ed.) *Genetic research strategies for psychobiology and psychiatry*. Pacific Grove, Calif.: Boxwood Press, 1981.

Rice, J., & Cloninger, C. R. Multifactorial inheritance with cultural transmission and assortive mating, 1. Description and basic properties of the unitary models. *American Journal of Human Genetics*, 1978, *30*, 618–643.

Rice, J. A., Reich, T., Wepte, R., & Cloninger, C. K. An approach to the multivariate normal integral: Application to multifactored qualitative traits. *Biometrics;* 1979, *35*, 451–459.

Rickels, K., & Cattell, R. B. The clinical factor validity and trueness of the IPAT verbal and objective batteries for anxiety and regression. *Journal of Clinical Psychology*, 1965, *21*, 257–264.

Rife, D. C. A research proposal for UNESCO. *Mankind Quarterly*, 1961, *2*, 128–131.

Roberts, D. F., & Harrison, G. A. *Natural selection in human populations*. New York: Pergamon Press, 1959.

Robinson, S. Experimental studies of physical fitness in relation to age. *Arbeitsphysiologie*, 1938, *10*, 279.

Robson, C. C., & Richards, O. W. *The variations of animals in nature.* London: Longmans, 1936.

Rodd, W. G. *A cross cultural study of Taiwan's schools.* Unpublished doctoral dissertation. Western Reserve University, 1958.

Roe, A. Children of alcoholic parents raised in foster homes. In *Alcohol Science and Society.* New Haven: Journal of Studies on Alcohol, 1954.

Rosanoff, A. J., Handy, L. M., & Plesset, R. R. The etiology of manic depressive syndromes with special reference to their occurrence in twins. *American Journal of Psychiatry,* 1935, *97,* 16–30.

Rose, R. J., Procidano, M. E., Conneally, M., & Yu, P. *Blood-group discordance and verbal intelligence: Analysis in offspring of MZ twins.* Paper presented at the 9th Annual Meeting of the Behavior Genetics Association, 1979.

Rosenthal, D. The offspring of schizophrenic couples. *Journal of Psychiatric Research,* 1966, *4,* 169–188.

Rosenthal, D. *Genetic theory and abnormal behavior.* New York: McGraw-Hill, 1970.

Rosenthal, D. *Genetic factors in behavioral disorders.* Paper presented in the Medical Genetics symposium at the 1972 Annual Meeting of the National Academy of Sciences, Washington, D. C., 1972.

Rosenthal, D., & Kety, S. S. (Eds.), *The transmission of schizophrenia.* Oxford: Pergamon, 1968.

Rosenthal, D., Wender, P. H., Kety, S. S., Welner, J., & Schulsinger, F. The adopted-away offspring of schizophrenics. *American Journal of Psychiatry,* 1971, *128,* 307–311.

Roubertoux, P. L'etiologie des depressions: Apport de l'analyse genetique. *Seminary Hopital de Paris,* 1981, *57,* 768–773.

Roubertoux, P., & Carlier, M. Analyse genetique des comportements. *L'annee Psychologique,* 1972, *72,* 391–441.

Roubertoux, P., & Carlier, M. Intelligence: Differences individuelles, facteurs genetiques et interaction entre genotype et environnement. *Annee Biologie Clinique,* 1978, *36,* 101–112.

Rowe, D., & Plomin, R. The Burt controversy: A comparison of Burt's data on I. Q. with data from other studies. *Behavior Genetics,* 1978, *8,* 81–83.

Royce, J. R. A factorial study of emotionality in the dog. *Psychological Monographs,* 1955, *69,* 1–91.

Royce, J. R. Factor theory and genetics. *Educational and psychological measurement,* 1957, *17,* 361–376.

Royce, J. R. Concepts generated in comparative and physiological psychological observations. In R. B. Cattell (Ed.), *Handbook of multivariate experimental psychology.* Chicago: Rand McNally, 1966.

Royce, J. R., & Covington, M. Genetic differences in the avoidance conditioning of mice. *Journal of Comparative and Physiological Psychology,* 1960, *53,* 197–220.

Royce, J. R., Holmes, T. M., & Poley, W. Behavior genetic analysis at mouse emotionality, III: The diallel analysis. *Behavior Genetics,* 1975, *5,* 351–372.

Ruggles-Gates, A. *Human heredity.* New York: Macmillan, 1946.

Rummell, R. J. Indicators of cross national and international patterns. *American Political Science Review,* 1969, *63,* 1–5.

Rummell, R. J. *Applied factor analysis.* Evanston: Northwestern University Press, 1970.

Rummel, R. J. *The dimensions of nations.* Beverly Hills, Calif.: Sage, 1972.

Salzano, F. M., & Rao, P. C. Path analysis of aptitude, personality, and achievement scores in Brazilian twins. *Behavior Genetics,* 1976, *6,* 461–466.

Sanders, E. M., Mefford, R. B., Jr., & Brown, O. H. Verbal–quantitative ability and certain personality and metabolic characteristics of male college students. *Educational Psychological Measurement,* 1960, *20,* 491–503.

Saunders, D. R. On the dimensionality of the WAIS battery for two groups of normal males. *Psychological Reports,* 1958, *5,* 529–541.

Saville, P. The British standardization of the 16PF (Supplement of Norma). Windsor, England: The National Foundation for Educational Research, 1972.

Scarr, S. Environmental bias in twin studies. In S. G. Vandenberg (Ed.), *Progress in human behavior genetics.* Baltimore: Johns Hopkins University Press, 1968.

Scarr, S. Effects of birth weight on later intelligence. *Social Biology,* 1969, *16,* 249–256.

Scarr, S. Resemblances in biologically-related and adoptive families. In S. Scarr (Ed.), *How families affect the personality development of their children.* Symposium presented at the 17th International Congress of Behavior Genetics, Lima, 1979.

Scarr, S. *I.Q., race, social class and individual differences: New studies of old problems.* Hillsdale, N.J.: Erlbaum, 1980.

Scarr, S., & Carter-Saltzman, L. Twin method: Defense of a critical assumption. *Behavior Genetics,* 1979, *9,* 527–542.

Scarr, S., Scarf, E., & Weinberg, R. A. Perceived and actual similarities in biological and adoptive families: Does perceived similarity bias genetic influences? *Behavior Genetics,* In press.

Scarr, S., Webber, P. L., Weinberg, R. A., & Wittig, M. A. Personality resemblance among adolescents and their parents in biologically-related and adoptive families. In press.

Scarr, S., & Weinberg, R. A. I.Q. test performance of black children adopted by while families. *American Psychologist,* 1976, *31,* 726–739.

Scarr, S., & Weinberg, R. A. Intellectual similarities within families of both adopted and biological children. *Intelligence,* 1977, *32,* 170–191.

Scarr, S., & Weinberg, R. A. The influence of "family background" on intellectual attainment. *American Sociological Review,* 1978, *43,* 674–692.

Schaie, K. W., & Strother, C. R. A cross sectional study of age changes in cognitive behavior. *Psychological Bulletin,* 1968, *68,* 10–25.

Schmidt, L. R., Hacker, H. O., & Cattell, R. B. *Objective Test batterie OA-TB, 75. Test Hefte.* Weinheim n. Basel: Beltz, 1973.

Schneewind, K. A. Entwicklung einer deutschsprachigen version des 16PF tests von Cattell. *Diagnostica,* 1977, *2,* 188–191.

Schoenfeldt, L. F. The hereditary components of the Project Talent two-day test battery. *Measurement and Evaluation in Guidance,* 1968, *1,* 130–140.

Schooley, C. Birth order effects. *Psychological Bulletin;* 1972, *78,* 161–175.

Schulz, B. Empirische untersuchungen uber die Bedeutung beidseitegen Belastung mit endogen Psychosen. *Zeitschrift fur Neurologie und Psychiatrie,* 1939, *165,* 97–108.

Schulz, B. Kinder schizophrener Elternpaare. *Zeitschrift fuer die Gesamte Neurologie und Psychiatrie,* 1940, *168,* 332–381.

Schwartz, M., & Schwartz, J. Comment on "I.Q.'s of Identical Twins Reared Apart." *Behavior Genetics,* 1976, *6,* 367–368.

Scott, J. P. Social genetics. *Behavior-Genetics,* 1977, *7,* 327–346.

Scott, J. P., & Fuller, J. L. *Genetics and the social behavior of the dog.* Chicago: University of Chicago Press, 1965.

Scott, J. P., Fuller, J. L., & King, J. A. The inheritance of animal breeding cycles in hybrid basenjicocker spaniel dogs. *Journal of Heredity,* 1959, *50,* 254–261.

Shaw, G. B. *Everybody's political what's what?* New York: Dodd Mead, 1947.

Shaw, G. B. *The complete prefaces of Bernard Shaw.* London: Hamlyn, 1865.

Sheldon, W. H., & Stevens, S. S. *The varieties of temperament.* New York: Harper, 1942.

Shields, J. Personality differences and neurotic traits in normal twin school children. *Eugenics Review,* 1954, *45,* 213–46.

Shields, J. *Monozygotic twins brought up apart and brought up together.* London: Oxford University Press, 1962.

Shields, J., Gottesman, I. I., & Slater, E. Kallman's 1946 schizophrenic twin study in the light of new information. *Acta Psychiatrica Scandinavica,* 1967, *43,* 385–396.

Shockley, W. Deviations from Hardy–Weinberg frequencies caused by assortative mating in hybrid populations. *Proceedings of the National Academy of Science USA*, 1973, *70*, 732–736.

Shuey, A. M. *The testing of Negro intelligence*. New York: Social Science Press, 1966.

Simpson, G. G. *Principles of animal taxonomy*. New York: Columbia University Press, 1961.

Skodak, M., & Skeels, H. M. A final follow-up study of one hundred adopted children. *Journal of Genetic Psychology*, 1949, *75*, 85–125.

Slater, E. The inheritance of manic-depressive insanity and its relation to mental defect. *Journal of Mental Science*, 1936, *82*, 626–633.

Slater, E. Psychiatry. In A. Sorsby (Ed.), *Clinical genetics*. London: Butterworth, 1953.

Slater, E. A review of earlier evidence on genetic factors in schizophrenia. In D. Rosenthal & S. S. Kety (Eds.), *The transmission of schizophrenia*. New York: Pergamon Press, 1968.

Slater, E., & Cowie, V. *The genetics of mental disorders*. London: Oxford University Press, 1971.

Smith, J. M. and Misiak, H. The effect of iris color on critical flicker frequency. (CFF). *Journal of General Psychology*, 1973, *89*, 91–95.

Smith, R. T. A comparison of sourenvironmental factors in monozygotic and dizygotic twins, testing an assumption. In S. G. Vandenberg (Ed.), *Methods and goals in human behavior genetics*. New York: Academic Press, 1965.

Sokal, R. R., & Sneath, P. H. A. *Principles of numerical taxonomy*. San Francisco: Freeman, 1963.

Sorensen, J. R. *Social aspects of applied human genetics*. New York: Russell Sage Foundation, 1971.

Spearman, C. General intelligence objectively determined and measured. *American Journal of Psychology*, 1904, *15*, 201–293.

Spearman, C. *The nature of 'intelligence' and the principles of cognition*. London: Macmillan and Co., 1923.

Spielberger, C. D. *Anxiety: Current trends in theory and research*. New York: Academic Press, 1972.

Spuhler, J. N. Assortative mating with respect to physical characters. *Eugenics Quarterly*, 1968, *15*, 128–140.

Stafford, R. E. Sex differences in spatial visualization as evidence of sex linked inheritance. *Perceptual and Motor Skills*, 1961, *13*, 428.

Stafford, R. E. New techniques for analyzing parent–child test scores for evidence of hereditary components. In S. G. Vandenberg (Ed.), *Methods and goals in human behavior genetics*. New York: Academic Press, 1965.

Stafford, R. E. Hereditary and environmental components of quantitative reasoning. *Review of Educational Research*, 1972, *42*, 183–201.

Stafford, R. E. *Hereditary components of musical aptitude: A review*. Paper presented at the 10th Annual Meeting of Behavior Genetics Association, June, 1980.

Stent, G. S. *Molecular genetics*. San Francisco: Freeman, 1970.

Sturtevant, A. H., & Beadle, G. W. *Introduction to genetics*. Dover: Constable, 1964.

Suzanne, C. On the relationship between psychometric and anthropometric traits. *American Journal of Physical Anthropology*, 1979, *51*, 421–424.

Sutton, H. E. *Genes, enzymes, and inherited diseases*. New York: Holt, Rinehart & Winston, 1961.

Sutton, H. E. *Introduction to human genetics*. New York: Holt, 1965.

Sutton, P. R. N. Association between colour of the iris of the eye and reaction to dental pain. *Nature*, 1959, *184*, 122.

Swan, D. A. The relation between mental ability and dermatoglyphic traits among South Mississippi Anglo-Saxon school children. In press.

Swan, D. A., Hawkins, G., & Douglas B. The relationship between ABO blood type and factors of personality among South Mississippi "Anglo Saxon" school children. *Mankind Quarterly*, 1980, *20*, 205–258.

Swan, D. A., Hawkins, G., & Douglas, B. The relations of 45 anthropological variables to ability and personality factors. In press.

Symonds, P. M. *Diagnosing personality and conduct.* New York: Appleton Century, 1937.

Tatro, D. F. The utility of source traits measured by the O–A (objective–analytic) battery in mental hospital diagnosis. *Multivariate Behavioral Research,* 1968. Special issue.

Taylor, C. C. Marriage of twins to twins. *Acta Genetica,* 1971, *1,* 96–113.

Teasdale, T. W. Social class correlations among adoptees and their biological and adoptive parents. *Behavior Genetics,* 1979, *9,* 103–114.

Tedford, W. H., Hill, W. R., & Hensley, L. Human eye color and reaction time. *Perceptual and Motor Skills,* 1978, *47,* 503–506.

Thiessen, D. D., Lindzey, G., & Owen, K. Behavior and allelic variations in enzyme activity and coat color at the *C* locus of the mouse. *Behavior Genetics,* 1970, *1,* 257–267.

Thompson, W. R. Some problems in the genetic study of personality and intelligence. In J. Hirsch (Ed.), *Behavior Genetic Analysis.* New York: McGraw-Hill, 1967.

Thomson, G. H. *The factorial analysis of human ability* (5th ed.). Boston: Houghton Mifflin, 1951.

Thurstone, L. L. *Primary mental abilities.* Chicago: University of Chicago Press, 1938.

Thurstone, L. L. *Multiple factor analysis.* Chicago: University of Chicago Press, 1947.

Thurstone, T. G., Thurstone, L. L., & Strandskov, H. H. *A psychological study of twins* (Rep. No. 4). Chapel Hill, N.C.: Psychometric Laboratory, University of North Carolina, 1955.

Tienari, P. Psychiatric illnesses in identical twins. *Acta Psychiatrica Scandinavica,* 1963, *39,* (suppl. 171), 1–195.

Tinbergen, W. *The study of instinct.* Oxford: Clarendon Press, 1951.

Tolman, E. C. The inheritance of maze-learning ability in rats. *Journal of Comparative Psychology,* 1924, *4,* 1–18.

Toynbee, A. *A study of history.* New York: Oxford University Press, 1947.

Tryon, R. C. Genetic differences in maze learning ability in rats. *Yearbook of the National Society for the Study of Education,* 1940, *39,* 111–119.

Tsujioka, B., & Cattell, R. B. Constancy and difference in personality structure and mean profile, in the questionnaire medium, from applying the 16P.F. test in America and Japan. *British Journal of Social and Clinical Psychology,* 1965, *4,* 287–297. (a)

Tsujioka, B., & Cattell, R. B. A cross-cultural comparison of second stratum questionnaire personality factor structures—anxiety and extraversion—in America and Japan. *Journal of Social Psychology,* 1965, *65,* 205–219.

Tuddenham, R. D. The constancy of personality ratings over two decades. *Genetic Psychology Monographs,* 1959, *60,* 3–29.

Undheim, J. D. Ability structure in 10–11 year old children and the theory of fluid and crystallized intelligence. *Journal of Educational Psychology,* 1976, *68,* 411–423.

Valen, L. van. Brain size and intelligence in man. *American Journal of Physical Anthropology,* 1968, *40,* 417–424.

Van Abeelen, J. H. F. (Ed.), *The genetics of behaviour.* Amsterdam, North Holland, 1974.

Vandenberg, S. G. The primary mental abilities of Chinese students: A comparative study of the stability of a factor structure. *Annals of the New York Academy of Science,* 1959, *79,* 257–304.

Vandenberg, S. G. The hereditary abilities study: Hereditary components in a psychological test battery. *American Journal of Human Genetics,* 1962, *14,* 220–237.

Vandenberg, S. G. *Methods and goals in human behavior genetics.* New York: Academic Press, 1965. (a)

Vandenberg, S. G. Multivariate analysis of twin differences. In S. G. Vandenberg (Ed.), *Methods and goals in human behavior genetics.* New York: Academic Press, 1965. (b)

Vandenberg, S. G. Hereditary factors in psychological variables in man, with a special emphasis on cognition. In J. N. Spuhler (Ed.), *Genetic diversity and human behavior.* Chicago: University of Chicago Press, 1967. (b)

Vandenberg, S. G. Hereditary factors in normal personality traits (as measured by inventories). *Recent Advances in Biological Psychiatry*, 1967, *9*, 65–104. (a)

Vandenberg, S. G. The primary mental abilities of South American students: A second comparative study of the generality of a cognitive factor structure. *Multivariate Behavioral Research*, 1967, *2*, 175–198. (c)

Vandenberg, S. G. Primary mental abilities or general intelligence? Evidence from twin studies. In J. M. Thoday & A. S. Parke, (Eds.), *Genetic and environmental influences on behavior*. New York: Plenum Press, 1968.

Vandenberg, S. G. Human behavior genetics. Present status and suggestions for future research. *Merrill Palmer Quarterly of Behavior and Development*, 1969, *15*, 121–154.

Vandenberg, S. G. Assortative mating: Or who marries whom? *Behavior Genetics*, 1972, *2*, 127–257.

Vandenberg, S. G., & Johnson, R. C. Further evidence on the relation between age of separation and similarity in I.Q. among pairs of separated identical twins. In S. G. Vandenberg (Ed.), *Progress in human behavior genetics*. Baltimore: Johns Hopkins Press, 1968.

Vandenberg, S. G., Meredith, W., & Keese, A. Two higher order factors in intelligence test scores. A partial replication of the Cattell–Horn theory of fluid and crystallized intelligence. In press.

Vaughan, D. S. The relative methodological soundness of several major personality factor analyses. *Journal of Behavioral Science*, 1973, *1*, 305–313.

Vaughan, D. S., & Cattell, R. B. Personality differences between young New Zealanders and Americans. *Journal of Social Psychology*, 1976, *99*, 3–12.

Vernon, P. E. Environmental handicaps and intellectual development: Part II and Part III. *British Journal of Educational Psychology*, 1965, *35*, 1–22.

Vernon, P. E. *Intelligence and the cultural environment*. London: Methuen, 1969.

Waddington, C. H. Genetic assimilation of an acquired character. *Evolution*, 1953, *7*, 118–126.

Waddington, C. H. *New patterns in genetics and development*. New York: Columbia University Press, 1967.

Wardell, D., & Royce, J. R. Relationships between cognitive and temperament traits and the concept of style. *Journal of Multivariate Experimental Personality and Clinical Psychology*, 1975, *1*, 244–266.

Wardell, D., & Yeudall, L. T. A. A multidimensional approach to criminal disorders, 1: The factor analysis of impulsivity. *Journal of Clinical Psychology*, 1976, *32*, 12–31.

Watkins, M. P. *Assortative marriage for physical characteristics in newlyweds*. Paper presented at the 10th Annual Meeting of the Behavior Genetics Association, June, 1980.

Watson, G. A comparison of the effects of lax versus strict home training. *Journal of Social Psychology*, 1934, *5*, 102–105.

Weiss, V. V. Empirische untersuchung zu einer hypothese uber den autosomal-rezessiven erbgang der mathematisch-technischen begabung. *Biologisches Zentralblatt*, 1972, *91*, 429–435.

Welch, W. B. On linear combination of several variances. *Journal of the American Statistical Association*, 1956, *51*, 1–9.

Wells, H. G. *Mankind in the making*. London: Chapman & Hall, 1903.

Weyl, N. The impact of famine on I.Q. *Mankind Quarterly*, 1974, *14*, 211–213.

Weyl, N. *Karl Marx—Racist*. New York: Arlington House, 1979.

Whitney, G., McClearn, G. E., & De Fries, J. C. Heritability of alcohol preference in laboratory mice and rats. *Journal of Heredity*, *61*(4), 165–169.

Willerman, L., Horn, J. M. & Loehlin, J. C. The aptitude–achievement test distinction: A study of unrelated children reared together. *Behavior Genetics*, 1977, *7*, 465–470.

Willerman, L., & Stafford, R. E. Maternal effects on intellectual functioning. *Behavior Genetics*, 1972, *2*, 321–325.

Williams, R. J. *Free and unequal*. Austin: University of Texas Press, 1953.

Williams, R. J. *Biochemical individuality*. New York: John Wiley & Sons, 1956.

Williams, T. Family resemblance in abilities: The Wechsler Scales. *Behavior Genetics,* 1975, *5,* 405–410.

Williams, T. Reply to Professor Lewoutin's letter. *Behavior Genetics,* 1977, *7,* 37.

Willoughby, R. R. Family similarities in mental test abilities. *Twenty-seventh Yearbook of the National Society for the Study of Education,* 1928, *27,* 55–59.

Willoughby, R. R. The relationship to emotionality of age, sex and conjugal condition. *American Journal of Sociology,* 1938, *6,* 920–931.

Wilson, E. O. *Sociobiology: The new synthesis.* London: Harvard University Press, 1975.

Wilson, R. S. Twins: Early mental developmental. *Science,* 1972, *175,* 914–917.

Wilson, R. S. Diagnosis of zygosity in twins: Reply to Lykken, *Behavior Genetics,* 1979, *9,* 245–246.

Witkin, H. A., Dyk, R. B., Paterson, H. F., Goodenough, D. R., & Karp, S. A. *Psychological differentiation.* New York: Wiley, 1962.

Woliver, R. *Assortive mating correlations of American and Japanese honeymoon couples on the MAT.* Unpublished doctoral dissertation, University of Hawaii-Manoa, Honolulu, 1979.

Wright, S. Coefficients of inbreeding and relationship. *American Nature,* 1922, *56,* 330–338.

Wright, S. The method of path coefficients. *Annals of Mathematical Statistics,* 1934, *5,* 161–215.

Wright, S. *Evolution and the genetics of populations (Vol. 1: Genetic and biometric foundations).* Chicago: University of Chicago Press, 1968.

Yen, W. M. Sex linked major gene influences on selected types of spatial performance. *Behavior Genetics,* 1975, *5,* 281–298.

Young, M. *The rise of the meritocracy.* Baltimore: Penguin, 1958.

Zerssen, D. Von. Premorbid personality and affective psychoses. In J. Burrows (Ed.), *Handbook of Studies on Depression.* In *Excerpta Medica,* 1977.

Zonderman, A. B., Vandenberg, S. G., Spuhler, K. P., & Fain, P. R. Assortative mating for cognitive abilities. *Behavior Genetics,* 1977, *7,* 261–271.

Glossary

Abstract variances The contributory genetic and threptic variances which the MAVA model sets up to explain the observed concrete family (intra- and inter-) variances.

Additive effects The contributions of genes, each in a particular allelic form, that in certain conditions, sum to the total genetic effect.

Adjustment process analysis A standard chart for analyzing a particular stage of dynamic adjustment.

Allele (from allelomorph) One of two or more alternative forms of expression from a given gene (multiple allelism is common).

Anxiety The anxiety factor as defined and measured by U.I. 24 in objective tests, or QI in questionnaires.

Assertiveness A personality–temperament unitary trait defined and measured in the O-A battery by U.I. 16.

Assortive (sometimes assortative) mating Mating in which parents are systematically more (or less) alike, phenotypically, threptically, or genetically, than would be expected from random pairing.

Asthenia A unitary trait found in objective test data (O-A battery) indexed as U.I. 28.

Autosomal Applying to chromosomes (or genes) not on the sex chromosomes.

Backcross   The cross of the heterozygotic offspring of two homozygotic parents with individuals of the same homozygosity as the parents.

Biometric genetics   The use of measurement and statistics in genetic research.

Chromosome   The DNA-histone thread in the nucleus of a cell, within which genes are located, and which is the basis of hereditary transmission.

Comparative MAVA   The use of MAVA in populations differing in culture, age, race, and measured environmental ranges, to reach further inferences on genetic structure and learning laws.

Conditioned dominance   The actions of an allele in heterozygotes as dominant or recessive according to environmental milieu.

Control   A unitary trait pattern in objective and questionnaire measures (U.I. 17, QVIII) of high social and ethical control of behavior.

Convarkin methods   A class of biometric genetic methods, which includes MAVA, in which abstract variances and covariances are deduced from *con*trasting *var*iances of *kin* groups, as observed variances.

Contributory model   Concept that most continuously variable traits are polygenic genetically (i.e., arising from the effects of many genes).

Cortertia   A unitary trait, probably temperamental, of "cortical alertness" appearing in objective tests as U.I. 22 and in questionnaires as the second order QIII.

Cultural species, in syntalities   Groupings of national cultures found from measured factorial syntality profiles for the world's nations.

Crossing-over   The exchange, in reduction division, of chromosomal parts between homologous chromosomes.

Cytoplasmic inheritance   Inheritance, non-Mendelian, not from chromosomes but from passing on of parental cytoplasm in the germ cell (organelles, plastids, etc.), and mainly on the maternal side.

Dizygotic (DZ) twins   Synonymous with fraternal twins, springing from two separate ova and distinct spermatozoa.

Dominance   Factor E in questionnaires, describing dominant assertive-versus-submissive behavior.

Dominant allele   An allele which masks to varying degrees (complete, partial, overdominance, etc.) the effects of another allele at the same locus, in the heterozygous combination. Opposed to *recessive allele*.

Econetics   The branch of psychology, covering psycho-physics as a branch, which systematically relates physico-social environmental characters to psychological impact variables.

Econetics matrix   A matrix attempting to cover the main features in the environment of the individual in terms of quantitative impacts or "presses."

Ego strength   The unitary trait, first recognized in psychoanalysis, but since verified and rendered measurable (C factor in the questionnaire domain) by factor analytic experiment.

Eidolon theory   The theory of a genothreptic interaction model in which the pattern of the threptic part follows closely that of the genetic loading pattern part.

Environmental variance   The variance in a population of people's exposure to measured *environmental characters* (e.g., temperature, years of schooling, nutrition, exercise). Not to be confused with *threptic* variance, the variance in a *trait* due to variance of the environment.

Epigenetics   The study of the intermediate processes (e.g., enzyme action) by which genes bring about their maturational effects.

Ergic tension, Q_4 in questionnaires   A primary trait of autonomic overactivity partly due to denial of ergic satisfactions.

Eugenics (positive or negative)   The planning of differential breeding to bring about better human adaptation.

Epistasis   The effect of one gene locus upon another where they do not stand in the relation of alleles. Usually one masks or modifies the expression of the other when they occur together.

Exvia–Invia   A second order questionnaire factor which defines the true core at the heart of the popular "extraversion–introversion" label. It appears as U.I. 32 in the objective test series.

Function fluctuation   The change in an individual's trait score from day to day with internal and external circumstances, such that scores need to be averaged across time to get a true trait score.

Gamete   Either of the halves produced by the cell division (meiosis) producing male or female germ cells. A haploid cell.

Gene frequency   For a particular gene in a population, the frequency of a given allele divided by the frequency with which it *could* occur.

Genealogical or pedigree method   Behavior genetic investigation by tracing syndrome occurrence across relatives and generations.

Genome   The set of genes in a single gamete (i.e., in a haploid split the haploid set).

Genothreptic correlations   The correlations in and between families and across the population of genetic and threptic deviations.

Guilt proneness   Factor O in questionnaires, measuring a sense of guilt and unworthiness.

Heterozygote   An individual with differing alleles at one or more loci.

Homogamy   The condition in a population in which like mates with like.

Imprinting   The acquisition of a learned behavior pattern in earliest life or some period of high internal susceptibility, so that it endures exceptionally well.

Inbreeding or consanguinity   In Wright's inbreeding coefficient the probability that two particular gene alleles appearing in a zygote are both descended from an ancestor common to the two parents. It is also defined as the proportion of a person's genes in which he is homozygous, first order in T-data, marked by independent and self-reliant behavior.

Inhibitory control, U.I. 17, QVIII   See **Control**.

Intelligence, crystallized, g_c   A general factor in abilities, distinct from fluid intelligence and considered due to investment of fluid intelligence in cultural learning.

Intelligence, fluid, g_f   The basic general ability among primary abilities, which appears most purely in unlearned performances with new data.

Karyotype   A statement of the number and nature of the set of chromosomes in species or individual, sometimes as a photomicrograph.

Least squares   Estimating (e.g., a line of best fit, a value) so that the sum of squares of departures from "true" values is minimized.

Linkage   A greater association of two or more non-allelic genes, in the inherited set, than would be expected from an independent, random shuffle. It occurs commonly from contiguity on the karyotype of a linkage group formed by a gene set located on the same chromosome.

Locus   The position of a gene on the chromosomes.

Meiotic process   The halving division of the zygotic chromosome number in producing a gamete so that the ensuing fertilized cell will have the ordinary number.

Modulator index   The index (s_k) which shows for the average person how much a particular situation, k, modulates the state proneness (L_i) score.

Mutation   The shift of a gene to a new allelic form.

Mutation rate   The frequency with which certain mutations occur in a given population.

Narcistic self   The unitary trait found in objective tests which is indexed as U.I. 26 and has a pattern of "spoiled" behavior.

OSES   A method of analysis of MAVA results which employs *O*verlapping *S*imultaneous *E*quation *S*ets.

Parmia, H   A questionnaire factor of bold insusceptibility to threat, hypothesized due to *Para*sympathetic predominance.

Path coefficient   A means of representing quantitatively the hypothesized causal action—in variance contribution—of one variable upon another.

Path learning analysis   A matrix method of determining the effect of environmental paths on personality learning.

Penetrance (or manifestation rate)   The fraction of individuals known to have a certain genotype (maybe some single gene allele) who show the expected phenotype (always relative to a specified environment).

Placement correlation   A correlation expressing the resemblance of a placed child to the parents or foster sibs in an adopting family.

Pleiotrophy   The action in which a single gene is responsible for diverse features of the phenotype that *a priori* would not be thought related.

Polygenic trait   A trait the genetic variation in which is due to the action of more than one gene, the various genes not necessarily contributing equal increments.

Polymorphism   Genetic presence of two or more genetically different classes in the same freely interbreeding population. Chromosomal polymorphism means the existence in the population of two or more structural forms for a chromosome.

Premsia-versus-Harria The unitary factor found in questionnaire response, indexed as *I*, characterized by emotional sensitivity versus toughness.

Primary ability An ability found as a first order factor from measures of actual performance variables.

Reality contact, U.I. 25 A unitary trait in objective tests (*T*-data) that seems to represent goodness of reality contact and is significantly low in neurotics and psychotics.

Recessive gene See **Dominant allele.**

Reciprocal crosses Comparison of offspring of male type X with female type Y and male type Y with female type X, valuable in recognizing sex linkage and cytoplasmic inheritance.

Recursive causal model A system of mutual interacting elements in which causal action goes in both directions.

Regression, U.I. 23 An objective test unitary trait along a dimension of capacity-to-mobilize versus regression. Higher (as regression) on neurotics and psychotics.

Sanguineness-versus-Discouragement A trait, possibly largely a state, factor found in objective test performance, indexed as U.I. 33, and involved in elation-versus-depression.

Second order (or stratum) factor A factor appearing from a factor analysis of correlated primary factor scores.

Self-sentiment A trait in questionnaire measurements of the organized attitudes to the self-concept. Indexed Q_3.

Self-sufficiency A unitary trait appearing in questionnaires, indexed as Q_2 and characterized by independence and lack of involvement in groups.

Sex chromosomes The homologous pair of chromosomes beyond the soma (general body) set, that are XX in the female and XY in the male. A gene on these is called sex linked.

Spiral action The theory that second order personality factors arise by mutual (spiral) interaction among the primaries that appear loaded in them.

Stub factor The specific part left in a primary factor when the variance due to a second or higher order factor is removed.

SUD (Stratified Unrelated Determiners) A factor model used in personality and elsewhere in which influences are admitted to operate at different strata levels but in which all are uncorrelated (unrelated).

Super ego A unitary trait recognized by psychoanalysis and since supported and rendered measurable by factor analytic experiment.

Surgency A temperament factor of high sociability and talkativeness—a primary within exvia—indexed as *F* in questionnaires.

Teachability (or Culturability) The extent to which the normal threptic component in a trait can be raised by intensive teaching time (at the cost of time normally given in the culture to other learning).

Threptic   That part of individual differences in a trait measurement which is due to environmental impact differences.

Triadic theory of abilities   The theory that ability performance arises from three structural levels—general factors, g's (e.g., intelligence); provincial factors, p's (e.g., visual capacity); and primary factors, or agencies, a's (e.g., numerical ability).

Validity, as concept validity (sometimes called construct)   The correlation of score on the given test with score on the factor (concept) it is supposed to measure.

Zygote   The cell, now diploid, from the union of the haploid male and female gametes.

Author Index

435

Thurstone, L. L., 3, 9, 39, 72, 75, 76, 78, 130, 228, 276, 278, 282, 283, 295, 297, 325, 344, *426*
Thurstone, T. G., 276, *426*
Tienari, P., 27, *426*
Tinbergen, W., 49, *426*
Tolman, E. C., 33, *426*
Tomson, B., 36, 360, *405*
Toynbee, A., *426*
Tryon, R. C., 3, 33, 180, *426*
Tsujioka, B., 35, 224, 342, 380, *407, 426*
Tuddenham, R. D., *426*
Tuttle, W. W., 197, *404*

U
Undheim, J. D., *426*

V
Valen, L. van, 37, 38, *426*
Van Abeelin, J. H. F., 295, *426*
Vandenberg, S. G., 3, 25, 60, 70, 71, 75, 76, 77, 79, 105, 129, 130, 134, 182, 210, 228, 232, 279, 292, 297, 298, 303, 304, 306, 313, 314, 331, *410, 418, 422, 426, 428*
Vaughan, D. S., 36, 277, 327, 329, 367, 371, 376, *408, 409, 427*
Vernon, P. E., 3, 181, 276, 280, *427*
Vogelmann, S., *409*

W
Waddington, C. H., 23, 25, *427*
Warburton, F. W., 327, 360, 368, 386, 398, *409*
Wardell, D., 360, 373, *427*
Watkins, M. P., 143, *427*
Watson, G. A., 3, 6, 17, 263, *427*
Webber, P. L., 114, 173, *423*
Weber, R. J., 197, *404*
Weinberg, R. A., 114, 152, 173, 177, *423, 424*
Weir, M. W., 33, *410*
Weise, C., 369, *408*

Weiss, V. V., *427*
Welch, W. B., 108, 293, *427*
Wells, H. G., 4, 5, *427*
Welner, J., 179, *423*
Wender, P. H., 179, *417, 423*
Wepte, R., 120, *422*
Weyl, N., 5, 23, 36, 158, 181, 182, 209, *418, 421, 427*
Whitney, G., *427*
Wilcock, J., 399, *412*
Willerman, L., 60, 86, 147, 173, 174, 175, 176, *415, 418, 422, 427*
Williams, E. L., 4, 32, 298, *405*
Williams, R. J., 4, *427*
Williams, T., 298, *427*
Willoughby, R. R., 128, 132, 146, *427*
Willson, J. L., 128, 133, 146, 245, *409*
Wilson, E. O., 4, *427*
Wilson, J. R., 130, *419*
Wilson, R. S., 4, 33, 75, *419, 427*
Winston, H. D., *418*
Witkin, H. A., 356, *413, 427*
Wittig, M. A., 114, 173, *423*
Woliver, R., 132, 133, 134, 135, 181, 182, 224, 226, 227, *407, 409, 427*
Wright, S., 35, 142, 152, 154, 155, 178, 251, 258, 265, 271, 393, *428*

Y
Yee, R., 369, *408*
Yee, S., 120, 217, 251, 258, 369, 399, *408, 413, 422*
Yen, W. M., *428*
Yeudall, L. T. A., 360, *427*
Young, H. B., 37, 39, 276, *409*
Young, M., 4, 5, *428*
Young, P. A., 4, 5, 7, 37, 39, 61, 276, 337, *411*
Yu, P., *422*

Z
Zerssen, D. Von, *428*
Zonderman, A. B., 129, 134, *428*

Subject Index

Genomics, 55
Genothreptic correlations, G
 discussion of values in g_f and g_c, 318
 environmental causes of r, 249
 for fluid intelligence, 306, 312
 genetic causes of between-family r, 248
 genetic causes of within-family r, 248
 "gestalt" effect in adoptive family r_{wgwt}
  and r_{bgwt}, 97
 joint causes of r, 250
 marked relation to H values, 338
 perspective on for personality traits, 350
 problematic $r_{wg_1wt_2}$ term, 96
 research methods and finding, 250
 rotation, sequence, path coefficients, 250
Genotype, defined, 15
Growth curves, methods of splitting into genetic and threptic components, 236
Guilt proneness, O, G
 heritability by OSES, least-squares and
  maximum-likelihood, by MAVA, 340
 rank in heritability, 349, 355
 theory of, support from H values, 339

H
Haploid, 14
Heritabilities (H_b and H_p) rank ordered
 for Q-data primary traits, 349
 for T-data (secondary) traits, 391
Heritability
 age changes in, 219
 broad and narrow, 142, 148, 269
 and the equity principle in, 219
 and genetically derived threptic variance,
  220
 gross and net, 273
 and imprinting, 219
 place of covariance in, 268
 and the stochastic process principle, 219
Heritability coefficient, H
 gross and net, 273
 initial calculation, 68
 role of covariance in, 268–270
 six different forms of, 267, 269
 standard error (twins), 107–108
 standard error of variances in (MAVA),
  108
 within, between, and population not
  equal, 273
Heterozygotic, meaning of, 12, G
Homogamy, 127, 135, G
Huntington's chorea, 22

I
Independence, QIV and U.I. 19
 as QIV
  heritability of, 346, 355
  secondary weights for, 243
 as U.I. 19
  fit of maximum likelihood solution, 363
  heritability by three analyses, 366
  impact on theory of U.I. 19, 367
  reliability and validity, 364
  subtests in measurement, 364
Inherited, meaning of, 11
Inhibitory control, QVIII and U.I. 17, G
 as QVIII
  heritability of, 346, 355
  secondary weights for, 343
 as U.I. 17
  abstract variances corrected and uncorrected, 365
  fit of maximum-likelihood solution, 363
  heritability by three methods, 366
  impact on theory of U.I. 17, 367
  subtests for measuring, 364
Innate, meaning of, 11
Intelligence
 age curves for, 197
 age curves for g_f and g_c, 214
 and cranial size, 38
 crystallized defined, 211, 283, G
 fluid defined, 211, 282, G
 heritability by twin method, 74–87
 investment theory of, 211
Interaction of genetic and threptic, 66, 85,
  157
 distinguished from covariance, 67, 189,
  243
Interaction of learning and genetic processes, 187–189
Intraclass correlation, 68, 267

L
Learning laws
 derivation from behavior genetics, 195,
  223, 237
 derivation from path learning analysis matrix, 195
Learning theory and genetics, 10
 evaluation of learning by maturational research, 205
 reciprocal assistance of learning theory
  and genetics, 401
 as structured learning theory, 188

PERSONALITY AND PSYCHOPATHOLOGY

A Series of Monographs, Texts, and Treatises

David T. Lykken, Editor

*Titles initiated during the series editorship of Brendan Maher.

PERSONALITY AND PSYCHOPATHOLOGY